MODERN PARASITE BIOLOGY

Cellular, Immunological, and Molecular Aspects

Edited by
David J. Wyler, M.D.
Professor of Medicine
Tufts University School of Medicine
and New England Medical Center Hospitals, Inc.

W. H. Freeman and Company
New York

Cover image: Scanning electronmicrograph of an erythrocyte infected with *Plasmodium falciparum*, showing the prominent excrescences (called knobs) on the erythrocyte surface. These knobs are associated with parasite-derived proteins that mediate attachment of the infected erythrocyte to the vascular endothelium. This attachment is important to the development and survival of the parasite and also plays a role in the pathogenesis of falciparum malaria (see Chapters 1, 9, and 16). (Photograph courtesy of Dr. Masamichi Aikawa, Case-Western Reserve University School of Medicine, reprinted by permission of the *Journal of Parasitology*, 1983, 69:435–437.)

Library of Congress Cataloging-in-Publication Data

Modern parasite biology: cellular, immunological, and molecular aspects/David J. Wyler, editor.
p. cm
Includes bibliographical references.
1. Parasitology. 2. Parasitic diseases—Immunological aspects.
3. Parasites—Cytology. 4. Parasites—Immunology. I. Wyler, David J.
QL757.M546 1990 89-23317
616.9′6—dc20 CIP
ISBN 0-7167-2038-8
ISBN 0-7167-2140-6 (pbk.)

Printed in the United States of America

1 2 3 4 5 6 7 8 9 0 VB 9 9 8 7 6 5 4 3 2 1 0

To my devoted wife, Deborah
and sons, Jonathan and Benjamin

Contents

MODERN PARASITE BIOLOGY

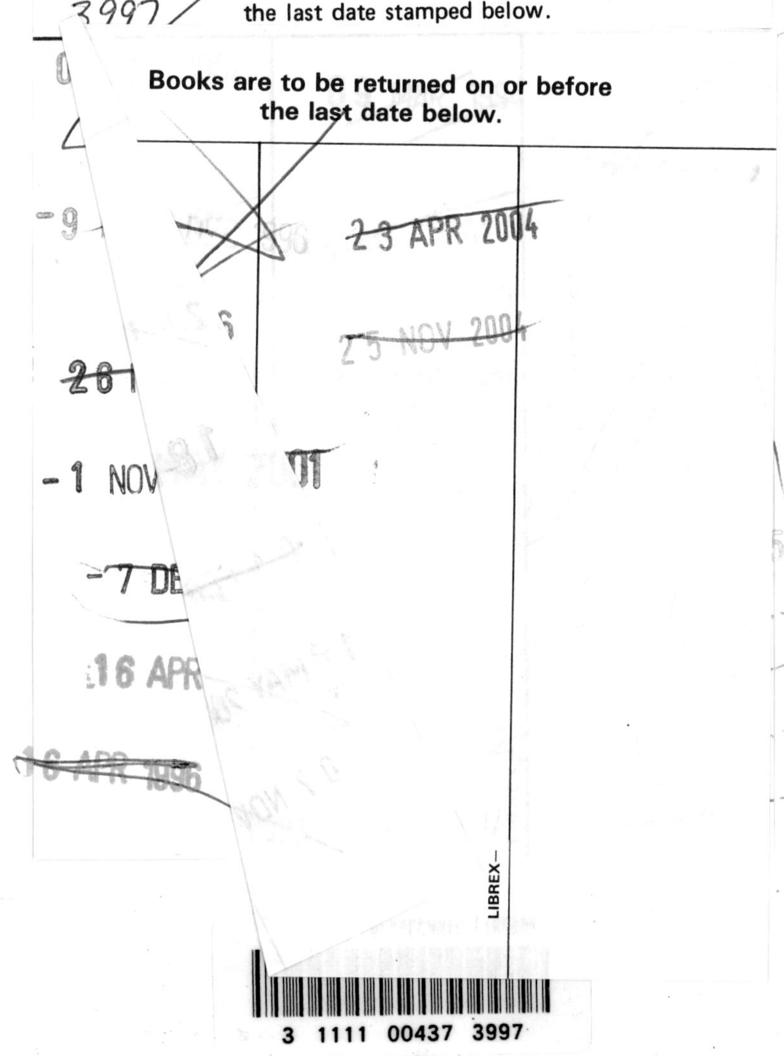

Preface

Medical parasitology, the study of parasitic diseases and parasites that infect man, has undergone remarkable developments in the last three decades. These developments, along with recent widespread application of the newest biomedical technologies, have imbued optimism that new strategies can be developed for controlling the important parasitic diseases that for centuries have been the scourges of mankind.

After the causative pathogens and their life cycles were identified (most in the late nineteenth and early twentieth centuries), control efforts relied heavily on case detection, invertebrate vector control, chemotherapy, and chemoprophylaxis. In some areas, impressive inroads were made. The use of DDT spraying, mosquito netting, and chloroquine prophylaxis reduced or eliminated malaria from many regions of the world. In parts of China, schistosomiasis was controlled by filling rice paddy irrigation ditches to kill the snail vectors; kala azar was controlled there by the elimination of the domestic dog reservoirs and by spraying houses against the sandfly vectors. New antihelminthic and antiprotozoal drugs were developed, largely through empirical screening of thousands of compounds. Each new success fed the fantasy that it was only a matter of time until control strategies and improved sanitation would erradicate the tropical parasitic diseases that prevailed in areas inhabited by over two-thirds of the world's population.

In reality, social, economic, and biological factors were radically undermining these disease control efforts. Economic development programs, such as new irrigation and housing projects, have facilitated expansion of certain vector populations and promoted greater human contact with the vectors. The construction of the Aswan High Dam swelled Lake Nassar and yielded flood waters needed for irrigation of new Egyptian farmland. However, the dam also created conditions that permitted the spread of the snails that transmit schistosomiasis, and thereby contributed significantly to the increased prevalence of this disease in Egypt. The application of green-revolution techniques in Asia and Latin America may have contributed to the resurgence of malaria in these areas because the irrigation projects provide new breeding grounds for the vector mosquitoes. Similarly, highway and rural development in Brazil is placing increasing numbers of workers and residents in the midst of areas where transmission of leishmaniasis occurs from forest animals to humans. Migration of human populations, often stimulated by wars and famine, also plays a role in the growing problem, as does political unrest which contributes to the disorganization of disease-control programs and of the health care infrastructure in general. The resurgence of malaria that followed the development of drug resistance by *Plasmodium falciparum* and the emergence of insecticide resistance in anopheline mosquitoes in the 1960s provided a stark reminder that biological systems, like ecological ones, can be dynamic. The recent emergence of toxoplasmosis and *Pneumocystis* pneumonia as opportunistic infections of patients with the acquired immunodeficiency syndrome is another reminder that hosts, as well as parasites, change.

By the late 1950s and in the 1960s, traditional disease control strategies for certain parasitic diseases became difficult to apply broadly or were ineffective and it became apparent that new methods were needed. However, unlike bacterial, viral, and fungal pathogens, helminths and protozoa could not be cultivated for potential vaccine production. Little was known of their unique metabolism, of the details of host cell-parasite interactions, or of immunological responses to parasite invasion. Morphology and taxonomy had long been the focus of basic medical parasitology research, and nosology and epidemiology dominated the attentions of clinical scientists in this field. To compound the difficulty, the parasitic diseases that plagued primarily third world nations were not accorded an appropriately high priority by the granting agencies in industrialized nations, especially during peacetime. Research funds were scarce, and training programs that could inspire young scientists to enter the field of medical parasitology were few. Despite its obvious relevance and fascinating subject matter, medical parasitology in the mid twentieth century was the stepchild of the microbiology family.

Circumstances began to change in the 1970s. The World Health Organization developed a new program to help fund applied and basic research and research training targeted at malaria, schistosomiasis, trypanosomiasis, leishmaniasis, filariasis, and leprosy. Given the major impact of these diseases on the social and economic development of the afflicted third world nations, it was fitting that the United Nations Development Programme and the World Bank joined WHO as partners in this project. Three private foundations (Rockefeller, Clark, and MacArthur) in the United States developed their own programs to support parasitology research. Government granting agencies continued their support. Communication and collaboration between scientists in industrialized and developing countries helped to ensure success of the undertaking. New technologies—notably, in vitro methods for parasite cultivation, improved methods of protein analysis, methods for production of monoclonal antibodies, creation of lymphocyte clones and hybrids, and cloning of parasite genes—were applied, and the scientific and lay press reported breakthroughs. Immunologists, biochemists, and molecular biologists were drawn to the challenges of modern parasitology both by fascination and opportunity. The possibility that studies of parasites might yield information with broad biological significance was also enticing, and within fifteen years the field began to hum.

The veritable explosion of new specialized knowledge that emerged in the 1980s from the new research resulted in a degree of overspecialization that sometimes interfered with communication between scientists working on the same parasite or disease. Understandably, scientists initially educated in one biological discipline come to parasitology with a perspective not easily shared with those from different scientific backgrounds and the ever-growing jargon of these disciplines has created a scientific Tower of Babel that further interferes with the cross-fertilization of ideas. Today, even review articles that provide broader perspectives to some scientific audiences may be intimidating to the non-initiated.

This book offers a compendium of reviews of the most active areas of research in medical parasitology. It provides an overview of scientific perspectives. By intent, the chapters are not exhaustive reviews. The contributors, all seasoned investigators in their areas, were invited to present their personal perspectives. They were also asked to identify lacunae in knowledge that deserve attention in the future, and were encouraged to synthesize concepts as coherently as possible, given the present state of knowledge. To this end, contributors organized information in their chapters without cumbersome citations. A list of more comprehensive reviews is provided by the contributors in the Additional Reading section at the end of each chapter and selected original scientific reports are also listed by chapter at the end of the book.

The chapters are written with the expectation that the reader will have a general understanding of the parasite's life cycle and the diseases the organism can cause, as well as a firm background in the major disciplines of modern biology. Course instructors may find it useful to coordinate readings from standard textbooks in medical parasitology and tropical medicine with chapters in this book. I have written introductions to the three sections to highlight certain perspectives that are appropriate to a particular discipline or to medically important parasites.

I am grateful for the guidance, patience, and support provided by the following people at W. H. Freeman and Company with whom I had the pleasure to work: senior editors Jim Dodd, Gary Carlson, Patrick Fitzgerald, and Kirk Jensen; project editors Diane Maass, Sonia Di Vittorio, and Jeanette Johnson; designers Nancy Field and Nancy Singer; and production coordinator Sheila Anderson. Without the considerable efforts of my colleagues who critically reviewed the concept of this book as well as specific chapters, this project would not have been possible: Richard Albach, Norma Andrews, David Arnot, Paul Bates, Clinton Carter, Richard Crandall, George Cross, Philip D'Alesandro, Raymond Damian, Dickson Despommier, Carter Diggs, Klaus Esser, Jacob Frankel, Bruce Greene, George Hill, John Hyde, Michele Jungery, Walter Kemp, Raymond Kuhn, Richard Locksley, Philip Lo Verde, Adel Mahmoud, Diane McLaren, Rima McLeod, Adolfo Martinez-Palomo, Steven Meshnick, Louis Miller, Geoffrey Pasvol, Edward Pearce, William Petri, David Sacks, Joseph Schwartzman, Philip Scott, Irwin Sherman, Kenneth Stuart, Herbert Tanowitz, Lex van der Ploeg, and R. Werk. Irene Doucette provided me with excellent secretarial assistance.

David J. Wyler

Contributors

Nina Agabian, Ph.D.
Director,
Intercampus Program in Molecular Parasitology;
Professor, Department of Pharmaceutical Chemistry
University of California, San Francisco
San Francisco, California

Paul F. Basch, Ph.D.
Professor,
Department of Health Research and Policy
Stanford University School of Medicine
Stanford, California

John C. Boothroyd, Ph.D.
Associate Professor,
Department of Microbiology and Immunology
and Department of Medicine
Stanford University School of Medicine
Stanford, California

Zigman Brener, M.D.
Professor,
Department of Medicine
University of Minas Gerais;
Chief, Laboratory of Chagas' Disease
Centro de Pesquisas Rene Rachou FIOCRUZ
Belo Horizonte, Brazil

Anthony E. Butterworth, M.B., B. Chir., Ph.D.
Associate Lecturer,
Department of Pathology
University of Cambridge
Cambridge, England

John P. Caulfield, M.D.
Associate Professor,
Department of Pathology,
Harvard Medical School
and Department of Tropical Public Health
Harvard School of Public Health
Boston, Massachusetts

K.-P. Chang, Ph.D.
Professor,
Department of Microbiology and Immunology
The Chicago Medical School
North Chicago, Illinois

John E. Donelson, Ph.D.
Professor,
Department of Biochemistry
University of Iowa;
Investigator,
Howard Hughes Medical Institute
Iowa City, Iowa

Juliet A. Fuhrman, Ph.D.
Research Associate,
Department of Tropical Public Health
Harvard School of Public Health
Boston, Massachusetts

Antoniana U. Krettli, Sc.D.
Professor,
Department of Parasitology
University of Minas Gerais;
Chief, Malaria Laboratory
Centro de Pesquisas Rene Rachou FIOCRUZ
Belo Horizonte, Brazil

John M. Mansfield, Ph.D.
Professor,
Department of Veterinary Science
University of Wisconsin-Madison
Madison, Wisconsin

Thomas F. McCutchan, Ph.D.
Senior Staff Fellow,
Laboratory of Parasitic Diseases
NIAID
National Institutes of Health
Bethesda, Maryland

George R. Newport, Ph.D.
Adjunct Assistant Professor,
Department of Pharmaceutical Chemistry
University of California, San Francisco
San Francisco, California

Richard D. Pearson, M.D.
Associate Professor,
Department of Medicine
and Department of Pathology
University of Virginia School of Medicine
Charlottesville, Virginia

Miercio E. A. Pereira, M.D., Ph.D.
Professor,
Department of Medicine and
Department of Molecular Biology and Microbiology
Tufts University School of Medicine
and New England Medical Center Hospitals
Boston, Massachusetts

Margaret E. Perkins, Ph.D.
Assistant Professor,
Laboratory of Biochemical Parasitology
The Rockefeller University
New York, New York

Elmer R. Pfefferkorn, Ph.D.
Professor and Chairman,
Department of Microbiology
Dartmouth Medical School
Hanover, New Hampshire

Willy F. Piessens, M.D.
Associate Professor,
Department of Tropical Public Health
Harvard School of Public Health
Boston, Massachusetts

Johnathan I. Ravdin, M.D.
Associate Professor,
Department of Medicine and
Department of Pharmacology
University of Virginia School of Medicine
Charlottesville, Virginia

John Samuelson, M.D., Ph.D.
Assistant Professor,
Department of Tropical Public Health
Harvard School of Public Health
Boston, Massachusetts

Somesh D. Sharma, Ph.D.
Co-Founder, Vice President and Director of Immunology
Biospan Corporation
Redwood City, California

Mervyn Turner, Ph.D.
Executive Director,
Basic Animal Science Research
Merck Sharpe & Dohme Research Laboratories
Rahway, New Jersey

Ann C. Vickery, Ph.D.
Associate Professor,
College of Public Health
University of South Florida
Tampa, Florida

Mary E. Wilson, M.D.
Assistant Professor,
Department of Internal Medicine
University of Iowa College of Medicine
Iowa City, Iowa

Dyann F. Wirth, Ph.D.
Associate Professor,
Department of Tropical Public Health
Harvard School of Public Health
Boston, Massachusetts

David J. Wyler, M.D.
Professor,
Department of Medicine and
Department of Molecular Biology and Microbiology
Tufts University School of Medicine and
New England Medical Center Hospitals
Boston, Massachusetts

PART I

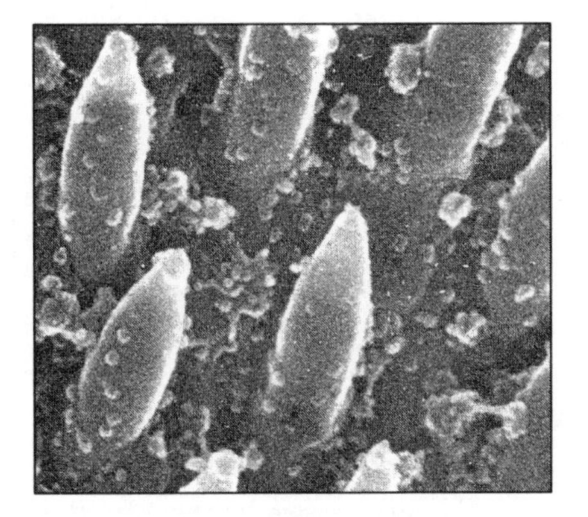

CELL BIOLOGY OF PARASITES

David J. Wyler

The definition of parasite life cycles has inspired numerous questions regarding parasite biology. Two particular concerns of biologists are to identify unique metabolic requirements of parasites that could define the basis for dependence on their hosts and to delineate the steps involved in the transformation from one stage of the parasite (such as that present in the invertebrate vector) to another (such as the stages present in the vertebrate host). Interest in parasite biochemistry has engendered substantial research into various aspects of parasite composition and metabolism; the results have been reviewed in a number of comprehensive texts.

The multidisciplinary field of cell biology, a field that has enjoyed remarkable growth in recent years, emerged with the introduction of electron microscopy and the development of powerful new techniques in biochemistry. Parasitologists began to apply the tools of cell biology and its structure-function perspective on the relationships of cells in an effort to better understand not only the parasite as the unit of interest, but particularly the cellular relationships between parasite and host. The breakthrough in the understanding of these relationships in the last decade has been truly exciting and many aspects are summarized in this section.

One particularly active area of parasite cell biology research explores the ultrastructural features and biochemical basis of invasion of host cells by obligate intracellular protozoa. An important realization in the general design of these studies is that some of these protozoa (*Plasmodia* and *Leishmania*) are highly selective in the host cells they parasitize, others *(Trypanosoma cruzi)* are somewhat less so, and some *(Toxoplasma gondii)* show little discrimination. Influenced by the important concept of specificity in the interaction of soluble ligands and cell-bound receptors, investigators have been attempting to identify the receptor-ligand systems that dictate specificity in parasite–host cell interaction. Most advanced is the research investigating how plasmodial merozoites attach to erythrocytes (Chapter 1). The results, which indicate a critical role for certain blood group substances (in the case of *P.vivax*) and components of glycophorin (in the case of *P. falciparum*) in the interaction, have established that intrinsic membrane components may act as receptors and ligands. In addition, the participation in the attachment process of soluble host proteins, such as complement and fibronectin, that can serve to bridge host cell and parasite has been appreciated in the case of *Leishmania* (Chapter 5) and *T. cruzi* (Chapter 4). The possibility that enzymes released by the protozoa influence invasion has also been suggested (Chapters 1 and 4). How parasites enter the host cells also has intrigued scientists. Invaders of mononuclear phagocytes—*Toxoplasma* (Chapter 2), *Leishmania* (Chapter 5), and *T. cruzi* (Chapter 4)—apparently gain entry into these cells passively by phagocytosis, although the molecular triggers for their ingestion remain to be identified. In contrast, the entry of merozoites of *Plasmodia* into erythrocytes (cells not recognized as phagocytic) entails the formation and movement of a tight junction between host and parasite plasma membranes and complex alterations in host membrane and cytoskeletal organization.

Once protozoa enter the host cell, the subcellular compartment they occupy has substantial implications with respect to their ability to evade intrinsic host defenses (in the case of inhabitants of mononuclear phagocytes such as *Leishmania*, *Toxoplasma*, and *T. cruzi*); different parasites display distinctive strategies for dealing with this challenge. Transport of nutrients and metabolic waste across parasite and host membranes is also under scrutiny, and modification in host membranes by the parasite, for example, the insertion of parasite-derived molecules that mediate the cytoadherance of *P. falciparum* – infected erythrocytes to endothelial cells (Chapter 1) has proved fertile ground for study. The investigation of the intracellular metabolism of protozoa has become possible more recently as cultivation methods have been developed and no doubt will be facilitated by application of new technologies for intracellular analysis, including fluorescence probe methods and nuclear magnetic resonance detection. Nucleic acid metabolism has proven of particular interest (Chapter 2) since the intracellular protozoa have been found incapable of de novo purine synthesis.

Helminths, by virtue of their multicellular construction and extracellular habitation, have engendered other lines of investigation. The ultrastructural analysis of different stages of schistosomes (Chapter 6) has provided an impetus to assess morphogenesis and various aspects of interstage transformation. The recognition that host cells can attack the schistosomular tegument has led to fascinating membrane research (Chapter 7). Because the biochemistry and pharmacology of schistosomes has been investigated for several years and the resulting voluminous literature is not easily summarized, it has not been included in this volume.

The application of a cellular-biology perspective to the study of medically important parasites has at least two virtues: (1) it may uncover critical elements in host–parasite interaction that could be amenable to interruption by pharmacological or immunological intervention, and (2) it may facilitate the discovery of biological principles perhaps most readily revealed in lower eukaryotic systems. Thus, for example, knowledge about the invasion of erythrocytes by merozoites has stimulated efforts by immunologists to prepare vaccines against molecules involved in the process (Chapter 9). On the other hand, insights into the membrane chemistry of antigenic variation in African trypanosomes (Chapter 3) has opened a whole new line of research in the molecular biology of trypanosomes (Chapter 17), parasites presently considered excellent models for uncovering principles of eukaryotic molecular biology.

Because the perspectives presented in the chapters of this section are based on the parasite life cycle, they provide the transition from classical medical parasitology to modern parasite biology. Inasmuch as progress in modern parasite biology depends heavily on an interdisciplinary approach, however, the reader will discover some overlap here with chapters in other sections. In several cases, the biological process investigated by the cellular biologist has defined the potential targets of host defense that the immunologists study, and the immunologists in turn may identify the antigens that provide specificity to this defense. The molecular biologist may then undertake efforts to clone the genes that code the antigens. The interplay of basic disciplines is particularly revealed in the chapter on ameba (Chapter 8) in which both cellular biology and immunology are discussed. While presentation of this combination may have heuristic value, it also reflects the fact that a systematic approach to the immunology of ameba is a relatively recent endeavor.

The challenges that remain in the cellular biology of parasites are substantial and should prove particularly appealing to investigators who enjoy the application of a variety of perspectives and techniques in addressing fundamental biological questions. At present, research in this area is generating many compelling questions whose answers may lead to novel disease-control strategies.

1

Cell Biology of *Plasmodium*

■

Margaret E. Perkins

Plasmodium, the protozoan parasite that causes malaria, exists in a variety of different forms which have successfully adapted to different cellular environments, in both the vertebrate host and the mosquito vector. The parasite develops in a highly regulated manner through three distinct cycles in the vertebrate host and two cycles in the insect. At certain stages of its life cycle, development is intracellular. In the vertebrate host this protective environment, together with the fact that each stage is antigenically distinct, provides the main strategies for evading the host's immune attack. This chapter reviews the complex process of host cell invasion, the parasite's ability to alter or subvert host cell functions to its advantage, and its ability to subsequently free itself from the intracellular environment, en route to a new host cell. In biochemical terms we know very little of these events, and experimental investigation of the processes has proved difficult in many cases. Therefore, questions about the suitability and the limitations of techniques applied to the study of the plasmodia will also be considered.

The genus *Plasmodium* is a member of the subphylum Apicomplexa. Members of this subphylum include genera of other intracellular protozoa: *Eimeria*, *Toxoplasma*, *Babesia*, and *Theileria*. A unifying and unique feature of the subphylum is the complex array of subcellular apical structures known as rhoptries and micronemes. These organelles are presumed to be involved in entry into host cells, which for the most part are nonphagocytic. Entry requires the active participation of the parasite. Rhoptries and micronemes are organelles that are present in both sporozoites and merozoites, the extracellular stages that invade host cells, but they are absent in gametes and ookinetes, stages that do not invade.

Over 100 species of *Plasmodia* infect rodents, birds, and primates. The sporozoite is the form of the parasite that the mosquito injects into the vertebrate host and that initiates the intrahepatic cycle. Sporozoites rapidly circulate to the liver, where they invade liver cells and develop into tissue schizonts within hepatocytes. In the course of 1 to 2 weeks each tissue schizont divides to form at least 10,000 merozoites. On release from the liver into the blood stream, the merozoites invade erythrocytes. The specificity of the interaction between the merozoite and erythrocyte has received considerable attention in recent years because this step is a crucial and potentially vulnerable one for the parasite's survival. Rings and trophozoites are intraerythrocytic asexual stages into which the merozoites develop. Trophozoites in turn develop into the multinucleated schizonts that give rise to the merozoites that are released by rupture of the infected red cell. These merozoites then invade other red cells. The cyclical development from merozoite to schizont can proceed many times in vivo or until the host succumbs, and it also can be sustained indefinitely in vitro. Some merozoites develop intracellularly to form male or female gametocytes, which if ingested by mosquitoes, change into male and female gametes in the mosquito midgut; they cause no disease in the vertebrate host. During the parasites sexual cycle, fertilization occurs when gametes unite in the gut to form a unicellular zygote. The fertilized zygote transforms to a motile ookinete and penetrates through the intestinal wall of the midgut, where it forms an oocyst. The oocyst undergoes a process called sporogony that results in the formation of 10,000–20,000 sporozoites, which migrate to the salivary gland. From here sporozoites emerge to infect the host when the vector mosquito takes a blood meal. In summary, there are five distinct extracellular stages: the sporozoite and the merozoite in the vertebrate host and the gamete, zygote, and ookinete in the mosquito. All are unicellular. The ookinete and the male gamete are capable of multiplication; the latter forms eight motile microgametes. There are three intracellular stages and in two of these, the intrahepatic schizont and intraerythrocytic schizont, asexual division occurs by binary fusion. Only the gametocyte remains uninucleate. Attempts have been made, with varying degrees of success, to culture all stages of *Plasmodium falciparum* and some stages of other species. In vitro cultivation has been most successful for the asexual erythrocytic stages of *P. falciparum*. Production of gametocytes and ookinetes during cultivation of the erythrocytic stages has also been possible. Recently the complete hepatic cycle of *P. vivax* and *P. falciparum* has been accomplished. The in vitro culture systems will be discussed in more detail in the relevant sections that follow.

The Hepatic Cycle

Ultrastructure of Sporozoites

The sporozoite is an elongated unicellular organism, 11 micrometers (μm) in length. It is covered by a thin surface coat overlying a double unit membrane and a row of subpellicular microtubules. The only surface protein identified to date as a component of the surface coat is the circumsporozoite protein (CS), which probably will prove to be the only one. Incubation of sporozoite with antisporozoite immune serum results in the precipitation and sloughing off of the surface coat material, in a process called the circumsporozoite precipitation (CSP) reaction. The CS protein was first identified in *P. berghei*. Homologous proteins have also been identified in other species: *P. cynamolgi*,

P. knowlesi, *P. falciparum*, and *P. vivax*. The CS protein varies in size from 40 kilodaltons (kD) in *P. knowlesi* to 60 kD in *P. falciparum*; it appears to be synthesized as a high-molecular-weight precursor and is processed during transport to the cell surface. The CS protein of *P. knowlesi* was the first plasmodial protein to be cloned in *E. coli*, and base pair sequence analysis of the cDNA revealed a remarkable feature, the presence of an extensive repeat domain. Subsequent analysis of other plasmodial antigens or the genes coding for them revealed that many parasite proteins share this property, some having even larger repeat domains than the CS protein. For the CS protein, the number and sequence of amino acids per repeat and the number of repeats per molecule vary between species and in some instances between strains; however, the repeat domain generally constitutes 40 percent and occupies the same location in the molecule. Flanking the repeat domain at the 3′ and 5′ end of the molecule are regions of highly charged amino acids that are conserved between species.

The sporozoite exists in the bloodstream for a maximum of 1 hour. Efficient invasion of hepatic cells is necessary to guarantee protection against host defenses, and generation of the subsequent stages is its major objective. It has been proposed that the function of the CS protein is to mediate the interaction (attachment and entry) with the hepatic cells although there has been little biochemical information available to support this hypothesis. Recently, however, it has been demonstrated that a region of the CS protein immediately 5′ to the repeat domain is conserved in different species and apparently is responsible for binding to hepatoma cells in vitro (a culture system believed to mimic in vivo conditions). Rhoptries and micronemes are also present in the sporozoite and probably have an important role in invasion as well (for further discussion of the molecular biology of the sporozoite protein, see Chapter 16).

Sporozoite Invasion of Hepatocytes

Electron microscopic studies of the in vivo entry of *P. berghei* sporozoites into hepatic cells have been performed after sporozoite attachment. The hepatocyte plasmalemma invaginates at the point of attachment and creates a parasitophorous vacuole in which sporozoites reside. An active involvement of the sporozoite during this invasion is postulated. Because the proteins of the sporozoite rhoptries have not yet been identified, it has not been possible to directly examine their fate during invasion. If a monoclonal antibody to a rhoptry protein were available, it would be possible to follow the movement of its contents by immunofluorescence during invasion. It has been assumed that the sporozoite recognizes specific structures on the hepatic cell surface and that the CS protein plays a role in this recognition, since antibodies directed against the CS protein block hepatoma cell invasion in vitro. No specific hepatic receptor has been identified, however. The strict species specificity characteristic of the merozoite invasion of erythrocytes is not apparently applicable to sporozoite entry of hepatocytes. For example, *P. berghei* sporozoites can invade human hepatoma cell lines in vitro, whereas *P. berghei* merozoites cannot enter human erythrocytes. Currently, investigators are working under the assumption that sporozoites do bind to specific hepatic surface molecules and are considering that these may be the same for all species of *Plasmodium*. Although we know the complete gene sequence for several CS proteins, we understand little of the biochemical events that regulate sporozoite invasion. The gaps in our understanding are revealed by several questions: How does the sporozoite recognize the hepatocyte cell surface, or does it recognize the Kupffer or endothelial cells first? What triggers the nonphagocytic cells to apparently endocytose the parasites? Do any plasma components contribute to the invasion process? How efficient is the invasion process (how many sporozoites enter and survive)? Are rhoptry contents involved in entry, and what then is their fate? What is the fate of the CS protein during invasion? The limited availability of sporozoites has hindered efforts addressing these questions. Recently, however, several in vitro systems have been developed which may be suitable to examine some of these basic questions.

In Vitro Hepatocyte Culture System

In vitro systems have been developed to examine *P. berghei* sporozoite entry with a cultured human lung fibroblast cell line (WI38) and a hepatoma cell line (HepG2-A16). The fact that sporozoites enter fibroblasts at all suggests that receptor specificity is not particularly rigid, although it is unclear if the same receptors are employed by the two cell types. Attachment and entry occur more efficiently in the hepatoma cell line and appear to be active processes on the part of the parasite, since entry is not inhibited by antiphagocytic drugs such as cytochalasins. It has been observed that the sporozoite sheds its surface coat during invasion. Antibody directed against the CS antigen dramatically inhibited invasion in vitro, suggesting that the protein is directly responsible for recognizing certain receptors on the hepatic cell. Recently it has been found that antibodies raised against the repeat sequence of the *P. falciparum* CS protein, Asn-Ala Asn-Pro-, significantly block sporozoite attachment and invasion of primary hepatocytes in culture. Furthermore, sporozoites that do invade in the presence of antibody develop abnormally and in some cases fail to mature. *P. falciparum* and *P. vivax* sporozoites also infect the hepatoma cell line HepG2-A16. Although the hepatoma cell line system is suitable for in vitro invasion studies, a drawback is that after invasion, sporozoites of these two species develop only as far as intrahepatic trophozoites; mature schizonts and invasive merozoites cannot be obtained from this culture.

More recently, the complete development of hepatic stages of *P. falciparum* and *P. vivax* in vitro has been reported. With the use of primary cultures of human hepatocytes, it was possible to obtain mature and infective merozoites released spontaneously from the cultured cells. Primary cultures were inoculated with sporozoites obtained from mosquitoes fed on gametocytes. After a 20-hour incubation with sporozoites, intracellular forms could be seen. After 64 hours, intracellular schizonts were observed, and they developed to maturity after 6–7 days in culture. Erythrocytes were added to hepatocyte cultures on day 7, and erythrocytic ring stages were subsequently identified. Although the infectivity rate was low (1 ring per 3×10^5 erythrocytes) such a system permits the examination of liver-derived merozoites. Because of the small number of parasites obtained in these cultures, the techniques to examine these stages in greater detail must be very sensitive. Immunofluorescence is a moderately sensitive technique that would allow one to discern when certain proteins of the sporozoite are lost and when blood-stage antigens begin to be synthesized. This in vitro system to examine sporozoite entry obviously has great potential and will be suitable for studies on schizont development and stage-specific gene regulation.

The Erythrocytic Cycle

Ultrastructure

The erythrocytic cycle is initiated after merozoites leave the ruptured hepatic cell. The extracellular existence of merozoites in the plasma is transitory, and immediately on the invasion of erythrocytes, it is transformed to the ring stage. The merozoite is oval in shape, about 1 μm in diameter, has a nucleus and various cytoplasmic organelles, and is surrounded by a double lipid bilayer membrane and thick surface coat. Beneath the inner plasma membrane is a row of microtubules that may function as a cytoskeleton responsible for maintaining cell shape. The apical end of the merozoite is distinguished by a cone-shaped structure called the conoid. The underlying inner membrane is discontinuous at this end. During invasion, the rhoptries and associated micronemes move to the apical end, and a ductule extending from each rhoptry forms a common duct that extends to the lip of the conoid structure. (Similarities with the ultrastructure of *Toxoplasma gondii* are described in Chapter 2.) Many merozoite proteins have been identified, and their probable location is shown in the diagram in Figure 1-1. In addition to several organelles common to the Apicomplexa, merozoites contain mitochondria. Their mitochondria have poorly developed cristae, and it is therefore assumed that certain biochemical components of the respiratory chain are not functional, although this has not been investigated in any detail. The major metabolic end product of the organism is lactate, indicating that glycolysis is the important pathway to energy production. Another structure, known as the multilamellae structure, is an organelle bounded by layered membranes and usually located close to the mitochondrion. This location has prompted speculations that the multilamellae structure serves as an energy reservoir. The remainder of the cytoplasm is filled with ribosomes apparently unbound to endoplasmic reticulum.

The merozoite undergoes many changes after invading erythrocytes. It loses its surface coat and appears to discharge contents from the rhoptries during invasion. It comes to reside inside an endosomal compartment called the parasitophorous vacuole (PVM) that is formed by the invagination of the erythrocyte plasma membrane. The transformation of *P. falciparum* merozoites to rings, but not their further development, has been observed in a cell-free system. The structures of the merozoite break down inside the erythrocytes; apical organelles and microtubules disappear. The rings are

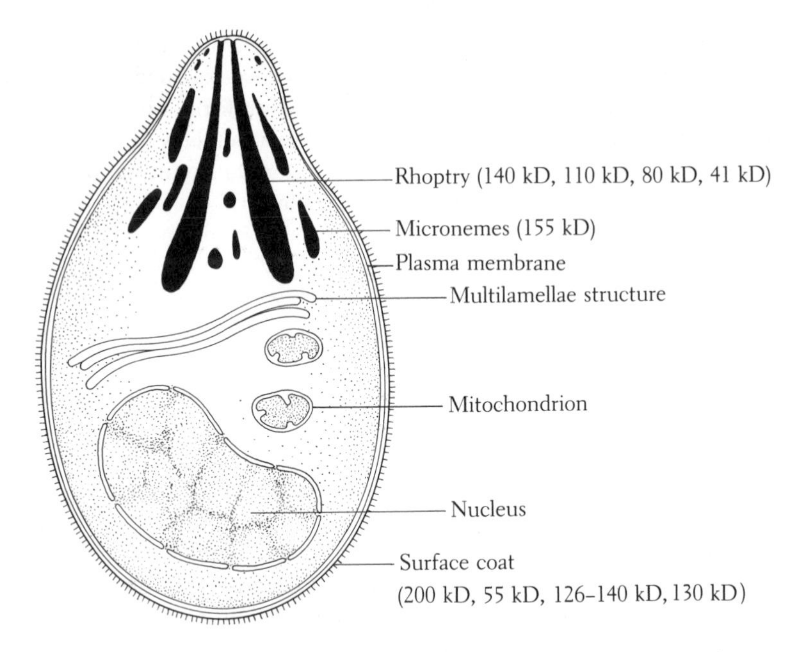

FIGURE 1·1 Localization of *P. falciparum* antigens. Major proteins of the merozoite stage that have been identified by monoclonal antibodies or biochemical properties. The surface-coat protein of 200 kD is processed to proteins of 90 kD and 50 kD at the merozoite stage.

biosynthetically quiescent stages, although ring stage–specific proteins are synthesized. Certain proteins that are present on other stages are absent on ring stages.

The uninucleate trophozoite is characterized by the loss of the regular shape of the ring and the appearance of large food vacuoles. At the trophozoite (feeding) stage, ingestion of the host cell cytoplasm occurs through a structure known as the cytostome. The cytostome invaginates into the parasite and is pinched off, fusing to form a food vacuole. The process of intracellular pinocytosis is found only in *Plasmodia* and a few other intracellular parasites. The contents of the food vacuole, principally hemoglobin, are digested by lysozymelike enzymes, the end product of hemoglobin being crystals of hemozoin. The antimalarial drug chloroquine appears somehow to disrupt the digestion of hemoglobin, or interferes with sequestration of one of the toxic end products, ferriprotophorphyrin (hemin). Parasites resistant to chloroquine may have developed an alternate pathway to process digested hemoglobin. More recent studies suggest that chloroquine-resistant strains of *P. falciparum* are more efficient in exporting chloroquine from the food vacuole. The proteases that regulate hemoglobin digestion have been studied with interest since their active sites could represent important targets for chemotherapeutic attack.

The intracellular schizont, by definition, has more than one nucleus. Schizogony (formation of the schizont) is the process of asexual nuclear division and is accompanied by considerable ultrastructural differentiation. The process of de novo formation of subcellular organelles such as the rhoptries and micronemes is uncommon in mammalian cells and therefore may be of particular interest to cell biologists. The mitochondria enlarge and bud, and the submembrane microtubules and rhoptries reappear, at first randomly, in the cytoplasm. The schizont membrane begins to invaginate and eventually pinches off, forming individual merozoites. The numerous nuclei, rhoptries, micronemes, and other organelles migrate into the developing merozoite. The only identified material remaining outside the merozoite is the hemozoin pigment, surrounded by its own membrane. Recently it has been possible to induce *P. knowlesi* trophozoites, stripped of the outer erythrocyte membrane, to divide to form multinucleated schizonts. At the time merozoites are ready for release, the host cell cytoplasm is fully digested and the parasitophorous vacuole membrane (PVM) has expanded to enclose the growing schizont. The mature merozoites are surrounded by their own double membranes in addition to the PVM and the erythrocyte membrane. Biological membranes do not readily disassemble; therefore, the nature of the signal that induces these outer two membranes to rupture, allowing the individual merozoites to be released, is of exceptional basic biological interest. It is known that the rupture of *P. knowlesi* schizonts is inhibited by chymostatin, suggesting that chymotrypsinlike enzymes may be involved in the release.

Structural differences that exist between different species may be noteworthy for future investigation. The most notable are structural differences in the modification of the cytoplasm and membranes of the infected erythrocyte. In erythrocytes infected with vivax-type parasites (for example, *P. vivax* and *P. cynomolgi*) caveolae-vesicle complexes located adjacent to the erythrocyte membrane have been observed. These complexes may correspond to Schüffner dots seen on Giemsa-stained blood films. The caveolae are typified by electron-dense clefts in the erythrocyte plasma membrane and closely associated vesicles in the host cytoplasm. In *falciparum*-type parasites (e.g., *P. falciparum*, *P. fragile*, and *P. malariae*), electron-dense excrescences or knob structures are observed and appear to be on the cytoplasmic face of the erythrocyte membrane. For *P. knowlesi*, electron-dense invagination or caveolae of the erythrocyte membrane of infected cells are observed. Since *P. vivax* and *P. falciparum* both infect human cells, the differences in cytoplasmic and membrane structures presumably are not related to the host red cell but appear to be dictated by the parasite. In cells infected with all types of primate malaria parasites, membrane-bound clefts in the erythrocyte cytoplasm are seen. These have been termed Maurer's clefts and appear under transmission electron microscopy as elongated vesicles. A parasite-synthesized protein of 50 kD has been identified as a component of the cleft membrane. The various cytoplasmic vesicles may be involved in intracellular transport from the intravacuolar schizont to the infected erythrocyte membrane. It has been proposed that knob structures in *P. falciparum*–infected cells are involved in the binding of infected erythrocytes to the endothelial cells of postcapillary venules during late schizogony. This sequestration process appears to be a protective measure against filtration of the cell by the spleen, and might provide a milieu of low oxygen tension favorable to terminal schizont development. However, *P.*

malariae–infected erythrocytes also have knobs but do not sequester in deep vascular beds; it therefore is unlikely that promotion of sequestration is the only function served by these excrescences. Once adapted to culture, *P. falciparum* rapidly looses the knob structures without consequence to its in vitro development. The chromosomal rearrangement of the gene for one of the knob-associated proteins has been shown to account for the loss of synthesis of this protein.

How may we investigate the function of some of the erythrocyte cytoplasmic-membrane structures, for instance, the intracytoplasmic vesicles and clefts that may be involved in cellular transport in the infected cell? Since certain parasite-synthesized proteins are known to be localized in the infected-erythrocyte membrane, it may be possible to follow their transport from the schizont to the outer membrane by immunocytochemistry, a technique that can localize proteins in imbedded sections of the cell. The visualization depends on binding of an antibody directed against the specific protein of interest, tagged with an electron-dense marker such as gold or ferritin. An example shown in Figure 1-2 uses a monoclonal antibody directed against a *P. falciparum* rhoptry protein.

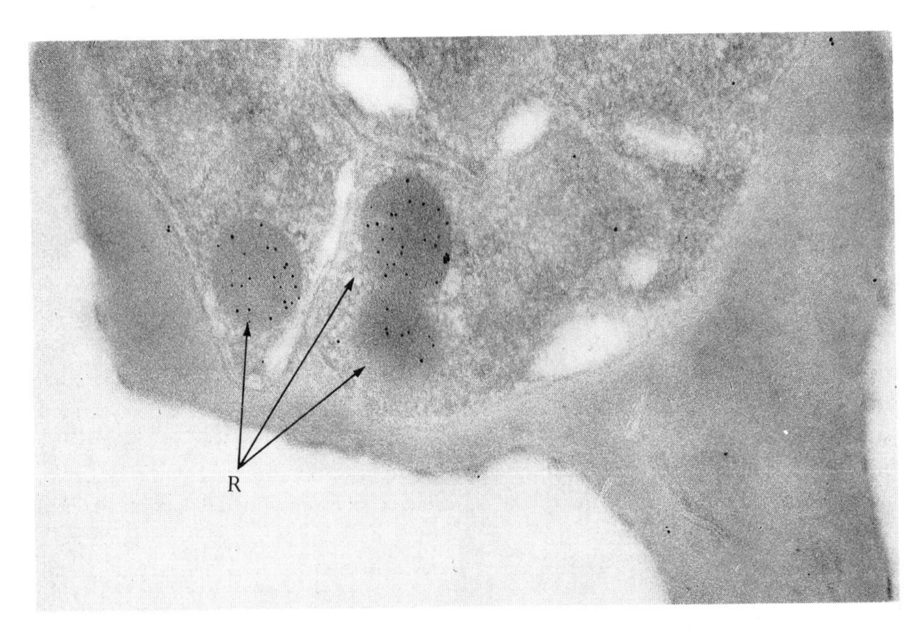

FIGURE 1·2 Immunocytochemical localization of a *P. falciparum* rhoptry protein. Merozoites were collected from in vitro cultures of *P. falciparum* and fixed in 1% glutaraldehyde. Cells were then incubated with monoclonal antibody 1B9 directed against a rhoptry protein of 110 kD. The gold particles (linked to the monoclonal antibody) are identified as small electron dense particles and localize the protein to the rhoptries. (Courtesy of T. Sam-Yellowe. Reproduced from the *Journal of Cell Biology*, 1988, 106:1507–1513 by copyright permission of the Rockefeller University Press.)

This brief description of morphology indicates that many novel processes of the parasite would be interesting to study at the molecular level. How does the merozoite invade the small, rigid erythrocyte? How is development in an intracellular environment regulated? What is the biochemical basis of the radical modification of the erythrocyte cytoskeleton? We shall discuss some of the processes in the following section.

Merozoite Invasion

Merozoite invasion of erythrocytes has long been a major interest of malaria biologists because it represents a process potentially vulnerable to inhibition by antibodies or drugs and because it is a unique event in terms of erythrocyte endocytosis. Its study can be traced to the early 1950s, when the species specificity of this interaction was first observed. Although the mechanism of merozoite entry appears to be similar in most species, considerable evidence suggests that different plasmodial species recognize different erythrocyte surface proteins as receptors. A consequence of this specificity is that for the most part merozoites in vitro invade only erythrocytes of their natural vertebrate host. For instance, *P. falciparum*, which infects humans and higher primates in vivo, invades only erythrocytes of humans, chimpanzees, and Aotus and Saimiri monkeys. *P. falciparum* merozoites do not invade erythrocytes of the rhesus monkey or avian erythrocytes. *P. berghei*, which causes natural infections in rodents, does not invade human erythrocytes. Such differences in host specificity must be very subtle, because many of the major erythrocyte surface proteins are highly conserved among mammals, although there are some interesting differences. Invasion requires at least two discrete steps: (1) receptor recognition and binding and (2) erythrocyte membrane deformation.

Receptor recognition: *P. falciparum, P. knowlesi,* and *P. vivax* Initial attachment between the merozoite and the erythrocyte can involve any area of the parasite's surface, but entry proceeds only after alignment of the apical end onto the erythrocyte surface. Attachment between the apical end and the erythrocyte is through fibrils of the merozoite surface coat. Whether the protein composition of the parasite's surface coat at the apical end differs from the rest of the surface coat is unknown. Attachment results in the formation of a tight junction between the two cells and subsequent invagination of the erythrocyte membrane. The tight junction formed moves back over the merozoite as the parasite proceeds to enter the forming parasitophorous vacuole created by the invaginating erythrocyte membrane. The thick surface coat does not enter with the merozoite; it is sloughed off during entry and may be deposited on the erythrocyte surface. It is possible that the shedding of the surface coat is an active process, and its accumulation at the neck of the vacuole might theoretically be associated with modification in the organization of erythrocyte surface receptors.

Early studies on merozoite invasion benefited from the extensive information available on the organization of the protein components of the erythrocyte membrane.

The two major surface proteins of the human erythrocyte are an anion transporter designated band 3 and glycophorin A, the most abundant in a family of related sialic acid–rich glycoproteins. The glycophorin family also consists of glycophorin B, C, and D. Glycophorin A and C, like band 3, span the lipid bilayer and therefore are said to be transmembrane. The unique feature of the sialo-glycoproteins, which accounts for their name, is their high (30 percent) content of sialic acid. All three glycophorins are highly substituted with the O-linked tetrasaccharide of the following structure: NeuNAc 2–3 Gal β 1–3 (NeuNAc 2,6) GalNAc. In addition, glycophorin A and C contain a single N-linked oligosaccharide. Considerable evidence has accumulated that *P. falciparum* merozoites depend on the glycophorins for invasion into human erythrocytes. Evidence comes from studies with abnormal erythrocytes having modified glycophorins and with enzyme-treated erythrocytes. The N-terminal tryptic fragment of glycophorin (containing 11 O-linked tetrasaccharides) has been found to inhibit invasion, as do Fab fragments of monoclonal antibodies to glycophorin, which provides additional evidence. En(-a) erythrocytes lack glycophorin A, S-s-U cells lack glycophorin B, and $Mk^{-}Mk^{-}$ cells lack glycophorin A, B. Tn cells lack the terminal (2–3) linked NeuNAc and Gal residues of the tetrasaccharide, and Cad cells contain an extra GalNAc residue in the tetrasaccharide. Invasion of *P. falciparum* merozoites into all these abnormal cells is considerably reduced, although it is entirely blocked only in the case of the Tn and Cad cells (in which all three glycophorins are altered). This finding illustrates that all three glycophorins can act as receptors independently. The feature common to all glycophorins is the O-linked tetrasaccharide, and other recent data suggest that this is the critical site recognized by the merozoites. Neuraminidase and O-glycanase–treated erythrocytes are also resistant to invasion, further supporting the model for direct involvement of the carbohydrates in the binding site. Invasion by *P. knowlesi* and *P. vivax* is normal in the glycophorin modified erythrocytes, underscoring the specificity of the interaction.

Recently it has been found that variability in receptor recognition exists between different geographic isolates or strains. This finding is based on studies of invasion of seven isolates into neuraminidase- and endoglycosidase-treated erythrocytes. Although some strains were found to be fully dependent on NeuNAc and glycophorin, in that invasion into neuraminidase and O-glycanase–treated cells was inhibited almost completely (5 percent invasion), other strains still invaded these same erythrocytes partially (25%–50%). It has been suggested that NeuNAc-independent invasion may reflect the ability of the merozoite to interact with a secondary low-affinity receptor other than glycophorin. This receptor appears to be a trypsin-sensitive component of the erythrocyte surface and therefore could not be band 3, which is trypsin- and neuraminidase-resistant. Soon after receptor binding, a tight junction forms between the merozoite and erythrocyte.

In contrast to the considerable detail known about the receptors for *P. falciparum*, information on the receptors for other species of *Plasmodium* has been elusive due in part to the difficulty in culture of these species in vitro. The Duffy blood group antigen, a 35–45kD glycoprotein, has been proposed as the receptor for both *P. knowlesi* and *P. vivax*. Neither species invades Duffy negative erythrocytes, analogous to the situation with *P. falciparum* and En(-a) erythrocytes. However, trypsin treatment of Duffy negative cells renders them susceptible to invasion by *P. knowlesi*, indicating the parasite is able to interact with a secondary receptor made available in these cells. Monoclonal antibody directed against the Duffy glycoprotein also blocks invasion by *P. vivax*.

Erythrocyte membrane deformation The events that induce deformation of the erythrocyte, resulting in its invagination and ultimately in the formation of the endocytotic vacuole membrane (PVM), have not been defined. Given the rigid nature of the erythrocyte membrane and the absence of any phagocytic ability of the cell, certain key processes are thought to occur. To appreciate these, a consideration of certain properties and components of the erythrocyte membrane is helpful.

Measurements on the deformability of erythrocytes indicate that the membrane exhibits properties similar to a semisolid with considerable elastic properties. The structural components responsible for the rigid–elastic property are the proteins of the membrane-cytoskeleton network. The major proteins of the cytoskeleton are the heterodimers of spectrin, ankyrin, band 4.1, erythrocyte actin, and band 4.9. Spectrin is linked to the major transmembrane protein band 3 by ankyrin and to the second major transmembrane protein, glycophorin A by band 4.1. These proteins form a highly ordered uninterrupted network that is extremely difficult to disrupt. Even in erythrocyte ghosts, (hypotonically lysed cells devoid of hemoglobin) the superstructure of the membrane-cytoskeletal network is maintained. Evidence that spectrin and its interactions with the other membrane proteins are central to the control of cell shape and rigidity comes from experiments in which the organization of spectrin is modified and from studies of abnormal erythrocytes with cytoskeletal defects. For instance, in hereditary elliptocytosis (a condition in which afflicted individuals have elliptical shaped erythrocytes) the biochemical alteration has been traced to an imbalance in the normal ratio of spectrin dimer–tetramer; more of the spectrin dimer is present in abnormal than in normal cells. In other abnormal cells, spherocytes, there is an absolute reduction in the amount of spectrin. In still other cells, the binding of spectrin to ankyrin may be impaired. Such abnormal erythrocytes are not uncommon and could be of considerable use in elucidating the role of spectrin and other cytoskeletal proteins in merozoite invasion.

After merozoite receptor binding and junction formation, the erythrocyte membrane invaginates and forms an endocytotic cup or space while the merozoite remains attached to the neck of the forming vacuole by the tight junction. The vacuolar membrane appears to expand rapidly until it is large enough to surround the invading merozoite. The expansion of the host membrane implies that there is probably a dramatic reorganization of the major transmembrane and cytoskeletal proteins spectrin and band 3. Indeed, freeze-fracture studies of the vacuolar membrane suggest that it is free of the transmembrane proteins, indicating that proteins may migrate to the tight junction at the entrance of the vacuole, creating the protein-poor, lipid-rich region. The critical question, however, is how this movement or redistribution of erythrocyte protein is regulated. Presumably the transmembrane proteins dissociate from the cytoskeletal proteins, and the individual components of the cytoskeleton are modified. We are left with no understanding of how the parasite may initiate these events. Two of several mechanisms that could account for erythrocyte membrane deformation appear the most accessible to investigate.

One mechanism, similar to that described for the receptor-mediated endocytosis of virus and certain ligands, could be that binding of the merozoite to the erythrocyte receptors results in their cross-linking or capping. The redistribution could result in disruption or uncoupling of the interactions of the transmembrane proteins with proteins of the cytoskeletal network. Capping of surface receptors in mammalian cells has been shown to be accompanied by rearrangements of the components of the cytoskeletal, which, by some unknown mechanism, regroup around the developing endocytotic vacuole and draw it into the cell cytoplasm. Studies on merozoite invasion with abnormally shaped erythrocytes (ovalocytes) supports a model in which dissociation of cytoskeletal components is essential for invasion. The ovalocytes are resistant to invasion by *P. falciparum* and *P. vivax*. This abnormal cell is at least one-tenth as deformable as normal erythrocytes. Specific alterations in biochemical associations between cytoskeletal membrane proteins contribute to this low deformability. Resistance to invasion may be due to an inability of the erythrocyte membrane proteins to dissociate and reorganize. Merozoites also do not invade erythrocytes fixed in low (1 percent) concentrations of glutaraldehyde. If receptor binding induces membrane-protein reorganization, there must be proteins on the surface of the merozoite that bind with high affinity to the erythrocyte receptor. Parasite proteins that bind to the correct receptor have been identified for *P. falciparum*, *P. knowlesi*, and *P. vivax*.

A second model to explain erythrocyte deformation involves the two organelles known as rhoptries and micronemes. Since these organelles are only found in the invasive stages of the parasite, the sporozoite and merozoite, it has been assumed that they are involved in invasion. Electron micrographs and immunofluorescence micrographs can be interpreted to suggest that some material is secreted from the rhoptries into or onto the erythrocyte membrane (Figure 1-3). The rhoptry proteins could be hydrophobic and thereby contribute to enhanced fluidity of the erythrocyte membrane. As a result, redistribution or destabilization of erythrocyte transmembrane proteins could occur. A diagram depicting a possible model of the various merozoite surface and rhoptry proteins during invasion is shown in Figure 1-4.

How may we examine the role of the proteins of the rhoptries and micronemes in merozoite invasion? During the course of production of monoclonal antibodies to *P. falciparum* merozoite antigens, common monoclonal antibodies that were selected in many laboratories were those directed against five proteins of the rhoptries; the proteins were 140 kD, 130 kD, 110 kD, 83 kD, and 41 kD. The antibodies gave a characteristic immunofluorescent pattern of two punctate dots in each merozoite. By immunoelectron microscopy they have been localized to the electron-dense rhoptry organelle, as shown for the 110-kD protein in Figure 1-2. With use of the antibodies it should be possible to purify the rhoptry proteins by immunoaffinity chromatography. It would then be necessary to devise a system to examine the insertion of such proteins in the erythrocyte membrane or in synthetic biomembranes containing erythrocyte-mem-

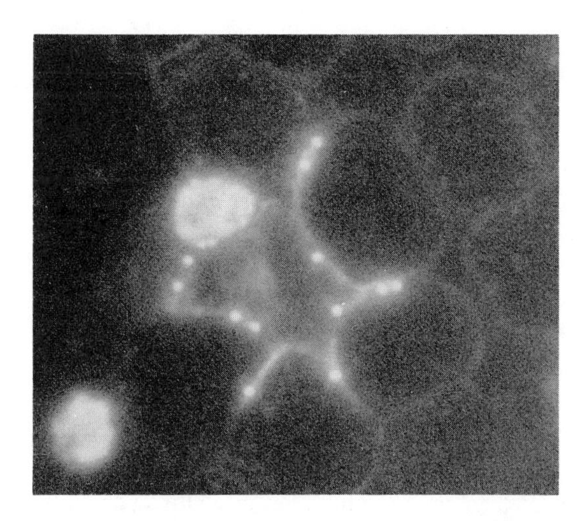

FIGURE 1·3 Immunofluorescent localization of rhoptries during merozoite entry. A thin smear of cells from an in vitro culture of *P. falciparum* was fixed in acetone and overlayed with monoclonal antibody 1B9 which precipitates a protein of 110 kD. Cells were washed and then incubated with FITC-rabbit anti-mouse antibody. (Courtesy of T. Sam-Yellowe. Reproduced from the *Journal of Cell Biology*, 1988, 106:1507–1513 by copyright permission of the Rockefeller University Press.)

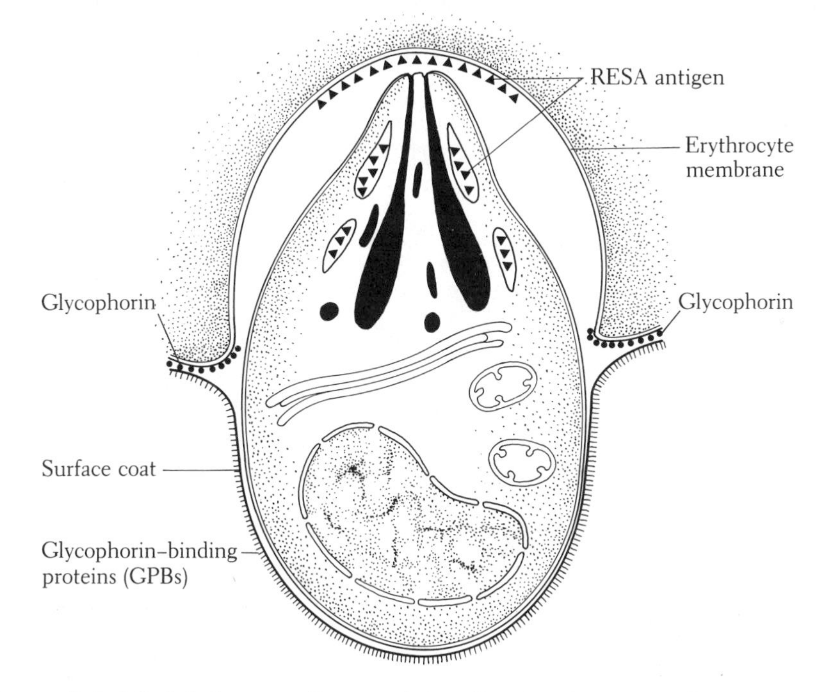

FIGURE 1·4 Model of merozoite protein localization during erythrocyte invasion. The glycophorin binding proteins bind to the erythrocyte receptor and remain external to the cell during invasion. The RESA antigen and perhaps other proteins of the rhoptries and micronemes may be inserted into the lipid bilayer after the receptor binding stage. From electron microscopic studies it appears that the merozoite surface coat is shed during invasion and that the electron-dense nature of the rhoptries changes, suggesting that material is secreted from this organelle during invasion.

brane proteins. Important questions should be addressed: Are the rhoptry proteins incorporated into the lipid bilayer? Do erythrocyte-membrane proteins—band 3 and glycophorin—alter their distribution or specific interaction with each other and with cytoskeletal proteins upon incorporation of the parasite proteins? This question could be examined by assessing changes in patterns of cross-linking with chemical cross-linkers. Biochemical analysis of protein organization could also include studies of phosphorylation patterns, since phosphokinase and phosphorylase reactions regulate interactions between spectrin and other cytoskeletal proteins.

Resealed erythrocyte ghosts provide a system that may be useful in studies of invasion. Ghosts are erythrocytes that have been hypotonically lysed and resealed. Red ghosts still contain some hemoglobin; white ghosts are devoid of hemoglobin. The amount of hemoglobin remaining after lysis depends on the procedure for lysis and resealing. During the resealing step, it is possible to include reagents that alter the internal (or cytoplasmic) face of the erythrocyte membrane. The resealed ghosts are susceptible to merozoite invasion. Preparation and properties of erythrocyte ghosts, susceptible to invasion by the malarial parasites, have been described in detail. How may the system be used to study cytoplasmic events involved in invasion? The intraerythrocytic levels of ATP, inorganic phosphate, and intracellular CA^{++} alter the association between spectrin, actin, and other red-cell proteins. In ghosts, it is possible to immobilize these states, whereas in intact erythrocytes, equilibrium is rapidly reestablished. In fact, reducing ATP levels in resealed ghosts was found to reduce merozoite invasion, suggesting that phosphorylation of certain membrane proteins was important to invasion. Interactions between cytoskeletal proteins and transmembrane proteins also can be modified with specific antibodies that could be included in the resealing step. Such modifications may result in organizational changes in the external domains of transmembrane proteins (band 3 and glycophorin) that may affect merozoite invasion. Understanding merozoite invasion will eventually give us greater insight into the structural–functional relationship of erythrocyte membrane proteins.

Protein Trafficking in the Infected Erythrocyte

In all eukaryotic cells, proteins are synthesized on ribosomes present randomly in the cell cytoplasm. Shortly thereafter proteins find their way to specific subcellular organelles, the plasma membrane, or a secretory structure. Although the exact mechanisms of intracellular trafficking are not yet known, there is good evidence to suggest that coded in the protein sequence are certain peptides or other components that are specifically recognized by membranes of the organelles to which the proteins are destined to travel. The mechanism for sorting of secretory proteins is known in considerable detail. A signal peptide on the N-terminus is recognized by a signal-recognition molecule that binds to a receptor or docking protein on the surface of the endoplasmic reticulum (ER). The emerging protein is then secreted into the lumen of the ER and eventually secreted when the ER fuses with the plasma membrane. A consensus-signal sequence, seemingly contained in all secreted proteins, has been established, and no doubt other consensus sequences or domains that define transport to other organelles such as mitochondria and lysozomes will also be found.

In the erythrocyte infected with a developing malarial parasite, parasite proteins are inserted into the developing membranes of the schizont. It is not difficult to imagine the information needed for export to this membrane, because the route would be similar to that in a free-living eukaryotic cell. However, proteins are located in the space between the schizont membrane and the PVM, and as the PVM expands, parasite proteins are no doubt inserted into this membrane as well. Such a protein may have two signal peptides, one to export it to the outside of the schizont and one to recognize the PVM, or it may have a hybrid signal that is sufficient to direct it to the second membrane. When a protein is translocated through the PVM, through the erythrocyte cytoplasm, and then inserted into the erythrocyte membrane or secreted, the signal sequences must be much more complex. Do such proteins have three or more signal peptides, with each one recognized at each membrane for translocation, or is there a single hybrid signal sequence that guarantees its final destination? Clearly these are questions of interest to scientists working on intracellular trafficking. In very few systems is there such complex trafficking as in the plasmodium-infected erythrocyte, and through study of this system we may learn some important principles governing protein recognition with respect to organelle and membrane biogenesis.

As yet only a few specific parasite proteins have been unequivocally localized by microscopy to the erythrocyte membrane and none have been identified in the PVM. These are the knob-associated proteins of 80 – 100 kD and a protein called MESA (merozoite-infected erythrocyte surface antigen) of 300 kD. However, there is indirect evidence that many parasite proteins probably exist in both membranes. By comparing sequences or tertiary structures of such proteins, we may be able to identify peptides or domains responsible for their export from the schizont and translocation across the vacuole space or erythrocyte cytoplasm.

Modification of erythrocyte structure and function Few alterations in the host-cell ultrastructure or function that occurs after the merozoite invades the erythrocyte have been identified. The density of the ring-infected erythrocyte, its morphology, and its deformability are almost indistinguishable from that of the uninfected cell. About 12 to 16 hours after invasion, when the ring stage begins to transform to the trophozoite stage, certain events occur. Parasite enzymes that specifically degrade hemoglobin are present. Since hemoglobin is the main source of amino acids for the developing schizont, it is very important to understand this pathway in detail. At the end of the intraerythrocytic stage, almost all the host hemoglobin has been digested.

Major changes have been described in the activity of ion transporters of the erythrocyte membrane as the parasite develops from a trophozoite to a schizont. Ca^{++} levels are increased 25-fold in *P. chabaudi* – infected cells, possibly the result of an altered Ca^{++}-ATPase in the erythrocyte membrane. There also may be newly synthesized Ca^{++} transporters inserted into the PVM. The presence of parasite-coded or -dependent transporters for monosaccharides and transferrin has also been reported. No new or modified transporters have been isolated or their protein structure analyzed in any detail, although they could represent model systems to examine membrane protein topogenesis. Theoretically, some might be targets for novel approaches to chemotherapy.

Probably the most dramatic change in the infected erythrocyte is the change in cell shape and deformability. The regular discoid shape and smooth surface of the normal erythrocyte is lost and the cell assumes an irregular shape with large protrusions, not unlike a sack of potatoes. It is assumed that the rigid erythrocyte cytoskeletal membrane network is radically altered. This is true of cells infected with all species of *Plasmodia* regardless of whether they do or do not synthesize knob structures or caveolae. One consequence of the decreased deformability is enhanced filtration in the splenic red pulp. What is the origin of these modifications? Does the parasite secrete proteases that cleave the cytoskeleton proteins? Does the parasite synthesize proteins that bind to erythrocyte membrane proteins and disrupt their organization? Perhaps only in virus-transformed mammalian cells are such dramatic reorganizations of the existing cell structures observed. The molecular explanation of the modifications may be relevant to general questions of maintenance of cell shape and morphology. How may we study these interactions? It may be possible to mimic the morphological changes by treating erythrocytes with reagents known to alter cytoskeletal or membrane protein interactions. Alternatively, the cytoskeleton from infected cells can be extracted with minimum perturbations, and the altered associations can be characterized by classical methodologies.

In *P. falciparum*–infected erythrocytes a major modification of the erythrocyte membrane is the appearance of knob structures, electron dense plaques on the cytoplasmic side of the membrane. Two parasite-synthesized proteins have been localized to the knobs. Sequestration of infected erythrocytes in the postcapillary venules is mediated by adherence of the erythrocyte surface located over the knob. The cell–cell adhesion appears to involve specific proteins on both the surface of the infected erythrocyte and the endothelial cell. An 88 kD glycoprotein recognized by the monoclonal antibody OKM5 has been suggested as possible endothelial cell surface receptor involved in adherence. A corresponding protein on the infected erythrocyte surface, a so-called cytoadherence antigen, has been somewhat elusive to define, although antibody inhibition of endothelial cell adhesion by immune serum was found to be strain-specific, suggesting such a molecule is antigenically variant. Antibodies against the 80 kD and 300 kD proteins of knobs do not inhibit cytoadherence, suggesting they are not directly involved in binding.

In vitro cultivation of *P. falciparum* asexual stages Many of the experiments proposed earlier can be performed with *P. falciparum* cultured in vitro. The practicality of using in vitro-cultured parasites rather than parasites taken from infected animals is becoming increasingly clear. The successful cultivation of *P. falciparum* in vitro was first reported in 1976, and most laboratories now use this method with minor modifications. Low oxygen tension is an important condition for cultivation. The parasite is cultured in human erythrocytes in a simple medium with 10 percent human serum. It is possible to culture 20–30 ml of erythrocytes at a time, generally 8%–15% of which are parasite-infected, so that after concentration with gelatin, approximately 1 ml (5×10^9 cells) of schizont-infected erythrocytes can be obtained. For some unknown reason, a parasite density greater than 20 percent in in vitro cultures is difficult to maintain. Parasite proteins or lipids can readily be labeled with radioactive amino acids or other metabolites by including these in the medium. Erythrocytes, which have no protein-synthesizing machinery, will not incorporate the label. There have been reports on the culture of other species of *Plasmodium*, *P. berghei* and *P. knowlesi*, although *P. falciparum* is still the most popular culture system. To date there has been no successful long-term culture of *P. vivax*, another important species that infects humans. This has retarded aspects of research on the cell biology of *P. vivax*, but with the availability of cDNA libraries, it has been possible to sequence some of the important antigens of this species. It is not understood why other species, such as *P. vivax*, do not grow in vitro under the conditions established for *P. falciparum*. This problem may be interesting to investigate, although past failures by other workers should be thoroughly sorted out before an attempt is made to initiate such a project. Another advantage of in vitro cultivation is that it is possible to synchronize the various stages, for example, rings, trophozoites, and schizonts, and identify when specific proteins are expressed.

Isolation of *P. falciparum* merozoites The in vitro culture technique has also made possible the collection of viable *P. falciparum* merozoites. Although it was relatively easy to obtain infectious *P. knowlesi* merozoites from short-term cultures of parasites taken from an infected monkey, the shorter viability time of the *P. falciparum* merozoite (estimated to be 60 minutes) made similar studies with this species difficult. The isolation of viable *P. falciparum* merozoites was dependent on the ability to obtain synchronous cultures in which a high percentage of schizont-infected cells burst simultaneously. In the first such method reported, highly synchronous cultures were produced by repeated cycles of concentrations of schizont-infected cells with gelatin, then the most mature population were allowed to reinvade in a short period (4–6 hours), and nonreleased schizonts were lysed with sorbitol. When merozoite release began, mature schizont-infected erythrocytes and uninfected erythrocytes were removed by low-speed centrifugation (600 *g*) and merozoites in the supernatant were collected by centrifugation at 2000 *g*. It was estimated that 25 percent of the schizont population released merozoites in a 2-hour period. A maximum of $2-5 \times 10^9$ merozoites could be obtained in 1 hour from 2 ml of parasitized blood, although only 10 percent of these were infective.

Several methods now exist for the isolation of spontaneously released *P. falciparum* merozoites from culture, and the modifications have resulted in greatly increased yields. One method for collecting spontaneously released merozoites over a long period involves the use of a nylon membrane seive to remove debris and schizont-infected cell membranes. Merozoites were collected in the supernatant from concentrated cultures over a 6-hour period. Schizonts were removed from culture by low-speed centrifugation (200 *g*). The supernatant, containing merozoites, broken membranes, and hemoglobin, was filtered through an acrylic nylon membrane sieve (3 μm pore size) and then a membrane of 1.2 μm pore size. The effluent containing the merozoites was centrifuged at 3000 g for 10 minutes. From cultures containing 5 × 10^9 schizont-infected erythrocytes, 10^{10} merozoites were released and over 70 percent were recovered. The infectivity of the merozoites was estimated to be 80 percent.

For certain types of studies on invasion, it has been and will continue to be preferable to start with isolated merozoites rather than mature schizonts. In studies with free merozoites, problems that arise because of the toxic effects of reagents on the mature schizont are eliminated. It is anticipated that any biophysical studies on the erythrocyte membrane during merozoite invasion will require merozoites, since it is only with merozoites that one can obtain a tightly synchronous invasion. However, if we want to detect biochemical and biophysical changes in the erythrocyte membrane during invasion, it will be necessary to have highly efficient invasion rates to permit measurement of subtle parameters of invasion. This will require large-scale production of invasive merozoites. The rapid loss of viability is the major obstacle to overcome, and establishing procedures to stabilize or preserve released merozoites will be helpful.

The Sexual Cycle

Gametocyte Development and Culture

The gametocyte is the stage that links the asexual blood stages of the vertebrate host to the sexual stages of the mosquito. For reasons that are not understood, during the course of an infection, a certain percentage of asexual ring stages will not develop to form schizonts but will form mature male or female gametocytes that are infectious to the mosquito. First observed in vitro in 1976, the process of gametocytogenesis occurs in in vitro cultures; development from rings to mature infective gametocytes requires 12–14 days. A small proportion of parasites develop into gametocytes at any given time. Strains differ in their propensity for gametocytogenesis in vitro; some strains rarely produce gametocytes in culture. Supplementation of the media for cultivation of the asexual blood stages with cyclic AMP or hypoxanthine promotes gametocytogenesis to variable degrees in different strains. Gametocytes produced in vitro are capable of completing the sexual cycle; mosquitoes fed on gametocyte-containing cultures produce sporozoites infectious to chimpanzees and *Aotus* monkeys. Various attempts have been made to improve the efficiency of the in vitro culture of gametocytes, to produce sufficient numbers of cells for biochemical and immunological studies. Two problems exist. One is to induce asexual stages to switch to gametocytes; the second is to promote gametocyte maturation. Once mature gametocytes have been produced, it is a relatively straightforward process to induce gamete, zygote, and ookinete formation.

A recent development using a semiautomated static culture may prove to be the most dependable system yet for the production of gametocytes. This system relies on propagation under prolonged uninterrupted periods of growth in the appropriate gas levels and temperature: 3 percent O_2, 4 percent CO_2, 93 percent N_2 at 37°C. In this system, the number of gametocyte-infected erythrocytes can reach about 1 percent, or about 50 percent of the total parasite (sexual and asexual stage) -infected cells. It is possible that exposure to ambient temperatures and gases inhibits transformation. Some isolates are better gametocyte producers than others, a fact that may be related to the density of the asexual stages from which they are derived; high densities of asexual stages appear to be correlated with low levels of gametocyte production. It is possible that conditions favorable to the asexual stages are unfavorable to gametocytes or that adverse conditions unfavorable to the blood stage induce gametocyte formation. When human serum is replaced with fetal bovine serum, an increase in infective oocyst formation or gametocyte formation is observed. Clearly the development of a dependable system for the production of large numbers of gametocytes in culture will be of great value. Major questions to be studied concern signals that regulate gene expression during gametocytogenesis and expression of male and female specific genes.

Gamete Development

While still in the bloodstream of the infected host, the gametocyte will not develop further. After reaching maturity, 10–14 days after invasion of the erythrocyte, it is fully infectious to mosquitoes. If not ingested it probably disintegrates, which may be the reason antibodies directed against gamete antigens are sometimes found in infected individuals. Gametocyte production in the host is thought to be staggered, which would increase the chances of a mosquito being infected at the appropriate stage for further development. Immediately upon being ingested by the mosquito, the gametocyte undergoes rapid transformation. In a process known as exflagellation, the mature gametocyte sheds the red cell membrane, and the macrogamete (female) and microgametocyte (male) are liberated in the mosquito midgut. The microgametocyte completes exflagellation by dividing into eight motile microgametes. The microgamete then fuses with a macrogamete to form the zygote. Exflagellation and fertilization can be controlled in vitro: by the regulation of the temperature and the pH of medium to between 7.7 and 8.0, it was possible to ensure complete exflagellation production in *P. gallinaceum*. The media used for transformation contained 10 millimole (m*M*) Tris, 155 mM NaCl, 10 mM glucose, and 25 m*M* $NaHCO_3$ pH 8.0. Gamete production was reduced if $NaHCO_3$ was not included in the buffer. Similar conditions induced gametogenesis in *P. falciparum*. Fertilization to form the zygote takes place immediately (10–20 min) after exflagellation. Fertilization has been described in detail in many electron microscopic studies, but little is known of the biochemical events governing fusion. Several methods exist for the separation of the large macrogametes and zygotes from the smaller microgametes. One method involves layering the blood on a Hypaque-Ficoll gradient and centrifuging for 10 minutes at 10,000*g*. Others have used metrizamide gradients or wheat germ agglutinin (lectin) chromatography to select macrogametes.

Ookinete Development

Conditions supporting the in vitro transformation of the fertilized zygote to the ookinete have been established for *P. gallinaceum*. Zygotes were incubated in Medium 199 supplemented with 17 mM dextran, 25 mM $NaHCO_3$, and 0.7 mM L-glutamine, plus antibiotics. After 24 hours, fully mature ookinetes were observed. These could be separated from unfertilized female gametes by centrifugation on a Hypaque/Ficoll gradient; mature ookinetes, being more dense, were collected in the pellet fraction. However, this step represents the end of the line for the in vitro cultivation of plasmodia. The last remaining link in the cycle, that of the development of the ookinete to the oocyst to the sporozoite, has not been established yet in culture. As a consequence, little is known of the biochemical steps involved.

As the zygote develops it becomes a motile ookinete, which penetrates the intestinal wall of the mosquito. This interaction has been studied morphologically but not biochemically. The ookinete develops into an oocyst when it reaches the outer layer of the wall, the basement membrane. The factors triggering ookinesis are unknown. The ookinete is similar in structure to the merozoites and sporozoite, having a double plasma membrane with a truncated cone-shaped apical end, except that it does not have rhoptries. The ookinete migrates through the epithelial cell layer, possibly between cells. There is no biochemical description of migration of the ookinete through the gut wall to the basement membrane and development to an oocyst. Initially the oocyst is a single cell surrounded by a double membrane, with a single nucleus. Nuclear division occurs, and as the oocyst grows, vacuoles appear to form large clefts. The clefts subdivide the cytoplasm to form sporoblasts, which give rise to sporozoites, in a manner morphologically similar to the development of merozoites from schizonts.

Identification of Surface Antigens of Sexual Stages

The availability of the in vitro system for production of *P. gallinaceum* gametes in culture has made possible the identification of their surface proteins and that of the fertilized zygote and ookinete. By immunizing rabbits with gametes and zygotes, antibodies against the surface antigens were obtained and found to block zygote formation, principally by binding to the gamete surface. Monoclonal antibodies blocking development of *P. gallinaceum* midgut stages were produced and found to immunoprecipitate antigens of 240 kD, 56 kD, and 54 kD present on both male and female gametes and zygotes. Other proteins of 48 kD, 19 kD, and 17 kD were on female gametes only, and a protein of 28 kD was on male gametes only. As zygotes undergo transformation to ookinetes, these proteins are shed from the parasite surface. Two new proteins of 28 kD and 26 kD appear on the ookinete surface; the 26 kD protein is immunoprecipitated by monoclonal antibodies that block ookinesis.

The molecular events that regulate gametocytogenesis and fertilization of the micro- and macrogamete are of considerable importance but little understood. Now that reliable in vitro culture systems are available, this question should be more accessible to investigate. Another problem that has received little attention is the migration of the ookinete through the mosquito gut mucosal cells. This could be examined by devising the appropriate endothelial-cell culture system and examining migration in vitro. Since each of these steps represents a stage for blocking parasite development, they all would be worthwhile for further study.

Coda

The complex transmission of plasmodia between vertebrate and invertebrate host involves a single hepatic cycle, multiple erythrocytic cycles, and a single sexual cycle. Continuous culture in vitro of the erythrocytic stage of *P. falciparum* has been achieved, and limited cultivation of other species is also possible. Gametocytes can be obtained from these cultures, and methods are available to induce gamete, zygote, and ookinete formation. It is also possible to grow the complete hepatic stages of *P. falciparum*, *P. vivax*, and *P. berghei* in vitro. Despite certain limitations on the quantities of parasites that are available from the in vitro systems, considerable progress in our understanding of the biology of the parasite has been made since their introduction. The expression in *Escherichia coli* of the genes of some plasmodial antigens has allowed for the purification of amounts of the proteins that have permitted characterization of their immunological and biochemical properties. Also of great interest in the near future will be questions concerning regulation of gene expression within each of the different cycles of the parasites, and the factors that regulate the transformation from one antigenically distinct stage of the parasite to the next.

Additional Reading

AIKAWA, M., and SEED, T. M. 1980. Morphology of plasmodia. In *Malaria*, vol. I., J. P. Kreir, ed. Academic Press, pp. 285–344.

BENNETT, V. 1985. The membrane skeleton of human erythrocytes and its implications for more complex cells. *Ann. Rev. Biochem.* 54:273–304.

BREUER, W. V. 1985. How the malaria parasite invades its host cell, the erythrocyte. *Int. Rev. Cytol.*, 96:191–238. Academic Press.

HOWARD, R. J. 1986. Malaria: antigens and host-parasite interactions. In *Parasite Antigens: Toward New Strategies for Vaccines*, T. W. Pearson, ed. Marcel Dekker, pp. 111–165.

NUSSENZWEIG, V., and NUSSENZWEIG, R. S. 1985. Circumsporozoite proteins of malaria parasites. *Cell* 42:401–403.

PASVOL, G. 1984. Receptors on red cells for *Plasmodium falciparum* and their interaction with merozoites. *Phil. Trans. R. Soc. Lond.* B307:189–200.

SHERMAN, I. 1979. Biochemistry of Plasmodium (malaria parasites). *Microbiol. Rev.* 43:453–495.

2

Cell Biology of *Toxoplasma gondii*

■

Elmer R. Pfefferkorn

Toxoplasma gondii, an obligate intracellular protozoan parasite, was first described by Charles Nicolle in 1908, one year before he did the work on typhus that earned him the Nobel Prize in Medicine. Nicolle and a colleague found *T. gondii* in the African rodent *Ctenodactylus gondi*, which supplied the parasite's species name. The genus name does not refer to any toxic property of the parasite but instead is derived from the Greek *toxon*, bow (as in bow and arrow), in reference to the crescent shape of the organism. Although *T. gondii* has been known for eight decades, it is the most recent of the major human parasites to have its life cycle understood. Not until 1970 did several laboratories independently establish the critical role of felines as the definitive host.

Infection of humans with *T. gondii* is common but generally asymptomatic. Toxoplasmosis poses a major clinical problem in two contexts. If a woman has a primary infection during pregnancy, her developing fetus may also become infected and suffer severe congenital defects. Recrudescent or primary toxoplasmosis in patients with a defect in cell-mediated immunity, as in AIDS, can result in fatal encephalitis.

Taxonomy and Life Cycle

Toxoplasma gondii has been assigned to the phylum Apicomplexa and class Sporozoa, which also includes the malaria parasites (genus Plasmodium). However, in contrast to the narrow vertebrate host ranges of the Plasmodia, *T. gondii* is a cosmopolitan parasite

The personal research reviewed in this chapter was supported by grant AI-14151 from the National Institutes of Health.

that infects a wide variety of birds and mammals. Acute toxoplasmosis in these hosts is characterized by intracellular growth in many tissues of the rapidly dividing tachyzoite stage. As the immune response (see Chapter 10) controls the acute infection, some of the tachyzoites become encysted, particularly in muscle and brain, in an essentially dormant stage termed the bradyzoite. These cysts contain hundreds of bradyzoites within a wall that is contributed, at least in part, by the host cell, and they are retained for years and possibly for the lifetime of the vertebrate host. If the tissue that contains these cysts is eaten by a nonfeline carnivore, the bradyzoites are released in the intestine and again yield an infection characterized initially by rapid growth of tachyzoites and ultimately by the persistence of encysted bradyzoites. However, when encysted bradyzoites are eaten by a member of the cat family, a different scenario ensues. In addition to yielding tachyzoites, the bradyzoites also differentiate within intestinal epithelial cells through several stages of schizogony to yield male and female gametes. These gametes fuse to form a zygote, which synthesizes an impervious protective wall for itself and is excreted in the feces as a unicellular unsporulated oocyst. At ambient temperatures the excreted oocyst sporulates and becomes infectious through three cell divisions yielding eight sporozoites, all contained within the oocyst wall. When a sporulated oocyst is ingested by a mammal or bird, the sporozoites are released and infect the intestinal epithelial cells. They become rapidly multiplying tachyzoites that produce first an acute and then a chronic infection with encysted bradyzoites, thereby completing the natural life cycle.

Since only the tachyzoite stage of *T. gondii* can be conveniently grown in

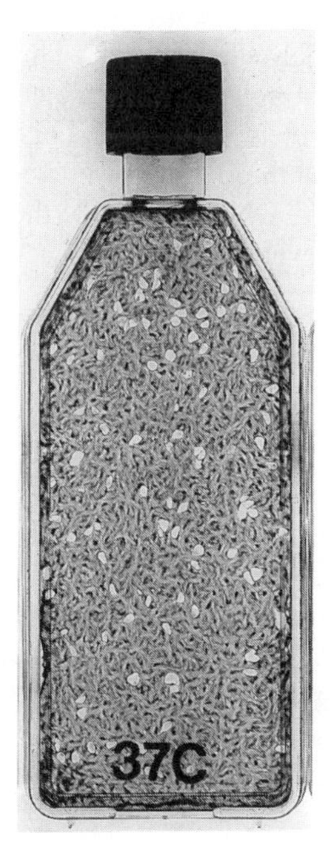

FIGURE 2·1 Plaques formed by *T. gondii* 7 days after infection of a confluent monolayer culture of human fibroblasts. The fibroblasts have been stained for photography. The plaques are the small, completely clear areas. (Reproduced from PFEFFERKORN, E. R., and L. C. PFEFFERKORN. 1976. *Toxoplasma gondii:* isolation and preliminary characterization of temperature-sensitive mutants. *Exp. Parasitol.* 39:365–376 by permission of Academic Press.)

cultured cells, most knowledge of the cell biology of *T. gondii* is drawn from studies of this rapidly growing asexual form. This stage offers an excellent model for the cell biologist interested in the complex system of one eukaryotic cell parasitizing another. Principal among the advantages of this parasite are the ability of the tachyzoite to grow in practically any cultured mammalian cell and the availability of a plaque assay (Figure 2-1) for infectivity that allows precise titrations, easy cloning, and simple selection of mutants. Although a good deal has been learned about the cell biology of *T. gondii*, major gaps in our knowledge remain. Fortunately, some of these gaps can be filled by reference to data from studies of *Eimeria tenella*, a related parasite of chickens that belongs to the same subclass, the Coccidia, as does *T. gondii*. Although extrapolation of data from this closely related organism may not always be reliable, potentially useful new experimental approaches to the study of *T. gondii* may be suggested by data from *E. tenella*.

Morphology and Composition

Tachyzoites of *T. gondii* are crescent-shaped cells, about 6 μm long and 2 μm wide. They share many cytological features with other eukaryotic cells, including a nucleus, mitochondria, Golgi apparatus, and endoplasmic reticulum (Figure 2-2A). Only a few studies of the RNA of *T. gondii* have been reported. The ribosomal RNA consists of the usual large and small molecules that are presumably associated with their respective ribosomal subunits. Recent reports demonstrate conclusively that the mRNA of *T. gondii* resembles that of other eukaryotic cells in bearing a 3′-polyadenylated tail. This mRNA has been translated in heterologous systems to yield polypeptide products serologically identifiable as *T. gondii* proteins. The DNA content of the sporozoite stage of *T. gondii* has been determined to be about 0.1 pg/cell through direct microfluorometric analysis of single cells. This value is the haploid DNA content of the parasite because all stages except the zygote are haploid. Several laboratories have prepared genomic and cDNA libraries for *T. gondii*.

As might be expected from the size of its genome, the tachyzoite contains a vast array of proteins. At least 1000 proteins have been observed by autoradiography of two-dimensional electrophoresis of biosynthetically labeled parasites, a value that undoubtedly underestimates the complexity of the organism. In contrast, cell surface specific radioiodination followed by SDS acrylamide gel electrophoresis reveals only a limited array of five to seven major radioactive bands with relative molecular weights (M_r) that range from 14,000 to 60,000. The most prominent protein (M_r 30,000) may represent as much as 3 percent of the total parasite protein. Like some surface proteins of the Plasmodia (Chapter 1), this protein contains an immunodominant repetitive epitope. The relative simplicity of the tachyzoite surface is reflected in the humoral immune response, which primarily recognizes four major surface antigens of M_r 14,000 to 43,000. Unlike most eukaryotic cells, the *T. gondii* tachyzoite is reportedly unable to

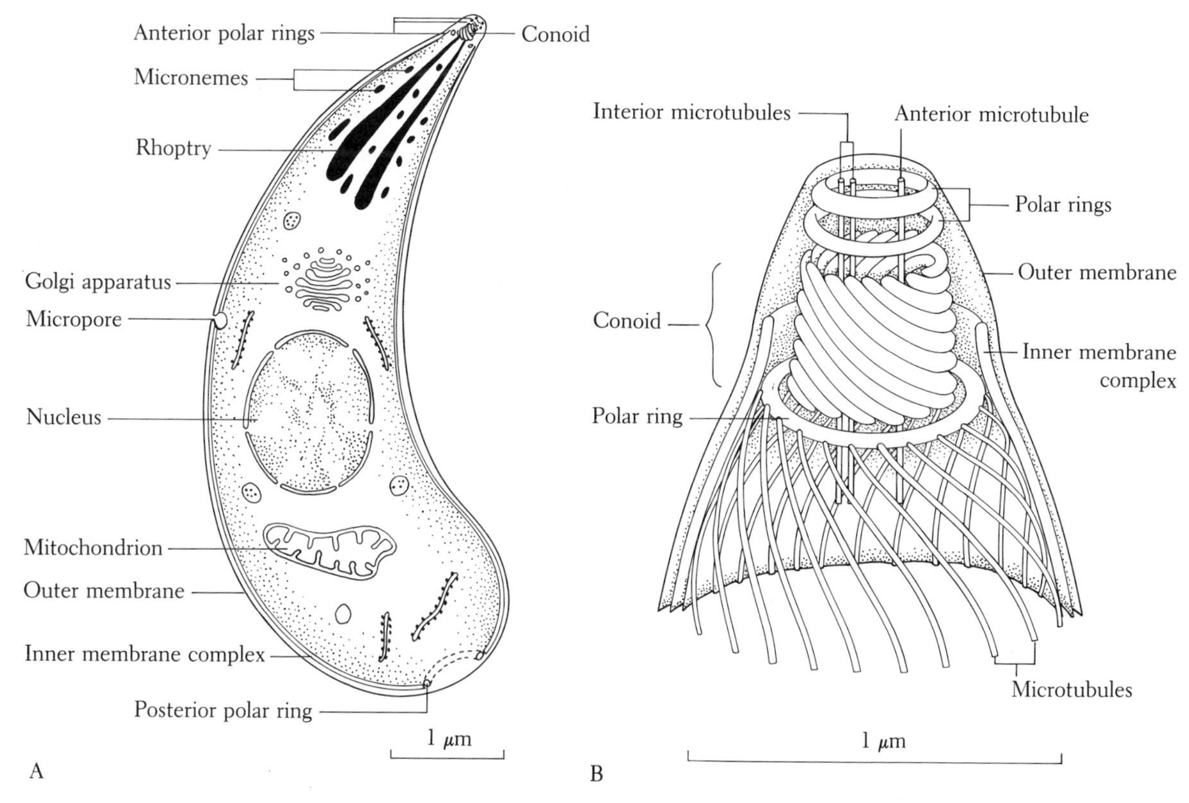

FIGURE 2·2 A. The ultrastructure of the tachyzoite of *T. gondii* (based on various electron microscopic studies). B. Detailed view of the structure of the anterior end of *T. gondii* (redrawn from NICHOLS, B. A., and M. L. CHIAPPINO. 1987. Cytoskeleton of *Toxoplasma gondii. J. Protozool.* 34:218–227).

bind any of the fluorescently labeled lectins that are generally used to identify surface glycoproteins. This difference from other eukaryotes is likely to be more apparent than real, for recent studies have detected low levels of concanavalin A binding both to the intact organism and to isolated proteins. It may simply be that for steric reasons, the glycosyl residues of the surface proteins of *T. gondii* are not readily accessible to lectins.

In addition to the general characteristics that define *T. gondii* as a eukaryotic cell, the tachyzoite also has a number of structures that mark it as a member of the Apicomplexa, and within that phylum as one of the subclass Coccidia. Many of these ultrastructural features were, in fact, first observed in *T. gondii* (Figure 2-2B). The tachyzoites are coated with not one but three unit membranes that together make up the pellicle. The outer membrane is continuous. The two closely apposed innermost membranes are discontinuous, ending at structures called polar rings, which are located at both the anterior and the posterior ends of the parasites. The innermost membranes also have small circular gaps called micropores that may be involved in the uptake of nutrients. The subpellicular cytoskeleton is composed of 22 microtubules that stretch from the anterior polar ring through most of the length of the parasite. The anterior end of the *T. gondii* tachyzoite is marked by the presence of the conoid, a hollow, truncated cone of spirally wound fibers that are also likely to be microtubules (Figure 2-2B). Protrusion of this structure during the early stages of entry into a host cell can be detected by both light and electron microscopy. Two intracellular organelles peculiar to the Apicomplexa are also thought to play a role in the penetration of host cells. The rhoptries are club-shaped, densely osmophilic structures whose narrow ends terminate in the conoid. Their structure suggests a secretory function that is probably carried out in the early stages of host-cell entry. Each parasite has only a small number of rhoptries, perhaps only two. Packed between the rhoptries are a large number of smaller osmophilic vesicles called micronemes. The relationship between these two types of anterior organelles remains unclear although the micronemes may also play a role in penetration.

Infection of Host Cells

A critical event in the life of any obligate intracellular parasite is the infection of its host cell. This process can conveniently be considered to occur in two steps, attachment and penetration, or invasion; unfortunately, practically nothing is known of the first, and the precise events of the second are disputed. The conventional view of the attachment of an intracellular protozoan parasite to its host cell requires recognition by the parasite of some specific macromolecule on the host cell surface. For certain Plasmodium species (see Chapter 1 on the Cell Biology of Plasmodia), substantial progress has been made in identification of the host cell receptor. The only clue as to the nature of the host cell receptor for *T. gondii* is the remarkably broad range of cells that can be infected, namely cells of a wide variety of histological types from all mammals, probably from all birds, and from at least one reptile. Such a broad host range all but rules out any specific protein as the receptor and forces the consideration of components that are common to the plasma membranes of most cells. The only two likely possibilities are lipids and the glycosyl residues that are present on both glycoproteins and glycolipids. Existing evidence argues against glycosyl residues because addition of a wide variety of sugars to the extracellular medium fails to block the infection of cultured cells by *T. gondii*. However, it may be that these sugars would have mimicked the true receptor more effectively had they been linked to a carrier protein. An alternative possibility, that *T. gondii* recognizes cellular membranes by attachment to some lipid component, should be considered. Cholesterol, a component of all animal cell plasma membranes, has been implicated as the receptor for certain rickettsia, which are obligate intracellular bacteria.

However it occurs, attachment of *T. gondii* serves to orient the parasite with its anterior end, which bears the conoid and rhoptries, in contact with the plasma membrane of the host cell (Figure 2-3). At this time the conoid is extended and the tachyzoites proceed rapidly to invade and soon come to lie within a cytoplasmic vacuole

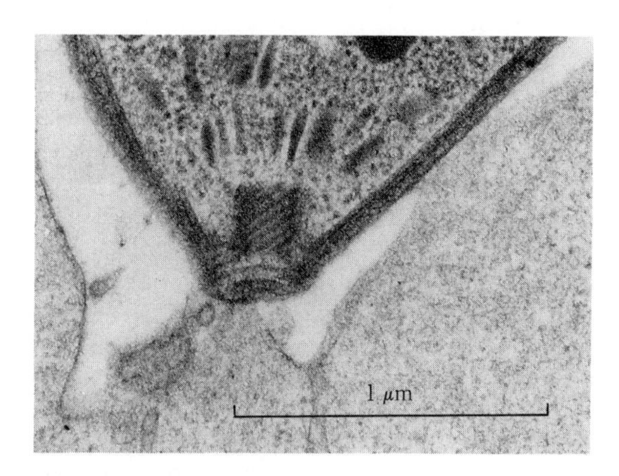

FIGURE 2·3 Extension of the conoid as a tachyzoite of *T. gondii* becomes attached to a host cell. (From AIKAWA, M., et al. 1977. Transmission and scanning electron microscopy of host cell entry by *Toxoplasma gondii*. *Am. J. Pathol.* 87:285–296.)

called the parasitophorous vacuole. Many ultrastructural studies of invasion by *T. gondii* have used mouse peritoneal macrophages as the host cell, probably because of the long-standing tradition of maintaining the parasite in vivo by serial intraperitoneal passage. In retrospect, however, this use of macrophages was probably unfortunate. The macrophage is capable of phagocytosing dead *T. gondii* through a mechanism unrelated to the active invasion that results in a productive infection. The two processes, phagocytosis and invasion, initially yield a superficially similar outcome, a parasite within a cytoplasmic vacuole. However, when dead parasites or even live parasites coated with antibody are phagocytosed, they are doomed to destruction by an assortment of agents, including lysosomal enzymes and active oxygen metabolites (see Chapter 10). The complications of studying a culture in which both phagocytosis and invasion are taking place can be avoided by using one of the many nonphagocytic cultured cells that allow penetration and growth of the parasite.

The penetration of a host cell by *T. gondii* probably requires the interplay of biochemical and biomechanical events, although their relative importance is uncertain. The anterior organelles, rhoptries and micronemes, are generally regarded as secretory in nature and are thought to produce substances that promote the penetration of *T. gondii* into its host cell. Both biochemical and ultrastructural evidence support this contention. A fraction of lysed *T. gondii* that increases the ability of the parasite to invade cultured cells has been partially purified. This material, termed "penetration-enhancing factor," is active at 1 μg protein/ml and seems not to be lysosomal in origin. Instead, immunofluorescence using antibody made against partially purified penetration-enhancing factor showed it to be preferentially located within the rhoptry region of intact parasites, suggesting that this factor is a secretory product. This conclusion is supported by ultrastructural studies showing that during the invasion of host cells, the membranes of the rhoptries fuse with the plasma membrane of the parasite in a manner that would release the contents of the rhoptries close to the host cell membrane. The biochemical role of the penetration-enhancing factor, or rhoptry secretion (if indeed they are synonymous), remains to be determined. It is likely to alter the properties of the host cell membrane, perhaps by increasing its fluidity, in such a way as to promote penetration of the parasite. At concentrations significantly higher than those required to promote invasion, the penetration-enhancing factor is actually lytic to cultured cells. The significance of these observations and elucidation of the precise role of the penetration-enhancing factor must await further study. The recent preparation of a monoclonal antibody against penetration-enhancing factor provides a promising direction for the future study of this factor.

Once attachment has occurred, secretion from the rhoptries and invasion by the parasite are remarkably rapid. Several investigators have noted that invasion is complete in less than 15 seconds. Various mechanisms have been proposed to explain this process, some claiming the major role for the host cell and others favoring active penetration by the parasite. The truth lies somewhere between the two extremes, although the parasite probably plays the predominant role. Before the model presented below can be considered, the related topics of capping and motility in the tachyzoite must be considered. Capping is the process by which polyclonal antibodies attached to the surface of *T. gondii* are transported to its posterior end and then shed into the medium. This process presumably reflects the action of an actin-based contractile system coupled to pellicular proteins and perhaps microtubules. Capping of antibody-linked pellicular proteins is thought to be closely related to the motility of *T. gondii* tachyzoites. Anyone who observes extracellular parasites in cell cultures is immediately struck by their rapid twisting motions, particularly just after an infected cell has lysed. Although it has no flagella or any of the other specialized organelles associated with motility, the *T. gondii* tachyzoite is capable of rapid twisting and gliding motions, provided that it is attached to some substratum, such as a monolayer of cultured cells or a plastic tissue-culture plate. This motility is likely to be related to the phenomenon of capping. In a reasonable model for motility, pellicle proteins attached to the substratum are transported to the rear of the organism, resulting in forward motion. The direction is determined by the subpellicular microtubules acting through linear arrays of intramembranous proteins.

How are these phenomena related to penetration of the parasite into a host cell? The hypothesis described below depends heavily on a model originally developed for *E. tenella*, in which the processes of capping and motility are assumed to be mirrored by that of penetration. In this model, the parasite first attaches through an unidentified receptor and immediately perturbs the host plasma membrane by secretion from its rhoptries (Figure 2-4). A close apposition of the parasite pellicle and the perturbed host cell membrane ensues. This junction is then "capped" (that is, transported away from the anterior end), which drives the parasite into the cell and at the same time generates the parasitophorous vacuolar membrane from the plasma membrane of the host cell. Finally, as the capping process reaches the posterior end of the parasite, the junction between the parasite and the host cell membrane is broken and the host cell plasma membrane fuses to yield a tachyzoite lying free within a cytoplasmic vacuole. Although the role of the host cell during this process is secondary to that of the parasite, the host cell is probably not wholly passive. For example, its cytoskeleton must somehow accommodate the intrusion of the parasite.

FIGURE 2·4 Model for the penetration of *T. gondii* tachyzoites into a host cell. (A) adsorption, (B) conoid extension and secretion of penetration-enhancing factor from rhoptries that perturbs the host cell membrane, (C–E) formation of a junction between the pellicle of the parasite and the plasma membrane of the cell and translocation of that junction to the posterior end of the parasite, after (F) sealing of the host cell membrane to form the parasitophorous vacuole.

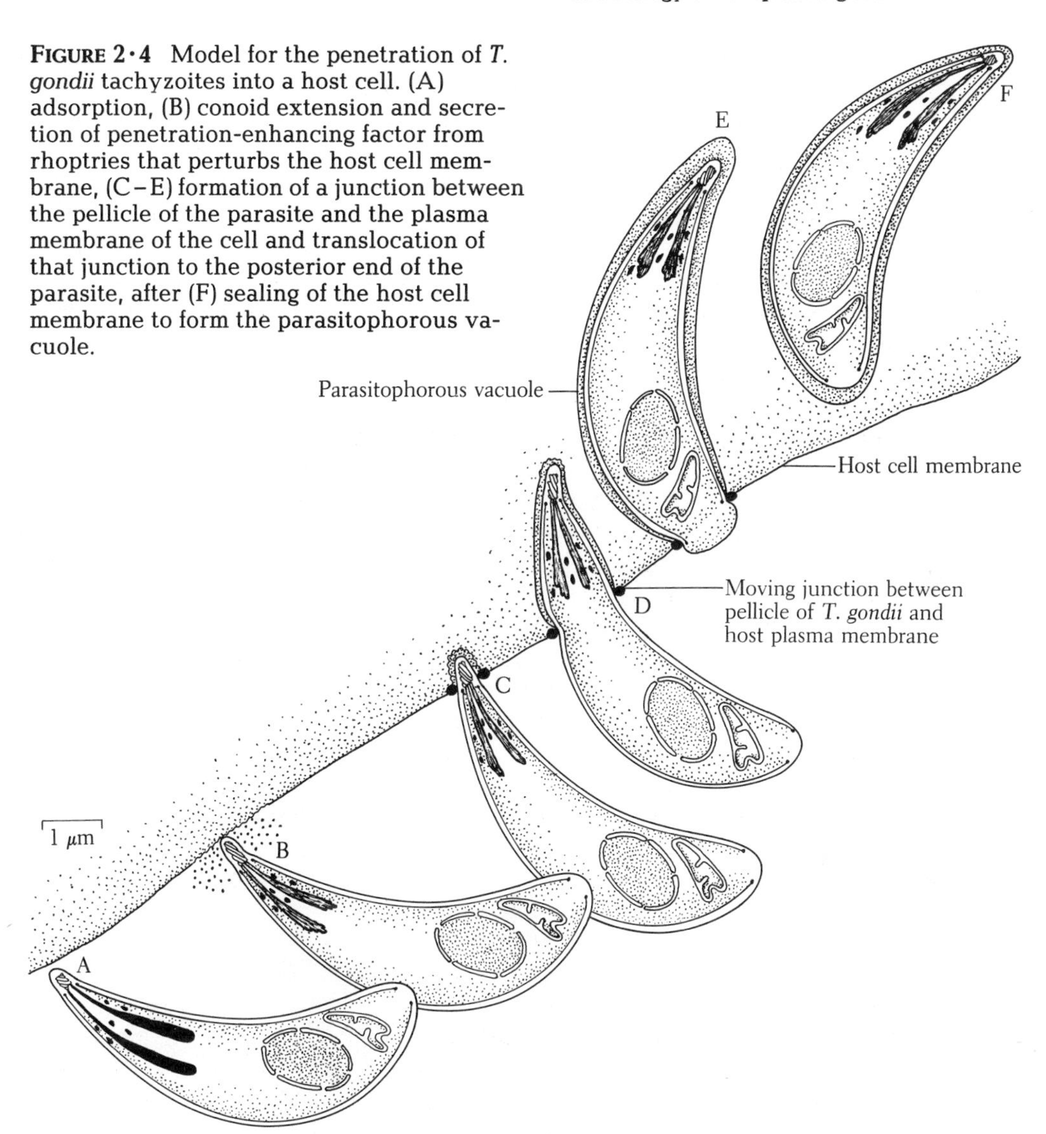

The evidence supporting the foregoing model is largely circumstantial. *T. gondii* will invade certain erythrocytes and indeed multiply within immature nucleated erythroid precursor cells. The invasion of mature erythrocytes provides a useful model system because these cells are incapable of phagocytosis. During their invasion the close apposition of the host cell and parasite membranes is readily observed. Further evidence for this apposition comes from a striking study of the invasion of cells by *T. gondii* that have been coated with monoclonal antibody directed against the principal surface protein (M_r 30,000) of the tachyzoite. Fortunately, this particular antibody is not capped and therefore serves as a marker for the entire surface of an extracellular parasite that has been treated with monoclonal antibody. Uniform coating of the surface of extracellular parasites can be demonstrated in electron micrographs with the immunoperoxidase method. As these surface-labeled parasites invade host cells, the attached antibody is stripped off at the point of the moving junction between host and parasite membranes, testifying to its tightness (Figure 2-5). At present it is not known if only the attached antibody is shed during entry or if its antigen is lost as well. The role of actin in the biomechanical events associated with the penetration of *T. gondii* is supported by the observation that cytochalasin D blocks penetration but not attachment. This observation does not exclude a role of actin in the host. Penetration of host cells by *T. gondii*

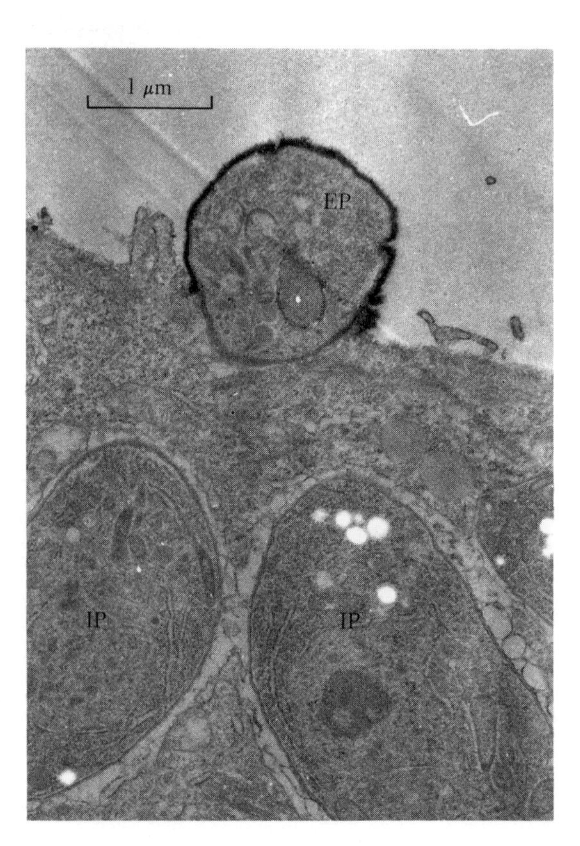

FIGURE 2·5 Stripping of an attached antibody (stained black) at the point of the parasite–host cell junction as a tachyzoite of *T. gondii* enters a host cell. Before entering the cell, the parasites had a uniform black-staining antibody coat. IP: intracellular parasites without antibody coat. EP: entering parasite losing its antibody coat. (Reproduced from DUBREMETZ, J. F., C. RODRIQUEZ, and E. FERREIRA. 1985. *Toxoplasma gondii:* redistribution of monoclonal antibodies on tachyzoites during host cell invasion. *Exp. Parasitol.* 59:24–32 by permission of Academic Press.)

also depends on energy metabolism by the parasite and possibly also by the host cell (compare the penetration of *Plasmodia* described in Chapter 1).

If this model for penetration is accepted, the principal unresolved issue becomes the severity of the effect on the host cell membrane at the site of entry. Several ultrastructural studies suggest that the host cell membrane is physically disrupted and that the newly intracellular parasite lies within torn membrane fragments but partially exposed to the cytoplasm of its host cell. The vacuolar membrane is then thought to be quickly knit together, perhaps through the addition of membrane components supplied by the parasite. In the alternative, simpler model adopted in Figure 2-4, the host cell membrane is assumed to remain intact during the entire process of penetration, and the breaks in this membrane observed during entry are considered artifacts ascribed to fragility induced by the secretion from the rhoptries.

The Role of the Parasitophorous Vacuole in the Intracellular Growth of *T. gondii*

However it is formed, the parasitophorous vacuole is a safe haven for the growth of *T. gondii*. In sharp contrast to the fate of a cytoplasmic phagosomal vacuole, the parasitophorous vacuole avoids both fusion with lysosomes and acidification. Instead, the outside of the parasitophorous vacuole becomes covered with host-cell mitochondria. How these two seemingly similar vacuoles come to have totally different fates remains a mystery that is central to our understanding the intracellular growth of *T. gondii*. The absence of lysosomal fusion and acidification of the parasitophorous vacuole does not reflect a global defect in the host cell. If a phagocytic cell infected with *T. gondii* also contains a phagosome, the phagosome is acidified and fuses normally with lysosomes. Thus, the inhibition of these functions is peculiar to the parasitophorous vacuole. The only likely way in which such a vacuole can communicate with the host cell is through the domains of vacuolar membrane proteins that are exposed to host cell cytoplasm. *T. gondii* could determine the fate of its parasitophorous vacuole in several ways: (a) host proteins that signal lysosomal fusion and acidification could be systematically excluded during formation of the vacuole, (b) host cell proteins that signal these cellular functions could be altered allosterically or in some other manner to negate the signal, or (c) parasite-specific proteins that override the cellular signals could be incorporated into the vacuolar membrane. One of the major difficulties in distinguishing among these explanations is that the vacuolar membrane has never been purified because of its fragility and lack of obvious identifying characteristics.

Multiplication of the Tachyzoite

Secure within its parasitophorous vacuole, the tachyzoite begins to multiply; the division time of 5 to 9 hours is characteristic of whichever strain one is studying. As might be expected, greater virulence in vivo is associated with more rapid growth in vitro. Within any one vacuole, the division of tachyzoites is remarkably well synchronized, at least through the first three or four divisions. In an infected culture as a whole, *T. gondii* multiplies asynchronously, probably because the infecting parasites are themselves in different stages of their cell cycle.

Tachyzoites divide through a type of binary fission termed endodyogeny. This process (Figure 2-6) begins as the crescent-shaped parasite becomes more spherical. Below the anterior pole in the cytoplasm of the mother cell, the two internal pellicular membranes of the daughter cells begin to be organized and soon acquire their own rudimentary rhoptries, micronemes, and conoids. As the internal membranes grow posteriorly, the nucleus of the mother cell divides and the daughter nuclei become enclosed within the new inner membranes as they extend to the posterior pole. In contrast to mitosis in higher eukaryotes, the nuclear membrane of *T. gondii* remains

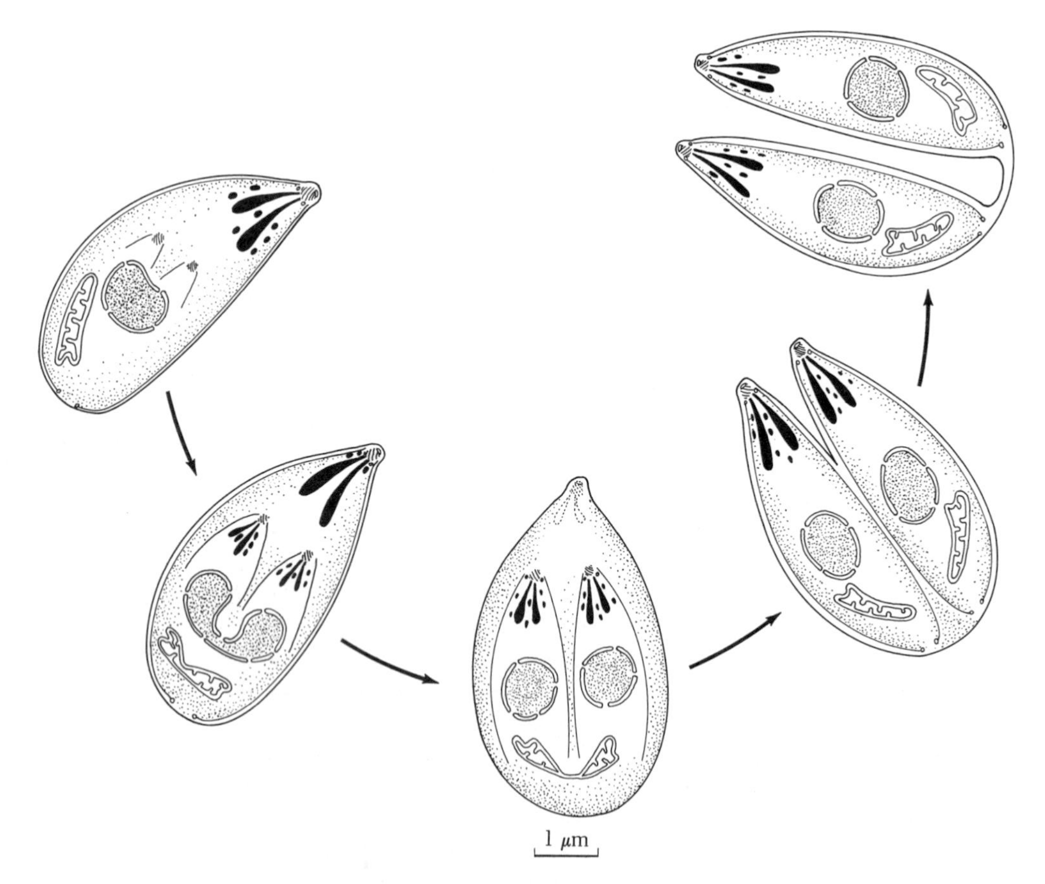

FIGURE 2·6 Stages in the division of tachyzoites of *T. gondii* by endodyogeny (based on various electron microscopic studies).

intact and its chromosomes do not condense at metaphase. The mitochondria of the mother cell are also divided between the daughter cells. As the daughter cells, now virtually complete, enlarge to fill the mother cell, the inner membrane complex of the mother cell disappears while the outer membrane is reused as the outer layer of the pellicle of the daughter cells. The daughter cells may remain temporarily attached at their posterior ends through a residual body of mother-cell cytoplasm. The tachyzoites produced by synchronous endodyogeny in individual cultured cells usually lie in the same plane and form the rosettes that are characteristic of *T. gondii*.

As the multiplication of tachyzoites progresses, the parasitophorous vacuole expands to fill more and more of the host cell cytoplasmic volume. The origin of the additional membrane required for the expansion of the vacuole is unknown. It is likely to be supplied by the host but a contribution by the parasite cannot be ruled out. Remarkably, the growth of tachyzoites has little apparent effect upon the host cell until it lyses to release a new crop of parasites. Autoradiography detects no marked diminution of incorporation of amino acids into host cell proteins or of nucleosides into host cell nucleic acids. Even the entry of cells into S phase, as detected by extensive nuclear incorporation of thymidine, or into mitosis—probably the most sensitive measure of the host cell's sense of well being—does not seem to be markedly affected by infection with *T. gondii*. Perhaps the fact that the growing parasites are insulated from the host cell cytoplasm by the parasitophorous vacuole diminishes their impact on host cell metabolism. Roughly 2 days after infection, both the vacuolar membrane and the host cell lyse to release 64 to 256 tachyzoites. As observed by phase-contrast microscopy, this release is rapid and the newly emerging parasites are highly motile.

Biochemical Communication Between Host Cell and Parasite

Understanding the biochemistry of the host–parasite relationship in cells infected with tachyzoites of *T. gondii* depends to a large extent on our knowledge of the function of the parasitophorous vacuolar membrane. All biochemical traffic between host cell and parasite must, of necessity, cross this membrane. There is no evidence that pyrimidine nucleotides or deoxynucleotides are exchanged between host cell and parasite. The experiments designed to detect any flow of uridine triphosphate (UTP) or thymidine triphosphate (TTP) from the host cell to *T. gondii* used a mutant parasite that is resistant to fluorodeoxyuridine (FUDR), as will be discussed. This mutant lacks an enzyme that plays the central role in pyrimidine salvage by *T. gondii*. Autoradiography shows that this mutant is incapable of incorporating either [^{3}H]uridine or [^{3}H]deoxyuridine. However, the host cell, as expected, incorporates these precursors efficiently. Thus, the infected host cell labeled with [^{3}H]uridine must contain an ample amount of [^{3}H]UTP, the immediate precursor for RNA synthesis. Since there is no doubt that the intracellular mutant parasites could use this [^{3}H]UTP if it were present in their own cytoplasm, the fact that they are unlabeled in autoradiography experiments proves that *T. gondii* has no direct access to the pyrimidine nucleotide pool of its host cell. The same argument can be applied to prove that the TTP pool of the host cell, which is efficiently labeled by [^{3}H]deoxyuridine, is unavailable to the parasite.

To examine the reverse possibility, that pyrimidine nucleotides of the parasite were available to the host cell, wild-type parasites could be used because human fibroblast host cells lack the salvage enzyme uracil phosphoribosyltransferase. When these cells are infected with wild-type *T. gondii* that are capable of incorporating uracil, autoradiography shows that incorporation of [^{3}H]uracil is confined to the intracellular parasites. Since this precursor is used to make UTP and TTP for use in nucleic acid synthesis, these nucleotides synthesized by the parasite must be unavailable to the host cell. If they were available, the cytoplasm and nuclei of infected host cells would be labeled while the uninfected cells would remain unlabeled.

The purine nucleotide pool of the parasite is also unavailable to its host cell. This conclusion is based upon autoradiography of infected Lesch-Nyhan fibroblasts labeled with [^{3}H]hypoxanthine. Although these mutant cells are incapable of incorporating [^{3}H]hypoxanthine, their intracellular parasites are efficiently labeled and must contain labeled ATP and GTP to which the host cell has no access. This series of experiments on nucleotide exchange leaves only the traffic of purine nucleotides from host cell to parasite unexamined. As described later, access to host cell purines is critical for the growth of *T. gondii*.

The energy (ATP) required to support the intracellular growth of *T. gondii* is undoubtedly produced by the parasite itself. Extracellular tachyzoites can incorporate label from exogenous ATP that contains [^{3}H]adenine, but only after the phosphate groups are removed. All stages of *T. gondii* contain cristate mitochondria that closely resemble those found in most eukaryotic cells. Since extracellular parasites have been shown to respire and to contain enzymes of the tricarboxylic acid (TCA) cycle, there is good reason to believe that these mitochondria are fully functional, in that the production of ATP is coupled to this respiration. However, there is no direct evidence that this is the case. It is clear that something unexplained happens to the mitochondria of *T. gondii* after the tachyzoite enters host cells. A fluorescent rhodamine dye, R123, is specifically taken up by the mitochondria of extracellular *T. gondii*. This mitochondrial staining is sensitive to agents that disrupt the mitochondrial membrane potential and presumably reflects this potential. Remarkably, the mitochondria of intracellular parasites fail to take up the rhodamine dye even when it is injected directly into the parasitophorous vacuole. Dye injected in this manner even crosses the vacuolar membrane and stains the mitochondria of the host cell. These important observations suggest a reduced mitochondrial membrane potential in intracellular parasites. Thus, the relative importance of oxidative phosphorylation and glycolysis in ATP production by actively growing intracellular parasites remains unknown. The fact that the host cell itself is actively engaged in both modes of ATP production makes this a particularly difficult problem to attack. One reasonable approach would be to study the respiration of intracellular tachyzoites growing in a mutant host cell that is incapable of oxidative phosphorylation. Such mutant cells are known to support the growth of *T. gondii*.

The Biochemical Role of the Host Cell

Perhaps the most fundamental cell-biological question that can be asked about *T. gondii*, and indeed about all obligate intracellular parasites, is why the host cell is required at all. Current knowledge allows no decisive answer. For the most part, all that we can do is to deprive the parasite of certain host cell functions, thereby narrowing the range of explanations for obligate intracellular parasitism that we need to consider. The most decisive experiment in this vein is the demonstration that *T. gondii* grows quite normally in enucleated cells. In a single stroke, de novo RNA and DNA synthesis by the host cell are excluded as essential for the parasite. This conclusion is supported by experiments with drugs. Blocking host cell DNA synthesis with fluorodeoxyuridine (FUDR), an inhibitor of thymidylic acid synthetase, has no effect on the growth of an FUDR-resistant mutant of *T. gondii*. Similarly, a specific inhibitor of host cell mRNA synthesis, alpha-amanitin, allows normal growth of wild-type *T. gondii*, whose mRNA synthesis is relatively insensitive to this drug.

The possible dependence of *T. gondii* on ongoing host-cell protein synthesis is more difficult to study because all drugs that inhibit host protein synthesis also block the growth of the parasite. Studies with enucleated cells cannot exclude a role for host cell protein synthesis because of the presence of long-lived cytoplasmic mRNA. Two lines of evidence argue against host cell protein synthesis as a significant factor in the intracellular growth of *T. gondii*. The first approach takes advantage of the fact that the mycotoxin muconomycin A is a very poorly reversible inhibitor of protein synthesis: human fibroblasts treated with muconomycin A and washed free of the drug remain incapable of incorporating amino acids when incubated in drug-free medium. When cells treated in this way are infected with *T. gondii*, the parasites grow normally, and autoradiography demonstrates that [^{3}H]leucine is incorporated by the growing parasites but not by the host cell. The second approach to examining a requirement for host cell protein synthesis used a mutant host cell in which protein synthesis is temperature-sensitive because of a thermolabile leucyl-tRNA synthetase. Mutant Chinese hamster ovary (CHO) cells with this thermolabile enzyme are incapable of protein synthesis at 40°C, yet support normal growth of *T. gondii* at this temperature.

Another biosynthetic activity of the host cell that may be important to *T. gondii* is lipid synthesis. Nothing is known of this aspect of the host–parasite relationship for *T. gondii*. However, because two other Apicomplexan parasites, *Plasmodium berghei* and *E. tenella* are thought to be incapable of cholesterol synthesis, *T. gondii* probably also depends on its host cell for cholesterol. Nothing is known of how host cell lipids might be transferred to intracellular parasites.

As noted earlier, the synthesis of nucleic acids and proteins by the host cell is not essential for the intracellular growth of *T. gondii*. Yet the host cell must contribute something critical to the parasite in order to account for its obligate intracellular growth. Only one such contribution has been conclusively identified: *T. gondii* is absolutely dependent upon its host cell for a supply of purines. The demonstration of this requirement illustrates the difficulty of studying the biosynthetic capability of an intracellular protozoan parasite and offers one useful solution. The nutritional requirements and the biosynthetic capabilities of a free-living protozoan are relatively easily defined: the requirement for a given nutrient can be detected by observing the lack of growth in a medium deficient in that nutrient. Alternatively, the ability to carry out a given sequence of biosynthetic steps can be demonstrated unequivocally by supplying an isotopically labeled precursor in the medium and tracing its radioactivity to the final product in the free-living organism. Neither of these simple strategies can be used to define the biosynthetic capabilities of an intracellular protozoan parasite because it is impossible to distinguish true biosynthesis on the part of the parasite from biosynthesis by the host cell followed by export across the parasitophorous vacuole to the intracellular parasite. One easy way to make this distinction is to use a mutant host cell with a biochemical defect in a specific pathway. When such a host is infected, any biosynthetic activity that is observed must be ascribed to the intracellular parasite.

This approach is illustrated by the demonstration that *T. gondii* is incapable of de novo purine synthesis. In these experiments, the mutant host cell is a CHO cell that is completely deficient in 5-amino-4-imidazolecarboxamide ribonucleotide formyltransferase. This mutant is incapable of growth in the absence of exogenous purines and cannot incorporate [^{14}C]formate into the purine ring. *T. gondii* grows normally in these cells even though the medium contains no purines. Two explanations could account for this growth. The parasite could be growing at the expense of preformed host-cell purines or could be synthesizing the purine ring de novo. To distinguish between these alternatives, infected mutant cultures are labeled with [^{14}C]formate, and the parasite nucleic acid bases are examined for radioactivity. None is found in the parasite's purines, although the methyl group of their thymine is labeled, showing that the parasites have access to the labeled precursor. *T. gondii*, like other parasitic protozoa, is therefore incapable of purine synthesis. The reason for this widespread biosynthetic deficiency among phylogenetically diverse organisms remains to be established.

Since *T. gondii* grows normally within cultured cells in medium that supplies no purines, the host cell is the only possible source for this essential nutrient. The forms in which the host purines enter the parasitophorous vacuole and are taken up by the parasite are unknown. A glimpse of purine salvage by *T. gondii* can be obtained because tachyzoites, freshly released from infected cells, are briefly capable of limited RNA synthesis. Under these conditions, the parasite can incorporate a variety of purine nucleotides, nucleosides, and free bases. Of particular interest is the ability of the extracellular parasites to incorporate exogenous ATP into their nucleic acids. This observation suggests that host cell ATP might be directly used by the parasite, provided that it could enter the parasitophorous vacuole. However, when doubly labeled ATP that contains ^{3}H in the purine ring and ^{32}P in the alpha phosphate is supplied, only the purine ring is incorporated into the RNA of the parasite. Thus the ATP must be extensively degraded before or as it enters the parasite. Two purine salvage enzymes have been demonstrated in *T. gondii*, adenosine kinase and hypoxanthine-guanine phosphoribosyltransferase. The relative roles played by these two enzymes are not completely resolved. However, adenosine kinase–less mutants, as will be seen later, grow entirely normally; therefore, hypoxanthine-guanine phosphoribosyltransferase, acting alone, can provide adequate purine salvage. Conversely, parasite mutants lacking hypoxanthine-guanine phosphoribosyltransferase have never been isolated. Such a mutant should be readily selected through its resistance to azaguanine in cultures of Lesch-Nyhan fibroblasts, which themselves lack this enzyme. This failure to select such a parasite mutant suggests that hypoxanthine-guanine phosphoribosyltransferase may be an essential enzyme for *T. gondii*.

The frequent observation that the parasitophorous vacuole, in which *T. gondii* grows, is lined on the host cytoplasmic side with host cell mitochondria has led to the reasonable suggestion that these mitochondria migrate there to supply ATP to the vacuole and thence to the parasite. Although this sequence of events may happen, it is by no means essential for the parasite: *T. gondii* grows entirely normally in a mutant host cell in which the mitochondria are incapable of oxidative phosphorylation and ATP synthesis.

The ability of intracellular *T. gondii* to synthesize pyrimidines can be studied in experiments analogous to those that demonstrated a requirement for purines. However, quite the opposite conclusion is obtained. The parasite is entirely capable of pyrimidine synthesis. This ability is demonstrated with a mutant CHO host cell with three defects in the pyrimidine pathway, the most profound being a complete absence of dihydro-orotase. Uninfected cultures depend on exogenous uridine for growth. These host cells are almost totally incapable of incorporating labeled aspartic acid, a pyrimidine precursor, into nucleic acid pyrimidines. However, when these mutant CHO cells are infected with *T. gondii*, the cultures show efficient incorporation of [^{14}C]aspartate into the pyrimidines of the intracellular parasites. This circumstantial proof that *T. gondii* is capable of pyrimidine synthesis was definitively confirmed by the systematic demonstration of each of the enzymes of the synthetic pathway.

Although *T. gondii* is capable of de novo synthesis of pyrimidines, it probably can, at least under certain circumstances, also employ salvage pathways to supply its needs. The physiologically important pyrimidine salvage pathways have been shown to converge on one enzyme, uracil phosphoribosyltransferase. Mutants lacking this enzyme are readily selected through their resistance to FUDR, as discussed later. These mutants are incapable of incorporating not only uracil but also uridine and deoxyuridine. Thus both of these nucleosides must be cleaved to produce uracil before the pyrimidine base can be salvaged by the parasite. Although uridine kinase has been demonstrated in *T. gondii*, it must play a minor role since the FUDR-resistant mutant described earlier is also resistant to fluorouridine, which is known to be a substrate of this kinase. The salvage pathways for pyrimidines in *T. gondii* and in human cells are compared in Figure 2-7. The relative roles of de novo synthesis and salvage of pyrimidines in the growth of wild-type *T. gondii* remain to be established. Apparently either can suffice to support growth. The FUDR-resistant mutant that is incapable of salvage grows normally, presumably through de novo synthesis. Conversely, the wild type grows well when de novo synthesis is blocked by pyrazofurin or by CO_2 starvation, both of which suppress growth of a mutant defective in pyrimidine salvage.

In summary, because *T. gondii* is an obligate intracellular parasite, the host cell must supply something that is not available in rich media that do not support parasite

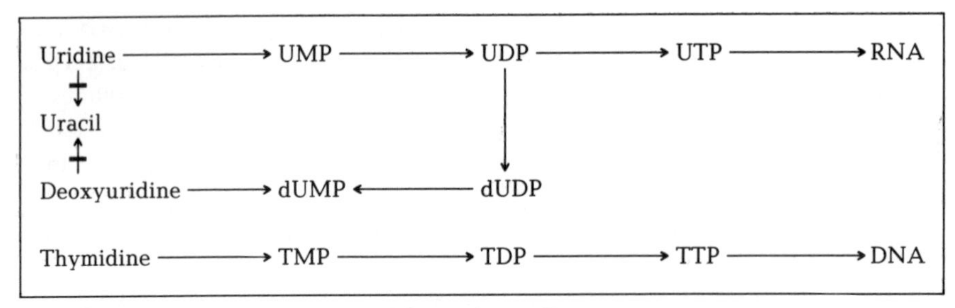

Mammalian cells

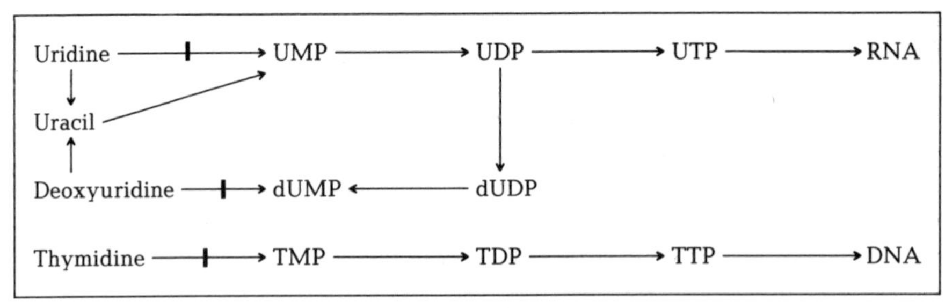

T. gondii

FIGURE 2·7 Comparison of pyrimidine salvage and nucleotide interconversion by *T. gondii* and by its mammalian host cell. Cytidine nucleotides are omitted for clarity. An arrow with a block indicates a reaction that is absent or biochemically unimportant. Abbreviations: UMP, UDP, UTP: uridine mono-, di-, and triphosphates, respectively; dUMP, dUDP, dUTP: deoxyuridine mono-, di-, and triphosphates, respectively; TMP, TDP, TTP: thymidine mono-, di-, and triphosphates, respectively.

growth. What the host cell supplies remains a mystery. The synthesis of pyrimidines, proteins, and nucleic acids by the host cell is not essential; lipid synthesis remains to be examined. Although the host cell must supply purines to *T. gondii*, this explanation is clearly not complete, since rich media that contain purines fail to support extracellular growth of *T. gondii*.

Other Stages of *T. gondii*

In contrast to the substantial body of knowledge of the cell biology of the tachyzoite of *T. gondii*, much less is known about the other stages of the parasite, because they can be studied only in vivo. However, we have some evidence of bradyzoite formation in cultured cells. These "bradyziotes" appear to be encased in cystlike structures; more important, they share two important biological properties with bradyzoites produced in vivo: resistance to pepsin and the induction of oocyst formation within 1 week after ingestion by cats. However, the in vitro production of encysted bradyzoites is seldom a consistent phenomenon, and the conditions that promote it are poorly understood. Bradyzoites produced in vivo are antigenically distinct from tachyzoites and also can be distinguished by their cytoplasmic granules, which probably represent stored carbohydrates that may serve as a source of energy during their long dormant phase and during their entry into intestinal cells when the dormant bradyzoite phase is interrupted as a carnivore eats tissue that contains cysts.

With the possible exception of the production of first-generation schizonts, none of the steps that lead to the formation of gametes in the feline intestine has been observed in vitro. The gametogeny of *T. gondii* therefore has been inaccessible to most of the tools of cell biology, with the exception of light and electron microscopy and Mendelian genetics. It should be noted that the problem of accomplishing in vitro gametogeny and oocyst formation by *T. gondii* may not be intractable since it has been accomplished with *E. tenella*. The process leading to the formation of male and female gametes in the feline intestine is relatively rapid, in that oocysts, the ultimate product of gamete fusion, begin to be excreted only 3 to 5 days after a cat ingests encysted bradyzoites. Nonetheless, it is a complex process in which the parasite population is expanded by extensive multiplication through five distinct steps of schizogony, which are distinguishable by light microscopy.

The basis for sexual differentiation in *T. gondii* is poorly understood, as it is for all Apicomplexans. However, it is well established that sex is not genetically determined in the sense that parasites in the asexual tachyzoite or sporozoite stages are not committed to make gametes of only one sex. This conclusion is based on the observations that tachyzoites cloned by plaque formation and sporozoites cloned by micromanipulation can be grown to populations capable of normal oocyst formation when their bradyzoites are fed to cats. Thus sexual differentiation in *T. gondii* gametogenesis is probably determined physiologically by some unidentified microenvironmental influence in the cat intestinal epithelium. At some point during the series of schizogonies that lead to the formation of gametocytes, the parasites must become committed to production of gametes of only one sex. Microspectrophotometric determination of DNA content suggests that this commitment is made at or before the third generation of schizogony, since the merozoites of this generation show the first evidence of the increased DNA content that is characteristic of the microgametocytes, which give rise to microgametes.

The mature microgamete (male gamete) of *T. gondii* is the only stage whose motility depends upon flagella. Its cytoplasm contains little more than a nucleus and a mitochondrion that undoubtedly supplies the energy to drive its two flagella. Although there is no evidence on this point, analogy with the well-studied *Plasmodia* suggests that the male gamete is likely to bear surface antigens, both on the flagella and on the plasma membrane, that are different from those characterized in the tachyzoite. The chemical signals, if indeed there are any, that allow a male gamete to seek and find a mature female gamete are unknown and cannot be readily studied until gametogenesis can be accomplished in vitro.

The macrogamete (female gamete) of *T. gondii* is a relatively large spherical cell that is also likely to be antigenically distinct from the various asexual stages. The macrogamete is distinguished primarily by the presence of two kinds of peripheral cytoplasmic granules. Because of their role in the construction of the oocyst wall, they are termed wall-forming bodies I and II. Shortly after gametes fuse to form a zygote, particulate material from these bodies, together with proliferating membranes, combines to form the outer and inner layers, respectively, of the rigid and impervious wall of the oocyst. The latter stages of this process, the final formation of the inner layer of the wall, become increasingly difficult for ultrastructural study because the impermeability of the developing oocyst wall prevents penetration of the usual fixatives.

The mature unsporulated oocyst of *T. gondii* can be purified from cat feces by a combination of velocity and equilibrium sucrose gradient centrifugation. Sporulation, which requires adequate aeration for 1 to 3 days at room temperature, results in the production of eight progeny sporozoites, packed four each into two sporocysts, all retained within the original oocyst wall. Both the sporozoites and the interior face of the oocyst wall bear stage-specific antigens. Since the oocyst wall is remarkably impermeable, sporulation can conveniently be carried out in 0.5 N H_2SO_4, which serves to kill any residual fecal bacteria. The oocyst wall is also impervious to NaOH and sodium hypochlorite. No analyses of the oocyst wall of *T. gondii* are available to explain this remarkable resistance to various harsh chemical treatments. However, studies of the oocyst wall of *E. tenella* offer data that are probably relevant to *T. gondii* (reviewed by Wang, 1982). The outer layer of this oocyst wall is composed primarily of fatty alcohols in which a C_{26} compound predominates. The inner layer is composed almost entirely of a single glycoprotein of M_r 10,000 with substantial sulfhydryl cross-linking.

In summary, although we know a good deal about the tachyzoite of *T. gondii*, the stimuli for gametogenesis and the antigenicity and metabolism of the gametes remain complete mysteries. The reason for this difference is clear: the tachyzoite multiplies readily in cultured cells, but the sexual stages are found only in the feline intestine. One of the current challenges in the research with *T. gondii* is to accomplish gametogenesis in cell or organ cultures.

Genetics of *T. gondii*

Although the study of Mendelian genetics is no simple task with any parasitic protozoan, *T. gondii* offers a reasonably useful system for several reasons. Most important, because the asexual forms are haploid, any potentially recessive mutation will be expressed. In addition, the ability of the tachyzoites to form plaques in cultured cells allows mutants to be cloned easily. Growth of parasites in cultured cells also allows application of the various methods for chemical mutagenesis that have become routine in the production of mammalian cell mutants. Since *T. gondii* grows well over the relatively broad temperature range of 33° to 40°C, it has been possible to isolate a large number of temperature-sensitive mutants that grow normally at the lower temperature but are incapable of plaque formation at 40°C. The principal advantage of temperature-sensitive mutations is that they can be found, at least in theory, in any gene that has an essential protein product. However, the disadvantage of this potentially wide array of mutations is that precise biochemical characterization of the defect is generally difficult. None of the temperature-sensitive mutants of *T. gondii* has been so characterized. However, these mutants, all isolated from the hypervirulent RH strain of *T. gondii*, are often remarkably avirulent for mice, and one of them has some potential as a vaccine.

The mutants that have been most useful in the study of *T. gondii* are drug-resistant. Two advantages of these mutants are that they can be directly selected by the drug and that their biochemical characterization is often relatively simple. One important constraint in the selection of drug-resistant mutants of *T. gondii* is that the drug used must not compromise the ability of the host cell to support the growth of the intracellular parasite during the week-long incubation required for the formation of plaques. The simplest way to avoid these complications is to use drugs that affect a host cell function that is not essential for the growth of *T. gondii*. Since DNA synthesis by the host cell is not essential for the parasite, drugs that inhibit eukaryotic DNA synthesis have proved to be of particular value in selecting mutants of *T. gondii*. An alternative approach is essential when using drugs that are incompatible with the long-term survival of cultured cells, such as inhibitors of protein synthesis. In this case the most useful tactic is to use host cell mutants that are themselves resistant to the drug in question, provided that resistance is not the result of impermeability to the drug.

Several of the *T. gondii* mutants resistant to inhibitors of DNA synthesis have been characterized biochemically. Adenine arabinoside (araA) is an analogue of deoxyadenosine. When incorporated into the nucleotide pool in the form of the nucleoside triphosphate, araA inhibits DNA synthesis by blocking the incorporation of deoxy-ATP. Two independently isolated araA-resistant mutants of *T. gondii* lack the salvage enzyme adenosine kinase, which catalyzes the first step in the incorporation of araA into the nucleotide pool of the parasite. This mutation is not lethal to the parasite, because its hypoxanthine-guanine phosphoribosyltransferase is sufficient for purine salvage.

Fluorodeoxyuridine (FUDR) is an analogue of deoxyuridine. Its mononucleotide is a potent noncompetitive inhibitor of thymidylic acid synthetase and thus of DNA synthesis. A mutant of *T. gondii* that is 100-fold more resistant to FUDR lacks the salvage enzyme uracil phosphoribosyltransferase. This mutation is not lethal, because de novo synthesis can supply the parasite's need for pyrimidines. Uracil phosphoribosyltransferase catalyzes the incorporation of fluorouracil into the nucleotide pool of the wild-type parasite after fluorouracil is produced from FUDR by the action of a nucleoside phosphorylase. What is probably the same enzyme produces fluorouracil from fluorouridine. Thus the FUDR-resistant mutant is also resistant to fluorouridine and fluorouracil. As noted above, this mutant has been useful in excluding pyrimidine nucleotide exchange between host cell and parasite and in elucidating the physiologically significant pathways for pyrimidine salvage by *T. gondii*.

Two other mutants of *T. gondii* resistant to other inhibitors of DNA synthesis are less well characterized. Hydroxyurea blocks the synthesis of deoxynucleotides by inhibiting ribonucleotide reductase. Because a resistant mutant does not have an enlarged pyrimidine deoxynucleotide pool, it is unlikely that it is resistant to hydroxyurea through the overproduction of ribonucleotide reductase. Hydroxyurea inhibits the incorporation of [^{3}H]uracil into the deoxynucleotide pool of the wild type but not of the mutant parasite. Thus the resistance of the mutant is probably a result of an altered ribonucleotide reductase that is less sensitive to the drug. Aphidicolin inhibits DNA synthesis by binding to and inhibiting DNA polymerase alpha at a site that is thought to overlap more or less completely with the binding site for deoxyCTP. A mutant of *T. gondii* that is fourfold more resistant than the wild type to aphidicolin overproduces deoxyCTP approximately threefold. Since this deoxynucleotide is known to reverse the action of aphidicolin in a competitive manner, it is likely that its overproduction is the basis for resistance in this mutant.

The drug-resistant mutants of *T. gondii* described earlier have been used in a series of genetic crosses. At present these studies of the Mendelian genetics must use in vivo experiments in cats because of the inability to accomplish gametogenesis in vitro. Two potential pitfalls should be noted. The most commonly used laboratory strain of *T. gondii*, called RH, has lost the ability to undergo gametogenesis in the feline intestine. This strain has been maintained by serial intraperitoneal passage in mice over many decades. Under these conditions, there is no selective advantage for the retention of genes that are uniquely expressed during gametogenesis. Indeed, if these genes represent a significant fraction of the genome, their continued maintenance during prolonged passage of the asexual form may be a sufficient burden to provide a small selective advantage to mutants in which they are deleted. This situation is not unique to *T. gondii*. Prolonged "syringe passage" of murine plasmodia has also been shown to yield strains incapable of gametogenesis. Even if a strain capable of gametogenesis is used to select mutants for use in genetic crosses, the possibility of inadvertently introducing a secondary mutation that blocks gametogenesis must be borne in mind. Such a secondary mutation could be introduced during the chemical mutagenesis required to induce the desired mutation or could be selected inadvertently during the prolonged asexual growth required during the mutant selection. Several drug-resistant mutants selected from a strain of *T. gondii* known to make gametes have been found to be incapable of forming oocysts. When the mutants that have lost the ability to make oocysts are crossed with a different mutant resistant to another drug and of proved fertility, no recombinants are obtained. This negative result suggests that the inability to make oocysts results from a defect in the production of both male and female gametes, because making either one alone should be sufficient for a successful cross with another mutant that can make both kinds of gametes.

Provided that both drug-resistant clones are by themselves able to make oocysts, recombination between them can easily be detected. The usual procedure is to produce encysted bradyzoites in chronically infected mice, using separate mice for each clone. Mouse brains, examined microscopically for cysts, are fed to seronegative kittens so that they ingest approximately equal numbers of cysts of each genotype. Oocysts purified from the cat feces are allowed to sporulate. Finally, sporozoites released from the oocysts are examined in cultured cells for recombinants resistant to both drugs. Such recombination is readily demonstrated in experiments in which the two parental strains were resistant to FUDR and to araA, respectively. The greatest frequency of doubly resistant recombinants observed is about 12 percent. The reciprocal doubly sensitive recombinant can also be detected.

In considering the results from genetic crosses with *T. gondii*, it should be noted that the term *recombinant* is used in its most general sense, to designate progeny parasites that have inherited genetic markers from two different parental strains; the term therefore does not imply crossing over between two genetic loci on the same chromosome. Recombinants could easily arise from the independent assortment of chromosomes at meiosis and, as will be noted, this may be the only way in which they can be formed.

In order to interpret the results of genetic crosses done with *T. gondii* and other coccidial parasites, it is necessary to speculate upon the possibility of a peculiarity in their meiosis (Figure 2-8). The meiotic reduction to the characteristic haploidy of *T. gondii* certainly occurs during sporulation of the oocyst. In contrast to their behavior in the mitotic divisions that take place during endodyogeny of the tachyzoite, the chromosomes of *T. gondii* and of other coccidia condense during the first division of sporulation. In most eukaryotic organisms the meiotic reduction to a haploid state requires two cell divisions. During metaphase of the first division the homologous chromosomes are found in the form of tetrads with each chromosome split into two chromatids. This is the stage at which meiotic crossing-over takes place and its morphological manifestations, the chiasmata, are observed. In contrast the coccidia are thought, on cytological grounds, to accomplish meiosis in a single cell division that immediately yields haploid progeny. Consistent with this model, there are no indications of chiasmata. Although these conclusions are based primarily on the cytology of the *Eimeria*, if correct, they are probably also valid for *T. gondii*. This model of single-step meiosis has important implications for an interpretation of recombination in the coccidia. If the meiotic reduction to haploidy is accomplished in a single step without the formation of tetrads and the opportunity that they afford for crossing over, genetic markers are locked into the chromosomes that bear them. Linkage therefore should be absolute for genetic markers on the same chromosome, whereas markers on different chromosomes should be randomly distributed among the progeny by independent assortment of homologous chromosomes during the single-step meiosis. Likewise, each oocyst resulting from a genetic cross should contain sporozoites of only two genotypes.

Tests for recombination with four drug-resistance markers in *T. gondii* show only independent assortment, with the exception of two mutants — one resistant to hydroxyurea and the other to FUDR — that yield no doubly resistant recombinants and thus appear to exhibit the absolute linkage predicted by the model described here. In this case, however, linkage is effectively ruled out because crosses between these two mutants yield the expected numbers of reciprocal recombinants that are sensitive to both drugs. The doubly resistant recombinants are not detected because the FUDR-resistance gene phenotypically suppresses the hydroxyurea-resistance gene, possibly by altering the deoxynucleotide pools of the parasite. However, in accord with the model of single-step meiosis, some genetic crosses between two mutants of *E. tenella* show evidence of absolute linkage.

In interpreting quantitative data from crosses involving unlinked markers in *T.*

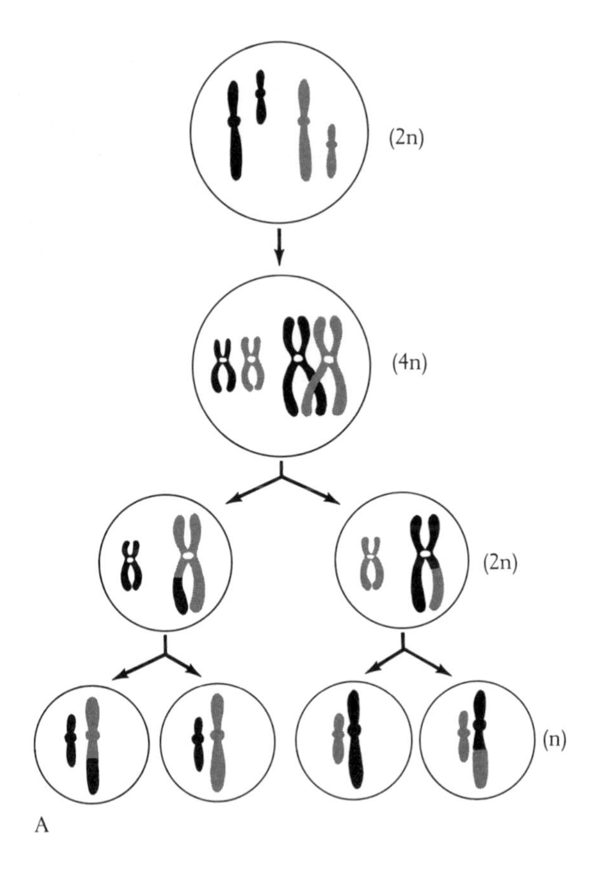

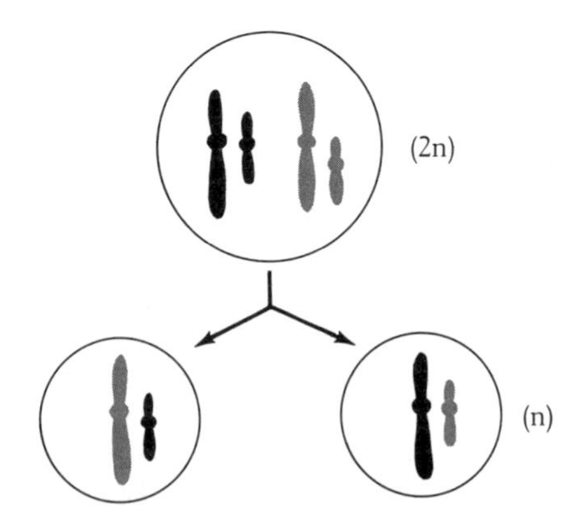

FIGURE 2·8 Comparison of (A) conventional meiosis and (B) the single step meiosis thought to be characteristic of the Coccidia. Only two different chromosomes are shown for clarity. The chromosomes donated by the male are shown in black; those donated by the females are shown in gray. The conventional meiosis shows a chiasma and a single meiotic crossover.

gondii, one feature of gametogenesis must be borne in mind. Since sexual differentiation in *T. gondii* is not genetically determined, each of the two parental strains used in a binary cross will produce both male and female gametes in the intestine of the infected cat. If both parents in the cross are used in exactly equal numbers, half of the matings will be self-crosses between gametes of the same strain. Thus, only half of the zygotes that are formed will be capable of giving rise to recombinants. These zygotes will yield 25 percent doubly drug–resistant recombinant sporozoites by independent assortment of genetic markers on different chromosomes. However, the apparent frequency of these recombinants will be halved, because an equal number of progeny will be produced from the self-crosses. Thus the highest possible frequency of recombinants that can be achieved is 12.5 percent, which is just larger than the greatest value actually observed with *T. gondii* and with *E. tenella*.

The maximal recombination frequency of 12.5 percent can be achieved only when the parental strains are present in exactly equal numbers. As the ratio of infecting parents deviates from 1 : 1, the expected maximal frequency of recombination between unlinked markers falls because self-crosses between gametes of the majority parent come to predominate. Thus, when genetic crosses with *T. gondii* are analyzed it is important to know the parental ratios that are actually involved. In theory, this ratio can be measured either at the input of encysted bradyzoites in the brains of mice fed to the kittens or in the output of sporozoites in the sporulated oocysts recovered from the kitten's feces. Although bradyzoites can be readily titrated in a plaque assay, their measurement may not reliably reflect the number of gametes produced in the cat intestine. Various strains of *T. gondii* and even individual preparations of encysted bradyzoites may not be uniform in their quantitative abilities to produce gametes. The most desirable measurement would be a determination of the numbers of gametes of each genotype in the feline intestine, but such assays are impossible. However, a close approximation of this measurement can be obtained, given the assumption that the gametes of each genotype have the same efficiency of mating and ultimately of producing sporulated oocysts. Under these assumptions, the ratio of sporozoites resistant to the two different drugs used as genetic markers should equal the ratio of the functional gametes of the parental genotypes actually present in the intestine of the infected kitten. This ratio then allows a simple calculation of the expected frequency of recombination between unlinked markers as one-half of the product of the frequencies of the two parental genotypes in the progeny sporozoites. Note that in a binary cross, if the parental genotypes are equal (each $= 0.5$), this equation predicts the maximal frequency of recombination noted above, 0.125.

An advantage of understanding the relationship between parental genotypes measured in the progeny sporozoites and expected frequency of recombinants is that a single kitten can be used for a large number of simultaneous crosses between various drug-resistant mutants. The only assumption required is that mating is random. This approach has been used in analyzing the data by infecting the same kitten with four different drug-resistant mutants, and the results were consistent with theory.

A good many more mutants of *T. gondii* will have to be used in crosses to make any inroads into the problem of defining linkage groups that presumably define the chromosomes of *T. gondii*. In resolving this problem Mendelian crosses are likely to be overtaken by molecular genetic techniques that can easily be coupled with the separation of parasite chromosomes by pulsed field electrophoresis.

Coda

From the viewpoint of a cell biologist, no obligate intracellular protozoan parasite is easy to study. The logistics of supplying adequate numbers of host cells and the problem of having to distinguish between biochemical activities of host and parasite are difficulties that do not have to be faced by those studying protozoa that grow extracellularly. These difficulties mean that the effort spent studying the cell biology of obligate intracellular parasites is probably best focused on the host–parasite relationship. *Toxoplasma gondii* is the best organism for studying this relationship because it is so easy to grow in a wide variety of cultured cells. All the major unsolved questions discussed in this chapter, the mode of entry into host cells, the structure and function of the parasitophorous vacuole, and the control of gametogenesis will have to be solved with cultured cells.

Additional Reading

CHOBOTAR, B., and SCHOLTYSECK, E. 1982. Ultrastructure. In *The Biology of the Coccidia*, P. L. Long, ed. University Park Press, pp. 101–165.

JOHNSON, A. M. 1985. The antigenic structure of *Toxoplasma gondii:* a review. *Pathology* 17:9–19.

PFEFFERKORN, E. R. 1981. Molecular genetics of Toxoplasma: *Toxoplasma gondii* as a model parasite for genetic studies. In *Modern Genetic Concepts and Techniques in the Study of Parasites*, F. Michal, ed. Schwabe and Co., pp. 195–212.

PFEFFERKORN, E. R., and SCHWARTZMANN, J. D. 1981. Use of mutants to study the biochemistry of the host-parasite relationship in cultured cells infected with *Toxoplasma gondii*. In *International Cell Biology*, H. G. Schweiger, ed. Springer Verlag, pp. 411–420.

RUSSELL, D. G. 1983. Host cell invasion by Apicomplexa: an expression of the parasite's contractile system? *Parasitology* 87:199–209.

WANG, C. C. 1982. Biochemistry and physiology of Coccidia. In *The Biology of the Coccidia*, P. L. Long, ed. University Park Press, pp. 167–228.

WERK, R. 1985. How does *Toxoplasma gondii* enter host cells? *Rev. Infect. Dis.* 7:449–457.

3

Cell Biology of African Trypanosomes

Mervyn J. Turner and John E. Donelson

Initially, cell biologists were attracted to African trypanosomes because of their fascinating antigenic variation — the process by which these protozoan parasites evade the immune response of their mammalian host by periodically changing their surface coat. As information about the structural and genetic aspects of this process became available, it revealed additional unexpected features of trypanosomes that were equally intriguing. Experimental investigations of these additional features are now among the most active areas of trypanosome research and provide the best hope for identifying new drugs to combat these parasites. This chapter summarizes our present understanding of the biochemical basis of antigenic variation and discusses two other distinctive features of trypanosomes: the organization of their microtubules and the presence of a unique cytoplasmic organelle called the glycosome, which contains the enzymes catalyzing glycolysis and other metabolic pathways.

Antigenic Variation

Antigenic variation is a consequence of changes in the chemical composition of the surface coat, which covers the entire surface of the trypanosome. This surface coat is visible in transmission electron microscopy as an electron-dense layer from 12 to 15 nanometers (nm) thick that lies outside of the lipid bilayer of the plasma membrane (Figure 3-1). It is composed of a monolayer of about 10^7 closely packed copies of a glycoprotein molecule called the variant surface glycoprotein, or VSG. By sequentially expressing genes that encode different VSG sequences, a trypanosome population is

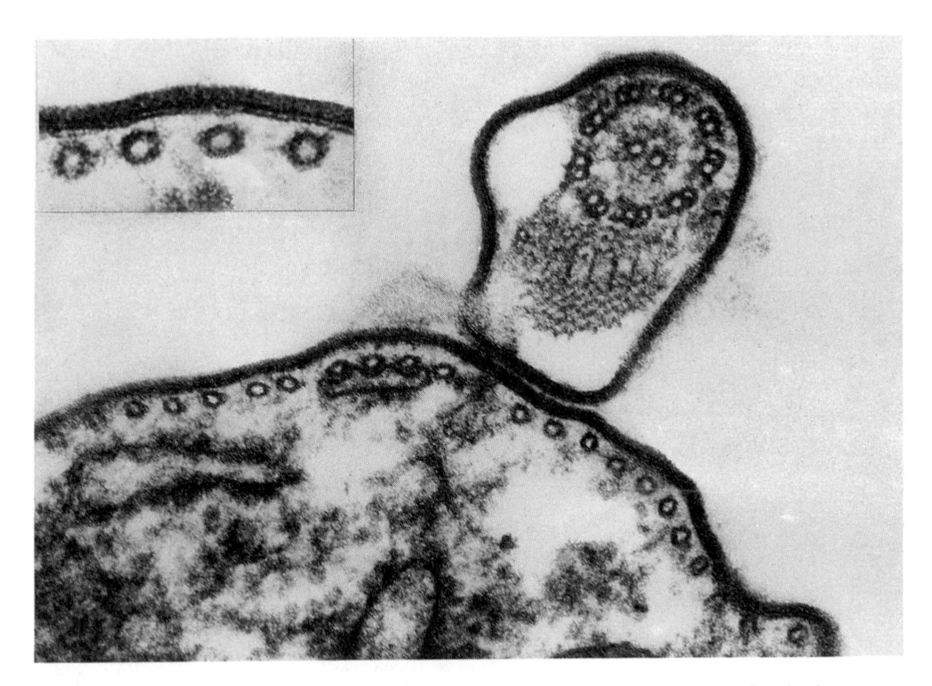

FIGURE 3·1 Transmission electron micrograph of a thin section through an African trypanosome shows its flagellum as well as the microtubules and the surface coat containing the VSG. Glycosomes occur in the cytoplasm but are not resolved in this photograph. Visible in the insert is the lipid bilayer below the surface coat. Magnification approximately ×200,000.

able to change its antigenic profile and keep one step ahead of the immune response against it.

The protein structure of VSGs, their arrangement on the cell surface, and their attachment to the plasma membrane have been subjects of much study in recent years, as has the immunochemical profile of the coat. Fully processed VSGs have molecular sizes between 55 and 60 kilodaltons (kD). They contain about 450 amino acids and have 7 percent to 17 percent of their weight as carbohydrate distributed between two different classes of oligosaccharide side chain. One class is composed of asparagine-linked oligosaccharides which, in most respects, are similar to the N-linked oligosaccharides found in many other eukaryotic glycoproteins. The number and positions of the asparagine residues to which these oligosaccharides are linked vary from one VSG to another. Curiously, although these carbohydrate moieties can be distributed throughout the polypeptide backbone, they are not accessible to plant lectins (which bind specifically to carbohydrates) when the VSG is on the parasite surface. Instead, they are apparently buried within the matrix of tightly packed VSGs.

The second class of oligosaccharide has a more unusual structure. Most membrane-bound glycoproteins of eukaryotes are anchored to the plasma membrane by a sequence of hydrophobic amino acids embedded in the membrane, which often completely spans the bilayer to link with a cytoplasmic domain. The trypanosome eschews such a conventional mode of attachment for VSG, instead favoring a more exotic, although not unique, linkage to a glycolipid molecule (Figure 3-2). The C-terminal carboxyl group of the VSG is in an amide linkage to an ethanolamine, which, in turn, is coupled through a phosphodiester linkage to an oligosaccharide containing three mannose residues and a side chain with a variable number of galactose molecules. At the reducing terminus of this glycan moiety is a molecule of glucosamine that is O-glycosidically linked to *myo*-inositol of phosphatidylinositol. In *Trypanosoma brucei* both acyl groups of the phosphatidylinositol are myristic acid ($C_{14:0}$), a relatively rare fatty acid in the trypanosome.

The trypanosome contains a phospholipase-C enzyme whose activity is specific for this glucosaminyl phosphatidylinositol. When activated, it cleaves the phosphate between the diacylglycerol and the inositol, resulting in the release of the surface coat from the plasma membrane. The role of this intriguing phospholipase in the life cycle of the trypanosome is not yet clear. Biochemists working on VSGs have used it unwittingly for many years, because its activation upon disruption of the plasma membrane leads to the release of the VSG in a water-soluble form that may be purified readily to homogeneity in tens of milligrams. Although free VSG can be detected in the circulation of infected animals, most of it is probably released from dying trypanosomes rather than from trypanosomes undergoing a VSG switch. Whether the lipase also has a role in this switch from the expression of one surface coat to another is not known. Antigenic variation of trypanosomes in the bloodstream is actually a rare event, occurring only

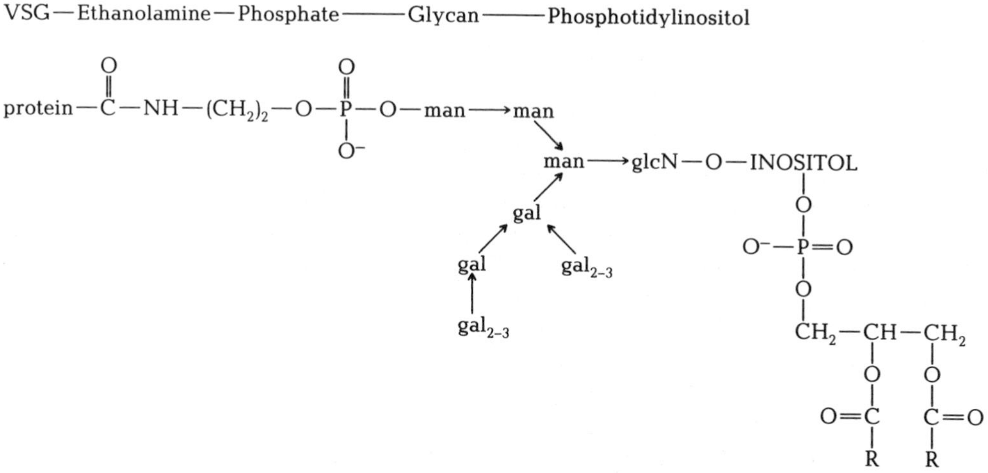

VSG—Ethanolamine—Phosphate——Glycan——Phosphotidylinositol

protein—C(=O)—NH—$(CH_2)_2$—O—P(=O)(O^-)—O—man⟶man⟶man⟶glcN—O—INOSITOL

man ← gal; gal ← gal ← gal_{2-3}; gal ← gal_{2-3}

INOSITOL—O—P(=O)(O^-)—O—CH_2—CH—CH_2

CH—O—C(=O)—R; CH_2—O—C(=O)—R

FIGURE 3·2 Structure of the glucosaminyl phosphatidylinositol anchor of a *Trypanosoma brucei* VSG; man, mannose; gal, galactose; glcN, glucosamine; R, $(CH_2)_{12}CH_3$, the side chain of myristic acid. The positions of the carbon atoms in the various sugar linkages are known but, for brevity, are not shown here.

about once in every $10^6 - 10^7$ trypanosomes per generation. Thus, it has not been possible to study the switch directly; we have been able to look only at the biochemical footprints of this switch many generations after it has happened. In contrast, when trypanosomes are ingested by a tsetse fly, they differentiate quickly to a form lacking the surface coat. This step can be mimicked in the laboratory by shifting a culture of trypanosomes from 37°C to 26°C in the appropriate media. This temperature shift is accompanied by activation of the phospholipase, and the subsequent release of the VSG also seems to involve loss of liberated diacylglycerol from the membrane. The enzyme has recently been purified to homogeneity, and it will be interesting to see whether specific inhibitors of its activity can be identified, and if so, what effect they will have on the life cycle of the parasite.

The VSG of trypanosomes is the best-characterized example of such a glycolipid linkage of a surface protein. However, many proteins in other organisms are known to contain similar or identical linkages, including the P63 surface protease of *Leishmania*, acetylcholinesterase, and decay accelerating factor (DAF) of red blood cells, the Thy-1 lymphoid and neuronal differentiation antigen, and the placental alkaline phosphatase. The malaria parasite, *Plasmodium falciparum*, contains a membrane protein linked to a similar glycolipid that contains no enzymatic activity when on the surface but becomes a protease when released in its soluble form. Many other membrane-associated proteins probably have similar glycolipid linkages, and their structures and biological significance will be an active area of investigation in the future. For the moment, however, trypanosomes stand out because of presence of the phospholipase-C activity that can quickly release the cell's entire surface coat. The abundance of both the substrate and the enzyme makes the trypanosome the organism of choice in searching for the biochemical and biological basis for this release phenomenon. Because immunization of rabbits with released VSG molecules stimulates production of many antibodies (some of which are directed at the C-terminal oligosaccharide), antibodies are available for the detection of such protein-bound glycolipids in trypanosomes and other organisms. This antigenic determinant on VSG that is also present on other eukaryotic cells is known as the cross-reacting determinant, or CRD. The combination of phospholipase treatment and detection with anti-CRD antibodies has been used to identify CRD-like structures not only on trypanosomes but also on red blood cell acetylcholinesterase and DAF, on P63 of *Leishmania*, and on the surface antigen of *Paramecium aurelii*. Nevertheless, the full extent of the glycan similarities in different organisms is not known. Some differences do occur because the Thy-1 anchor, in contrast to the VSG glycan anchor, contains no galactose and has one mole of N-acetylgalactosamine.

The special properties imparted to a membrane protein through a glucosaminyl-phosphatidylinositol anchor, in contrast to a more conventional transmembrane hydrophobic sequence, have not as yet been established. One obvious possibility is that the protein can be released rapidly when the anchor is enzymatically disrupted. Another possibility is that different membrane anchors affect the enzymatic activity of a surface protease (as has been shown for the surface protease of *Plasmodium*). Alternatively, this anchor might provide some conservation of space in the lipid bilayer or perhaps permit greater lateral mobility of the protein within the membrane's lipid bilayer.

The amino acid sequence diversity within the entire family of VSG molecules is so great that the presence of the CRD on the VSG is responsible for virtually all the immunological cross-reactions among different VSG molecules. Side-by-side comparison of different VSG sequences shows a hypervariability far more pronounced than that found in immunoglobulin variable region domains. This diversity is most pronounced within the N-terminal two-thirds of the molecule, which is believed to contain the parts of the VSG that are exposed on the surface of the living trypanosome (see Chapter 12). Within this portion of the molecule, the only significant sequence similarity among VSGs seems to be the occurrence of a cysteine residue between positions 14 and 17, although there is some tendency toward positional conservation of hydrophobic and small side groups.

In the last 50–100 residues at the C-terminus, some sequence similarities do emerge that are sufficient to allow classification of VSGs into two sequence families, class I and class II (Figure 3-3). Class I VSGs terminate with an aspartate or asparagine residue onto which the glycolipid is bound, whereas class II VSGs have a C-terminal serine. In each class the positions of several cysteines and some charged residues are generally conserved, suggesting that they contribute to a common structure of the C-terminal similarity domain that is class-specific, because the two classes are different.

The VSG mRNA encodes a C-terminal tail of 23 (class I) or 17 (class II) amino acids that is absent from the mature glycoprotein (Figure 3-3). These additional amino acids are largely hydrophobic, except for one lysine, and probably function to temporarily anchor the VSG to a membrane until the glycolipid can be transferred onto the nascent polypeptide. The classification into class I and class II includes these additional amino acids and, indeed, the most sequence conservation occurs within this region. At the extreme C-terminus, the class I and class II sequences converge, since all the C-terminal hydrophobic tails terminate in either leucyl-leucyl-phenylalanine or leucyl-leucyl-leucine.

In addition to these C-terminal sequence similarities, further structural similarity is indicated by the fact that all VSGs can be cleaved into two domains by the limited action of trypsin or other proteolytic enzymes. These enzymes produce an N-terminal fragment comprising two-thirds to three-quarters of the molecule, which is sometimes termed the hypervariable domain, and a C-terminal fragment containing one-third to one-quarter of the sequence, known as the similarity domain. VSGs associate to form homodimers both in solution and on the trypanosome surface. In addition, it appears that each of the two proteolytic domains can also form a dimer. Crystallization of two different N-terminal hypervariable domains, one isolated from a class I VSG and one from a class II VSG, has allowed a search for higher orders of structural similarity within VSGs that are only distantly related at the amino acid–sequence level. Interestingly, these two VSGs were found to share a common structural motif. Each hypervariable-domain dimer contains a symmetrical bundle of alpha-helices distributed around the longitudinal axis of the molecule. The bundle, which is approximately nine nm long, compared with the 12–15-nm thickness of the surface coat, has sixfold symmetry at the C-terminal end of the domain, fourfold symmetry in the central region, and sixfold symmetry again at the N-terminus. Although the presence of large amounts of alpha-

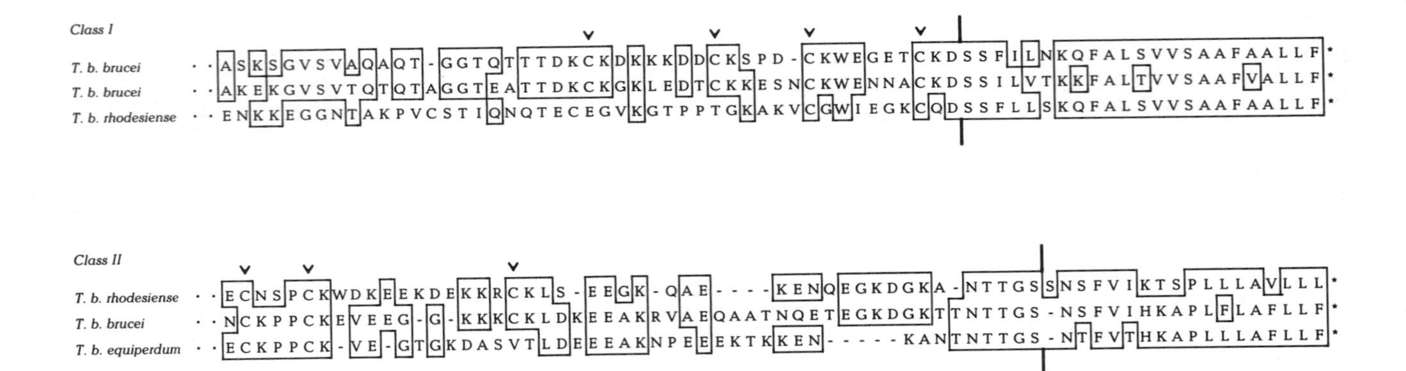

Figure 3·3 Comparison of the C-terminal amino acid sequences of three immunologically distinct VSGs in each of the two C-terminal sequence families, class I and class II. These VSGs, from the indicated trypanosome species, are representative of the 20 or more VSGs in several trypanosome species whose complete or partial sequences have been determined. Boxes indicate the most highly conserved positions, both among these VSGs and the others for which sequence information is available. Dashes have been placed in the sequences to maximize sequence similarity; arrowheads indicate cysteine residues; the vertical lines show the cleavage sites for removal of the hydrophobic tail of 23 amino acids (class I) or 17–18 amino acids (class II).

helix in VSGs has been inferred from spectroscopic measurements, secondary structure predictions failed to identify the locations of these alpha-helices in the VSG for which the most structural information is currently available. This failure points to the inadequacy of the rules currently available for secondary structure prediction and highlights the fact that with a database of approximately 1000 VSG genes in the trypanosome genome, the VSG system offers an excellent opportunity for the generation of new and better rules for the correlation of sequence and structure. Clearly such correlations would be of interest to a much wider audience of protein biochemists than just trypanosome specialists.

The eventual elucidation of the three-dimensional structures of VSGs to high resolution should provide answers to some interesting questions. First, what are the locations of the antigenic determinants that are exposed on the surface of the living trypanosome? Surely, the changes in this portion of the molecule provide the effective limit to the size of the VSG repertoire. Second, and of almost equal interest, what are the locations of the epitopes that are not exposed on the living trypanosome? Why do VSGs vary so much in these cryptic antigenic determinants, which would seem to have no survival value for the trypanosome? If the trypanosome does have a system for shuffling segments of VSG genes to create entirely new VSG sequences, as seems increasingly probable, then it may be that selection by the immune system for antigenically distinct variants is a secondary event, and that the primary selection pressure is for the ability of such new sequences to form a particular structure for the surface coat, impermeable to complement and other components of the immune system. Third, does the tendency of VSGs to form dimers suggest that packing compatibility of the new VSG with the old VSG is important in maintaining the integrity of the surface coat during the crucial switch period and plays a role in the order of VSG appearance?

Other interesting questions include where the buried asparagine-linked carbohydrate moieties are positioned relative to the VSG peptide backbone, since these structures are not exposed on the surface of the living trypanosome. The oligosaccharide component of the C-terminal glycolipid also is not exposed, because of its location underlying the surface coat. Furthermore, there is currently no information on the structures of the C-terminal similarity domains since, to date, only N-terminal hypervariable domains have been crystallized. The intact VSG molecule has proved refractory to crystallization, and no successful attempt to crystallize the C-terminal domain independently has yet been reported. Finally, the successful determination of the three-dimensional structure of an entire VSG structure should provide information about how individual VSG molecules are packed onto the trypanosome's surface.

From this consideration of antigenic variation and the properties of the surface coat, it seems likely that attempts to produce a vaccine for trypanosomiasis are doomed. Therefore, most current studies on VSG focus on its synthesis, membrane integration, and membrane release as potential targets for chemotherapy. Of obvious interest is the role of the VSG-specific phospholipase C in antigenic variation, and the extent to which interference with correct functioning of this enzyme could assist the infected host.

The biosynthetic pathway of the precursors for the glycolipid tail is another potential chemotherapeutic target, even though such glycolipids are not confined to the trypanosome. The glycan structure shown in Figure 3-2 suggests that its synthesis involves several novel enzymes not yet described whose inhibition would prevent VSG attachment to the membrane. Another consideration is the unusual and absolute specificity for myristic acid as the acyl groups in the phosphatidylinositol of *T. brucei* VSGs.

Perhaps furthest in the future, if VSGs of different sequences do share a common structural motif, then this structure may reflect a common protein-folding pathway. Agents that interfere with this pathway could prevent proper assembly of the surface coat, with fatal consequences for the trypanosome.

The Microtubules

All eukaryotic cells contain tubular protein structures known as microtubules. Microtubules have many roles within the cell, including chromosome segregation, motility, and maintenance of shape. Microtubules are formed by polymerization of heterodimers of α- and β-tubulin. Tubulins have been purified from many organisms and have been extensively studied. However, the microtubules of the trypanosomes have received less attention than they deserve, probably because of the overwhelming interest in the VSG. The hemoflagellates, of which trypanosomes are an example, have a unique arrangement of microtubules in the cell body. Underlying the plasma membrane, and in close contact with it, is a helical array of singlet microtubules (Figure 3-1). This subpellicular array is responsible for the characteristic shape of the parasite. The microtubules are laterally interconnected and form tight membrane contacts, the precise nature of which is not understood. In addition, the flagellum of the organism contains the conventional axoneme, which in cross section contains a circle of nine microtubules with an additional central pair. This arrangement is common to all flagellated organisms and the axoneme is responsible for cell locomotion. Finally, trypanosomes contain a transient intranuclear spindle, present only during nuclear division.

Tubulins are very abundant proteins in trypanosomes, second only to the VSG. Comparatively little work has been done on these proteins despite their ready availability. Tubulins are generally a heterogeneous group of proteins. Although both α- and β-tubulins have molecular weights of about 50,000, expression of different tubulin genes produces proteins that differ slightly in amino acid sequence. Differences in posttranslational modification also can lead to some heterogeneity. In trypanosomes, pellicular and flagellar microtubules may be distinct because a monoclonal antibody has been found that recognizes tubulin only in the flagellum. In addition, evidence for heterogeneity within the subpellicular microtubules comes from the related kinetoplastid organism *Crithidia fasciculata*. Three different forms of α-tubulin are found in the pellicular microtubules, only two of which occur in the cytoplasmic pool, whereas the third form appears in the axoneme only. Some of this heterogeneity is generated by posttranslational modification of a tyrosine residue. In all eukaryotes, this reaction occurs only on the α-tubulin chain. Heterogeneity in the β-tubulins of trypanosomes has not yet been detected. *Trypanosoma b. rhodesiense* contains about 15 α-tubulin and 15 β-tubulin genes arranged predominantly as tandem arrays of α-/β-tubulin gene repeats. One of these repeat units has been cloned and sequenced, allowing a comparison of the trypanosomal tubulins with those of their vertebrate hosts. The β-tubulin gene encodes a protein of 442 amino acids; the α-tubulin gene specifies a protein of 451 amino acids. Both contain a tyrosine as the C-terminal amino acid. Thus, it seems likely that the posttranslational modification of α-tubulin involves both a tubulin tyrosine carboxypeptidase, which removes this terminal tyrosine, and a tyrosine transpeptidase, which adds it back on. This apparently redundant action seems to occur in other organisms as well, since C-terminal tyrosines are encoded by both the rat and human α-tubulin genes. However, in contrast to higher eukaryotes, the trypanosome β-tubulin gene also encodes a C-terminal tyrosine, and yet there is no evidence for posttranslational tyrosinolation of β-tubulin in trypanosomes. This difference indicates that trypanosomes must have a highly specific system for tyrosinolation and detyrosinolation.

About 40 percent sequence identity is present in the genes for trypanosome α-tubulin and β-tubulin, indicating that they have diverged from each other to an extent similar to that found in tubulins of other organisms. More interesting is the extent to which these α- and β-tubulin sequences have diverged from the α- and β-tubulin sequences of other species. Comparison of the predicted sequences of *T. b. rhodesiense* α- and β-tubulin with those of rats, pigs, and humans shows about 84%–85% amino acid sequence identity, compared with 70%–71% identity with the corresponding tubulins of yeasts. The sequence differences are not evenly distributed throughout the molecules. For β-tubulin, the most variable regions are between positions 1 and 80, 340 and 390, and 430 and the C-terminus. For the α-tubulins, the corresponding variable regions are between positions 128 and 154, 168 and 204, and 437 and the C-terminus. In both the α- and β-tubulins there are also strongly conserved regions in which amino acid substitutions are generally conservative.

What is the significance of such differences? Microtubules have proved to be very good targets for drugs. In particular, the benzimidazole antihelmintics probably act by depolymerizing microtubules in nematodes. The benzimidazoles, however, have no effect on protozoan parasites such as the trypanosomes. Similarly, other drugs such as colchicine are potent inhibitors of microtubule formation in higher eukaryotes, but have no effect on protozoan parasites or, indeed, on other lower eukaryotes. In studies with isolated tubulins, benzimidazoles have a much higher affinity for nematode tubulin, whereas colchicine is more specific for tubulins of higher eukaryotes. As yet, there are no agents that bind selectively to microtubules of protozoan parasites. A careful sequence comparison of tubulins from the protozoa, from nematodes, and from vertebrates might aid both in the identification of the residues involved in drug binding and in the development of a new generation of compounds selective for protozoa.

In addition to the α- and β-tubulins, microtubules contain other proteins called microtubule-associated proteins (MAPs). The best-characterized MAPs have molecular weights of about 300,000 daltons and about 55,000–65,000 daltons. These latter proteins are generally known as the Tau MAPs. MAPs probably mediate the associations of microtubules with other cellular organelles and the interactions among different microtubules. The high-molecular-weight MAPs also seem to promote the polymerization of tubulins into microtubules. One study using porcine brain α- and β-tubulins concluded that amino acid residues 434–440 of β-tubulins are crucial for the binding of both classes of MAPs. This region corresponds to one of the variable portions in the comparison of different β-tubulin sequences. Should this conclusion be correct, then the interaction between MAPs and tubulins in protozoa may offer another advantage for selectivity in the development of novel drugs. Although no work has been performed directly on MAPs in trypanosomes or any other protozoan parasite, some indirect evidence exists for the presence of still other proteins that perturb microtubule function.

A number of tricyclic antidepressants, most notably the phenothiazines, have been shown to disrupt the contact between the subpellicular microtubules of trypanosomes and the plasma membrane, which leads to a loss of cell motility and death. A phenothiazine binding protein of 60,000 daltons has been purified from *T. brucei* and has been shown to be unique to trypanosomatids. Indeed, monoclonal antibody binding studies show that the protein is not well conserved even among different trypanosome subgenera. The function of this protein is unknown, but one possibility is that it is a microtubule-associated protein whose function is to connect the subpellicular microtubules to the plasma membrane. This region of the trypanosome cytoskeleton will be interesting to investigate further.

The Glycosome

In the vertebrate bloodstream, where they have ample access to glucose, trypanosomes derive ATP predominantly from anaerobic glycolysis through the Embden-Myerhoff pathway. Indeed, trypanosomes have the highest known glycolytic rate of any eukaryote. Most of the enzymes of the glycolytic pathway are confined to an organelle unique to the trypanosome, called the glycosome. The glycosome contains the first seven enzymes of the glycolytic pathway and two additional enzymes involved in glycerol metabolism. These enzymes are hexokinase, glucose phosphate isomerase, phosphofructokinase, fructose 1,6-bis-phosphate aldolase, triosephosphate isomerase, D-glyceraldehyde-3-phosphate dehydrogenase, phosphoglycerate kinase, glycerol 3-phosphate dehydrogenase, and glycerol kinase.

These enzymes differ in interesting ways from those found in the mammalian host of the parasite. In particular, their properties in solution, following disruption of the glycosome, indicate that they associate with each other to form multienzyme complexes. One attractive suggestion to account for such associations is that the substrate, glucose, is channeled through a chain of enzyme-active sites to provide the impetus for the observed rapid rates of glycolysis. Recent experiments, however, indicate that this is not the case, and the advantage to the parasite of compartmentalization of the glycolytic pathway remains uncertain.

It is intriguing to ask how the glycosome itself is assembled. The organelle contains no DNA and no ribosomes; therefore, the enzymes must be synthesized elsewhere in the cytoplasm and transported into, or assembled simultaneously with, the basic glycosome structure. Translation of the mRNAs for several of the enzymes has been shown to occur on free, rather than membrane-bound, cytosolic ribosomes, indicating that the nascent polypeptides are imported into the glycosome. It is likely, therefore, that each polypeptide found in the glycosome contains a unique signal to facilitate transport into the glycosome, and this signal is likely to be common to all proteins of the glycosome but absent from proteins found exclusively in the cytoplasm or elsewhere in the parasite. Much recent work centers on the search for such a signal; if identified, it could provide a target for the rational design of new trypanocidal drugs.

Comparison of the physical properties and amino acid sequences of glycosomal enzymes with their counterparts in the cytosol of both trypanosomes and their mammalian hosts has revealed a few differences. In general, the trypanosomal enzymes are more basic and are slightly larger in size. The three-dimensional structures of several of the mammalian enzymes are known, and the sequences of the trypanosome enzymes are sufficiently similar to suggest that they adopt similar folding patterns. When the sequences of the glycosomal phosphoglycerate kinase, triosephosphate isomerase, or glyceraldehyde-3-phosphate dehydrogenase are compared with the known structures of their vertebrate counterparts, the additional positive charges found on the trypanosomal enzymes are clustered at two spots on the surface of the enzyme. It has been suggested that these positively charged "hot spots" may function as the postulated glycosome recognition sequences. The distance between the two hot spots is about 12 angstroms (Å), which is about the distance between the two negatively charged groups on a well-known trypanocidal drug, suramin, whose mode of action is uncertain. It will indeed be fascinating to determine if an interaction between the suramin molecule and the glycosomal enzymes is responsible for this drug's activity.

Eventually, a knowledge of the three-dimensional structures of each of the enzymes of the glycosome will be necessary to allow comparison with the homologous enzymes of the vertebrate host. This information should be extremely helpful in devising trypanosome-specific inhibitors of glycolysis.

Other Significant Features

Other features of the trypanosome interest the cell biologist and biochemist. When in the vertebrate host, all trypanosomes obtain their ATP entirely from anaerobic glycolysis, although some trypanosomes in the bloodstream population prudently preadapt to aerobic life in the tsetse fly midgut by switching on some of the enzymes involved in mitochondrial respiration. These trypanosome forms, when ingested, probably propagate the population in the tsetse fly. The environmental signals controlling these changes in the trypanosome's metabolism have not yet been identified, nor have the mechanisms by which such signals are transduced to the mitochondrion. As discussed in Chapter 17, the trypanosome's mitochondrial DNA is remarkably complex, and although much is known, many questions remain to be answered about the functioning of this organelle in the trypanosome.

Other interesting puzzles are posed by the innate resistance of humans to several trypanosome species that readily infect domestic and wild animals. For example, *Trypanosoma brucei brucei* infects cattle but not humans because it is lysed by normal human serum. On the other hand, *Trypanosoma brucei rhodesiense* and *T. brucei gambiense*, both of which are morphologically indistinguishable from *T. b. brucei*, grow well in the human bloodstream and are the main causes of human trypanosomiasis. The mechanism by which *T. b. brucei* is killed by human serum is not completely understood, nor is the mechanism of the acquired resistance trait to *T. b. rhodesiense* or *gambiense*. The high-density-lipoprotein (HDL) fraction of human serum has been implicated in the killing, which seems to involve ion leakage leading to osmotic death. Intriguingly, although it was thought that *T. b. brucei* and *T. b. rhodesiense* were distinct subspecies, it now appears that *T. b. brucei* can, itself, occasionally switch to a form resistant to the action of human serum and is therefore infective to human beings, behaving in all respects like *T. b. rhodesiense*. Again, the nature of this switch and the signal that triggers it are unknown.

Trypanosomes lack one feature that would otherwise make them an ideal system for study by molecular or cellular biologists: the absence of a clearly defined sexual stage in the life cycle. The corollary is an absence of literature on the genetics of trypanosomes, which has hindered progress in other areas, such as in the study of the molecular biology of antigenic variation. This deficiency might be remedied if a system could be developed for transfection of trypanosomes by exogenous DNA whose expression was properly regulated. However, to date the trypanosome has proved refractory to transfection in which the introduced DNA is transcribed with fidelity.

By looking at the frequency of homozygous and heterozygous forms of different isoenzyme markers, it has been established that bloodstream trypomastigotes are diploid for housekeeping genes and that the frequency distribution for homozygotes and heterozygotes is consistent with the occurrence of a sexual stage in the life cycle. This stage probably occurs in the tsetse fly, perhaps within the salivary glands, and it has been possible to genetically cross different strains of trypanosomes. Such experiments should provide the foundation for the study of trypanosome genetics. Rapid developments in this area will likely occur over the next few years.

Coda

Trypanosomes have properties of interest to investigators working on protein, lipid, or carbohydrate molecules, on cell structure and form, on gene structure and regulation, on intermediary metabolism, and on immune regulation. The main challenge now is to channel that interest in the cell biology of trypanosomes toward the development of effective new procedures for the control of this fascinating, yet devastating, parasite.

Additional Reading

DONELSON, J.E., and M. J. TURNER. 1985. How the trypanosome changes its coat. *Sci. Am.* 252:44–51.

FERGUSON, M. A. J., S. W. HOMANS, R. A. DWEK, and T. W. RADEMACHER. 1988. Glycosyl-phosphatidylinositol moiety that anchors *Trypanosoma brucei* variant surface glycoprotein to the membrane. *Science* 239:753–759.

HERELD, D., J. L. KRAKOW, J. D. BANGS, G. W. HART, and P. T. ENGLUND. 1986. A phospholipase C from *Trypanosoma brucei* which selectively cleaves the glycolipid on the variant surface glycoprotein. *J. Biol. Chem.* 261:13813–13819.

METCALF, P., M. BLUM, D. FREYMANN, M. J. TURNER, and D. C. WILEY. 1987. Two variant surface glycoproteins of *Trypanosoma brucei* of different sequence classes have similar 6 A resolution X-ray structure. *Nature* 325:84–86.

MISSET, O., O. J. M. BOS, and F. R. OPPERDOES. 1986. Glycolytic enzymes of *Trypanosoma brucei*. Simultaneous purification, intraglycosomal concentrations and physical properties. *Eur. J. Biochem.* 157:441–453.

RUSSELL, D. G., and J. F. DUBREMETZ. 1986. Microtubular cytoskeletons of parasitic protozoa. *Parasitol. Today* 2:177–179.

TURNER, M. J. 1985. The biochemistry of the surface antigens of the African trypanosomes. *Brit. Med. Bull.* 41:137–143.

4

Cell Biology of *Trypanosoma cruzi*

■

Miercio E. A. Pereira

Trypanosoma cruzi is the causative agent of Chagas' disease, which affects millions of people in Latin America; the disease is often lethal and virtually incurable. One of the main reasons for trying to determine the molecular basis of interactions between *T. cruzi* and its mammalian hosts is the likelihood that such knowledge may provide rational strategies for controlling Chagas' disease. In addition, learning the cell biology of *T. cruzi* should be rewarding because it may provide a model to dissect molecular mechanisms of differentiation and of cell–cell interaction.

T. cruzi reacts with and infects most mammalian cells, and the parasite is highly pleomorphic, exhibiting several distinct forms in its life cycle. Each form apparently displays unique properties. Transformation of developmental forms occurs in response to simple variations in environmental conditions. Furthermore, morphologically similar stages derived from diverse strains, or even from the same strain, display remarkable differences in biological properties. *T. cruzi* is clearly an organism well suited to the study of the structure–function relationship at the cellular level.

The molecular mechanisms governing the cell biology of *T. cruzi* remain largely unknown, although some progress has been made in the last few years. The aim of this chapter is to review some of the principles learned and to propose a novel model of *T. cruzi*–host cell interaction based on current evidence. Emphasis is placed on the structure–function relationship at the molecular level, and no attempts are made to be exhaustive in phenomenological descriptions and in morphological aspects of the parasite per se, since these concerns are covered in recent reviews referred to at the end of the chapter.

Developmental Forms

The identification of the developmental stages of *T. cruzi* is based on such morphological criteria as the general form of the cell, the position of the kinetoplast and nucleus relative to one another, and the region where the flagellum emerges from the flagellar pocket. The morphologically distinct stages show unique biological properties. Accordingly, three distinct stages are identified: trypomastigotes, amastigotes, and epimastigotes. (1) *Trypomastigotes* have the kinetoplast located posterior to the nucleus and are about 20 μm long and 2 μm wide (Figure 4-1). In mammalian hosts, they are found intracellularly in tissues and extracellularly in the circulation, and in invertebrate hosts (reduviid bugs), they are always extracellular and located in the posterior intestine. Because researchers assumed that the infection starts in mammals and ends in insects, they named the trypomastigotes found in invertebrates metacyclic trypomastigotes. (2) *Amastigotes* are the intracellular replicating forms of *T. cruzi* in mammalian cells (Figure 4-2). They are round forms possessing a short flagellum, and they multiply by binary fission with a doubling time of 7–14 hours. This doubling time was determined in tissue culture cells and was found to depend on several parameters, such as parasite strain and temperature of culture. (3) *Epimastigotes* have the kinetoplast located anterior to the nucleus, are spindle-shaped, and are 20–40 μm long. They are located in the midgut of invertebrate hosts, where they multiply and maintain the infection for as long as the insects live.

The life cycle of *T. cruzi* can be summarized as follows. When an insect infected with *T. cruzi* bites a mammalian host, it may release trypomastigotes in the feces or urine, and after entering the new host through skin abrasion or mucous membranes, trypomastigotes can infect neighboring cells. Inside the cell, trypomastigotes differentiate into amastigotes, some of which are capable of multiplying by binary fission and of redifferentiating into trypomastigotes. Newly transformed trypomastigotes leave the infected cell to initiate a new cycle in adjacent cells, or they can enter the circulation. From the circulation they can propagate the infection throughout the body. They also can enter reduviid bugs when ingested during a blood meal and initiate infection in the

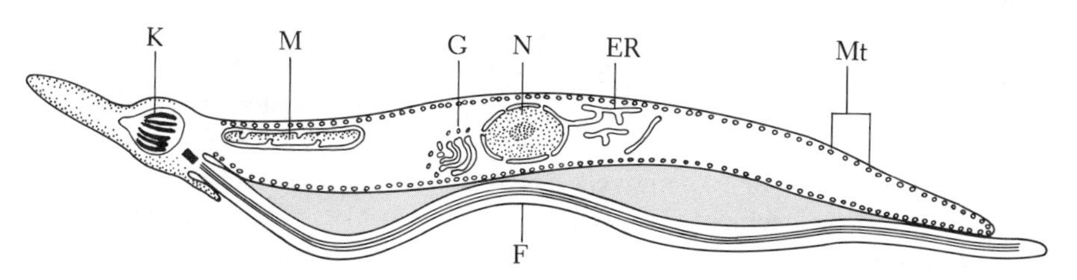

FIGURE 4·1 Representation of *T. cruzi* trypomastigote ultrastructure. K, kinetoplast; M, mitochondrion; G, Golgi apparatus; N, nucleus; ER, endoplasmic reticulum; Mt, subpellicular microtubules; F, flagellum.

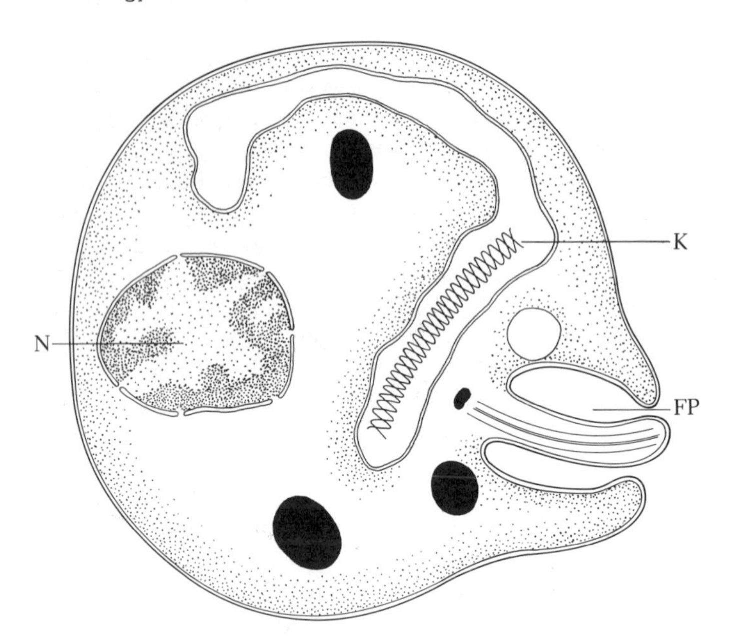

FIGURE 4·2 Representation of *T. cruzi* amastigote ultrastructure. K, kinetoplast; N, nucleus; FP, flagellar pocket.

invertebrate host. In reduviid vectors, trypomastigotes transform into epimastigotes, which then migrate to the midgut, where they multiply; they then migrate to the posterior intestine, where they attach to rectal glands. Attached epimastigotes differentiate into metacyclic trypomastigotes, which in turn are released and eliminated with the feces during a blood meal.

The complete differentiation process of *T. cruzi* in the invertebrate host remains largely unknown and the description just given is probably an oversimplification. However, an insect analogue can be reproduced in vitro by seeding epimastigotes into liquid medium containing fetal calf serum, liver or brain extracts, and glucose (LIT medium). In this medium, epimastigotes multiply and differentiate into trypomastigotes. But since LIT medium is quite different from the internal milieu of insects, parasites harvested from this complex medium may show qualitative and quantitative differences in biological properties from the insect-derived trypanosomes. For example, in stationary-phase liquid cultures containing LIT medium, attachment of epimastigotes to the substratum does not seem to be necessary for metacyclogenesis, whereas attachment is required in the reduviid bug. However, the degree of epimastigote-trypomastigote differentiation achieved under such in vitro conditions is generally very low. An exciting new development is the ability to grow epimastigotes in an insect urine analogue (a buffered salt solution containing a high concentration of L-proline), a condition that promotes epimastigotes to readily adhere to the substratum and favors substantial differentiation into trypomastigotes. These trypomastigotes become detached and swim in the medium. It therefore appears that growth of *T. cruzi* in the urine analogue in vitro simulates the events in the insect rectal gland. The in vitro–induced metacyclic trypomastigotes display biological properties similar to those of the insect-derived parasites, although they are less virulent to experimental animals. This system therefore should provide a useful tool to study the mechanism of differentiation of *T. cruzi* in the insect vector.

Amastigotes and epimastigotes are the parasite stages that multiply and maintain infection in the vertebrate and invertebrate hosts, respectively, but they are not capable of initiating infection in either host. In contrast, trypomastigotes do not replicate but instead are the infective forms responsible for transmission of the infection from mammals to insects and from insects to mammals. Despite the importance of trypomastigotes in the infection of mammalian hosts, several important questions related to their biology in mammalian hosts remain unanswered. First, it is generally assumed that trypomastigotes leave infected cells after the cells are ruptured by heavy parasitism. Although this process is readily observed in tissue culture cells, there is no evidence that trypomastigotes exit host cells in vitro passively; in fact, we could very well say that they do so by a well-timed and active mechanism. Second, after entering host cells, trypomastigotes are usually surrounded by host cell membranes (phagolysosomes), whereas amastigotes multiply freely in the cytosol. Thus, trypomastigotes must rid themselves of host membranes to differentiate into amastigotes. It is interesting that when epimastigotes get into macrophages under in vitro conditions, they are also surrounded by phagolysosomes; however, in this case, because they can not escape from the enveloped host membranes, they degenerate and die. It is not known why trypomastigotes and epimastigotes differ in their ability to escape from phagolysosomes; if the reason were understood, it might be possible to interfere with the trypomastigote escape. Third, an interesting problem surrounds the interaction of trypomastigotes with the vascular endothelium before they encounter myocardial and other potential target cells. For the parasite to enter or exit tissues, it must interact with two structures: the basement membrane, or glycocalyx, and the plasmalemma itself. Although little is known of this interaction, it is critical to parasite survival because in the bloodstream trypomastigotes do not infect blood cells and, unlike the African trypanosomes, do not divide. Bloodstream trypomastigotes must traverse the vascular endothelium in order to infect peripheral tissues and maintain infection in the mammalian host.

Cytoplasmic Organelles

Most organelles of higher eukaryotes are found in *T. cruzi*; the developmental forms of the parasite contain mitochondria, endoplasmic reticulum, ribosomes, peroxisomes, lysosomes, and the Golgi complex. The function of the organelles is apparently similar to that of the mammalian counterparts. For example, trypomastigotes have a linear mitochondrion and other stages have a branched one, but discoid cristae occur in all stages, as does glycolysis, a tricarboxylic acid cycle, proline oxidation, and partial cyanide sensitivity of oxygen consumption. However, there may be differences in the metabolic pathways and final synthesized products. In mammalian, avian, and higher eukaryotes, N-glycosylation of proteins involves the transfer of an oligosaccharide containing two N-acetylglucosamine, nine mannose, and three glucose units from a dolichol-P-P derivative to an asparagine residue. The presence of the three glucose units appears to be necessary for the transfer reaction to occur. The oligosaccharides transferred to the protein are then processed by loss of some of their monosaccharide constituents and, in some cases, by addition of other sugar residues. Transfer and processing of the N-linked sugar moiety of glycoproteins is in the endoplasmic reticulum and Golgi apparatus. It is therefore not surprising that proteins are N-glycosylated in *T. cruzi;* however, the trypanosomes (as well as other trypanosomatids) are peculiar in that unglucosylated oligosaccharides are transferred from dolichol-P-P derivatives to asparagine residues and they are strikingly different from higher eukaryotes. Another interesting peculiarity relates to the sugar composition of *T. cruzi* glycoproteins, which may be different from that of mammalian glycoproteins. For example, the glycoprotein GP-72 of epimastigotes contains relatively high concentrations of ribose, an unusual finding because eukaryotic glycoproteins are not known to contain ribose except for ADP-ribosylated proteins. It is important to identify differences in the metabolic pathways and in the composition of endogenous products of *T. cruzi* (and of other parasites) with host cells since this difference may provide a feasible strategy to control infection, in that it may be possible to find drugs that will inhibit unique parasite enzymes without significantly affecting the host-enzyme composition. In addition, parasite constituents with unusual composition should be highly immunogenic and therefore could induce protective immunity. Accordingly, the epimastigote glycoprotein GP-72 mentioned earlier was found to be highly immunogenic in experimental animals, and the anti-GP-72 antibodies reacted with the sugar moiety of the glycoprotein. The animals immunized with GP-72 were partially protected against infection with metacyclic trypomastigotes.

T. cruzi and other trypanosomatids have cellular structures unique to the protozoa. For example, one of their most distinguishing features is the subpellicular microtubules located below the plasma membrane; these microtubules, which are connected to one another and to the plasma membrane, confer rigidity to the parasite. Experimental animals infected with *T. cruzi* produce antibodies to tubulin, the main microtubule protein, and these antibodies cross-react with the host tubulin. It therefore would be of interest to determine whether these autoantibodies participate in the pathogenesis of Chagas' disease. Subpellicular microtubules are located on the entire plasma membrane, except in an area where the flagellum emerges from the cell (flagellar pocket). It is generally assumed (but with no direct evidence) that endocytosis takes place in the flagellar pocket and not on the rest of the cell. Certainly *T. cruzi* is able to pinocytose proteins such as peroxidase through the flagellar pocket, but whether it exhibits receptor-mediated endocytosis remains to be determined. Another specialized structure of *T. cruzi* and of other trypanosomatids is the kinetoplast, which is a complex array of DNA fibrils in the mitochondrial matrix. Changes in the position of the kinetoplast relative to the nucleus define the morphological stages of *T. cruzi*. Digestion of kinetoplast DNA with restriction endonucleases, followed by the electrophoretic analysis of the fragments on polyacrylamide gels, has been used for the characterization of stocks, strains, and clones of *T. cruzi*.

Despite the obvious importance of learning the biology of the organelles of *T. cruzi* (and of other parasites) at the molecular level, we know very little compared with what we know of higher eukaryotes. For example, understanding the signals that control the targeting of newly synthesized proteins or glycoproteins to specific organelles represents one of the most important and vexing problems facing modern cell biology. But most of what we know we learned using higher eukaryotes as model. It is known that primary translation products of many membrane and secreted proteins contain a hydrophobic amino-terminal peptide that "signals" the vectorial insertion of the nascent protein into (or across) the membrane of the rough endoplasmic reticulum after translation. These peptides are cleaved by a specific protease to yield the mature form of the protein. However, there are several variations on this theme: not all signal peptides are cleaved; many are not amino-terminal but internal; many proteins exhibit multiple membrane-spanning segments with their amino termini facing the cytosol, and the amino acid sequence of signal peptides is not highly conserved. Thus, protein compartmentalization in higher eukaryotes is complicated and studies with more primitive organisms such as *T. cruzi* might yield simple and rewarding results.

Cell-Surface Glycoproteins

It is generally assumed that proteins located on the outer membrane of parasites are important in host–parasite interaction and therefore might serve as targets for vaccination trials. Some of these surface proteins are potent immunogens during natural infections of mammalian hosts and appear as major bands in immunoprecipitation and immunoblotting assays that employ sera from immunized hosts, as discussed later in this chapter. In some cases the assumption that major surface proteins might be important vaccine candidates (e.g., the circumsporozoite antigen of malaria parasites) is warranted; however, major surface proteins do not invariably induce protective immunity. The argument can be made that if a surface protein naturally induces strong cellular and humoral immunity in a host that is nonetheless incapable of ridding itself of the infection, the "major" antigen is not a good candidate for a vaccine. On the other hand, "minor" surface proteins (as defined by surface-labeling technique and immunoprecipitation assay) that have important functions for parasite interaction with the host (e.g., enzymes involved in transport of essential nutrients and receptors for host-derived molecules) and that do not elicit strong antibody response could be valuable for artificially inducing host immune response and protective immunity. Although it is believed that surface proteins might be important for immunoprophylaxis, it is also generally assumed that cytoplasmic materials are not. However, it has been recently reported that immunization of animals with paramyosin, an endogenous protein from *Schistosoma mansoni*, induces protection against schistosomal infection. And paramyosin is a cytoplasmic protein. Clearly, while "major" surface proteins can exhibit functions critical for parasite survival in the host, it is possible that "minor" surface and even cytoplasmic proteins are important as well and that strategies other than simple identification and characterization of surface proteins must be sought.

The properties of several surface proteins described in *T. cruzi* are summarized in Table 4-1. The identification of surface proteins is generally accomplished by surface radioiodination of live parasites followed by lysis in detergent, protein separation by sodium dodecyl sulfate polyacrylamide gel electrophoresis, and exposure to x-ray film. To determine immunogenic proteins, the lysate is generally immunoprecipitated with sera of patients with Chagas' disease or of experimental animals infected with *T. cruzi* before electrophoresis. In other cases, monoclonal antibodies derived from mice immunized with glutaraldehyde-fixed parasites or from mice infected with *T. cruzi* have been used to identify surface proteins. To determine if the surface protein is glycosylated, lectins have generally been used as probe. Some surface proteins that have been identified are GP-25, GP-60, GP-90, and GP-100 and the complex glycolipid LPPG. GP-60, a glycoprotein of molecular weight (MW) 60 kilodaltons (kD), is highly immunogenic in man, and it may be useful for immunodiagnosis of Chagas' disease; GP-25 may or may not be a degradation product of GP-60; GP-90 is the major glycoprotein of *T. cruzi*, and it is composed of several proteins of isoelectric points

TABLE 4·1
Cell-surface glycoconjugates, from trypanosoma cruzi

Glycoprotein	Weight (%)	Carbohydrate Type	Stage distribution	Properties and proposed function
Neuramindase (set of glycoproteins ranging from 160–200 kD		N-linked	E,T	Negative control of infection; HDL receptor
GP-90	19	High mannose	E,A,T	Partial protection in experimental animal infection
GP-85	—	—	T	Host cell–parasite interaction; fibronectin receptor
GP-72	49	Phosphorylated fucose, pentoses, high galactose	E,MT	Differentiation complement receptor, partial protection against metacyclic challenge
GP-60	40	High galactose	E,T	Diagnostic antigen
GP-25	—	—	—	—
LPPG	42–63	Lipopeptidophosphoglycan, galactofuranose	E	—

Abbreviations: E, epimastigotes; T, glood-borne trypomastigotes; MT, metacyclic trypomastigotes; LPDG, lipopeptidophosphoglycan; GP, glycoprotein.

ranging from 5.0 to 7.5. A complex glycolipid can be isolated from the surface of epimastigotes together with three glycoproteins (GP-37, GP-31, and GP-24). The complex has been named lipopeptidophosphoglycan (LPPG), and it has not been detected in bloodstream trypomastigotes. Although the function of LPPG is unknown, antibody to the complex reacts to antigen present in the sera of acutely infected animals and with amastigotes, suggesting that LPPG may be shed or secreted during infection in vivo. These proteins will not be discussed further, because their function is totally unknown. A recently discovered surface protein of MW 160 kD appears to be involved in the resistance of humans and experimental animals to *T. cruzi* infection (see Chapter 13). We will analyze three surface glycoproteins implicated in various biological functions of *T. cruzi*: GP-72, GP-85, and neuraminidase.

GP-72

GP-72 is a major surface glycoprotein of MW 72 kD present on epimastigotes and trypomastigotes of *T. cruzi* grown in liquid medium. Because differentiation and multiplication of the parasite are analogous to these processes in the insect vector, it is assumed that GP-72 is present in insect-derived parasites as well. GP-72 is not detected in amastigotes or in bloodstream trypomastigotes. It is the most extensively characterized protein of *T. cruzi*. Several functions are believed to be associated with the glycoprotein.

Polymorphism The glycoprotein was identified and purified by means of a monoclonal antibody (designated WIC 29.26) and found to contain 49 percent carbohydrate, of which ribose and xylose were major constituents. The sugars were also phosphorylated. Because of its unusual sugar composition, GP-72 is highly immunogenic, as noted earlier. Detection of the epitope by the monoclonal antibody depends on the assay employed. Thus, not all strains and clones of *T. cruzi* react with the monoclonal antibody in a radioimmunoassay using intact living parasites as antigen. However, when parasites were examined by immunoprecipitation reaction, virtually all exhibited the epitope. Electrophoretic variants of GP-72 have been noted, some isolates expressing the molecule in the form of 72 kD – 59 kD or 79 kD – 66 kD doublets. In addition, WIC 29.26 epitope expressed on the surface of epimastigotes is correlated with the isozyme pattern (zymodeme) of *T. cruzi*, with certain zymodemes having distinct geographic distributions, which are distinguished on the basis of their epitope expression. Thus, genetic variations in the biochemistry, physiology, and pathogenicity of *T. cruzi* could be related to differences in the surface exposure of WIC 29.26 epitope.

Metacyclogenesis With an in vitro system, transformation of epimastigotes into trypomastigotes (metacyclogenesis) was found to be inhibited by WIC 29.26 but not by other monoclonal antibodies that recognize non-GP-72 surface molecules, suggesting that GP-72 might be involved in differentiation of *T. cruzi* in the insect vector. Because GP-72 is so heavily glycosylated, it is possible that interaction of the glycoprotein with insect gut lectins might regulate the extent of parasite morphogenesis. In support of this hypothesis, a lectin from the midgut of *Rhodinius prolixus* an insect vector of *T. cruzi*, was found to bind GP-72, as determined by hemagglutination inhibition and radioimmunoassay.

Regulation of complement activation Epimastigotes are very sensitive to lysis in fresh normal serum, whereas trypomastigotes are insensitive. This distinction appears to be due to a difference in the relative ability of these stages to activate the alternative complement pathway. GP-72 is the major C3 acceptor on epimastigotes during complement activation. During this activation process, lytic C3 is deposited on the parasite. In contrast, trypomastigotes fail to activate the alternative complement pathway, and only the inactive fragment iC3b can be found deposited on the metacyclic forms. GP-72, however, does not appear to be the acceptor of C3 in trypomastigotes. The ability of metacyclic trypomastigotes to evade the alternative complement pathway depends on the developmentally regulated synthesis of surface proteins and N-linked carbohydrate; resistance to serum lysis is ablated by pretreating metacyclic trypomastigote with pronase and other proteases, and with neuraminidase or N-glycanase. A stage-specific 90 kD to 115 kD glycoprotein present on the surface of trypomastigotes is suspected of being the component responsible for the alternative complement activation in the infective metacyclic stage. It remains an interesting question why GP-72, which serves as an efficient acceptor on epimastigotes, can no longer function as such when expressed on trypomastigotes.

GP-85

Recent efforts are now converging to identify, characterize, and probe the biological function of GP-85. GP-85 is a glycoprotein of relative molecular weight (M_r) 85 kD present on the surface of trypomastigotes but absent from amastigotes and epimastigotes. It was first identified by lectin binding studies, and its isolation was achieved by affinity chromatography with the use of wheat germ agglutinin immobilized on agarose. The sugar composition of GP-85 is not known, but it may interact with wheat germ agglutinin through N-acetyl-glucosamine residues since lectin binding is not abolished by neuraminidase treatment of the glycoprotein. However, it is not directly known if GP-85 contains N-acetyl-glucosamine or if this sugar, if present, is accessible to the lectin. The gene coding for GP-85 has been cloned and found to contain a nonapeptide unit that is tandemly repeated five times. The role that the repeat unit may have in the biological function of the protein is unknown. Studies from several laboratories suggest that GP-85 may be necessary for efficient host–parasite interaction in the process of invasion. One of the pieces of evidence is that binding and interiorization of *T. cruzi* into tissue culture cells is inhibited by antibodies specific for GP-85. Coincidentally, fibronectin has been found to bind to an 85 kD putative fibronectin receptor on the plasma membrane of *T. cruzi* trypomastigotes. Fibronectin promotes association of *T. cruzi* trypomastigotes and host cells, and parasite fibronectin receptors isolated by immobilized fibronectin affinity chromatography can inhibit infection of tissue culture cells by *T. cruzi* trypomastigotes in vitro. It is possible then that the GP-85 isolated by wheat germ lectin affinity chromatography is the fibronectin receptor on trypomastigotes, and if so, GP-85 should mediate *T. cruzi*–host cell interaction by binding to fibronectin. However, since the work on GP-85 was done by various groups who did not use identical methodologies or reagents, the various properties described for GP-85 still need to be confirmed to be due to the same molecule.

Neuraminidase

Neuraminidase is an enzyme that splices sialic acid from glycoconjugates in solution or on cell membrane. *T. cruzi* exhibit developmentally regulated neuraminidase activity. The enzyme is present at high levels in trypomastigotes, at 10- to 50-fold lower levels in epimastigotes, and is undetectable in amastigotes. Effects of neuraminidase activity have been observed in in vitro experiments: live trypomastigotes, but not amastigotes, remove sialic acid from rat myocardial and human endothelial cells during infection or from red blood cells exposed to the parasite. The degree of desialylation of red blood cells observed in mice infected with *T. cruzi* directly correlated with the magnitude of parasitemia. Strains and clones of *T. cruzi* are heterogeneous with respect to enzyme activity. The presence of high neuraminidase activity in these populations correlates with myotropism.

The enzyme is present on the outer membrane of trypomastigotes. Live parasites desialylate cells and macromolecules, and enzyme activity is not detected in the cell-free supernatant under the assay conditions. Live intact trypomastigotes digested with trypsin release all the enzyme activity into the supernatant (along with other surface proteins). Antibodies to the enzyme react with live parasites as determined by immunofluorecence microscopy. Of great interest is the finding that the enzyme is present on a subpopulation of trypomastigotes (15 percent to 40 percent in the Y strain of *T. cruzi*) as determined with the antineuraminidase antibodies. This subpopulation is morphologically distinct from parasites that are not bound by the antibodies. Complement-mediated lysis by the antineuraminidase antibody destroys the trypomastigotes that have the enzyme (NA^+) but not intact parasites that fail to be recognized by the antibody and that lack neuraminidase activity (NA^- trypomastigotes). NA^+ and NA^- parasites have different biological properties. The NA^+ trypomastigotes may exert a negative influence on the infectivity of NA^- parasites, since NA^- trypomastigotes are more infective to culture cells in vitro than is a mixed population composed of NA^+ and NA^- parasites. The presence of these two subsets of trypomastigotes has also been verified with a specific inhibitor of the neuraminidase (high density lipoprotein or cruzin) labeled with a fluorescence probe; only NA^+ parasites bind the inhibitor.

Immunoprecipitation of metabolically labeled trypomastigotes with monospecific antineuraminidase antibodies followed by sodium dodecyl sulfate polyacrylamide gel electrophoresis show that the neuraminidase is comprised of a set of proteins with high molecular weights (M_r 165–200 kD), some of which exhibit neuraminidase activity. The bands are glycoproteins as they bind to lectin-agarose columns and change mobility in polyacrylamide gel electrophoresis after digestion with N-glycanase.

Reaction with Nonimmune Soluble Host Molecules

It is known that serum proteins (such as fibronectin) bind to blood parasites, but it is generally assumed that the binding is nonspecific. The effect of nonimmune host factors on the biology of *T. cruzi* and of other parasites in their vertebrate and invertebrate hosts has just begun to be scrutinized. This clearly is a neglected area of *T. cruzi* research; most investigators are interested in determining the influence of the immune system on parasite behavior. Yet, as summarized below evidence is accumulating to indicate that nonimmunological host factors might contribute to successful host-parasite interaction.

High-density lipoprotein (HDL) HDL has been recently found to bind live, intact *T. cruzi* in a typical ligand-receptor type of interaction; the binding is saturable and specific (it is not inhibited by proteins unrelated to those that make up HDL). The binding occurs through the apoprotein(s) portion of HDL. HDL reacts strongly with the infective trypomastigotes and weakly with epimastigotes. The HDL receptors on *T. cruzi* appear to be parasite neuraminidase, since HDL is a potent inhibitor of the *T. cruzi* enzyme; HDL concentrations as low as 10^{-9} *M* inhibit the neuraminidase activity present in 10^8 trypomastigotes. The addition of HDL to a defined medium at concentrations that inhibited *T. cruzi* neuraminidase augments the ability of the parasite to penetrate and infect tissue culture cells in vitro. The infection-promoting effect of HDL depends on its neuraminidase inhibitory activity, since the effect is reversed by addition of *Vibrio cholera* neuraminidase. Interestingly, the activity of this bacterial neuraminidase is not inhibited by HDL. In addition to aiding infection, specific binding of HDL to *T. cruzi* may be used by the parasite to obtain host cholesterol for membrane biosynthesis. (HDL is a major scavenger of cholesterol in humans and in other mammals.) These findings may have implications for the pathogenesis of Chagas' disease.

Serum sialoglycoproteins Recently it has been found that some serum glycoproteins like fetuin, fibrinonogen, and transferrin can promote the penetration of trypomastigotes into tissue culture cells. However, because binding of the sialoglycoconjugates to the parasites could not be demonstrated it is thought that an enzymatic transfer of sialyl residues from the host sialoglyco-conjugates to the trypomastigotes might have a positive effect on the parasite invasion of host cells.

Alpha-2-macroglobulin (α2M) is a glycoprotein found in the plasma of vertebrates that has inhibitory activity against various proteases. The infection of macrophages and fibroblasts by bloodstream trypomastigotes can be inhibited by α2M and other protease inhibitors, whereas the ingestion of epimastigotes by macrophages is not. These results suggest that trypomastigotes might contain proteases on their plasma membrane that are involved in the infection of host cells. However, in order for this hypothesis to be validated it is essential to demonstrate α2M binding to intact trypomastigotes, which has not been done. In any case, it is an interesting suggestion that host protease inhibitors might negatively influence parasite infectivity.

These four examples suggest that host molecules may have important influences on host–*T. cruzi* interaction and underscore the need for additional related studies.

Interaction with Host Cells: A Proposed Model

Evidence has accumulated in the past few years to indicate that the interaction of *T. cruzi* trypomastigotes with host cells is mediated by specific recognition molecules. Similar evidence is accumulating for other protozoa–host cell interaction, namely leishmania and plasmodia (see Chapters 1 and 5).

Treatment of parasite or host cell membranes with oxidants, lectins, or enzymes (proteases, glycosidases, and phospholipases) can alter host cell–parasite association in vitro and suggests a role for surface ligands in the association. A role for carbohydrate–protein interactions in the binding of parasites to host cells is suggested, for example, by the reduction of infection following periodate treatment of host cells. Periodate normally hydrolyzes covalent bonds between carbon atoms having free hydroxyls, such as those from the glycerol moiety of N-acetylneuraminic acid and from vicinal hydroxyls of the pyranose ring of hexoses (Figure 4-3). Thus, alterations of a biological activity by periodate oxidation suggest carbohydrate involvement in the activity tested. However, periodate also hydrolyzes amino acids in peptide bonds and is therefore not specific for sugar. But even if periodate oxidizes cell-surface carbohydrate, it is possible that the decrease in infection resulted from physical modifications of the membrane, such as a change in fluidity, rather than from an alteration of the sugar per se. In addition to periodate, other treatments that affect carbohydrate–protein interaction also seem to influence *T. cruzi* infection. Several investigators found that the addition of plant lectins, such as concanavalin A (Con A), wheat germ, and ricin I, to trypomastigote or host cells substantially affected *T. cruzi* infection in vitro. These findings were generally assumed to imply glycoconjugate involvement in *T. cruzi*–host cell interaction. However, although this interpretation may be correct, it is equally likely that the lectins (which by definition are multivalent and have more than one sugar-binding site per molecule) could bring the interacting cells together at a faster rate by reacting with sugar residues on both parasite and host cell, with resultant increase in infection. This result

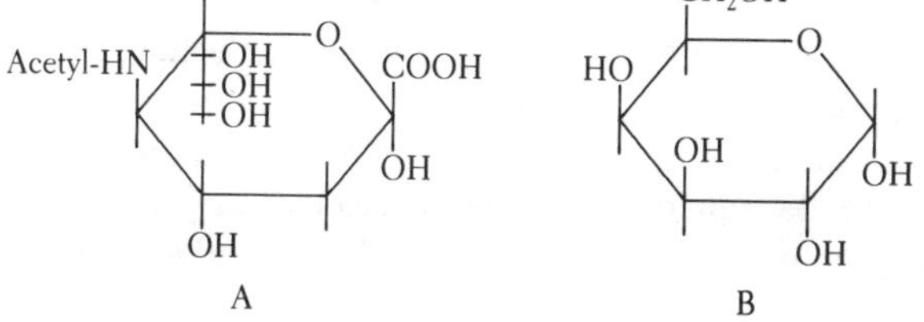

Acetyl-HN, OH, OH, OH, O, COOH, OH, OH — A
CH_2OH, O, HO, OH, OH, OH — B

FIGURE 4·3 Structure of *N*-acetylneuraminic acid (A) and galactose (B). Vicinal hydroxyl groups are susceptible to periodate oxidation.

was observed in the incubation of macrophages and trypomastigotes with Con A. If this latter interpretation is correct, the sugar residues do not necessarily participate actively in the parasite interaction with the host cell under natural circumstances, in which Con A is absent. On the other hand, if plant lectins inhibit infection, as was found in the interaction of macrophages with trypomastigotes in the presence of Con A, the lectin, once bound to the host cell or parasite, could sterically block the access of adjacent active ligands other than Con A receptors. Another suggestion for carbohydrate–protein interaction in *T. cruzi* infection has come from experiments in which monosaccharides such as methyl α-mannoside and N-acetyl-D-glucosamine were found to inhibit *T. cruzi* infection. The results were interpreted to suggest that a lectin molecule on the parasite mediated the interaction; however, another interpretation for monosaccharide inhibition is that the sugars used were toxic to cells and therefore inhibited infection by nonspecific perturbation of the parasite metabolism. The putative lectin of *T. cruzi* as yet has not been isolated.

Another approach used to verify the participation of sugar residues on *T. cruzi*–host cell interaction is to treat the interacting cells with glycosidases and determine whether the deglycosylated cells have an altered capacity to infect or to be infected. Treatment of the parasite with α-mannosidase increased their ability to associate with host cells, whereas digestion of the host cells with the same enzyme preparation decreased association. A similar finding was obtained with α-galactosidase treatment of trypanosomes and host cells. Because glycosidases are very specific in their reaction with sugar residues, these results strongly support the idea of carbohydrate–protein recognition in *T. cruzi* binding to host cells. Alternatively, the effects induced by the glycosidases could be due to contaminants in the enzyme preparations (commercially available glycosidases are not homogeneously pure) or to nonspecific perturbations on the interacting cells by the artificial removal of surface carbohydrates. Inhibitors of N-glycosylation, such as the plant alkaloid tunicamycin, have been found to affect *T. cruzi* infection, but here, too, the results cannot be unequivocally interpreted on the basis of sugar involvement in parasite penetration, because alkaloid inhibitors are toxic to cells and the alterations observed therefore might be caused by unknown mechanisms.

The findings just discussed nevertheless implicate surface molecules in the interaction of *T. cruzi* with host cells and suggest that recognition of carbohydrate by protein is one of the mechanisms involved. As suggested in earlier sections, more direct evidence for specific mediators of cell–parasite interaction is provided with the use of antibodies specific for cell surface components. In this way, antibodies to GP-85, a surface glycoprotein of trypomastigotes, was found to abrogate parasite binding to and penetration of host cells, indicating that GP-85 may be important in the parasite infection. Likewise, fibronectin receptors, which have been isolated from trypomastigotes and found to have a molecular weight (MW) of 85 kD seem to inhibit the infection of host cells by *T. cruzi*. Antibodies to the receptors inhibited infection as well, whereas soluble fibronectin increased parasite penetration, presumably by binding to receptors on trypomastigotes and on host cells, thereby bridging the two types of cells. It is well documented that the tripeptide sequence Arg-Gly-Asp is the recognition site in the binding of fibronectin to mammalian cells; accordingly, the tripeptide also was found to inhibit cell invasion by *T. cruzi*. These results argue in favor of GP-85 or of fibronectin receptors as a ligand in the pathogen binding to host cells; GP-85 or fibronectin receptors therefore promote infection. As a consequence, blocking such receptors with specific antibodies or with haptens and ligands should decrease parasite infection. Identification of receptors or ligands that exert positive effect on infection is therefore important because it might lead to a vaccine.

The underlying assumption is that the equilibrium achieved when a parasite infects a host results from two opposing forces: one is developed by the parasite to promote infection; the other, by the host through its immune system to prevent it. There is no doubt that these two mechanisms are critical for parasitism to occur; otherwise the parasites would quickly kill the host or the host would readily eliminate the parasites, and in either case parasitism would not develop. It has not been appreciated that the parasites themselves might have evolved mechanisms to down-regulate infection; paradoxically, parasitism would be even more efficient than if negative modulation of parasite growth depended solely on the host immune response.

Current evidence suggests that the neuraminidase of *T. cruzi* controls infection by a negative mechanism, as follows. Antibody to the enzyme, at concentrations that inhibit neuraminidase activity, augments infection of tissue culture cells by *T. cruzi*, a finding expected for an activity that exerts a negative control of host–parasite interaction. Accordingly, the specific inhibitor of *T. cruzi* neuraminidase, HDL, enhances infection as well. If the neuraminidase has a negative effect in infection, then one would predict that strains displaying high neuraminidase activity should be less virulent than those with low enzyme activity, a correlation that is verified experimentally. How could the neuraminidase exert a negative control of infection? If it indeed does, presumably it does so through its effect on sialic acid, since antibody and HDL augment infection in vitro only when present in the monolayers at concentrations that inhibit enzyme activity. Furthermore, the effect of antibody and HDL is completely abrogated by *V. cholera* neuraminidase, an enzyme that is not inhibited by the two reagents. It therefore follows that if neuraminidase has a negative control, then sialic acid, whether on the parasite or on the host cell, should exhibit a positive effect. Recent evidence supports this notion. Thus, as already discussed, fetuin and other sialoglycoproteins have been found to increase *T. cruzi* infection. These glycoconjugates do not inhibit but rather are substrates of the trypanosome neuraminidase, so they could not enhance infection by inhibiting the parasite neuraminidase activity. In addition, the glycoconjugates also serve as substrate for the trans-sialyltransferase of *T. cruzi*. Thus, the interpretation that fetuin enhances infection through the parasite sialyltransferase is consistent with the hypothesis that sialic acid is a key element in the control of *T. cruzi*–host cell interaction, the neuraminidase and the transferase having reverse effects in the process.

Coda

In conclusion, the model put forward here is that the interaction of trypomastigotes with host cells is governed by mechanisms evolved by the parasite to exert antipode effects on the interaction. Positive control would be through the parasite fibronectin receptor,

GP-85, trans-sialyltransferase, or protease, whereas negative control would be by the neuraminidase. Sialic acid would be pivotal in the recognition process. The extent of infection of host cells therefore would be the net result of the two opposing forces. Host molecules that react with the parasite receptors should decrease or increase infection, depending on whether the receptors up- or down-regulate infection. For example, fibronectin peptides and protease inhibitors (α2M) should inhibit infection, whereas neuraminidase inhibitors should have a reverse effect, as is observed experimentally. Antibodies to the receptors should affect parasite binding as well and might have positive and negative effects, depending on the kind of receptor being bound. *T. cruzi* should not be unique in having negative and positive mechanisms of infection; for example, influenza virus infects by a receptor-mediated endocytosis through a hemagglutinin that recognizes sialic acid residues on host cells. The virus has a neuraminidase that cleaves host sialic acid and thereby decreases the number of available hemagglutinin receptors. The viral hemagglutinin and neuraminidase therefore can be seen as exerting positive and negative effects on infection, with sialic acid having a critical role in virus–host cell interaction, a process somewhat analogous to what we propose for *T. cruzi*–host cell recognition. Finally, because of the negative effect of *T. cruzi* neuraminidase on infection, the action of the enzyme may help to keep amastigote replication in check and possibly allow the trypanosome to remain latent.

Additional Reading

BRENER, Z. 1973. Biology of *Trypanosoma cruzi*. *Ann. Rev. Microbiol.* 27:349–381.

DE SOUZA, W. 1984. Cell Biology of *Trypanosoma cruzi*. *Int. Rev. Cytol.* 86:197–283.

DVORAK, J. A. 1984. The natural heterogeneity of *Trypanosoma cruzi:* biological and medical implications. *J. Cell. Biochem.* 24:357–371.

SNARY, D. 1985. Biochemistry of surface antigens of *Trypanosoma cruzi*. *Br. Med. Bull.* 41:144–148.

VICKERMAN, K. 1985. Developmental cycles and biology of pathogenic trypanosomes. *Br. Med. Bull.* 41:105–114.

5

Cell Biology of *Leishmania*

K.-P. Chang

Many species of trypanosomatid protozoa of the genus *Leishmania* are parasites of man. They are transmitted by blood-sucking female phlebotomine sandflies. Figure 5-1 presents schematically various forms of leishmanias reported in the insect and mammalian hosts. The parasite lives extracellularly in the alimentary tract of the fly, largely as motile promastigotes, or cells each with a flagellum at their anterior end. When delivered into the mammalian host by the vector, they infect macrophages of the reticuloendothelial system and therein transform into nonmotile amastigotes. Subsequent multiplication of amastigotes is believed to lyse the infected macrophages, and the released amastigotes infect additional macrophages, resulting in the chronic diseases collectively known as leishmaniasis. There are simple cutaneous, diffuse cutaneous, mucocutaneous, and visceral forms of the disease, the last being often fatal in untreated cases. The variations in the clinical manifestations of leishmaniasis are due largely to the different species of parasites involved and partly the status of host immunity.

Leishmaniasis was first recognized as an important disease of humans at the turn of the century. It remains widespread in all continents except Australia. The epidemiological perspective on *Leishmania* organisms and leishmaniasis is in a state of flux; several endemic foci have been discovered frequently in recent years—especially in Africa and South and Central America, where different species of parasites, sandfly vectors, and reservoir animals are involved in transmission. There also have been epidemic outbreaks, resulting in part from inadequate vector control programs and the emergence of drug-resistant parasites. Today, hundreds of millions of people in new and old endemic foci are at risk of contracting leishmaniasis.

Leishmanias are a parasite of paradox. The evolutionary adaptation of organisms such as leishmanias to a parasite's way of life presumably proceeds from an extracellular to an intracellular mode. Intracellular parasitism in such cases is viewed generally as an evolutionary strategy of parasites to escape from humoral immunity of the host. Yet, the pathogenicity of leishmanias hinges completely on their ability to infect macrophages,

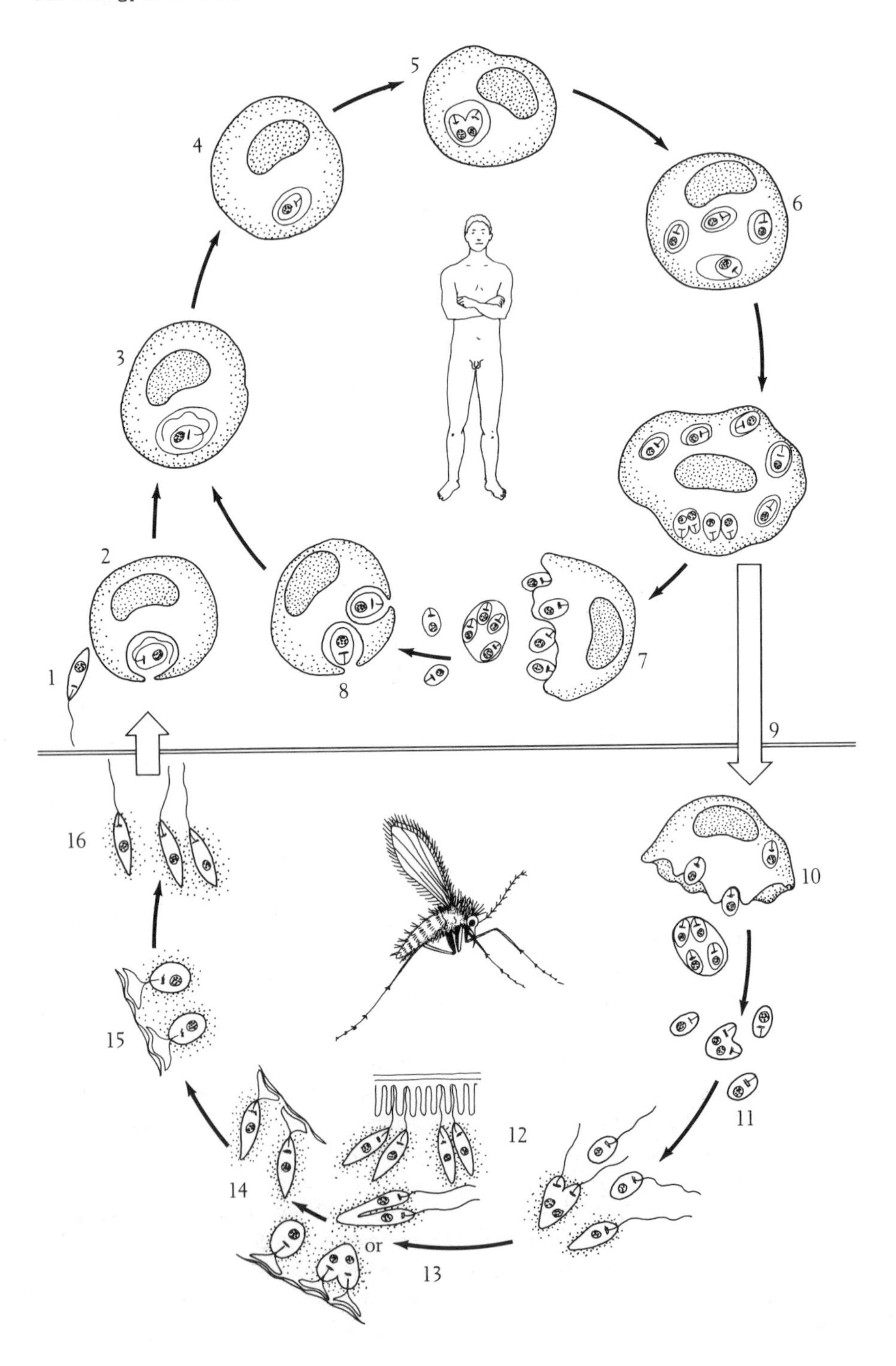
1
2
3
4
5
6
7
8
9
10
11
12
13
or
14
15
16

an archetype of phagocytes important in the cellular defense of the host against invading microorganisms. How leishmanias manage not only to escape from destruction by these cells but to survive, differentiate, and multiply intracellularly is a question of academic and practical importance, relevant to the pathogenic mechanism of leishmanias as well as to the immunobiology and cytopathology of leishmaniasis.

Like all cases of intracellular parasitism, leishmanial infection of macrophages follows multiple steps of host–parasite cellular interactions. The sequential events of such cell–cell interactions are (1) attachment of leishmanias to macrophages; (2) entry of leishmanias into macrophages; (3) intracellular survival of leishmanias; (4) intracellular differentiation of leishmanias from promastigotes into amastigotes; and (5) intracellular multiplication of amastigotes. How do these events occur at the cellular level? What is the biochemical or molecular basis of such cell–cell interactions? These questions have been studied more intensively during the last decade, following the advent of technical improvement for the isolation and cultivation of macrophages and both stages of leishmanias in vitro. Leishmanial infection of macrophages is studied in vitro to facilitate the analysis of the sequential steps involved. The macrophages used are largely derived from the peritoneal cavity of rodents and, to some extent, from human peripheral blood monocytes. Leishmanias used are amastigotes isolated from infected animals or macrophages in culture and promastigotes grown in cell-free media. Promastigotes are grown to late-log or stationary phase when infective cells corresponding to those in the vector become more abundant. This chapter summarizes and discusses pertinent points in the sequence of five events, emphasizing recent developments in cell biology in this area.

Attachment of Leishmanias to Macrophages

The adhesion of leishmanias to the macrophage surface is a prerequisite for gaining intracellular entry. Binding must occur for both stages of the parasite during evolution of the disease, in the initial primary infection by promastigotes deposited by the sandfly

---

FIGURE 5·1 Leishmanial infection in sandfly and in mammalian hosts. (1) Delivery of promastigotes (proboscis form) into human skin by the bite of sandfly vector; (2) attachment and engulfment by phagocytosis of promastigotes by a macrophage; (3) fusion of phagosome containing a promastigote with lysosome in a macrophage; (4) differentiation of promastigote into amastigote in the phagolysosome of the infected macrophage; (5) multiplication of an amastigote in a parasite-containing or parasitophorous vacuole; (6) formation of large parasitophorous vacuole and continuing replication of intravacuolar amastigotes; (7) rupture of heavily parasitized macrophage and release of amastigotes; (8) phagocytosis of released amastigotes by a macrophage; (9) ingestion of parasitized macrophage by sandfly after a blood meal taken from infected person or reservoir animal; (10–16) development of promastigotes in the sandfly vector.

into animal skin and in the subsequent infection at a later stage by the amastigotes released from heavily parasitized macrophages.

In vitro study of leishmania-macrophage binding indicates molecular compatibility at the host–parasite membrane interface. Leishmanias are clearly predisposed to bind to macrophages: promastigotes bind avidly to these phagocytes, less well to fibroblasts, poorly to lymphocytes, and hardly, if at all, to erythrocytes. The existence of specific membrane-binding sites during leishmania-macrophage interactions is suggested by ultrastructural observations of points of attachment separated by gaps of nonadhesion. Moreover, promastigote-macrophage binding follows a saturation kinetics suggestive of ligand–receptor interactions. This proposal has gained increasing support by recent findings that such binding can be blocked by presumed ligands of low molecular weight or complex molecules purified from the parasite surface and by the antibodies or their Fab fragments specific to either macrophage surface receptors or to leishmania membrane molecules. The list of macrophage surface molecules involved includes their receptors for antibody, complement, mannose, and fibronectin. All these receptors are known to have biological importance in cell–cell interactions. The surface molecules of leishmanias implicated in their binding to the macrophages appear to be glycoconjugates.

Leishmanias possess antigenic molecules on their surface. It has long been noted by a number of investigators that opsonization of living promastigotes with antisera prepared against them in laboratory animals or those from infected animals increases their uptake by macrophages. For example, prior treatment of *Leishmania donovani* promastigotes with rabbit antisera doubles the rate of their subsequent binding to hamster macrophages. Phagocytosis of *L. donovani* by mouse macrophages is increased when the host cells are pretreated with sera from infected mice. Similar findings in other systems leave no doubt that this binding is mediated by opsonic or cytophilic antileishmania antibodies, which bridge Fc receptors of macrophages and the surface antigens of leishmanias.

Leishmanial promastigotes of different species have been shown to fix complement. Evidence indicates the adsorption of C3 to and its conversion into C3b and iC3b by protease (gp63) on the surface of promastigotes. Their phagocytosis by macrophages is increased several times after complement fixation, and this is significantly inhibited by monoclonal antibodies or their Fab fragments specific to the iC3b receptor (CR3) of the macrophages. This iC3b receptor-dependent uptake of promastigotes also occurs under serum-free conditions, suggesting that it is mediated by the complement endogenously produced by the macrophages. That complement activation may be a means employed by promastigotes to facilitate their entry into phagocytes is thus suggested.

Monosaccharides, such as mannose, N-acetylglucosamine, N-galactosamine, galactose and/or sialic acid have been detected by specific probes (e.g., plant lectins) as terminal residues of glycoconjugates on the surface of various leishmania species. The possible participation of mannose residues on the surface of promastigotes as one of the determinants in their binding to macrophages was first proposed on the basis of competition-inhibition types of experiments under serum-free conditions using zymosan and mannose. Recent work provides strong support for this notion. The binding of promastigotes to macrophages is greatly reduced when the latter cells are cultured on mannan-coated glass coverslips to sequester their mannose receptors, thereby making them unavailable for binding. Promastigote-macrophage binding through other monosaccharide-specific pathways is also possible, but remains to be explored. Such monosaccharide specificity of host-parasite interactions at the cellular level may bear on the well-known tissue tropism of *Leishmania* species. There are indeed quantitative and qualitative differences on the externally disposed sugar residues detected in different *Leishmania* species.

Other known receptors of macrophages that may be involved in their binding of leishmanias include those for fibronectin. This molecule has been well characterized biochemically and is instrumental in several cell–cell adhesion interactions between different host cells on the one hand, and host cells and microbes on the other. There is evidence for the adhesion of serum fibronectin to leishmanias, leading to their increased binding to macrophages. Host fibronectin may serve simply as a bridge to enhance adhesion or also facilitate phagocytosis of *Leishmania* by cooperation with the macrophage complement receptor.

Which endogenous surface molecules of leishmanias are responsible for their binding to the receptors of macrophages? All leishmania surface antigens that react to natural antibodies or that can elicit opsonic and cytophilic antibodies are potentially involved as participants in the Fc receptor-dependent binding. The two species of surface glycoconjugates that have been identified recently as candidates for the macrophage-binding molecules of promastigotes function apparently independently of the mechanism mediated by Fc receptors. A surface antigenic glycoprotein of 63 kD (gp63) has been purified from *Leishmania amazonensis* to homogeneity by monoclonal antibody affinity chromatography. Preadsorption of these molecules to macrophages inhibits their binding of promastigotes by 50 percent. Since this antigen contains terminal mannose residues, the mannose receptors on the surface of macrophages may well be those that bind this saccharide. A glycoprotein of similar molecular mass was also isolated from *Leishmania mexicana* by concanavalin A and anionic exchange column chromatography.

The Fab fragments of a polyvalent antibody prepared against this antigen inhibit promastigote-macrophage binding. The purified antigen incorporated into liposomes binds complement and competes with mannan for binding to macrophages. It is suggested that gp63 is the leishmania ligand responsible for binding of promastigotes to both mannose receptors and iC3b receptors of the macrophages. It has been shown that gp63 is a surface protease present in all pathogenic species of *Leishmania* examined. Conservation of gp63 and its enzymatic activity indicate a universal importance for this molecule in the cleavage of complement by leishmanias for binding to macrophage receptors. Recently, analysis of the gene encoding gp63 revealed the presence of an RGD sequence, a known ligand for a family of receptors including CR3 of macrophages. An RGD-containing peptide was subsequently synthesized on the basis of the gp63 gene sequence and found to inhibit significantly the binding of promastigotes to macrophages. This finding, coupled with the inhibitory activity of the antibodies specific to either CR3 or the synthetic peptide, suggests that the binding of promastigotes to macrophages may be mediated also directly by the interaction of the RGD sequence in the gp63 to the receptor of concern. Another parasite surface molecule of interest is a species of glycolipid, which has been purified from *L. major* by monoclonal antibody affinity chromatography and from *L. donovani* biochemically, known as lipophosphoglycan. The Fab fragments of the antibody inhibit promastigote–macrophage binding by 80 percent. The receptors of macrophages for this leishmanial antigen are not CR3 but do belong to the integrin family.

Recent work outlined in the foregoing paragraph lends credence to the proposal that leishmanias use preexisting receptors of macrophages for binding, probably through multiple pathways. The binding of cells in this host–parasite system may occur through the direct interactions of their native surface molecules or through the intervention of soluble mediators released by the host cells, for example, antibodies, complement, and fibronectin. It is possible that there are binding mechanisms that do not use specific macrophage receptors. Different leishmanias may also use different binding mechanisms. These are additional areas that require investigation.

Topics of interest concerning receptor-mediated mechanisms are the cooperation or competition among different binding pathways identified and their relative importance in the natural infection. Under serum-free conditions, there appears to be cooperation among different receptors described earlier. Of interest is the possibility that the binding mechanism may change completely or partly, involving other pathways mediated, for example, by Fc receptors in the presence of antibodies generated against leishmania surface antigens. The macrophage-binding ligands of leishmanias can also be "masked" by other molecules derived either from the host or from the vector and "denuded" or clustered by the capping phenomenon of membrane molecules mediated, for example, by antibodies and lectins. Conceivably, all these events may affect the leishmania-macrophage binding under natural conditions and possibly alter the intracellular fate of leishmanias. This hypothesis is supported by a recent finding indicating that the entry of leishmanias into macrophages through iC3b receptors results in only a minimal respiratory burst in the host cell. Since promastigotes are susceptible to the toxic effects of oxygen radicals, the reduced burst may favor parasite survival. Detailed analysis of leishmania surface molecules and their interactions with macrophage receptors may yield useful information on the relative importance of the different binding pathways.

Intracellular Entry of Leishmanias into Macrophages

Leishmanias bound to the surface of the host cells must be interiorized to achieve intracellular parasitism. Phagocytic activity of macrophages plays a major role in this process. As noted, blocking phagocytosis with cytochalasins prevents parasite entry.

Under scanning and transmission electron microscopes, the cytological events have been well described for the uptake of leishmanial promastigotes by macrophages.

The attachment of leishmanias through either their anterior or posterior end induces an extensive membrane ruffling formation of the macrophage surface. The parasites are subsequently engulfed by cellular pseudopodia, as found during phagocytosis of other microorganisms or particles by macrophages. The process requires the expenditure of energy on the part of the macrophages and the participation of their cytokinetic elements, such as actin. The attachment of leishmanias to the surface of macrophages presumably triggers phagocytosis by mechanisms similar to those under investigation in other systems, including a sequential binding of molecules on host cell and parasite surfaces, in a zipperlike mechanism.

Intracellular Survival of Leishmanias in Macrophages

After gaining entry into macrophages, leishmanias must find ways to defend themselves against a battery of powerful microbicidal mechanisms known to exist in these phagocytes. How leishmanias survive intracellularly is a matter of considerable interest, given our understanding of the major roles played by these phagocytes in host defense generally. Discussion here will be limited to the cellular and molecular aspects of this event. Issues of immunological importance are covered in Chapter 11.

After entry into macrophages, leishmanias are lodged in phagosomes to which lysosomes of the macrophages fuse, creating phagolysosomes. Surprisingly, leishmanias survive in this hostile environment, which normally destroys most microorganisms ingested by macrophages. Leishmanias share this capability with few other species of intracellular parasites, *Mycobacterium leprae* and *Coxiella burnettii* being the other exceptional pathogens that survive in this subcellular compartment.

How do leishmanias resist microbicidal factors in the lysosomal compartments of macrophages? Since many potential microbicidal properties have been reported for macrophages but have not been systematically studied for leishmania specifically, this question is not easily answered.

Phagocytosis of many microorganisms triggers a respiratory burst by mononuclear phagocytes, resulting in the generation of microbicidal oxygen radicals such as hydrogen peroxide, hydroxyl radicals, and singlet oxygen. There has been report of an unusual iron-containing superoxide dismutase of trypanosomatid protozoa, including leishmanias, that allows them to dismutate superoxide anions into hydrogen peroxide and molecular oxygen. More recently "trypanothione," similar to glutathione of mammalian cells, has been discovered in these parasites. In contrast to mammalian glutathione "trypanothione" possesses additional molecules of spermidine. Leishmanias have little or no catalase and may rely on trypanothione peroxidase as the major scavenger for the detoxification of hydrogen peroxide. It is possible that these and similar enzyme systems of leishmanias work in concert to dispose of the oxidative metabolites generated not only from their own aerobic respiration but also from the respiratory burst of the mononuclear phagocytes. Of interest is the finding that an acid phosphatase purified from the plasma membrane of *L. donovani* diminishes the respiratory burst of neutrophils induced by zymosan. The implication of this finding is that the binding of leishmanias to phagocytes may result in the generation of oxygen metabolites in amounts less than usual, thereby relieving the burden of detoxification for the parasites.

Mononuclear phagocytes possess a number of "nonoxidative" microbicidal factors, such as channel- or pore-forming molecules (polyperforins) that might result in perforation of microbial cell walls or membranes, and lysosomal hydrolytic enzymes. Surface membrane glycoproteins of *L. m. amazonensis* are refractory to degradation at pH 5.0 by lysosomal preparations purified from rat liver. Leishmanias may be protected by a surface layer of molecules resistant to acid hydrolases, enabling them to survive in the acidic environment of the macrophage's phagolysosome. Leishmania surface molecules may also play an active role in detoxification of microbicidal factors in this environment; gp63 has been identified as a protease active at pH 4.0 and conceivably could destroy host enzymes that normally have microbicidal properties. Additional protection against lysosomal enzymes might be conferred on leishmania by their metabolites, which are released into their immediate surroundings. It has been suggested that ammonium ions and proteases released by *L. m. mexicana* might interfere with lysosomal functions by raising the intralysosomal pH and by directly interacting with lysosomal enzymes, respectively. Soluble factors excreted by *L. major* (called "excreted factors") are reported to inhibit the activity of beta-galactosidase and to partially protect erythrocytes from degradation by macrophages; other lysosomal enzymes tested are not affected. The "excreted factors" appear to contain glycolipids and glycopeptides possibly shed from the leishmania cell surface, but their precise composition and chemical structure have not been well characterized. There may be multiple factors at the disposal of leishmanias to disarm the degradative actions of lysosomes.

"Molecular mimicry" has been proposed as a means of intralysosomal survival of leishmania, on the assumptions that mechanisms must exist whereby hydrolytic enzymes in the lysosome avoid mutual digestion and that leishmania surface or secreted molecules, by virtue of their resemblance to lysosomal enzymes, may somehow fit into such a "scheme of protection." The acid phosphatase (a usual lysosomal marker enzyme) and protease found on leishmania surface may represent examples involved in this phenomenon. Additional possibilities have been suggested. Surface glycoconjugates of leishmanias are anchored to their plasma membrane via glycosyl phosphotidylinositol, analogous to that reported for the variant-specific glycoproteins of African trypanosomes (see Chapter 3). Furthermore, a promastigote surface glycoprotein is released on their entry into macrophages and becomes demonstrable on the macrophage surface. The release of the leishmania surface glycoproteins may be triggered by activation of a phosphotidylinositol-specific phospholipase C endogenous to the parasites or to the lysosomes of macrophages. Do one or more of these molecules released from leishmanias inhibit lysosomal enzymes directly or shut off their "trigger" and "activator" of respiratory burst, if they indeed work in sequence as a cascade of reactions? Further investigation into the molecular structure of leishmania membrane molecules (e.g., gp63) and their interactions or the interactions of their components after release with macrophages will undoubtedly provide insight into questions relating to the intracellular survival mechanisms of leishmanias. Of relevance is the recent finding that leishmania surface lipophosphoglycan inhibits proteinkinase C, which may be involved in the activation of microbicidal mechanisms during phagocytosis by macrophages.

Cellular Differentiation of Leishmanias

In the phagolysosome of macrophages, leishmanias differentiate from promastigotes into amastigotes, an adaptation necessary for their survival during the transition from an extracellular to an intracellular environment. Which factors trigger this differentiation? Some leishmania species differentiate from promastigotes into forms resembling amastigotes in nonlysosomal compartments of human fibroblasts. This finding suggests that transformation probably is not triggered by events dependent on the activation of lysosomal enzymes. The body temperature of the mammalian host and as yet undefined factors in the intracellular milieu are involved, but may not be absolutely required. Forms that resemble amastigotes morphologically have been observed in aged cultures of promastigotes grown continuously at 27°C. It has been possible to obtain such "amastigote forms" of leishmanias from promastigotes of some species incubated in a serum-containing medium at 33 – 35°C in the absence of macrophages or other mammalian cells. These forms have some morphologic, antigenic, and metabolic similarities to intracellular amastigotes but nonetheless may not be identical to tissue-derived amastigotes.

Profound morphological and biochemical changes accompany leishmanial differentiation from promastigotes into amastigotes. Regression of external flagella occurs, as does shortening of the body length and reduction in the number of pellicular microtubules and other organelles. There is a substantial reduction in the biosynthesis of macromolecules and in oxygen consumption. The diminution of leishmanias in size, growth rates, and metabolic activities during this differentiation may represent a generalized biological "attenuation," perhaps occurring in the face of a very hostile microenvironment. This notion may be incorrect, however, because cyclic nucleotides, certain nucleoside metabolic enzymes, sterol and fatty acid biosynthetic enzymes, membrane antigenic molecules, and heat-shock proteins appear to be synthesized by amastigotes either as new products or at a level higher than those in promastigotes. Indeed, cross hybridization between genomic and cDNA libraries prepared from amastigotes and promastigotes provides evidence for the selective expression of genes specific to these different leishmanial stages. The regulatory mechanisms of gene expression for biosynthesis of specific proteins during leishmanial differentiation may vary with different species. For example, tubulin biosynthesis occurs at a higher rate in promastigotes than in amastigotes. This is regulated under either posttranscriptional control *(L. m. amazonensis)* or transcriptional control *(L. enriettii)*. The regulatory mechanisms of gene expression during interstage transformation of leishmania are clearly a fertile area for research, one that could divulge unique biological control mechanisms.

Intracellular Multiplication of Leishmanias

Leishmanias must multiply to cause all the sequelae of leishmaniasis, as only a very small number of promastigotes are delivered into the mammalian host by the sandfly vector. Shortly after the differentiation from promastigotes into amastigotes, leishmanias have been seen to divide in cultured macrophages. Cell division of leishmanias and other trypanosomatid protozoa is akin to binary fission of bacteria. In general, division of the flagella, flagellar pockets, kinetoplasts, and nuclei precedes that of the cell; a division furrow forms usually at the anterior end and extends posteriorly, splitting each cells lengthwise into two daughter cells.

Where do leishmanial amastigotes multiply intracellularly in macrophages? They do so in the parasite-containing vacuoles, which have properties of secondary lysosomes. Leishmania residence in this site is not unexpected since lysosome-phagosome fusion occurs shortly after parasite entry into macrophages. Continuous residence of leishmanias in this cellular compartment is indicated by the following evidence: (1) electron microscopy and UV fluorescent microscopy have shown that the tracer labels (e.g., thorotrast or FITC-dextran) added to the culture medium of infected cells accumulate in the parasite-containing vacuole, and (2) this vacuole has an acidic pH of 5.0 to 5.5, as determined by fluorometric assays of FITC-dextran-labeled macrophages infected with leishmanias. This compartment represents a portion of the lysosome-phagosome vacuolar system of the macrophages and is accessible to the molecules present in the immediate environment of the infected cells. Amastigotes may occupy only a portion of the vacuolar space, and some adhere to the vacuolar membrane; the remaining portion of the vacuole is fluid-filled.

The nutritional implications of the intraphagolysosomal residence for leishmanias in macrophages are that they must be acid-tolerant in their metabolism and that they might use nutrients that are lysosomotropic or are taken into the phagosome-lysosome vacuolar system by endocytic or other transport activity of the macrophages. Oxygen consumption and utilization of energy substrates for respiration by *L. donovani* proceed optimally for amastigotes at an ambient pH of 5.0 and for promastigotes at neutral pH, respectively. A proton pump system reported in their plasma membrane presumably assists in compensating for the proton concentration gradient that exists between their intracellular milieu and extracellular environment. The transport of certain substrates required for macromolecular biosynthesis appears to be driven by this proton pump.

Macrophages may not supply all the essential substrates needed by the leishmania. There is evidence that intracellular amastigotes acquire heme from the culture medium instead of from that synthesized by macrophages; leishmanias cannot synthesize this compound de novo. Furthermore, success has been reported in growing some leishmania species as amastigotelike forms at 34°C in cell-free medium containing additional factors, especially water soluble vitamins and nucleosides. It is not known if these are nutrients normally supplied to the parasites by macrophages, enabling them to grow intracellularly. Ectoenzymes of leishmanias, such as acid phosphatase and nucleotidases, have been postulated as serving a nutritional role in scavenging purine nucleotides by leishmanias, since these and related protozoa lack the pathway for de novo purine synthesis. The major surface glycoprotein (gp63) of leishmanias as a protease may function similarly by breaking down proteins into peptides and amino acids that can then be used by the parasite. Lysosomes have long been considered the "digestive bag" of mammalian cells. Whether leishmanias exploit the degradative capacity of the lysosomal enzymes for their own nutritional gains (much like some parasitic helminths do in the intestinal tract of their host) is an interesting question that awaits further assessment.

Coda

Leishmanial infection of macrophages in vitro has much to offer for studying a variety of fundamental aspects of cell–cell interactions in the general area of cell biology. The events of host–parasite cellular interactions in leishmaniasis bear directly on several basic issues in cell biology, such as (1) receptor-mediated endocytosis; (2) functional aspects of endosomes and lysosomes; (3) intracellular targeting and traffic of interiorized molecules; (4) molecular signals and regulation of gene expression in cell differentiation; and (5) membrane transport and metabolic compartmentalization in cell division.

Host–parasite surface interactions are often crucial for the outcome of microbial infections. In the case of leishmanial infection, the surface molecules of the parasites must interact with soluble immune factors in the tissue fluids of the host as well as with its macrophages to achieve intracellular parasitism. Understanding such interactions at the molecular level requires additional efforts to further study the expression, structure, and posttranslational modifications of leishmania ectoenzymes and other surface glycoconjugates.

Of specific interest is the recent work with gp63 and protein glycosylation in relation to leishmanial virulence; gp63 is a predominant surface glycoprotein, ubiquitous on the surface of all pathogenic species so far studied. This molecule is a unique endopeptidase, capable of acting on several substrates and active over a broad range of pHs. The proposed biological functions of gp63 are (1) degradation of antibodies by leishmanias to evade host humoral immunity; (2) conversion of C3 to iC3b for complement receptor-mediated endocytosis of promastigotes by macrophages; (3) neutralization of lysosomal microbicidal factors; and (4) degradation of proteins for nutritional benefits of the parasites. The potential functional roles of this molecule in multiple steps during leishmanial infection of macrophages might even qualify it as a virulence factor. Indeed, gp63 is more abundant in virulent than in avirulent promastigotes. This appears to be regulated in part by posttranslational modifications of gp63, such as *N*-glycosylation.

Although many questions remain to be answered regarding the molecular basis of leishmania survival in phagolysosomes, the observation of leishmanias as intralysosomal parasites has already been applied to experimental chemotherapy of leishmaniasis, based on the principle of lysosomotropic therapy. The therapeutic efficacy of the classic antimonial drugs used for leishmaniasis can be increased several hundredfold when they are prepared in liposomes for targeting to macrophages. Lysis of amastigotes also has been observed in macrophages in vitro following the addition of amino acid methyl esters (which are cleaved by lysosomal esterases, resulting in a local accumulation of amino acids and osmotic disruption of leishmania lysosomes). Such possible applications of basic research in parasite cell biology help to motivate continued investigation.

Theoretically, blockading any one of several sequential events that attend intracellular parasitism might abort the infection. Intracellular survival, differentiation, and multiplication may prove more vulnerable targets than the initial steps of parasite attachment and entry, because the latter involve a multiplicity of preexisting pathways essential to the well-being of the hosts. Novel practical solutions to the control of leishmaniasis may ultimately depend on detailed understanding of the molecular basis of host–parasite interactions at the cellular level.

Additional Reading

BORDIER, C. 1987. The promastigote surface protease of *Leishmania*. *Parasitol. Today* 3:151–153.

CHANG, K.-P. 1983. Cellular and molecular mechanism of intracellular symbiosis in leishmaniasis. *Int. Rev. Cytol.* 14(Suppl.):267–302.

CHANG, K.-P., and R. S. BRAY, eds. 1985. *Leishmaniasis* vol. 1 of *Human Parasitic Diseases*. Elsevier Science Publishers.

CHANG, K.-P., C. A. NACY, and R. D. PEARSON. 1986. Intracellular parasitism of macrophages in leishmaniasis: *In vitro* systems and their applications. *Methods Enzymol.* 132:603–625.

DWYER, D. M., and M. GOTTLIEB. 1983. The surface membrane chemistry of *Leishmania:* its possible role in parasite sequestration and survival. *J. Cell. Biochem.* 23:30–45.

HANDMAN, E. 1986. Leishmaniasis: antigens and host-parasite interactions. In *Parasite Antigens*, T. W. Pearson, ed. Marcel Dekker, pp. 5–48.

RABINOVITCH, M. 1985. The endocytic system of *Leishmania*-infected macrophages. In *Mononuclear Phagocytes: Characteristics, Physiology, and Function*, R. Van Furth, ed. Martinus Dorcrecht, pp. 611–619.

RUSSELL, D. K. and P. TALAMAS-ROHANA. 1989. Leishmania and the macrophage; a marriage of inconvenience. *Immunol. Today* 10:328–333.

TURCO, S. 1988. Proposal for a function of the *Leishmania donovani* lipophosphoglycan. *Biochem. Soc. Trans.* 16:259–261.

6

Cell Biology of Schistosomes

I. Ultrastructure and Transformations

■

Paul F. Basch and John Samuelson

As trematodes, schistosomes share the phylum of flatworms (Platyhelminthes) with such organisms as tapeworms and planaria. Although these are generally considered among the "lowest" bilaterally symmetrical animals, they are profoundly complex creatures and provide an exceptional challenge to the cell biologist. Each stage of their life cycle consists of numerous cell types arranged into tissues, organs, and organ systems showing great diversity in cell size, structure, and function. Analysis of the membrane biology and the physiology of transformation between stages of schistosomes has been carried out in recent years and has revealed some exciting new information. This chapter is devoted primarily to interstage transformation and other general aspects of cell biology. Chapter 7 deals mainly with schistosome membrane biology.

Life Cycle of the Parasite

Humans are hosts to about a half dozen species of schistosomes although other mammals, such as cattle, horses, and dogs, may carry these parasites in endemic regions. The intermediate hosts are various species of freshwater snails. Cercariae, the stage infectious to man, are released from the snail into fresh water, penetrate the host skin, and become schistosomulae. After a few days, schistosomulae enter the systemic venous return, pass to the lungs, and in most species migrate to the portal system between the intestine and liver. There the parasites mature as adult male and female schistosomes, and the latter produce thousands of eggs, which lodge in the liver and cause granulomas and fibrosis. Other eggs rupture through the bowel wall and are passed with stool, which may contaminate lakes or rivers. Miracidia hatch from eggs deposited in fresh water and penetrate snail vectors where they transform into sporocysts. The sporocysts subsequently give rise to the cercariae, which are released into the water and are capable of infecting other susceptible hosts.

The Miracidium and Intramolluscan Stages

The miracidium, which infects the snail intermediate host, is 200 micrometers (μm) in length and 40 μm in diameter and is passed in the feces within an ovoid egg made of chitin, measuring about 300 μm by 100 μm (Figure 6-1). On the miracidial surface are 21 ciliated and anucleate epithelial plates, attached by septate junctions to narrow syncytial ridges connected to cell bodies deep within the parasite. Ciliated sensory nerve cells lie within the syncytial ridges and connect to a central nerve plexus. There is a basement membrane overlying two concentric layers of smooth muscle that envelop internal structures, which include secretory glands involved in penetrating the snail; an excretory system marked by flame cells; and primitive germ cells that will develop into daughter sporocysts. Within the snail, the miracidium sheds the epithelial plates, expands the syncitial ridges to cover the entire parasite surface, and transforms into a mother sporocyst. Mother sporocysts reproduce asexually, making daughter sporocysts which migrate to the hepatopancreas of the snail where they produce cercariae, completing the cycle.

For a miracidium to infect the snail it must hatch from the egg, swim through fresh water, find and penetrate the snail, and shed the epithelium and remodel its surface. Manipulation of the environmental conditions that might trigger these events and observation of the parasites by video microscopy have revealed some of the stimuli and processes involved in miracidial infectivity. Miracidia swim at about 2 millimeters (mm) per second by the synchronous beating of the cilia on the epithelial plates. The ciliary beating stops when the miracidium is placed in physiological saline but continues when the miracidium is placed in fresh water. This finding indicates that when the miracidium is in the egg within the body of the mammalian host, the cilia do not beat and the parasite remains within the egg. However, when the eggs are passed into fresh water, the miracidium "swims" by the beating of the cilia within the egg. When the shell fractures, the miricidium can escape. A swimming miracidium turns not by an asymmetric beating of the cilia as does a *Paramecium*, but, rather by angling the body (like a rudder) and using the musculature beneath the epithelial plates. Miracidia accumulate around the snail or in a drop of snail-conditioned water (water in which snails have been placed for a period of time) by turning more frequently and swimming around in circles.

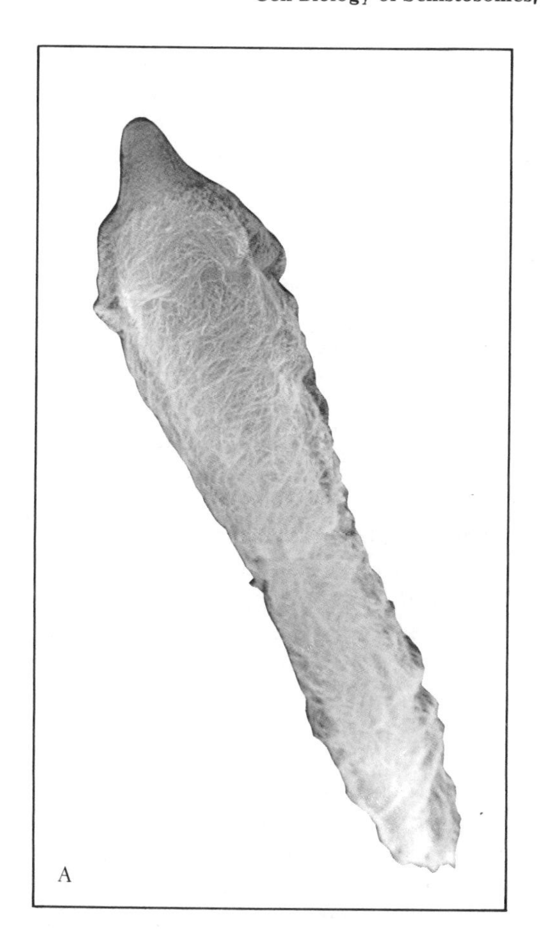

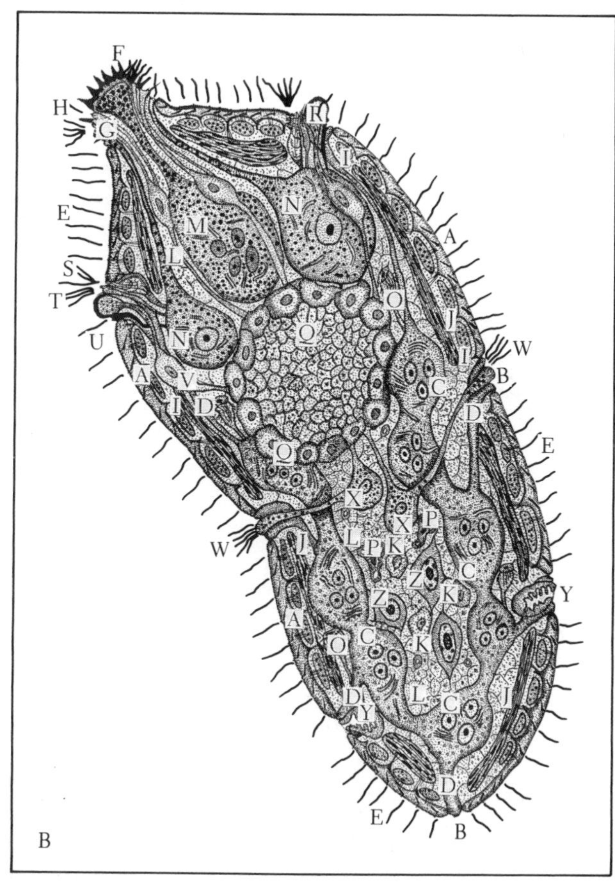

FIGURE 6·1 Miracidium of *Schistosoma mansoni.* A. Scanning electron micrograph of miracidium showing cilia (original ×975). B. Schematic representation of miracidium based on electron micrographs. A, epidermal plate; B, epidermal ridge; C, ridge cyton; D, cytoplasmic bridge; E, cilium and its rootlet; F, terebratorium and the profile of cytoplasmic expansion; G, multiciliated, deep-pit sensory papilla; H, uniciliated sensory papills; I, outer circular muscle fiber; J, inner longitudinal muscle fiber; K, interstitial cyton; L, processes of interstitial cells; M, apical gland and secretory duct; N, lateral gland and secretory duct; O, flame cell with excretory tubule; P, cyton of common excretory tubule; Q, neural mass with peripheral ganglia; R, internal papilla; S, multiciliated saccular sensory organelle; T, multiciliated shallow-pit sensory papilla; U, uniciliated sensory papilla; V, perikaryon of the neuron to internal papilla; W, multiciliated sensory papilla; X, perikaryon of the multiciliated sensory papilla; Y, execretory vesicle; Z, germ cell. (Part A from SAMUELSON, J. C., and J. P. CAULFIELD, 1985. Role of pleated septate junctions in the epithelium of miracidia of *Schistosoma mansoni* during transformation to sporocysts *in vitro. Tissue Cell* 17:667; Part B from PAN, S. C.-T. 1980. The fine structure of the miracidium of *Schistosoma mansoni. J. Invert. Pathol.* 36:307–372 by permission of Academic Press.)

The accumulation of miracidia within a drop of snail-conditioned water can be inhibited by eserine sulfate (physostigmine). This acetylcholinesterase inhibitor paralyzes the muscles of miracidia (and all other stages of the parasite) so that they cannot turn. Conversely, miracidia accumulate in a drop of serotonin, which transiently increases the muscular activity of miracidia. Miracidia stop swimming and may shed their epithelial plates when saline is added to fresh water. In the presence of serotonin, miracidia transferred from fresh water to saline stop swimming but do not transform, suggesting that the nervous system of the parasite is involved in sensing or transducing the osmotic signal for transformation.

Within 2 hours after penetration into a snail host, the ciliated plates are shed from the miracidium. The epidermal ridges gradually broaden until by the second day a continuous new surface layer appears around the newly formed mother sporocyst. At the outer surface of this layer numerous microvillus-like folds appear, covered by a fuzzy electron-dense coat which constitutes the actual interface for metabolic exchange with the hemolymph and for challenge by amebocytes of the snail host. The new syncytial tegument contains mitochondria, ribosomes, endoplasmic reticulum, glycogen, lipid, and other normal constituents of eukaryotic cells, and it remains connected with the nucleated cell bodies further beneath the surface. This arrangement is retained in the tegument of the mother sporocyst.

During subsequent development the muscular layers and parenchymal cells degenerate, the scattered germinal cells become more prominent, and the central portion of the sporocyst differentiates into a brood chamber. Now a remarkably rapid and efficient round of asexual reproduction occurs. Less than a week after infection, daughter sporocyst embryos begin to form by subdivision and differentiation from germinal cells that bud from the inner wall of the saccular mother sporocyst. Starting 5 to 6 days later, they escape, and proceed rapidly through hemolymph vessels and sinuses to the hepatopancreas and ovotestis of the snail. These active vermiform daughter sporocysts, up to a quarter of a millimeter long, contain muscle fibers, protonephridial and nervous systems, and a complement of their own germinal cells to make the next generation. As in other stages, the tegument is typically syncytial and anucleate, resting on a basement membrane, with connections to the nucleated cell bodies below the muscle fibers.

Production by germinal cells of early cercarial embryos ("germ balls") normally begins around the time that the daughter sporocysts exit from the mother. By 3 weeks after infection, the first cercariae are almost fully formed within the daughter sporocysts of *Schistosoma mansoni*. They begin to escape a few days later, first into the hemolymph of the snail, and then through the snail's integument into the surrounding water, where they swim away to find their definitive host. In *S. japonicum* the process is much slower, with cercarial embryos developing some 2 months after infection of the *Oncomelania* snail. All cercariae arising from a single miracidium are presumably identical and will produce adults of the same sex. Therefore, a snail releasing both male and female cercariae had been penetrated by at least two separate miracidia.

It has been shown recently that daughter sporocysts transplanted surgically into new snails will give rise to additional daughter sporocysts and that this process can be repeated through numerous passages. It is not yet clear whether the transplanted daughter sporocysts merely proliferate by growing and budding in the plenitude of their new environment or if the contained germ balls become diverted from production of cercariae and actually produce a further generation of sporocysts. What is apparent is that the extent of development of sporocysts is naturally limited by the volume and nutritive resources within the original snail host, and when artificially freed from such constraints, the amount of sporocyst tissue potentially arising from a single miracidium is essentially limitless. We can take advantage of this capability to maintain genetically constant clones of schistosomes by repeated surgical implantation from a single unimiracidially infected donor into generations of fresh snails.

The Cercaria

Cercariae are composed of a body about 125 μm long and 25 μm in diameter, to which is attached a 200-μm-long tail. Video microscopy has revealed that a cercaria swims from the snail to the mammalian host by alternating side-to-side rhythmic contractions (at 20 cycles per second) of the musculature of the body and tail.

The swimming cercariae (Figure 6-2) are covered by a single continuous syncytial tegument about 0.5 μm thick on the body and 0.2 μm thick on the tail. The tegument is virtually devoid of ribosomes, secretory structures, and other organelles that, along with the nuclei of the syncytium, are found in cellular extensions into the subtegumentary region. A trilaminate plasma membrane over a trilaminate basement membrane forms the outer surface of the tegument, which is covered by a dense, labile glycocalyx about 1 to 2 μm thick. This material, probably a complex carbohydrate, consists of numerous fibrils, each about 8 to 15 nanometers (nm) thick, which may protect the cercaria. It is shed once the cercaria has penetrated the skin of its vertebrate host. The highly antigenic glycocalyx can combine with antibody to produce the "cerkarienhüllen reaktion" (CHR), sometimes used as a serological test for schistosomiasis. In this procedure, cercariae are incubated with serum from a suspected case of schistosomiasis. If antischistosomal antibodies are present, the cercarial surface undergoes a blebbing reaction. The cercarial glycocalyx also activates complement by the alternative pathway, so that fresh human serum (even from an individual never exposed to schistosomes) kills cercariae in vitro.

The cercarial body contains unicellular glands (Figure 6-3) whose secretion products are probably involved in facilitating penetration into the skin of the host. Several proteinases produced by these glands can digest elastin, gelatin, laminin, fibronectin, keratin, type I (interstitial) collagen and types IV and VII (basement membrane) collagen. Presumably degradation of these extracellular proteins of the host destroy a potential barrier to penetration. Other elements within the cercaria include a primitive "brain" anlage, a well-developed flame-cell (osmoregulatory) system, and primordial digestive and reproductive systems that are destined to function only in later stages. Reproductive structures are not developed in the cercarial body, and it is impossible to

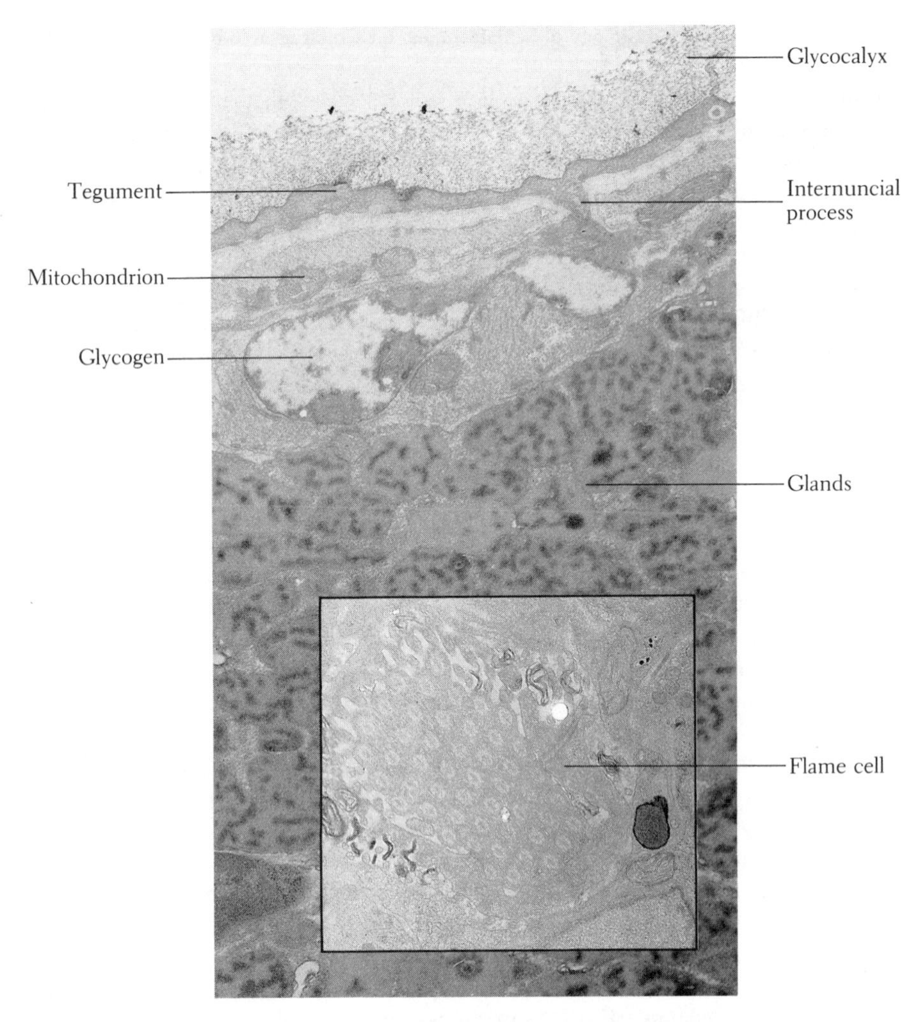

FIGURE 6·2 Electron micrograph of cercaria of *Schistosoma mansoni.* G, glycocalyx; T, tegument; M, mitochondrion; gE, glycogen; I, internuncial process; gl, glands. Inset: flame cell (original ×5000). (Electron micrographs courtesy of C. -P. Chiang and J. P. Caulfield, Harvard Medical School.)

FIGURE 6·3 Cercaria in process of secretion from acetabular glands. A–C, progressive eversion of head region during secretion (Nomarski optics, original ×800); D–F, same process viewed by scanning electron microscopy (original ×4000); G, transmission electron micrograph showing secretion (original ×4000); h, head region; n, nerve endings surrounding opening of duct of acetabular gland; gs, globular secretions. (A–C, E, F from SAMUELSON, J. C., J. J. QUINN, and J. P. CAULFIELD. 1984. Video microscopy of swimming and secretion cercariae of *Schistosoma mansoni. J. Parasitol.* 70:996; D, G from J. C. Samuelson and J. P. Caulfield, unpublished.)

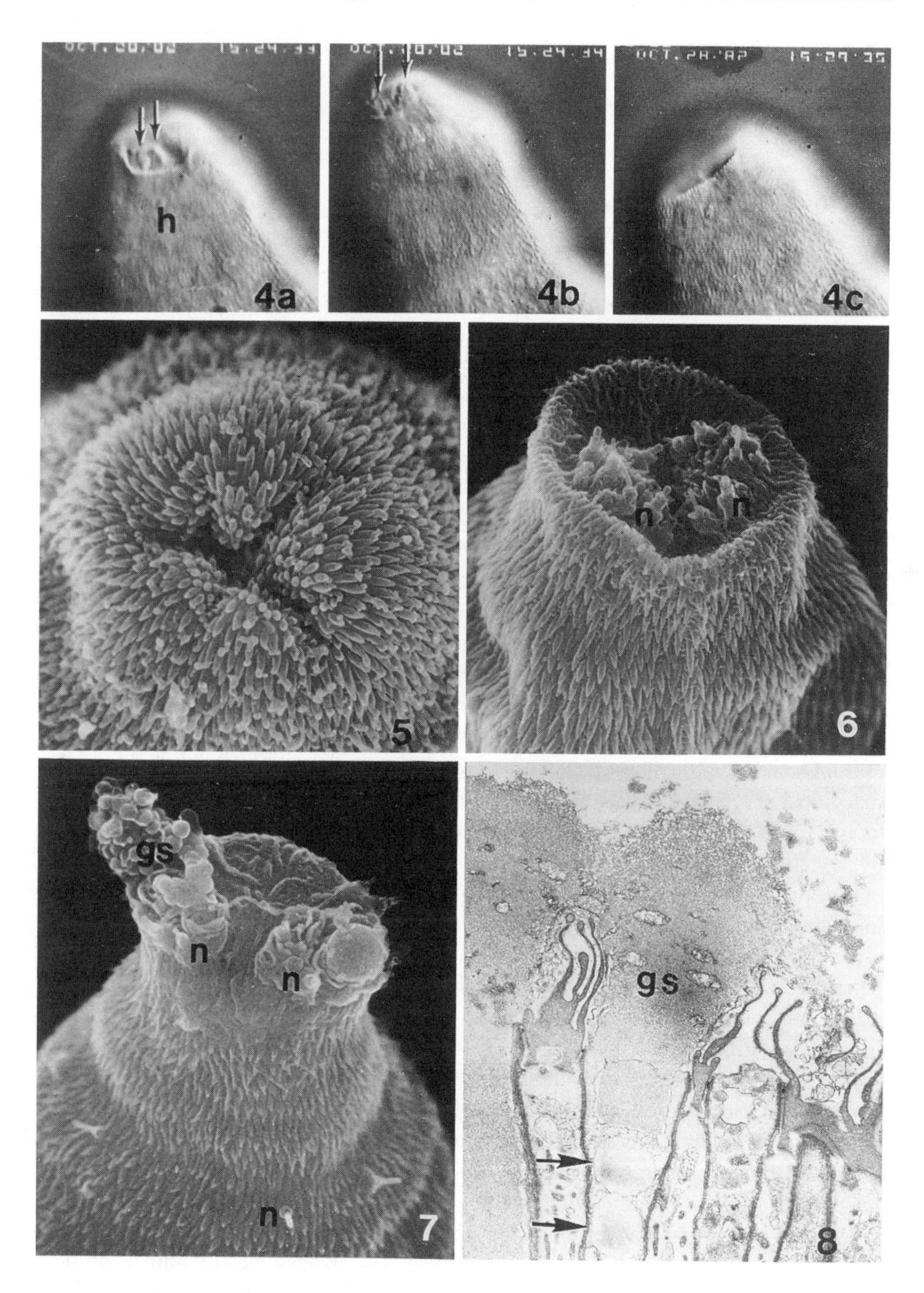
h
4a
4b
4c
5
n
n
6
gs
n
n
n
7
gs
8

tell males from females morphologically until perhaps 3 weeks after host penetration (in the schistosomulum stage).

Development from the Cercaria to the Schistosomulum

At the moment of cercarial penetration, the tail is cast off and the body enters the skin of the definitive host. Sometimes penetration is attempted into a host that cannot support further parasite maturation. Cercarial dermatitis ("swimmer's itch") of humans represents an example of such an aborted infection. When nonhuman schistosomes (commonly, schistosomes of birds) enter the skin of human bathers, they die in the subdermal connective tissue and there elicit a hypersensitivity reaction that gives rise to the dermatitis.

A cercaria penetrates the host skin and transforms to a schistosomulum by losing the tail, emptying the secretory glands, shedding the glycocalyx, and forming a new, double-unit membrane on the syncytial surface. These changes occur in vitro when cercariae (1) penetrate through an excised skin into physiological medium or (2) have their tails removed mechanically and are incubated in saline. The environmental trigger for surface transformation appears to be a step increase in the saline concentration rather than the absolute saline concentration, because cercariae transform to schistosomula when they go from fresh water to 120 mOsm saline (the osmolarity of the snail) or to 300 mOsm saline (the mammalian osmolarity). Similarly, cercariae dissected from the snail hepatopancreas and placed into 120 mOsm saline transform when they are shifted to 300 mOsm saline, indicating also that immediate prior experience in fresh water is not necessary for triggering transformation. The nervous system is apparently involved in sensing or effecting transformation, because the acetylcholinesterase inhibitor, eserine sulfate, inhibits in a reversible manner cercarial surface transformation to schistosomula. Eserine sulfate also inhibits cercarial swimming, glandular secretion, and skin penetration. This inhibition permits investigators to radiolabel the surface of eserine sulfate–treated cercariae and study its fate with subsequent transformation under permissive conditions. In addition, cercariae, which are prevented from transforming into schistomula, can be assessed for their interaction with antibodies, complement, or host effector cells. Such studies can shed light on the nature of the tegument of transformed schistosomula, particularly as it relates to a target of host defense. Because cercariae transform to schistosomula and live for days in completely defined medium, it is possible to study changes in parasite antigenicity and susceptibility to immune killing in the absence of host antigen acquisition (see Chapter 14).

Scanning electron microscopy has revealed the appearance—within 5 minutes of transformation—of a single set of microvilli over the entire tegumental surface. These microvilli elongate to 3–5 μm and fall off into the medium within 40–60 minutes. With the microvilli, some two-thirds of the fibrillar glycocalyx is shed, but much of the glycocalyx remains on the surface of schistosomula. This residual glycocalyx on recently transformed schistosomula may account for the binding of antibody or of alternative pathway complement components to these organisms. Preformed multilaminate membranous vesicles, which are secreted onto the parasite surface from cell bodies below the muscle layer, provide the membrane that is added onto the tegument surface of the transforming parasite to replace the shed microvilli. These membranous vesicles change the outer tegumental membrane from the single-unit membrane of the cercariae to the new, double-unit membrane found on the surface of schistosomula and adult schistosomes. The outer and inner components of the double-unit membrane are different (as shown by freeze-fracture microscopy). The parasite membrane has been the focus of much study because it can fuse with the plasma membranes of adherent effector cells in vitro, presenting an intriguing biological phenomenon (see Chapter 7).

Radiolabeling of the schistosomular surface shows at least 12 new glycoproteins not identified on the surface of cercariae. In addition, schistosomula bind the lectin concanavalin A, a lectin that does not bind to the cercarial surface. An interesting consequence of events that occur during the period when the cercarial glycocalyx is lost and the double-unit membrane is formed is the loss of resistance of schistosomula to the hypoosmolar stress of fresh water.

After 3 to 4 days in the subdermal layers of an appropriate host, the young schistosomula enter the bloodstream where they transit the circulation briefly, passing through the right side of the heart into the pulmonary artery. They arrive in the pulmonary capillaries on or soon after the fourth day and remain at least 2 or 3 days. By then, subtle and poorly understood surface changes have rendered the schistosomula resistant to the lethal action of the humoral or cellular mechanisms of immune attack.

In the pulmonary capillaries, the fibrous interstitial layer beneath the tegument, as well as some of the anterior musculature, disintegrates and disappears. This process may be a prerequisite for the extreme changes in shape necessary for the schistosomula to traverse the narrow vessels. The parasites elongate and lose much of their fine body spination, except for broad bands at either end, which seem to help them in crawling through the capillaries. The young schistosomula now measure only 8 μm in diameter, the approximate size of the smallest pulmonary capillaries through which the schistosomula must crawl.

After passing through the lungs, schistosomula of most species come to rest in the hepatic and portal venous plexus. Exceptions are *S. haematobium*, which migrates to the venules surrounding the bladder, and the Indian cattle parasite *S. nasalis*, which migrates to the nasal mucosa. In *S. mansoni*, the schistosomula are first observed in the portal system 6 days after infection in mice (7 days in hamsters) but they may take up to 40 days to wander to this site. In *S. japonicum* they arrive in the rodent liver within 3 days of infection.

Subsequent development is best understood in the species parasitizing humans. In *S. mansoni* and *S. japonicum* (and presumably in most others) immature schistosomula find partners of the opposite sex (when available) within the portal vein, where pairing occurs. The young pairs migrate against the portal blood flow into the venous branches, where maturation is completed and egg laying commences about 5 weeks after infection. Worms may live for 10 years or more, although shorter life spans are the rule. The mechanism by which male and female schistosomes recognize each other has been a topic of interest in recent years, since strategies to prevent pairing could represent a novel approach to disease control.

The Tegument of the Schistosomulum and Adult Schistosome

The tegument of schistosomes consists of a syncytium having cytoplasmic connections with underlying nucleated cell bodies (cytons) (Figure 6-4). The outer tegumentary surface is often marked by folds and is pitted with numerous openings of tubular canals readily seen with the scanning electron microscope. These presumably provide added surface for absorption of nutrients.

In mature male worms, the dorsal surface is marked by warty tubercles bearing many heavy spines which help to maintain the position of the worm pair against the blood flow. The surface of the male worm is characterized by thousands of spines in the suckers, throughout the gynecophoral canal and elsewhere. All the spines are invested by the tegumental membranes. The female surface is generally wrinkled annularly, and is smooth except for dense spination in the posterior end surrounding the aperture of the osmoregulatory system (excretory pore). A scattering of sensory organelles of various types is found on the surface of both sexes. There is some evidence for regional differentiation of the surface from one part of the worm to another, but this has not been well explored.

The tegumental syncytium itself, about 4 μm thick, is poor in organelles, having only scattered mitochondria and various types of membrane-bound inclusion bodies. Its outer plasma membrane has a unique heptalaminate structure that represents two fused lipid bilayers, each of which contains inner and outer leaflets.

Much attention has been given to the turnover of the adult surface. The rapid renewal of surface components is generally considered a crucial process in protecting schistosomes from immune damage. However, some early studies based on biosynthetic metabolic labeling or the binding of various materials to the surface apparently failed to distinguish between normal membrane renewal and shedding of tegument under adverse conditions, and some workers reported very rapid rates of surface turnover. Recent evidence shows that adherent immunological markers, specifically antibodies to the human complement component C3b may remain for several weeks on the surface of worms maintained in vitro. It has been suggested also that different areas of the worm surface may turn over at different rates.

Underlying the tegument is a basal lamina, from which microtubules extend into the tegumentary matrix. The nucleated (sometimes multinucleated) tegumental cell bodies are found beneath the muscle fibers and are the source of the granular inclusions found in the overlying syncytium. These cells also contain mitochondria and the Golgi apparatus.

The surface may be disrupted or released by freeze-thawing or by incubation in dilute digitonin, in similar detergents, or merely in phosphate buffered saline. The

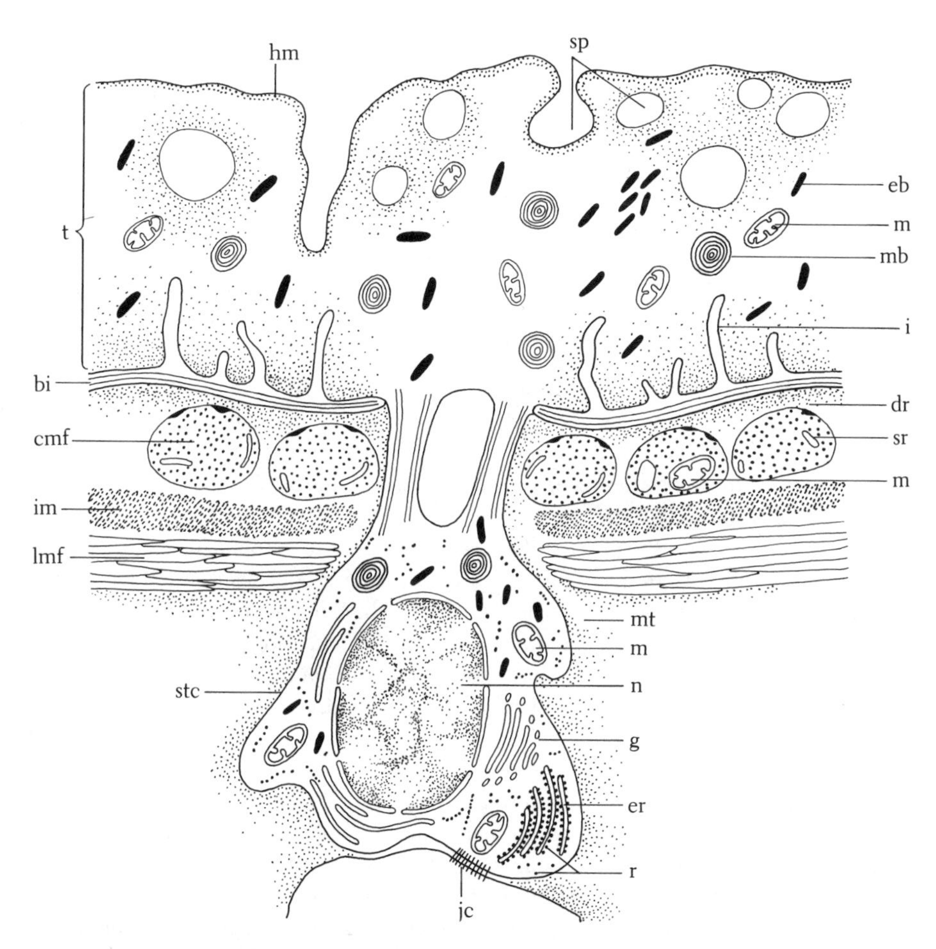

FIGURE 6·4 Representation of adult schistosome tegument and underlying structures: bi, basal lamina; cmf, circularly arranged muscle fibers; dr, peripheral dense region of muscle fibre; eb, elongate body; g, Golgi region; hm, heptalaminate outer membrane; i, invagination of basal membrane; im, interstitial material; jc, junctional complex with parenchymal cell; lmf, longitudinal muscle fibre; m, mitochondrion; mb, membraneous body; mt, microtubule; n, nucleus; r, ribosomes; sp, surface pits; sr, sarcoplasmic reticulum; stc, subtegumental cell; t, tegument. (Adapted from HOCKLEY, D. J. 1973. Ultrastructure of the tegument of *Schistosoma. Adv. Parasitol.* 11:233–305 by permission of Academic Press.)

component layers can be enriched or purified by gradient centrifugation and by confirmation that a particular subcellular component has been isolated, for example, by assessing specific enzyme markers, detecting radioactive surface label (commonly ^{125}Iodine), or analyzing interaction with surface-binding ligands such as lectins or specific antibodies.

Some investigators perceive the outer surface of schistosomes as capable of unidirectional transport of solutes and water by virtue of morphological, enzymatic, and functional asymmetry. For example, if the two bilayers are removed by sequential incubations in 0.1% digitonin, only the inner layer possesses alkaline phosphatase and a sodium and magnesium ATPase, analogous to the apical plasma membrane of other animal epithelia. In contrast, the outer bilayer, which is predominantly lipid, is thought to function primarily to maintain a permeability barrier to soluble proteins.

The adult schistosome tegument is rich in lipids, containing sphingomyelin, phosphatidylcholine, phosphatidylserine, phosphatidylinositol and phosphatidylethanolamine. There is a high content of eicosanoic acid (barely present in the mammalian host), synthesized by the worms from fatty acid precursors in the blood. Portions of the tegument contain specialized structures, particularly spines of various shapes and sizes, which are especially prominent on the ventral sucker. Small bulbous nerve endings, with or without attached cilia, are scattered on the surface of both sexes. Presumably these are sensors to detect and transmit environmental messages to other parts of the organism. How these or other receptors may act in such life functions as migration and mate finding is still unknown.

Continuous with the outer surface is the lining of the esophagus and intestinal ceca. The transition from the tegument to the anterior esophagus is accompanied by increasing complexity and folding of the surface. In this region, esophageal glands produce and transport secretory granules (presumably containing proteolytic enzymes) to the lumen. The cecal lumen itself is similar in adults of both sexes. Its lining is thrown into complex, irregular projections that serve for pinocytosis of hemoglobin breakdown products. A potent proteinase, whose activity is stimulated by sulfhydryl-containing compounds, acts in digestion of host blood.

As acoelomate animals, schistosomes and other Platyhelminthes lack a body cavity and a circulatory system. The internal organs are embedded within a solid cellular parenchyma. This tissue is composed of several cell types, including the cell bodies of the syncytial tegument and at least two kinds of muscle cells. One of these has large, electron-lucent nuclei, large cytoplasmic vacuoles containing a finely granular material, and abundant free ribosomes. The other, much smaller, has scanty cytoplasm but also gives rise to muscle fibers. In addition there are nerve cells, at least some of which contain electron-dense, membrane-bound vesicles reminiscent of neurosecretory granules. The solid tissue is traversed by a network of osmoregulatory (or excretory) tubules emanating from flame cells and collecting through ducts into the excretory opening at the posterior tip. The parenchyma occupies the bulk of male schistosomes from just behind the testes to the posterior end of the body. The long, slender females of the common schistosomes, filled to a large extent with reproductive structures, also have parenchyma filling the intervening space.

The Adult Male

The male reproductive system is far simpler than that of the female and is relatively uniform in the commonly considered *Schistosoma* species infecting humans and domestic animals. The system consists usually of about four to eight testes, located dorsal to the anterior part of the gynecophoral canal, with a collecting duct leading to the genital pore. Around each testis is a dense heterogeneous basal lamina within a layer of circular muscle fibers. The germinal cells were examined during the heyday of descriptive cytology in the early decades of this century but have not been the subject of recent scrutiny. They arise from the lining of the testicular wall, and after undergoing spermatogonial divisions, the daughter cells separate. Meiotic divisions lead to spermatids and eventually to mature sperm cells in a conventional manner. At least two types of nongerminal cells have been described within the testes, closely associated with spermatids and maturing sperm cells. These may help regulate the production of spermatozoa, or may possess nutritive, supportive, or phagocytic functions.

Some recent studies have followed the incorporation of tritiated thymidine into *S. mansoni*. In young worms, thymidine incorporation was found throughout the tissues. After about 35 days of age, when worms were sexually mature, incorporation occurred principally in the testes, indicating a continuing high degree of DNA synthesis in reproductive tissues with only isolated sites of activity in somatic tissues. At 45 days after infection, autoradiography indicated that the densely labeled cells were always in a peripheral position in the testes, with the central area generally unlabeled.

In late spermatocytes of *S. mansoni*, the Golgi complex may be absent, which perhaps is associated with the absence of an acrosome in the mature sperm. In fine structure, the sperm of most schistosomes appear typical of animal sperm, although an acrosome is lacking. In *Schistosomatium douthitti*, sperm cells are biflagellated, with a thin head about 30 to 35 μm long, sometimes appearing ribbonlike, and two short flagella less than half the head length. Sperm cell biology therefore suggests a phylogenetic split between these and other schistosomes. In *S. mansoni* the head measures 2 by 8 μm, rounded anteriorly and tapered posteriorly. A prominent anterior mass of undifferentiated mitochondria lies beneath the plasma membrane in a cuplike depression in the nucleus. The nucleus itself is largely electron-opaque, but contains electron-lucent patches of Feulgen-negative material. A single layer of microtubules is found between the periphery of the large nucleus and the plasma membrane of the sperm head. The single flagellum, about 20 μm long, may be of the $9 + 0$ tubule configuration or some variant.

The Adult Female

Female schistosomes, particularly in the species that parasitize birds, demonstrate considerable variation in shape, from elongated and filiform to flattened and leaflike. Mammalian species in general are less variable but still exhibit a range of forms, from the familiar cylindrical females of *Schistosoma* to the broad, flat females in *Schistosomatium*.

The external surface of female schistosomes that have been observed by scanning electron microscopy is simpler and more uniform than that of the heavily spined male. For the most part, females have a relatively featureless tegument with delicate transverse ridges bearing both ciliated and nerve papillae. Under higher magnification, tegumental pores may be seen throughout. Near the posterior end commonly lies the area of long, sharp spines previously mentioned.

The schistosome female reproductive tract was described in general terms by Theodor Bilharz, the German physician who first saw adult schistosomes while performing autopsies in Cairo in the early 1850s. This reproductive tract is much more complex than the male system. There is always a single ovary, filled with cells in various stages of development, neatly displayed in linear progression from the smaller and more immature oogonia in the blind anterior area to the more loosely packed oocytes ready for discharge into the oviduct. Sperm cells that have come upstream from the female gonopore congregate in the dilated proximal oviduct to rendezvous with and penetrate the newly discharged oocytes.

Oogonia have a large, irregularly shaped nucleus with clumped chromatin masses and a nucleolus, and a narrow rim of cytoplasm. Cytoplasmic organelles vary with the stage of maturation and degree of metabolic activity. In the mature oocyte, there is usually a well-defined nucleolus and little nuclear chromatin; ribosomes are often associated with the nuclear membrane. The cytoplasm also contains ribosomes, granular endoplasmic reticulum, mitochondria, and abundant Golgi material. Dense cortical granules, formed from the Golgi complexes, are located under the peripheral membrane.

In female *S. mansoni* from unisexual infections (in which no males are present) the ovary is small, often coiled, and relatively inconspicuous, and the entire reproductive system is greatly retarded in development.

Little is known in general about the control of cellular development and movement in the ovary of parasitic flatworms. When adults of *S. mansoni*, *S. japonicum*, and *S. haematobium* are incubated in tritiated thymidine and surgically implanted into the circulatory system of hamsters, nuclear label is taken up rapidly into oogonia at the anterior periphery of the ovary. After 3 days, labeled cells appear in the central anterior region of the ovary. Label is seen in the nuclei of primary oocytes in *S. japonicum* in 6 days, and in the other species in 7 days. Electron microscope autoradiography shows extranuclear DNA associated with mitochondria in the oocytes of *S. mansoni*. Some investigators who have labeled *S. mansoni* females with tritiated thymidine find label only in the anterior oogonial cells.

The ovary has understandably been a favorite potential target of chemotherapy. Antimonials and miracil, for example, lead to complete destruction of ovarian tissues. Niridazole causes a massive extrusion of ovarian cells into the uterus, which are then passed out by the worms. Emetine, although very destructive to the vitelline gland, has little effect on the ovary.

The early egg consists of the oocyte (normally with contained sperm cell) plus mature vitelline cells. In *S. japonicum* the number of vitelline cells is usually about 20; in *S. mansoni*, about 30 to 40. As a general rule in trematodes, eggshell formation begins in the ootype, which contains a mucoid secretion that forms a thin coat over the cells destined to make up the egg. Granules freed from the vitelline cells migrate to the exterior of the package. As the granules coalesce, the shell gradually thickens in a process that may involve a quinone tanning system. Then the cells may switch to production of lipid globules, which provide nourishment for the miracidial embryo within the eggshell. The autofluorescence of vitelline material and eggshells is well known and readily demonstrable by fluorescence microscopy. The shell consists of sclerotin, a tanned protein derived from phenolic precursors present in the vitelline granules. Phenol oxidase activity, reported in *S. mansoni* and *S. japonicum*, is depressed by many drugs that inhibit protein tanning. The resulting eggs are distorted and lack normal shell and contents.

The vitelline gland occupies the posterior half to two-thirds of the body. Many follicles connect through ducts to the large vitelline duct, which opens into the oviduct. Vitelline cells are haploid, suggesting a common origin of the vitelline gland and ovary. In *S. mansoni* cultured in vitro from cercariae to adults, vitelline cells are essentially normal, demonstrating that specific host influences are not necessary for their development.

The high demands of egg production dictate an active metabolic role for the vitelline gland. The intimate relation between the nutrient-providing intestinal ceca and the vitelline gland is shown by the thin border between these organs. Assuming that each female *S. mansoni* produces 300 eggs per day and there are 38 vitelline cells per egg, the vitelline gland in this species must release about 11,000 mature cells daily. In *S. japonicum* this figure is higher perhaps by a factor of 10. Various authors have estimated that a female *S. mansoni* converts from 10 percent to 100 percent of her own body weight into eggs every day.

Many workers have studied the effects of antischistosomal drugs on the vitelline gland because of its importance in egg production and its high biosynthetic turnover. Fuadin, Astiban, lucanthone, hycanthone, miracil, niridazole, and other drugs all have deleterious effects on vitelline cells.

Chromosomes

We may wonder how it is that the peculiar sexual separation in this family developed from what were certainly hermaphroditic ancestors. What is known now is that male and female schistosomes have distinct chromosomal patterns. Sex chromosomes have been found in a dozen species of schistosomes, in which the female appears to be heterogametic. That is, sex is determined—as in birds—through the ovum, rather than through the sperm. It is very difficult to obtain chromosome preparations from adult schistosomes, but relatively simple to find metaphase figures in sporocysts taken from the intermediate host snail. Therefore most karyology of schistosomes has been done from the larval stages. Studies have been conducted in 11 species from mammals and 5 from birds. Of the mammalian species, those from Africa, Asia, and North America are distinct. Each genus and species has a different karyotype, distinguishable by C-banding if not by chromosome morphology. All of the African and Asian mammalian schistosomes have diploid numbers of 16, but the two species from North

America, *Schistosomatium douthitti* and *Heterobilharzia americana*, have 14 and 20, respectively.

Information about schistosome cytogenetics, although still very crude, should be compatible with emerging knowledge about the molecular biology of this group, discussed in Chapter 19.

Coda

The cell biology of schistosomes is vastly more complex than that of protozoa. Numerous cell types, many of them sex- and stage-specific, characterize these metazoan parasites. The descriptive ultrastructural studies of recent decades are giving way to analyses of physiology and structural biochemistry, with particular reference to interstage transformation. Membrane biology, discussed further in Chapter 7, represents another active area of research. Additional knowledge of the reproductive biology of schistosomes could lead to novel chemotherapeutic or immunoprophylactic strategies.

Additional Reading

CRABTREE, J. E., and R. A. WILSON. 1986. *Schistosoma mansoni:* an ultrastructural study of pulmonary migration. *Parasitology* 92:343–354.

ERASMUS, D. A. 1977. The host-parasite interface of trematodes. *Adv. Parasitol.* 15:201–242.

HOCKLEY, D. J. 1973. Ultrastructure of the tegument of *Schistosoma*. *Adv. Parasitol.* 11:233–305.

SILK, M. H., I. M. SPENCE, and J. H. S. GEAR. 1969. Ultrastructural studies of the blood fluke-*Schistosoma mansoni*. *S. Afr. J. Med. Sci.* 34:(1) 1–10; (2) 11–20; (3) 93–104.

STIREWALT, M. A. 1974. *Schistosoma mansoni:* Cercaria to schistosomule. *Adv. Parasitol.* 12:115–182.

7

Cell Biology of Schistosomes

II. Tegumental Membranes and Their Interaction with Human Blood Cells

■

John P. Caulfield

The previous chapter presented the structural organization of schistosomes, showing that the surface of the parasite facing its host was always a syncytium in either the mammalian host or the snail. Here we extend the discussion into the functional properties of the syncytium, particularly its surface membranes. The structural, molecular, and functional properties of the schistosome surface are critical in explaining the parasite's long survival in the human host, because the surface is a significant potential target of host defense. Unlike some protozoa that hide within host cells, schistosomes are exposed to the circulating components of the immune system. To survive, the parasite surface presumably possesses mechanisms for evading immune attack. We wish to explain how the parasite can be attacked and killed by some cells, such as eosinophils, but evade the attack of other cells. The chapter will focus on *Schistosome mansoni* because most of the experimental work has been done on this species. Presumably, the paradigm constructed for *S. mansoni* will be applicable to the other species of schistosomes that infect humans. We will consider initially knowledge of the structure and composition of the surface membranes and then review functional studies performed on the membranes themselves. The chapter will conclude with a discussion of the interactions between various human blood cells and the parasite.

Structure of the Parasite Surface

During its life span in its mammalian host, the parasite is covered by a syncytium called the tegument (see Figure 6-4). In all stages, the tegument is connected to cell bodies that are located deep within the organism below the large circumferential bands of muscle that are responsible for such gross movements of the whole parasite as contraction, elongation, and turning. The cell bodies contain the basic machinery for protein synthesis, namely, nucleus, endoplasmic reticulum, and Golgi apparatus. The surface of the tegument has a varied topology consisting largely of protruding spines, which contain crystalline actin, and invaginating pits. The distribution of the spines and pits is altered during larval development. During the first 4 days after transformation from cercariae to schistosomula, the spines are lost over much of the body, there is more pitting at the surface, and the surface area increases 30 percent. In adults, different regions of the parasite have characteristic patterns of folding, pitting, or spines. Despite these large variations in structure, the ultrastructure of the surface membranes that face the host tissues or bloodstream remains relatively similar from the completion of cercarial transformation through adulthood.

Transmission electron microscopy reveals that the parasite surface consists of adjacent lipid bilayers that appear as two apposed membranes by transmission electron microscopy (Figure 7-1). These membranes are anisotropic, however, as shown by freeze-fracture technique. In newly transformed larvae, the inner of the tegumental membranes, that is, the one that is closest to the cytoplasm of the tegument, contains many intramembrane particles and resembles the plasma membrane of most mammalian cells. Intramembrane particles are associated with membrane proteins, and membranes devoid of these particles usually contain little or no protein. The outer membrane which faces the host in vivo or culture medium in vitro has few intramembrane particles and resembles a simple protein-poor lipid bilayer (Figure 7-2). Interestingly,

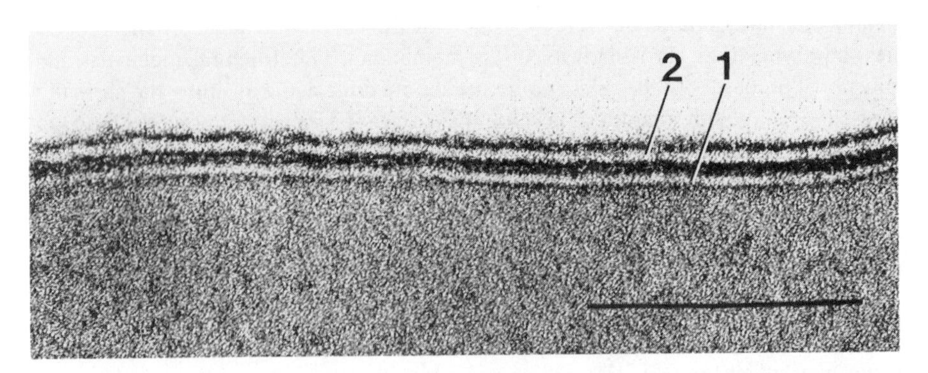

FIGURE 7·1 Transmission electron micrograph of schistosomular tegumental membrane. The membrane is composed of two closely apposed trilaminar membranes. The inner membrane is designated 1; the outer 2. Bar, 0.1 μm. ×400,000. (Reproduced from the *Journal of Cell Biology*, 1980, 86:46–63 by copyright permission of the Rockefeller University Press.)

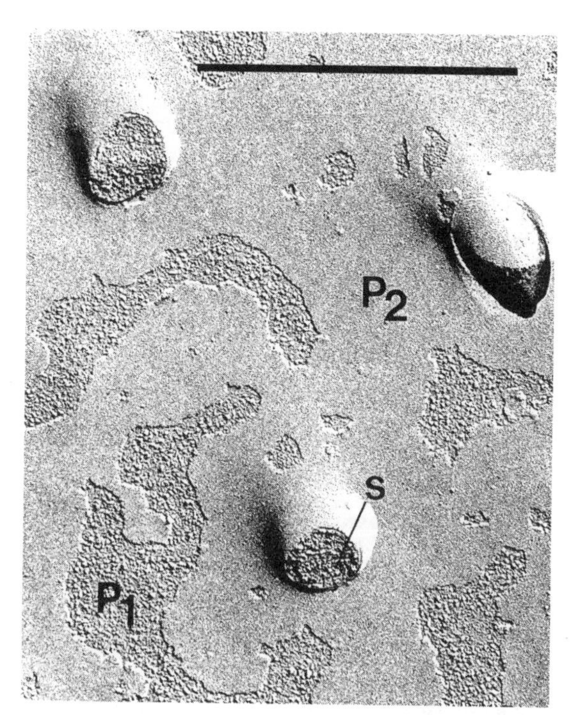

FIGURE 7·2 Freeze-fracture of tegumental membranes. P-face view, showing large areas of the IMP-rich inner membrane (P_1) and the IMP-poor outer membrane (P_2). The spines (s) are cross-fractured. Bar, 1 μm. ×41,500. (Reproduced from the *Journal of Cell Biology,* 1980, 86:46–63 by copyright permission of Rockefeller University Press.)

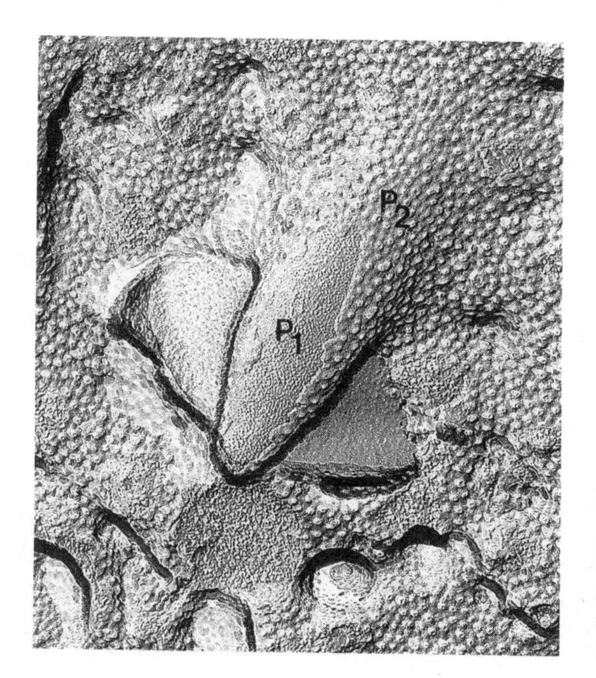

FIGURE 7·3 Freeze-fracture of the tegumental membranes of a schistosomulum that was incubated with the polyene antibiotic filipin. The outer membrane (P_2) has large numbers of punctate dimples that are characteristic of membranes containing cholesterol, whereas the inner membrane (P_1) is unperturbed (compare with Figure 7·2). ×55,000.

the outer membranes of worms isolated from the lungs of mice about a week after infection contain intramembrane particles, whereas outer membranes of worms cultured in vitro for the same period do not. The parasite may acquire these particles or proteins from its host. However, the host may induce the parasite to synthesize intramembrane particles in its outer membrane. Electron microscopy has also suggested that the outer membrane contains cholesterol, since the membrane structure is perturbed by digitonin and the polyene antibiotic filipin, both of which characteristically alter cholesterol-containing membranes (Figure 7-3). These structural observations suggest that the schistosomula are covered by two different membranes: an inner one that probably can function in ion transport much like mammalian cell plasma membranes and an outer one that is simply a lipid bilayer composed of sterol and phospholipid. The schistosome surface thus bears a strong resemblance to gram-negative bacteria, which also have a double membrane system.

Functional Studies of the Surface Membranes

When the plasma membrane of most mammalian cells is exposed to a molecule that binds specifically to a receptor on the membrane, for example, low-density lipoproteins or transferrin, or to reagents that bind to specific chemical groups on the membrane, for example, lectins or polycations, the bound molecules are usually internalized by the process of endocytosis and transported to internal compartments within the cell, such as phagosomes or lysosomes. Some mammalian cells, notably neoplastic cells, also can shed bound ligands into the culture medium. Schistosomes bind antibodies and complement at various stages of their life cycle in the mammalian host but most extensively shortly after transformation. However, these multivalent ligands bound to the surface are not endocytosed by the parasite. Furthermore, low-density lipoproteins, which also bind to the tegument, and fluid-phase markers such as horseradish peroxidase are also not endocytosed by the tegument. Thus, the tegumental membranes appear incapable of endocytosis and are very different from typical normal mammalian cell plasma membranes.

Although complement and antibody bind and are not endocytosed, they do not remain on the parasite surface but are slowly lost. This loss has been explored by applying the lectin concanavalin A (Con A) to schistosomula. The lectin binds to a unique population of molecules that are present for 24–48 hours after transformation but are not present on cercariae or schistosomula older than 48 hours. Thus, the Con A-binding molecules are lost from the surface independent of the binding of ligand, and the loss therefore is physiological and not induced by Con A. Furthermore, the kinetics of the loss of Con A binding is the same as that of the loss of bound rat and human antibody and bound complement, suggesting that the mechanism of loss is not confined to this lectin; a more general mechanism of membrane turnover is probably at work. The kinetics of the loss of these three molecules—antibody, complement, and Con A—follows that of a decay curve and indicates that the loss of bound molecules is a random event. In addition, Con A can be bound to the parasite and recovered in the culture medium over time. The recovered Con A is the same molecular weight as the Con A originally bound. These studies suggest the parasite is sloughing or shedding the bound ligand and attached membrane component into the culture medium.

In mammalian cells, multivalent ligands such as Con A or antibody bound to the plasma membrane at 0°C are capped when the temperature is raised, so the ligand becomes localized to a region of the cell surface. The studies with Con A also show that there is no capping of the bound ligand on the parasite surface, suggesting that the contractile proteins of the tegument do not functionally interact with the outer membrane. As suggested by the kinetics of loss of Con A that follows a decay curve, bound Con A is randomly distributed on the parasite surface. However, after the addition of hemocyanin that is rich in mannose and binds to the Con A, clustering of the hemocyanin and hence the lectin is observed (Figure 7-4). This clustering occurs after the parasites have been killed with azide, demonstrating that contractile proteins are irrelevant in cluster formation. Instead, the hemocyanin itself induces aggregation of the Con A in the membrane. These experiments also demonstrate movement in the plane of the membrane by the moieties to which Con A is bound.

The loss of bound Con A and its recovery in the culture medium suggest that the outer membrane or portions of it are being sloughed into the medium. This hypothesis can be tested by labeling the parasite surface and examining the culture medium over time to see if labeled membrane components appear there. Several general labeling schemes are available, particularly iodination and labeling oxidized sugars on glycoproteins and glycolipids with tritium. The sugar-labeling techniques have the advantage of being two-step procedures, permitting the nonspecific binding of the radiolabel to be determined in the absence of oxidation. Further, the oxidation step can be performed with an enzyme, such as galactose oxidase, which limits the specific labeling to the parasite surface, since the parasite does not endocytose. Experiments of this type as well as subsequent experiments employing monoclonal antibodies show that many surface glycoproteins and glycolipids are being shed into the culture medium with a half-time of approximately 12 – 14 hours. Given that the surface area of a schistosomulum is 20,000 square micrometers (μm^2), this figure represents losses of approximately 800 μm^2/hour, an astronomical rate by mammalian cell standards. Furthermore, the rate of synthesis must be double that figure, since the surface area is increasing by 100 percent per day. Studies of this type are usually not carried out on adult schistosomes, largely because such studies would be complicated by the presence in adults of a functioning gut, which can ingest sloughed proteins.

Although the surface membrane proteins and lipids are clearly being sloughed into the culture medium, some experiments do not fit this model and there are serious methodological limitations in all experiments of this type. In particular, studies with monoclonal antibodies have suggested that some proteins may not be sloughed and may persist for long periods of time on the surface. More generally, the interpretation of the turnover experiments is clouded by the fact that schistosomes are covered by two membranes and that the distribution of labeled or antigenic membrane components within the double-membrane system is usually unknown. In particular, it is unclear whether only components of the outer membrane are labeled or those of the inner

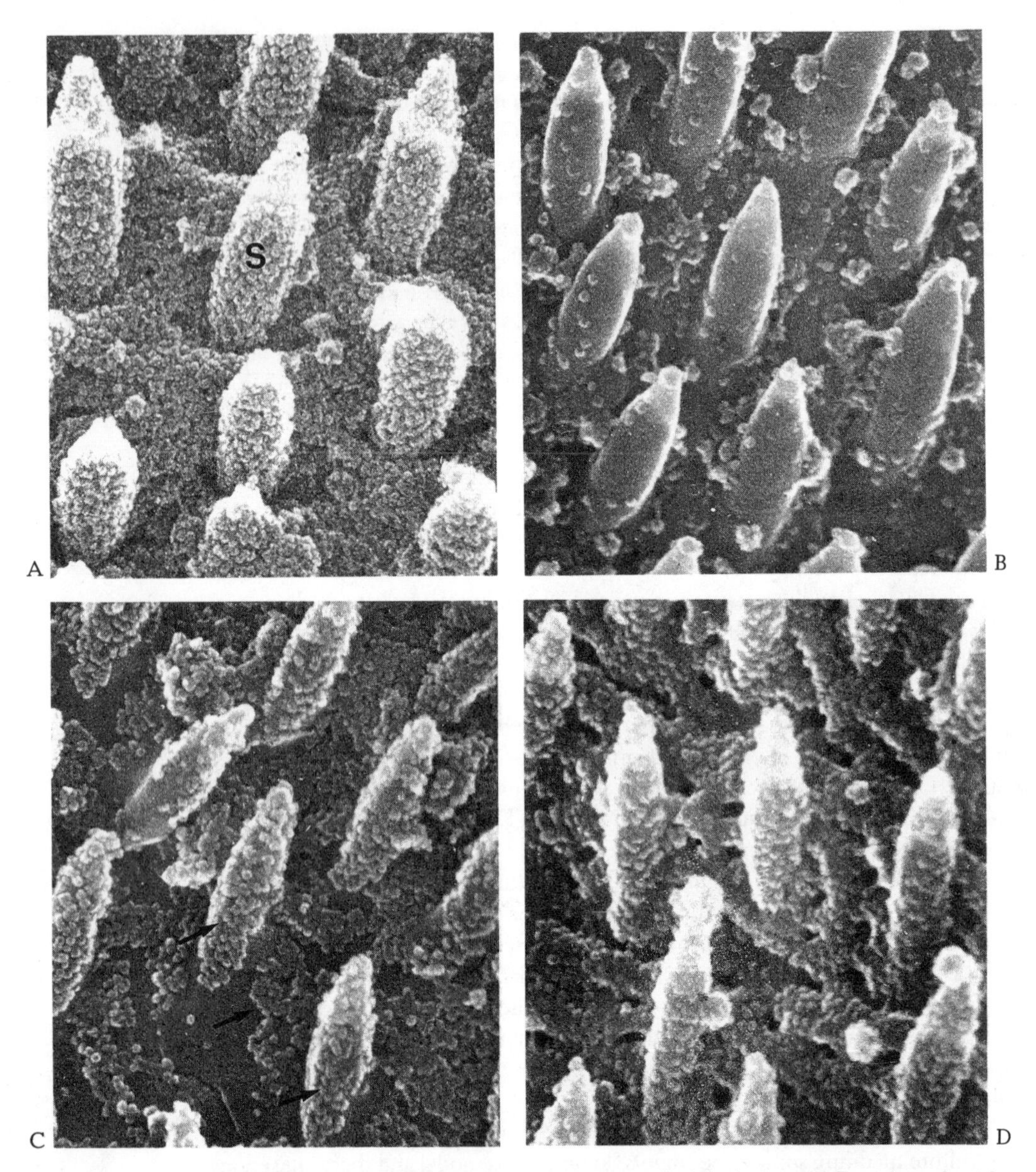

FIGURE 7·4 Scanning electron micrographs of skin schistosomula labeled with Con A and with hemocyanin immediately before fixation. A. High-power view of a fresh schistosomulum shows rectilinear hemocyanin molecules binding uniformly and densely over spines (S) and areas between the spines. B. A control organism incubated with FITC-Con A in the presence of α-methyl mannoside shows little hemocyanin binding to the schistosomular surface. C. A schistosomulum cultured for 12 hours and then labeled at 37°C has aggregates of hemocyanin molecules (arrows) over the schistosomulum surface between bare areas. D. In contrast, a schistosomulum cultured for 12 hours and then labeled at 4°C has an even, diffuse distribution of hemocyanin molecules on its surface. ×40,000. (Reproduced from the *Journal of Cell Biology,* 1982, 94:355–362 by copyright permission of the Rockefeller University Press.)

membrane are also labeled. Definitive experimentation in schistosomes is complicated by difficulties in obtaining large numbers of organisms and ultimately in determining the composition and distribution of the components of the two membranes.

Composition of Surface Membranes

Many laboratories are now actively investigating the composition of the surface membranes, but a definitive statement cannot yet be made about which molecules constitute the two membranes. Investigations have employed isolation of the membranes, surface labeling, selective extraction, and analysis of biosynthetically labeled materials shed into the medium. Taken together, they yield a general picture. The outer membrane is primarily composed of phospholipids. (Note that phospholipids are made up of a glycerol backbone attached to two fatty acids, called acyl chains, and phosphate. The phosphate is linked to another molecule, such as choline or serine, called the head group.) The major phospholipids in the outer membrane are phosphatidylcholine and phosphatidylethanolamine, with small amounts of phosphatidylserine, phosphatidylinositol, sphingomyelin, and lysophosphatidylcholine. Cholesterol is also present as well as glycolipids, some of which are antigenic. There are a variety of antigenic and nonantigenic proteins, most of which are glycoproteins as well (see Chapter 14 for a more extensive discussion of the antigenic proteins). The inner membrane probably contains the various enzymes known to be associated with the tegumental membranes, namely acetylcholinesterase, Na-Ca ATPase, and alkaline phosphatase as well as phospholipids and cholesterol.

Schistosomes do not synthesize cholesterol but instead use host cholesterol; they do not modify host acyl chain length during phospholipid biosynthesis but perform a limited number of head-group modifications. Presumably, the lipid components are assembled into membranes in the cell bodies and transported to the tegument in multilamellar bodies. There they fuse with the base of the pits and insert the new membrane onto the parasite surface.

Interactions with Cells

Schistosomes dwell in the bloodstream from about 4 days after they penetrate through the skin until they die years later. Being in the circulation exposes them to cells and plasma components that have the potential to destroy the parasites. On the other hand, one of the ways the parasite can protect itself from immune attack is to acquire host antigens from serum or blood cells themselves. In this way, the parasite mimics the host, is not recognized as foreign, and in a sense becomes the wolf in sheep's clothing. The attack of cells and parasite defense can be studied in vitro by mixing purified leukocytes, schistosomula, and infected patient's serum that contains antibodies against the parasite. Let us consider four blood cells: eosinophils, neutrophils, monocytes, and erythrocytes.

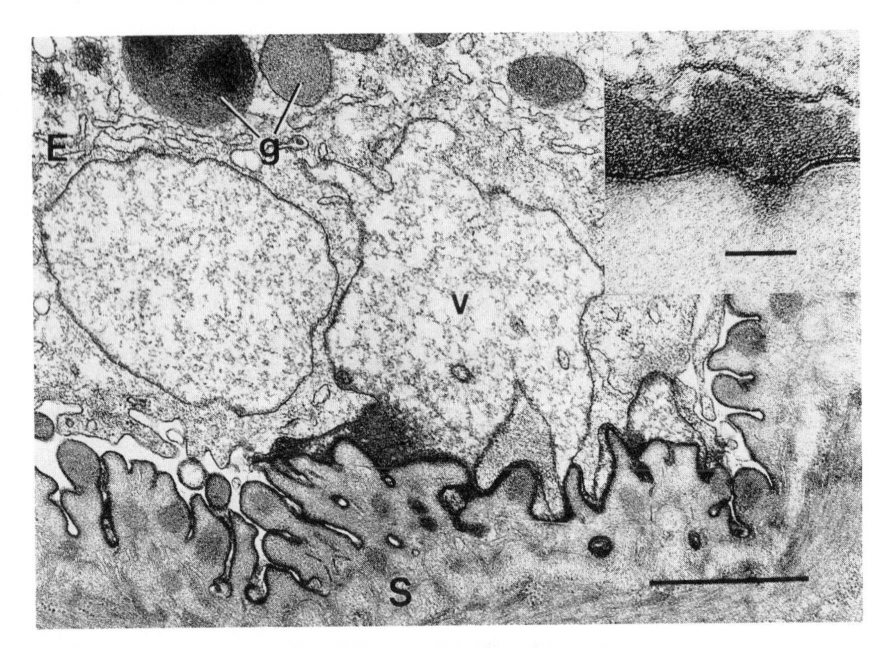

FIGURE 7·5 Eosinophil (E) discharging granule (g) contents onto the surface of a schistosomulum (S) preincubated with antibody. Note the large invaginations (v) into the eosinophil cytoplasm. These invaginations are formed by multiple granules discharging at the same site. Electron-dense material adheres to the surface of the schistosomulum and has penetrated into the pits. The inset shows an eosinophil adhering to the electron-dense material on the parasite surface. Bar, 1 μm. ×24,000. Inset bar, 0.1 μm. ×110,500. (Reproduced from the *Journal of Cell Biology*, 1980, 86:46–63 by copyright permission of the Rockefeller University Press.)

FIGURE 7·6 A. Eosinophils (E) treated with a monokine that activates eosinophils adhering to an Ab-coated schistosomulum (S) after 3 hours incubation. The dead cell in the center is clearly an eosinophil because some of its distinctive granules (gr) are intact. Note that electron-dense discharge material (d) is present both between the cell and the worm and on the side of the cell opposite the worm. The lucent areas on the other two eosinophils are glycogen (gl). ×21,000. B. Surface of an Ab-coated schistosomulum (S) after 2 hours incubation with monokine-treated eosinophils. Electron-dense discharge material (d) mixed with vesicular membrane fragments (arrows) is adherent to the tegumental membrane of the worm. ×65,000. C. Surface of an Ab-coated schistosomulum (S) after 3 hours incubation with monokine-treated eosinophils. There is a large mass of electron-dense discharge material (d) adherent to the tegumental membrane (tm) of the parasite. On the side of the discharge material opposite the parasite, there are the remnants of an eosinophil plasma membrane (arrows) and some membrane fragments. ×53,000. (Reproduced from the *Am J. Path.*, 1985, 120:380–390 by permission of J. B. Lippincott and Company.)

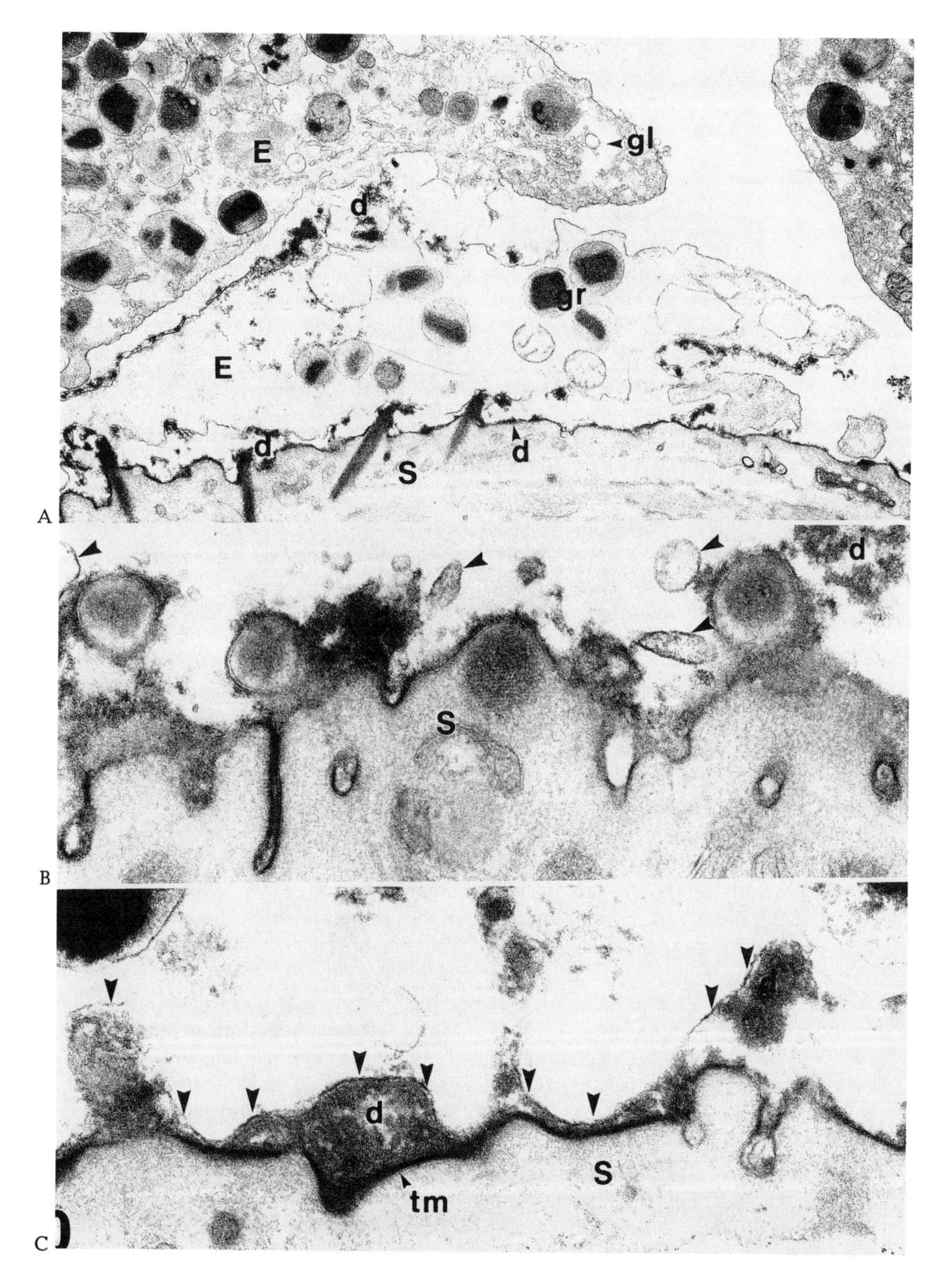
E
gl
d
gr
E
d
S
d
A
d
S
B
d
tm
S
C

Eosinophils adhere to the parasite surface and degranulate within minutes (Figure 7-5). The parasites die about 18 hours later when most of the surface is covered by material discharged from the granules. Since a single eosinophil discharges over an area of about 80 square micrometers (μm^2) and since about 1800 μm^2 of the parasite are covered by discharge material after 12 hours, several hundred eosinophils have to discharge onto each parasite. The successful attack of the eosinophils should be regarded as a mass onslaught directed broadly against the entire surface of the parasite that is carried out over the course of a half a day or more.

Eosinophil degranulation is mediated through antibody bound to the parasite surface. The cells bind to the worm by means of an Fc receptor in the plasma membrane binding to the Fc portion of the IgG on the worm. The granules move through the cytoplasm to the zone where the plasma membrane is bound to the antibody. The granule membrane then fuses with the plasma membrane, thereby releasing the granule contents at the site of eosinophil–parasite contact.

During the attack, eosinophils die and lyse on the parasite surface. Cells can be seen with electron lucent cytoplasm and granules that are still intact and undischarged (Figure 7-6). Presumably the cells died before they can exocytose all of their granules. Furthermore, fragments of the plasma membranes of eosinophils are seen covering degranulated material on the parasite surface (Figure 7-6), perhaps assisting in the attack by confining exocytosed material to the target and preventing diffusion of material from the surface. Whether the eosinophils are killed by the parasite or by material released by their own or other eosinophil granules remains to be determined.

Unlike eosinophils, neutrophils do not kill schistosomula in the presence of antischistosomal antibodies alone, but they will kill if complement is also present (see Chapter 14). Although neutrophils adhere to the worm in the presence of antibody alone, they do not degranulate, which probably explains their failure to kill the worms. The lack of degranulation is in turn explained by experiments examining how neutrophils process multivalent ligands such as antibodies and lectins on the parasite surface. Fluorochrome-conjugated antischistosomal antibodies or lectins promote cell–parasite adherence and can be observed simultaneously by fluorescence microscopy. Immediately after labeling, the ligands are evenly distributed on the parasite. Soon after neutrophils are added, dark areas appear on the parasite and fluorochrome can be identified in the lysosomes of cells (Figure 7-7). Neutrophils endocytose antibody or lectin from the parasite surface. Thus, unlike the interaction of the Fc portion of the antibody with the Fc receptor on the eosinophil, which promotes fusion of the granule membrane with the bound membrane at the cell surface, the neutrophil membrane receptor ligand complex is internalized and fuses with the granule membrane in the cell. Parasite antigens are also endocytosed along with the antibody, as demonstrated by autoradiography. Removal of both antigen and antibody from the parasite by neutrophils reduces the density of recognition signals available to eosinophils.

About 10 percent of adherent neutrophils fuse their plasma membranes to the outer of the two tegumental membranes of the worm when worms are first coated with antibody or lectin (Figures 7-8, 7-9). A hybrid membrane is derived from the plasma membrane of the neutrophil and the outer tegumental membrane of the parasite is

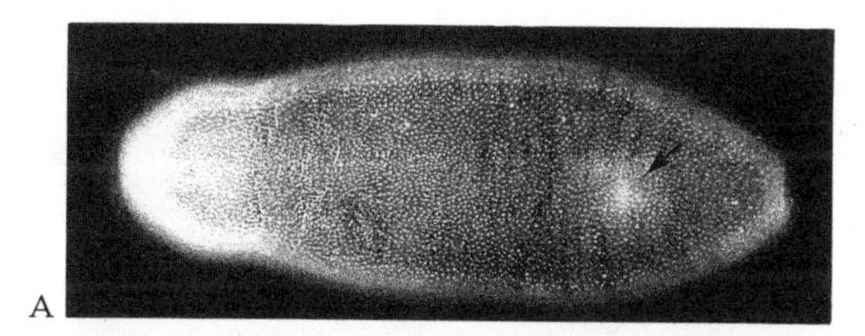
A

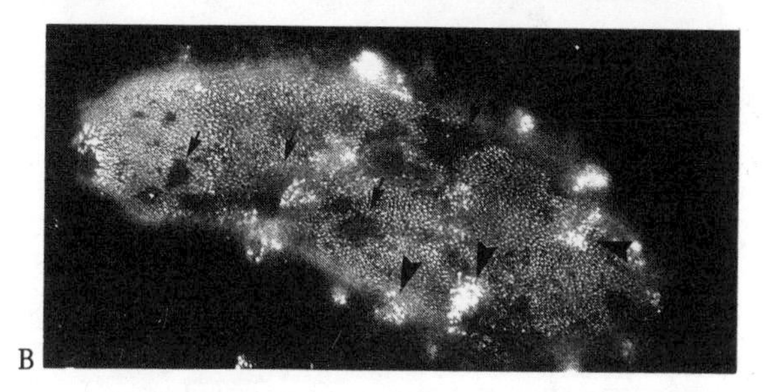
B

FIGURE 7·7 A. Skin schistosomulum at the end of the preincubation with fluorescein-Con A. The label is evenly distributed over the parasite. The anterior end, at the top, is brighter because there is a higher density of spines on this portion. The bright area on the body (arrowhead) is due to staining of the acetabulum or ventral sucker located on the other side of the worm. B. Mechanically prepared schistosomulum preincubated in fluorescein-Con A and incubated with buffy coat cells for 3 hours. The label is no longer uniformly distributed on the worm. Dark, nonfluorescent areas are present on the parasite (arrows). The fluorescein is also present in granules in the cells (arrowheads). ×680. (Reproduced from the *Journal of Cell Biology,* 1982, 94:370–378 by copyright permission of the Rockefeller University Press.)

formed as a result. No antibody or lectin is detectable by electron microscopic autoradiography in the fusions, suggesting that the ligand is removed by endocytosis before fusion occurs. These fusions are probably the largest (up to 10 μm) to be described between biological membranes. They may be a mechanism for acquiring host membrane components, particularly integral membrane proteins, such as histocompatibility antigens, which are thought to defend the parasite against being recognized as foreign by the host immune system. The fusions are not reversible and ultimately result in lysis of the fused neutrophil. After this lysis, the fused membrane appears to mix with the normal worm outer membrane (Figure 7-10).

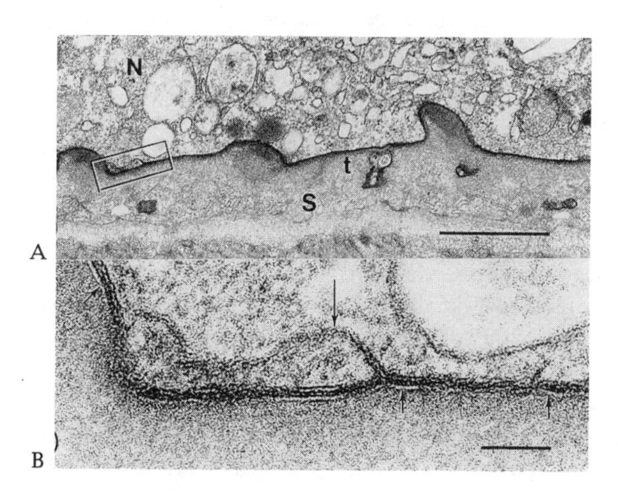

FIGURE 7·8 Neutrophil (N) adhering tightly to the surface of a schistosomulum (S) preincubated in antibody and complement. Note that the cell is flattened and tightly adherent to the tegument (t). B is a high-power view of the rectangle in A. Note that the membrane between the cell and the schistosomulum is pentalaminar (short arrows). Some membrane appears to be lifted off the surface (long arrow). A. Bar, 1 μm. ×26,500. B. Bar, 0.1 μm. ×171,600. (Reproduced from the *Journal of Cell Biology,* 1980, 86:46–63 by copyright permission of the Rockefeller University Press.)

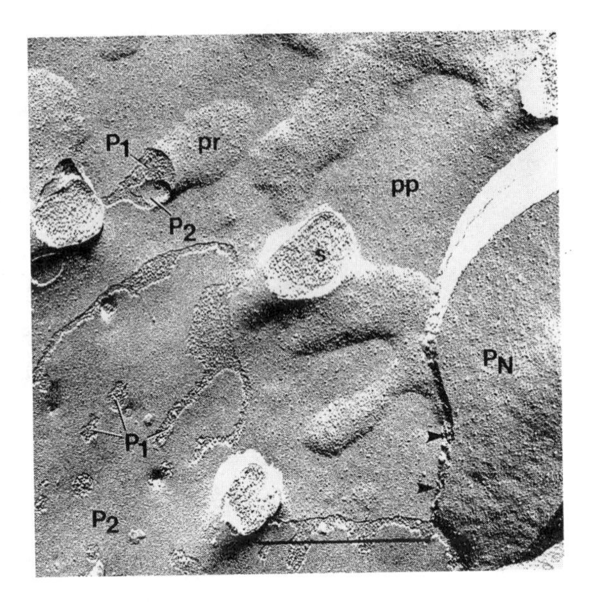

FIGURE 7·9 High-power view of neutrophil adhering to a schistosomulum preincubated with antibody and complement. Normal schistosomular P_1 and P_2 faces are seen at the bottom of the picture. An area of attachment with raised IMP-rich (pr) and IMP-poor (pp) areas is seen at the top and is separated from the normal membrane by a zone of P_1. The IMP-poor membrane and the P_2 faces are the same distance from the P_1 membrane. In the IMP-rich area where the fracture steps out of the membrane, normal P_2 and P_1 membrane faces are seen. Note that the concentration of IMPs in the rich area is approximately equal to that of the P face of the neutrophil (P_N) on the right. The P_N face is on the side of the cell not attached to the schistosomulum and is separated from the fused membrane by the cytoplasm (arrowheads). s, spine. Bar, 1 μm. ×41,000. (Reproduced from the *Journal of Cell Biology,* 1980, 86:46–63 by copyright permission of the Rockefeller University Press.)

The third major effector cell in human blood is the monocyte (Figure 7-11). Human monocytes are not as effective as eosinophils in killing schistosomes. Monocytes that have been cultured for several days are called monocyte-derived macrophages because of a certain resemblance to tissue macrophages. Quiescent monocyte-derived macrophages do not effectively kill tumor cells or schistosomula that have been preincubated with antibody. However, after exposure to interferon-γ (a mediator that activates the host cells), the monocyte-derived macrophages acquire the capacity to kill tumor cells; their ability to kill schistosomula is inconsistent. Human monocyte–derived macrophages fuse with the outer tegumental membrane (Figure 7-12). Murine macrophages, in contrast to human monocyte-derived macrophages, consistently kill schistosomes and apparently are major host defense effector cells in murine schistosomiasis infections.

Parasite interactions with erythrocytes are potentially important, particularly since adult worms can acquire human ABH blood group antigens in vitro. After coculture of schistosomula and erythrocytes, an electron-dense plaque is seen between the parasite outer membrane and the plasma membrane of the attached cell (Figure 7-13). Membrane proteins and glycolipids are not transferred to the parasites from adherent cells. However, carbocyanine dyes, which are anchored in the red-cell membrane by acyl chains, are transferred rapidly in 30 minutes to 3 hours. These studies also demonstrated that erythrocytes lyse on the parasite surface and that the lysed membrane fragments remain attached to the parasite (Figure 7-14). This mode of host antigen acquisition and the resultant masking of parasite antigens differs from the process of host antigen acquisition in which the antigens are presumably inserted into or absorbed onto the outer tegumental membrane. Additional modes of host antigen acquisition could

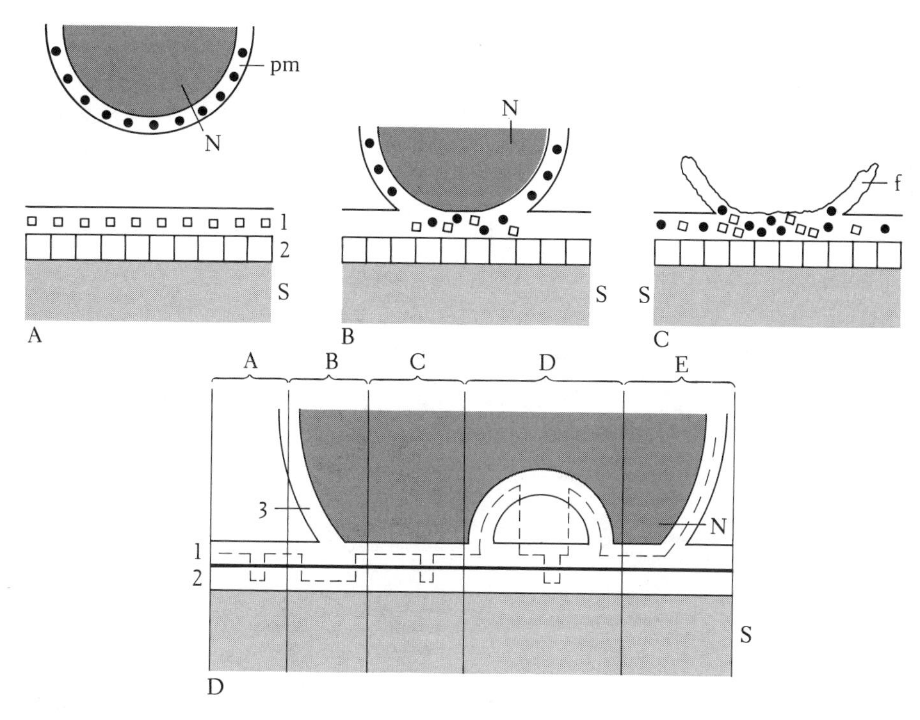

FIGURE 7·10 A hypothetical model of the transfer of membrane components from the plasma membrane (pm) of a neutrophil (N) to the tegumental membrane of a schistosomulum (S). In panels A–C, the cell approaches the worm, and the membrane components are seen as open or closed circles in the outer tegumental membrane (2) or cell plasma membrane, respectively. In B, the neutrophil has fused with the worm, and a hybrid membrane that contains components of both original membranes has formed. The worm inner membrane (1) and the unfused neutrophil membrane are both intact. In C, the neutrophil has lysed, and a fragment (f) of the neutrophil membrane remains attached to the worm by the fusion. The neutrophil membrane components have moved from the hybrid membrane into the unfused portions of the outer membrane. Panel D corresponds to the freeze-fracture micrograph in Figure 7.9. Host molecules acquired in this manner may protect the parasite from recognition as foreign by the immune system.

exist. Since adult worms ingest red blood cells, they also might be able to transport intact glycolipid antigens of erythrocytes from their gut to the tegumental surface. Furthermore, glycolipid antigens can be acquired from serum.

Each cell type examined—eosinophils, neutrophils, monocytes, and erythrocytes—is lysed on the surface of the parasite. With the first three cell types, studying the mechanism of lysis is difficult because the cells themselves have cytolytic molecules. It therefore is not clear whether lysed effector cells represent a form of suicide or homo-

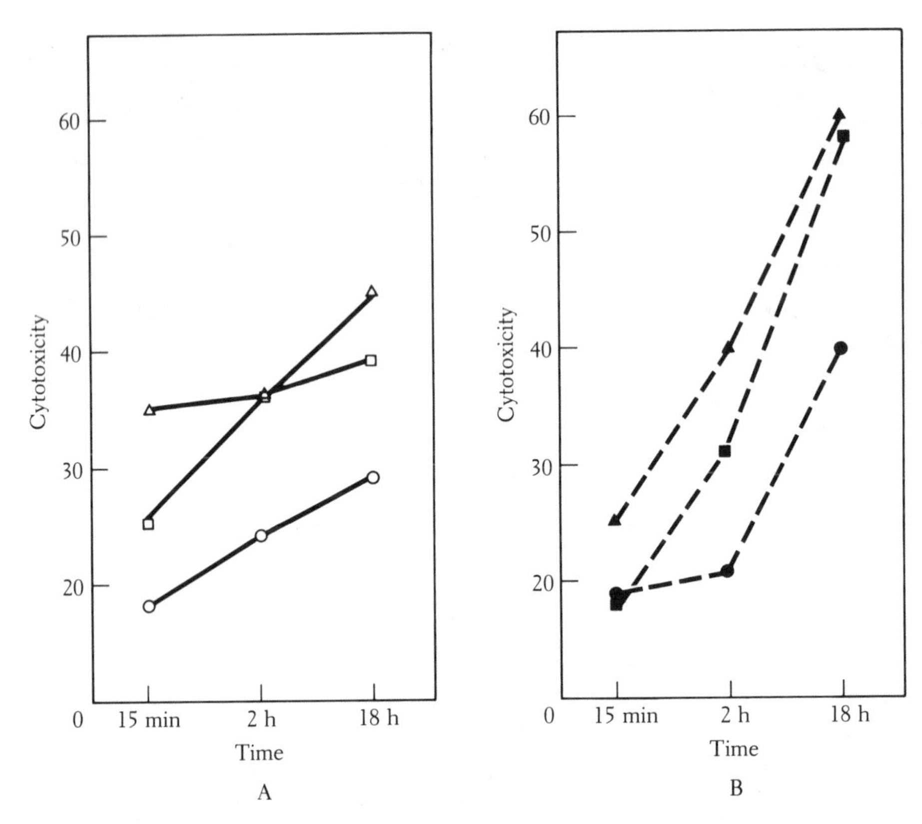

FIGURE 7·11 Demonstration of the cytotoxic effect of schistosomula on human monocyte-derived macrophages. The ordinate shows percentage of specific killing of the cells in the presence of schistosomula after total killing was corrected for spontaneous death of the cells alone. At each point for each experiment the cell death in cultures without schistosomes was subtracted from the cell death in cultures with schistosomes. The abscissa indicates the incubation time. The left panel (A) shows three experiments performed in the absence of antischistosome antibodies; the right panel (B), three experiments performed in the presence of antibodies. Each point is the mean of 12 counts, the standard deviation for each point is less than 5 percent. (△, ○, and □ each represent an experiment done on a different day; ▲, ●, and ■, experiments performed on the same day as the corresponding open marker.) The recovery of monocyte-derived macrophages in the absence of schistosomula compared to the starting number of cells was 100 percent at 15 minutes and 82 ± 4 (S.E.) percent at 2 and 18 hours. (Reproduced from *Am. J. Path.* 1988, 131:146–155 by permission of J. B. Lippincott and Company.)

cide, with the parasites as the murderer. Erythrocytes, on the other hand, do not have such autocytolytic capacities. Erythrocyte membranes bound to parasites have been studied by fluorescence photobleaching recovery (FPR) methods to further address the role of the parasite in host cell lysis. In this technique, a fluorochrome is inserted into the red cell membrane. For example, either glycophorin or band 3 can be selectively

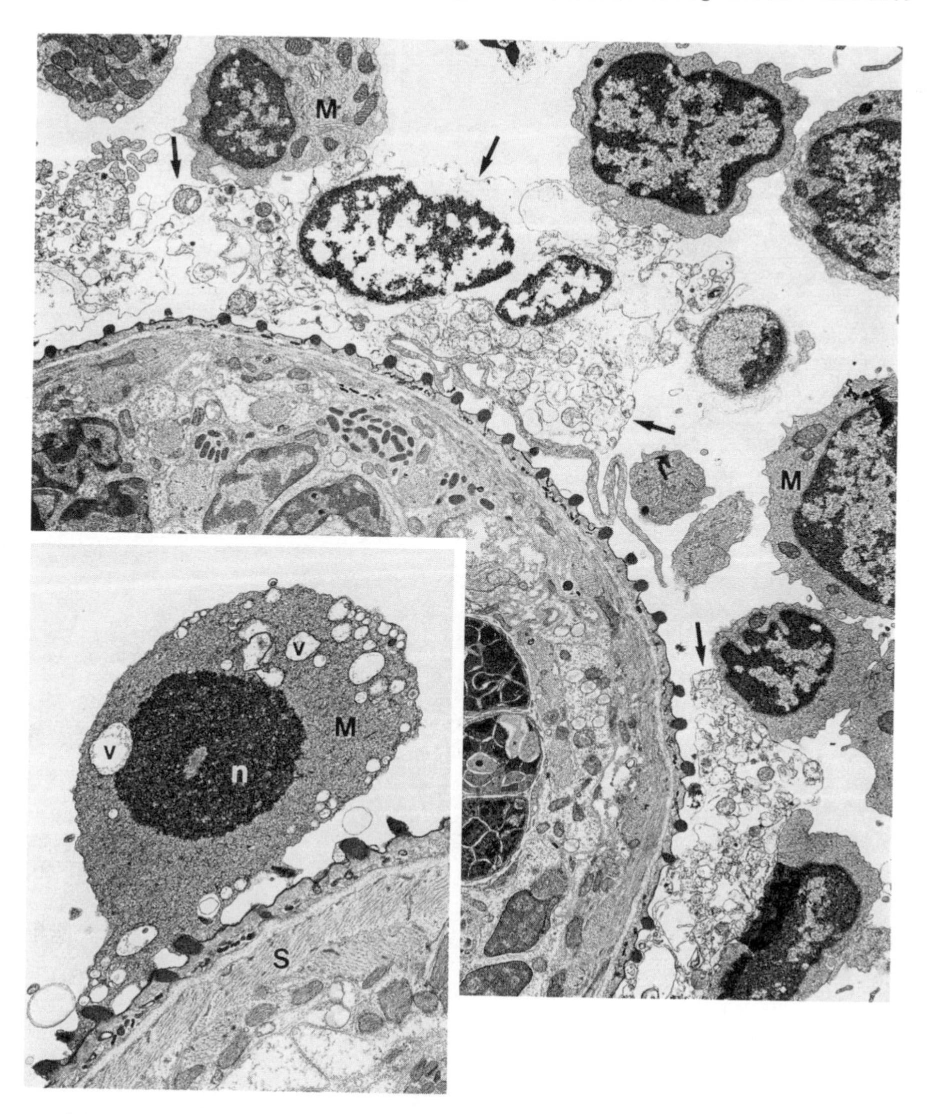

FIGURE 7·12 Monocyte-derived macrophages (M) incubated with antibody-coated schistosomula (S) for 15 minutes. Note that most of the cells adjacent to the worm are lysed (arrows). The inset shows a mononuclear cell attached to a schistosomulum. Note the pyknotic nucleus (n) and vacuoles (v) that indicate the cell is dying. ×7,000. Inset ×10,500. (Reproduced from *Am J. Path.* 1988, 131:146–155 by permission of J. B. Lippincott and Company.)

fluorescinated or the phospholipid analogue, fluorescein phosphatidylethanolamine (Fl-PE), can be partitioned into the membrane. After the labeled erythrocyte adheres to the worm, a laser is used to bleach the fluorochrome in a small spot on the surface of the cell. If the molecules bearing the fluorochromes are able to diffuse in the plane of the membrane, then the fluorescence will reappear at the bleached spot. Two measure-

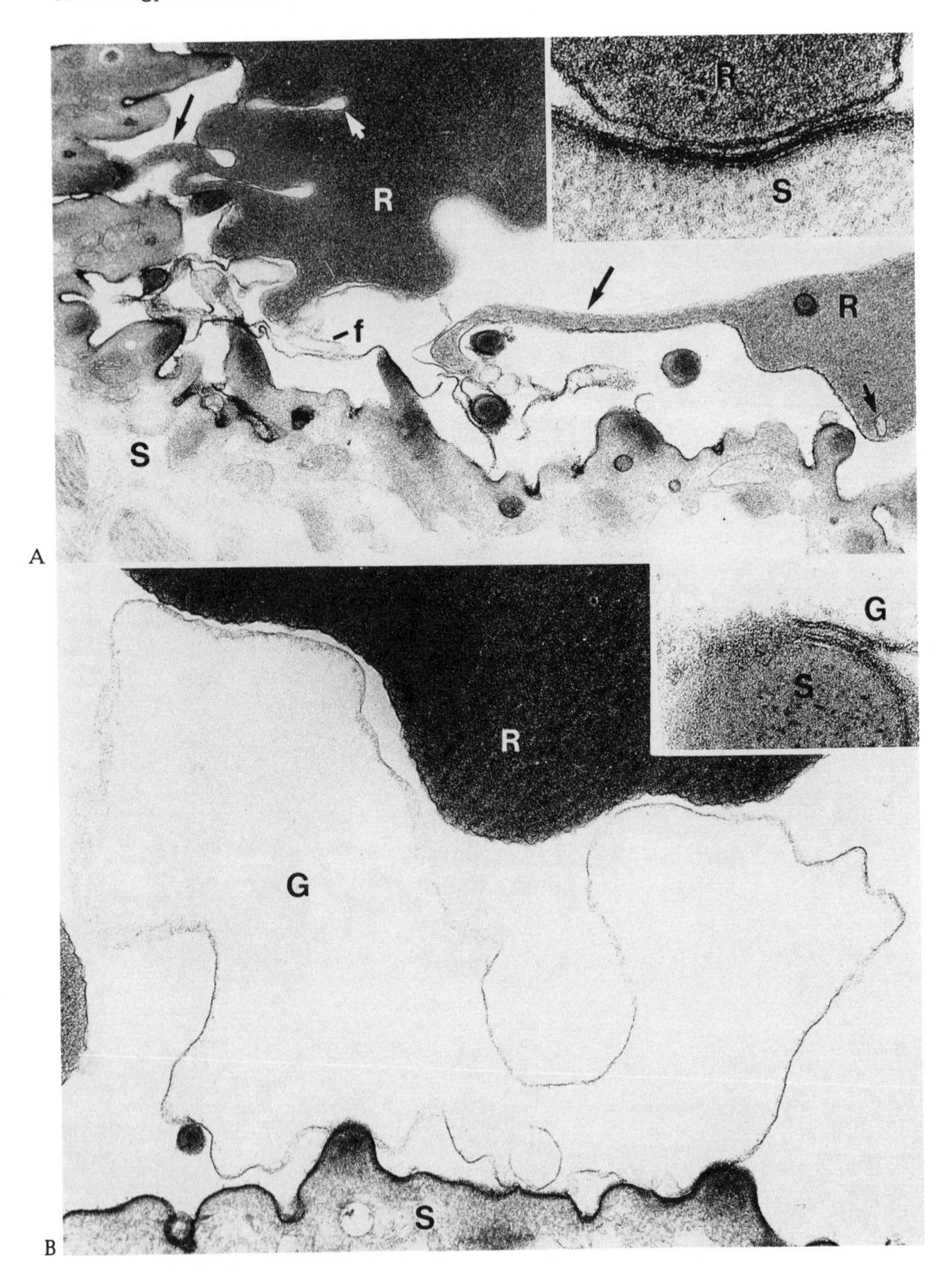
R
S
R
f
R
S
A
G
S
R
G
S
B

ments of recovery can be obtained: the mobile fraction, or percentage of labeled molecules that are able to move, and the diffusion coefficient, or rate of movement of the fluorochrome into the bleached area.

In such experiments, both glycophorin and band 3 are totally immobilized in lysed red cell membranes adherent to the worm (i.e., they do not move into the bleached spot). The phospholipid probe, Fl-PE, is partially immobilized (50 percent mobile fraction). These results obtained with red-cell membranes adherent to worms contrast with what is observed in studies of intact red-cell suspensions, in which the mobile fraction of glycophorin and band 3 is about 60 percent and that of Fl-PE is 100 percent. Red-cell-membrane ghosts produced by hypotonic lysis exhibit enhanced membrane movement, with mobile fraction of the proteins nearly 100 percent. Relative immobilization of the proteins observed on the parasite-associated erythrocyte membrane can be reproduced by cross-linking the protein in intact erythrocytes with lectins, polylysine, or diamide. Under these conditions there is no decrease in Fl-PE mobile fraction, however.

The only treatment of labeled erythrocytes that causes red cell lysis and alterations of the FPR measurements of all three test probes in a manner similar to those found in red cells adherent to the worm is the addition of exogenous lysophosphatidylcholine (lyso-PC). Interestingly, lyso-PC is also produced by worms metabolically labeled with palmitate or choline. Furthermore, the parasite releases lyso-PC into the culture medium in quantities sufficient to account for the FPR measurements. Together, these observations and the results of photobleaching experiments strongly suggest that the parasite is capable of lysing erythrocytes attached to its surface by the generation and release of lyso-PC. Perhaps the lysis of effector cells (eosinophils, neutrophils, and macrophages) is also due to the action of lyso-PC.

Coda

The schistosome appears to operate according to three general principles: (1) I'll take anything I can get; (2) If you want it back, you can have it; (3) If you get too close, I'll

---

Figure 7·13 A. Transmission micrograph of RBCs (R) incubated with the lectin wheat germ agglutinin (WGA)-coated schistosomula (S) for 22 hours. The RBC membrane is distorted into long projections (long arrows). Membrane also appears internalized into the RBC (short arrows), and membrane fragments (f) are present on the parasite surface. The inset shows the adherence of an RBC to a WGA-coated schistosomulum after 3 hours of incubation. The RBC plasma membrane and the double tegumental membrane of the parasite are separated by an ~200 Å gap containing electron-dense material. ×29,000. Inset, ×130,000. B. Transmission micrograph of an RBC ghost adherent to a WGA-coated schistosomulum (S) after 22 hours of incubation. The inset shows a high-power view of the attachment of a ghost to the parasite. (Compare with A inset: R, RBC, ×33,000. Inset, ×130,000.) (Reproduced from the *Journal of Cell Biology,* 1985, 101:158–166 by copyright permission of the Rockefeller University Press.)

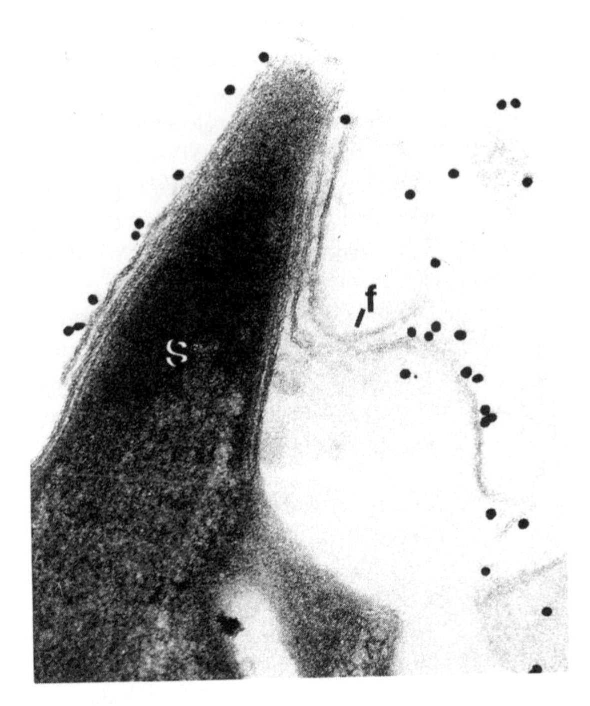

FIGURE 7·14 Transmission micrograph of the distribution of rabbit antihuman red blood cell membrane (RαHRBC). RBCs (R) were incubated with the lectin wheat germ agglutinin-coated schistosomula (S) for 22 hours and reacted with RαHRBC and gold-conjugated goat antirabbit IgG. The colloidal gold is seen on red cell membrane fragments (f) but not on the parasite membrane. The distribution of gold on the fragment does not lead to any conclusions about the sidedness of the antigens in the membrane because of the polyclonal nature of the RαHRBC. ×100,000. (Reproduced from the *Journal of Cell Biology,* 1985, 101:158–166 by copyright permission of the Rockefeller University Press.)

bite. The parasite outer membrane appears to be deceptively simple, containing mainly phospholipids and cholesterol derived from the host as well as adsorbed host proteins and lipids. The true parasite molecules, proteins and antigenic glycolipids, are probably only a small fraction of the outer membrane. The outer membrane can be shed rapidly because it is energetically cheap to make. It is also poorly antigenic because so much of it is derived from the host.

The presence of lyso-PC in the outer membrane or the ability to generate this molecule explains many of the membrane properties. Lyso-PC is a detergent very similar to sodium dodecyl sulfate. When its critical micelle concentration is reached in the membrane, proteins are solubilized in lyso-PC micelles or mixed micelles formed from lyso-PC and the other phospholipids and cholesterol. This property of lyso-PC explains the shedding of both membrane proteins and ligands from the parasite. Endocytosis by neutrophils of ligands such as antibody and lectins and the membrane components to which they are bound can be seen as an extension of this process. Lyso-PC is a fusigen, which accounts for the fusion of neutrophils and monocytes with the outer membrane. Finally, as suggested by its name, lyso-PC is a powerful lytic agent and is most likely responsible for the lysis of various cells on the parasite surface. The success of eosinophils in attacking the parasite probably owes to the high levels of lysophospholipase on their surface and in their granules. This enzyme cleaves lyso-PC into a fatty acid and glyceryl phosphorylcholine, thereby inactivating it.

Many important questions about the function and molecular architecture of the tegumental membranes remain unanswered. In particular, it is not clear whether the outer membrane has high steady-state concentrations of lyso-PC or generates it enzymatically in a controlled manner or in response to stimuli. It is also not clear how the inner membrane is separated from the outer membrane, how the membranes maintain their identity, and how the inner membrane functions as a true biological membrane when covered by an apparently intact separate lipid bilayer. Finally, the biogenesis of these membranes needs further study to determine if the models based on morphological data — namely, assembly and transport in multilamellar bodies — are correct.

For Additional Reading, see Chapter 6.

8

Cell Biology of *Entamoeba histolytica* and Immunology of Amebiasis

Jonathan I. Ravdin

Entamoeba histolytica is an enteric protozoan infecting 10 percent of the world's population, resulting in approximately 50 million cases of invasive colitis or liver abscess and 50,000 to 100,000 deaths per year. Considering that *E. histolytica* is the third leading parasitic cause of death, behind malaria and schistosomiasis, most scientists focus their efforts on the aspects of cell biology, biochemistry, and host immune response most relevant to understanding how this organism causes disease and the development of strategies to prevent it. This chapter will do the same.

Epidemiology and Clinical Disease Syndromes

Infection with *E. histolytica* is acquired by ingestion of the acid-resistant cyst form, with a single cyst being a potentially infective dose. Transmission of amebic infection occurs by any means that allows fecal-oral contamination. Examples are fecal contamination of water supplies or food, poor levels of hygiene in the institutionalized mentally retarded population, oral-anal sexual practices, or infected food preparers with poor hygienic standards.

In the vast majority of individuals, infection of the large bowel by *E. histolytica* does not result in invasive disease as manifested by symptomatic colonic ulceration, a serum antiamebic antibody response, or liver abscess. The fraction of infected individuals who suffer from amebiasis is not uniform throughout the world. In Mexico City up to 25 percent of infected individuals have symptomatic invasive disease, a figure probably 25 times greater than the incidence of colitis or liver abscess in infected sexually active male homosexuals in New York City, San Francisco, or London. The parasite and host factors that determine the occurrence of amebic invasion are unclear and an area of great interest and controversy. Apparently, most individuals spontaneously eradicate noninvasive *E. histolytica* infection over a period of 6 to 12 months. It is unknown if clearance of asymptomatic amebic infection is followed by host resistance to rechallenge; it probably is not, as there is an increasing prevale ce of intestinal infection with increasing age, in contrast to infection with *Giardia lamblia*, where prevalence peaks at age 1 – 5 years and then declines.

Invasive colonic disease usually presents with bloody diarrhea and abdominal pain, and less often with fever. The parasite is found in flask-shaped ulcerations penetrating down to muscle tissue (Figure 8-1). Amebae can form localized chronic lesions (amebomas) or can present in a fulminant form with total colonic involvement. An indolent presentation of amebiasis is very difficult to diagnose and can persist for years.

Amebic liver abscess is actually a misnomer, as the so-called abscess is filled with proteinaceous debris rather than pus (polymorphonuclear cells). The amebae are found at the periphery of the lesion; multiple lesions are frequently seen. Patients who develop liver abscess usually do so within 3 months of acquisition of intestinal infection. Interestingly, less than 20 percent of individuals have *E. histolytica* within their intestine when they manifest symptoms of a liver abscess, and they usually do not develop antiamebic serum antibodies until they have had symptoms for 7 days. How the parasite penetrates the gut and gains access to the portal venous system and the liver without immediately causing symptoms or eliciting a host humoral response is a mystery.

Life Cycle and Phylogeny

In the life cycle of *E. histolytica*, the infective cyst form can reside outside the host for weeks to months, depending on environmental conditions. After ingestion, excystation occurs in the small bowel. The factors that induce excystation are not totally defined; in vitro excystation requires low oxygen tension, inorganic salts, and appropriate osmotic conditions for culture of trophozoites. A mature quadranucleate cyst forms a metacystic stage, and eight trophozoites result after cytoplasmic division. The trophozoite resides exclusively in the large bowel; again, apparently most often as a commensal organism. Invasion is limited to the trophozoite form; alternatively, the parasite can encyst with excretion of up to 45 million cysts per day per individual. The cyst wall contains chitin (a glucosamine and N-acetyl-D-glucosamine polymer) as shown by intense staining with the fluorescent dye Calcofluor M_2R and binding by glucosamine-recognizing plant lectins, such as wheat germ agglutinin. A potential strategy for interrupting the life cycle

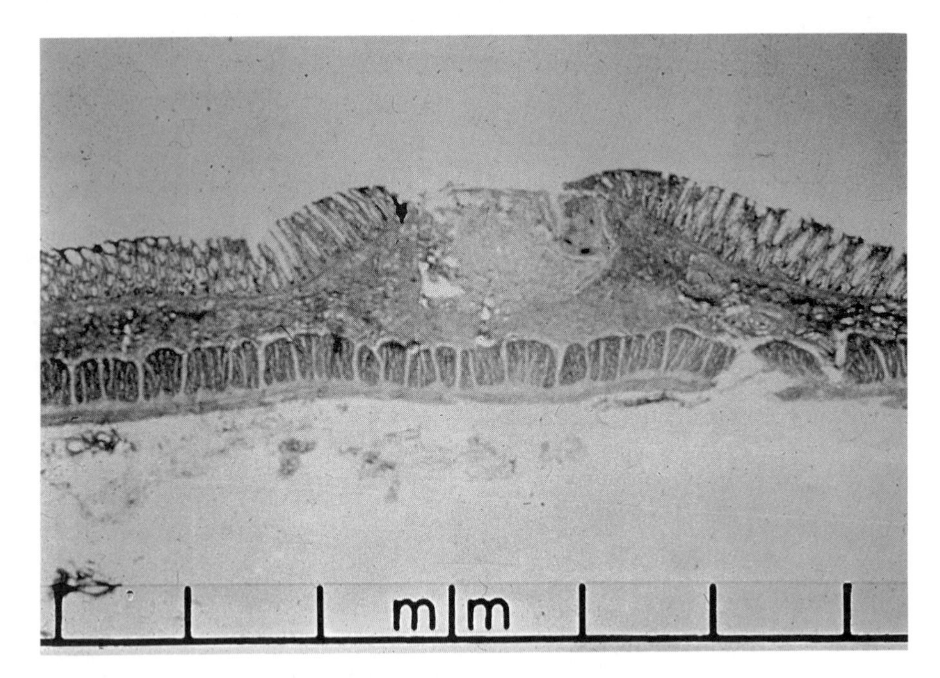

FIGURE 8·1 Light micrograph demonstrating a flask-shaped ulceration in a pathologic specimen from a patient with severe colonic amebiasis.

of *E. histolytica* is use of chitin synthase inhibitors such as Polyoxin D and Nikomycin, which are structural analogs of the chitin precursor UDP-N-acetylglucosamine.

Entamoeba histolytica belongs to the family Entamoebidae of the order Amoebida in the subphylum Sarcodina of the pseudopod-forming protozoan superclass Rhizopoda. Within the genus *Entamoeba* are two recognized species that infect the large bowel of humans, *E. histolytica* and *Entamoeba hartmanni*; other *Entamoeba* include *E. moshkovskii*, *E. invadens*, *E. terrapinae*, *E. knowlesi*, *E. aulostomi*, and *E. gingivalis*. At present, taxonomy of *Entamoebae* is based on morphology, in vitro growth characteristics, virulence, antigenic distinctness, amino acid content, drug susceptibility, and host specificity. The degree of DNA hybridization or sequence homology between species has not been determined.

A presently controversial area is the use of isoenzyme analysis to identify *E. histolytica* strains with distinct pathogenic abilities. Starch gel electrophoresis of *E. histolytica* of clinical isolates, with associated bacterial flora, from patients with invasive amebiasis has been found to have characteristic mobility patterns for hexokinase, phosphoglucomutase, L-malate, $NADP^+$ oxidoreductase, and glucose phosphate isomerase when compared to amebae cultured from asymptomatic individuals who have noninvasive intestinal infection. Some investigators interpret this data to indicate distinct pathogenic and nonpathogenic strains of *E. histolytica* and to indicate that individuals infected with nonpathogenic strains are not at future risk for invasive amebiasis and do not pose a risk to others. I believe there are insufficient data to support this assumption. The long-term follow-up of individuals with "nonpathogenic" infection has been inadequate, cross-infection with stability of zymodeme pattern has not been evaluated, and recent studies with a closed nonpathogenic isolate demonstrated changes in zymodeme to a pathogenic pattern during in vitro association of the ameba with irradiated bacterial flora obtained from a patient who had amebic colitis. The utility of zymodeme analysis and designation of strain specificity awaits rigorous studies of DNA homology and a more complete understanding of the molecular regulation of parasite pathogenicity.

Structure

E. histolytica trophozoites are 10 to 60 micrometers (μm) long and contain a single nucleus of 3 – 5 μm; cysts average 12 μm with one to four nuclei, which are morphologically identical to the nucleus of the trophozoite. The nuclei contain chromatin granules that are evenly distributed on the interior of the nuclear membrane. Trophozoite cytoplasm contains numerous vacuoles; amebae that are highly virulent in vivo or cytopathogenic in vitro have a greater degree of vacuolization. Cysts contain glycogen

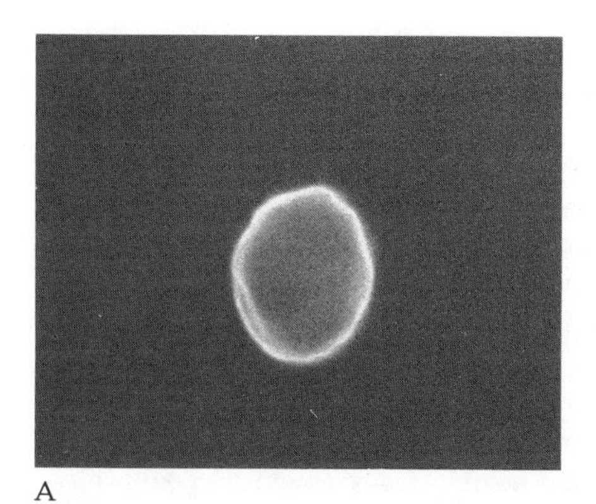

A

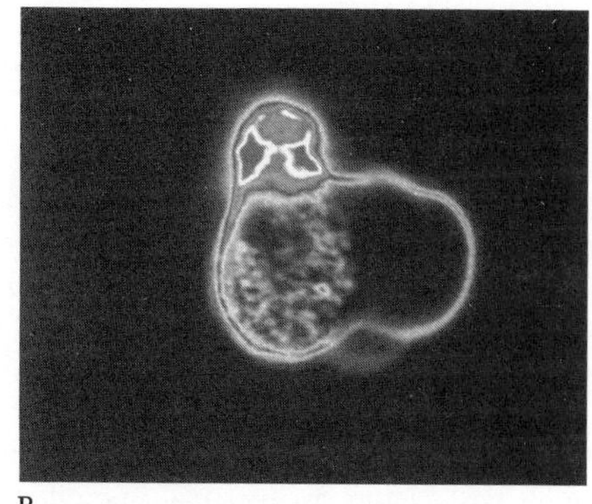

B

FIGURE 8·2 Cell-surface capping of the adherence lectin of *Entamoeba histolytica* detected by indirect immunofluorescence with the F14 monoclonal antibody; (A) 4°C; (B) 20°C, 5 minutes. Combined phase contrast and epifluorescence was computer enhanced from a Zeiss Axiomat with a Quantex 9210 image processor with the use of background subtraction, frame averaging, and grey scale expansion. (Reprinted from the *Journal of Clinical Investigation* 1987, 80:1238–1244 by copyright permission of the American Society of Clinical Investigation.)

granules or aggregates that stain with iodine and chromatoidal bars, rounded rod-shaped masses of ribosomes.

Scanning and transmission electron microscopy reveal that *E. histolytica* trophozoites have a surface glycocalyx that binds Concanavalin A and ruthenium red. At the uroid area extension of multiple filopodia is noted. The cell ribosomes within exist in a free or aggregated form; the typical Golgi apparatus is absent, and it is unclear how the parasite processes proteins. Microfilament cytoskeletal structures have been identified; the gene for actin is the first and presently the only gene cloned from *E. histolytica*. Trophozoites rapidly aggregate and "cap" attached Concanavalin A or antiamebic antibodies (Figure 8-2); microfilament inhibitors such as cytochalasin D prevent surface "capping." Mitosis has not been observed in *E. histolytica* despite numerous studies of in vitro cultivated organisms.

Pathogenicity

Pathogenesis can be defined as the origination and development of a disease. The concepts of disease causation by *E. histolytica* include (1) colonization of the gut by a virulent amebic strain with adherence to the intestinal mucus layer, (2) disruption of intestinal barriers by enzymes or toxic products, and (3) lysis of intestinal cells and host inflammatory cells, leading to mucosal interruption, colonic ulcers, and deep-tissue and/or distant-organ (liver) invasion (Figure 8-3).

Intestinal Colonization and *E. histolytica* Adherence Mechanisms

Intestinal infection with *E. histolytica* is species specific, with humans and Old World primates the only reservoir for natural infection. Although various animal models have been studied, including those of dogs, cats, rats, guinea pigs, and gerbils, no model of intestinal disease directly parallels the course of human disease, that is, ingestion of trophozoites, followed by colonic disease, encystation, and transmission. Host species specificity may be related to intrinsic amebic properties (i.e., adherence proteins) or to such host factors as intestinal microflora and colonic redox potential. Younger age increases susceptibility of animals to infection with axenic trophozoites, a finding that is consistent with the rapidly fatal invasive amebiasis reported in infants. The mechanisms underlying increased pathogenicity in younger animals are unclear; perhaps host factors (e.g., secretory antibody, intestinal mucosal changes) are important.

During intraluminal intestinal infection there is a close association and probable nutritional dependence of *E. histolytica* on the gut bacterial flora. In models of intestinal disease the in vivo virulence of *E. histolytica* depends on the presence of viable bacteria. As previously summarized, recent experiments suggest that in vitro association of *E.*

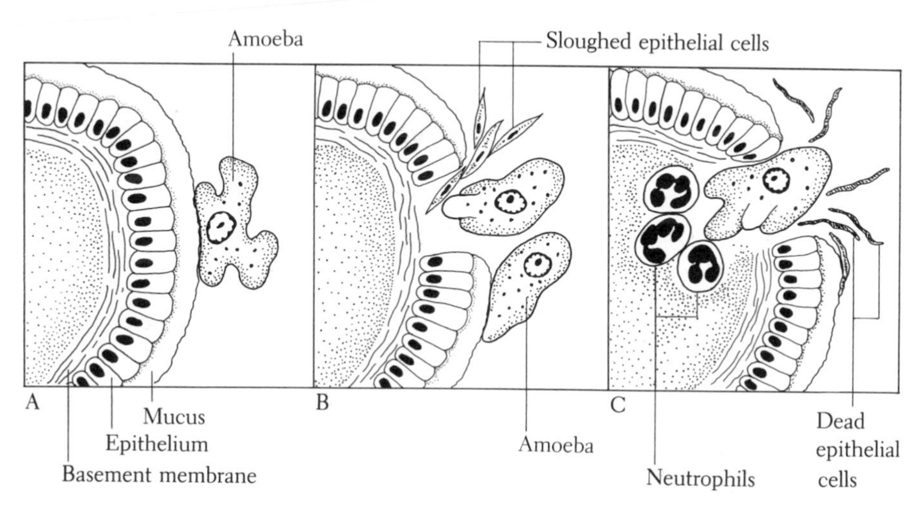

FIGURE 8·3 Pathogenesis of invasive amebiasis. Discrete steps in the disease pathogenesis include adherence to the intestinal mucus barrier (A), depletion of the mucus layer and disruption of the integrity of the intestinal mucosa (B), penetration of the trophozoite with lysis of adherent epithelial cells and responding host inflammatory cells (C).

histolytica trophozoites with bacteria from a human host with invasive colonic amebiasis results in a change of pathogenic phenotype. The mechanisms by which bacteria alter parasite pathogenicity are unclear.

Bacteria with mannose-binding substances on their cell surface adhere to and are ingested by trophozoites; amebae also adhere to non-mannose-binding bacterial species opsonized with specific antibacterial antibody. In addition, *E. histolytica* trophozoites adhere to and ingest nonopsonized bacteria that contain galactose or N-acetyl-D-galactosamine (GalNAc) residues on their surface, such as *Escherichia coli* 055 or *Salmonella greenside* 050. An increase in the in vitro virulence of axenic *E. histolytica* trophozoites follows their brief (30-minute) incubation with live adherent bacteria. Bacteria inactivated by heating, fixed in glutaraldehyde, or exposed to inhibitors of protein synthesis do not augment parasite cytopathogenicity or in vivo virulence; in addition, metronidazole inhibits bacterium-mediated augmentation of virulence. One possible explanation for this phenomenon is that bacteria function as scavengers of oxidized molecules and stimulate the parasite's electron transport system. Whether bacteria induce gene switching or alter mRNA transcription or translation in *E. histolytica* is unknown. Bacteria may also lower the oxidation-reduction potential in the gut lumen, thereby facilitating amebic growth and pathogenicity.

In sequentially studied animal models, *E. histolytica* trophozoites adhere to the colonic mucosa before invasion. Colonic biopsies of patients have demonstrated amebae in the bowel lumen or adherent to the mucus layer. Depletion of the colonic mucus layer is apparently necessary for amebic penetration to the epithelial barrier. In vitro studies of adherence of *E. histolytica* trophozoites to target cells, colonic mucosal-organ culture preparations, and purified colonic mucins have identified the relevant parasite adherence molecule.

A galactose or N-acetyl-D-galactosamine (Gal/GalNAc)-inhibitable *E. histolytica* lectin mediates in vitro adherence. Millimolar concentrations of galactose or GalNAc, but not numerous other monosaccharides, inhibit adherence of axenic trophozoites to Chinese hamster ovary (CHO) cells. The galactose terminal glycoprotein, asialoorosomucoid (ASOR), is more effective than asialofetuin and 1000-fold more effective than galactose in inhibiting amebic adherence to CHO cells (Figure 8-4). The amebic Gal/GalNAc lectin also mediates parasite adherence to erythrocytes, mononuclear cells, and numerous other mammalian cells in tissue culture and to opsonized bacteria or bacteria with galactose-containing lipopolysaccharide.

Adherence of [^{3}H]thymidine-labeled axenic *E. histolytica* trophozoites to fixed rat and human colonic mucosa in vitro (Figure 8-5) is specifically inhibited by GalNAc, galactose, and asialofetuin and human (10%) and rabbit (5%) immune sera, but not

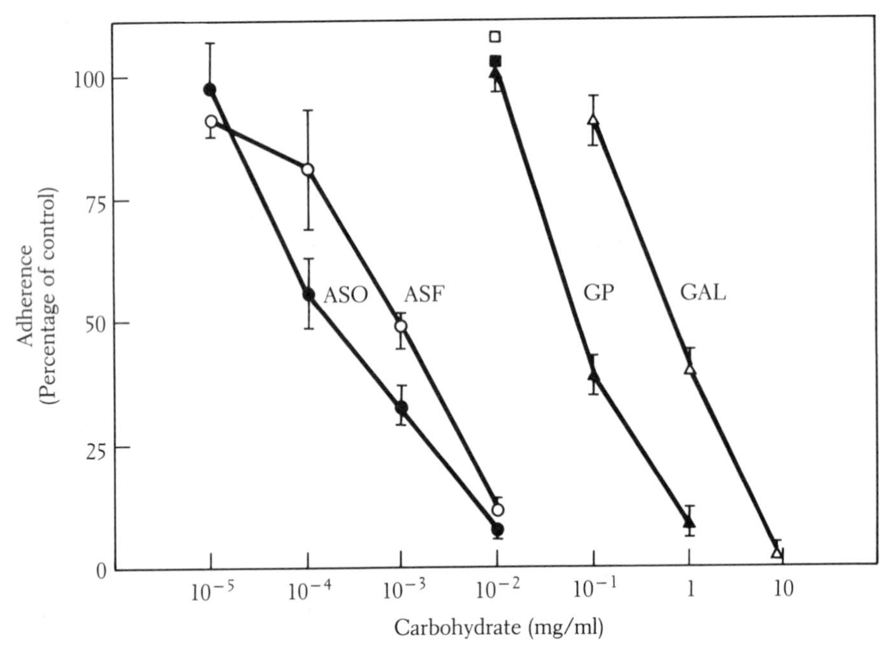

FIGURE 8·4 Carbohydrate inhibition of amebic adherence to target cells. Adherence was expressed as percentage of adherence in paired studies performed in control medium (where 60 percent of amebae had at least three adherent CHO cells upon suspension of the pellet by vortexing). Bars are mean ± S.E. of 6–11 determinations. (Gal, △) galactose; (GP, ▲) asialofetuin glycopeptide; (F, ◇) fetuin; (OR, ◆) orosomucoid; (ASF, ○) asialofetuin; (ASOR, ●) asialoorosomucoid. (Reprinted from the *Journal of Clinical Investigation* 1987, 80:1238–1244 by copyright permission of the American Society of Clinical Investigation.)

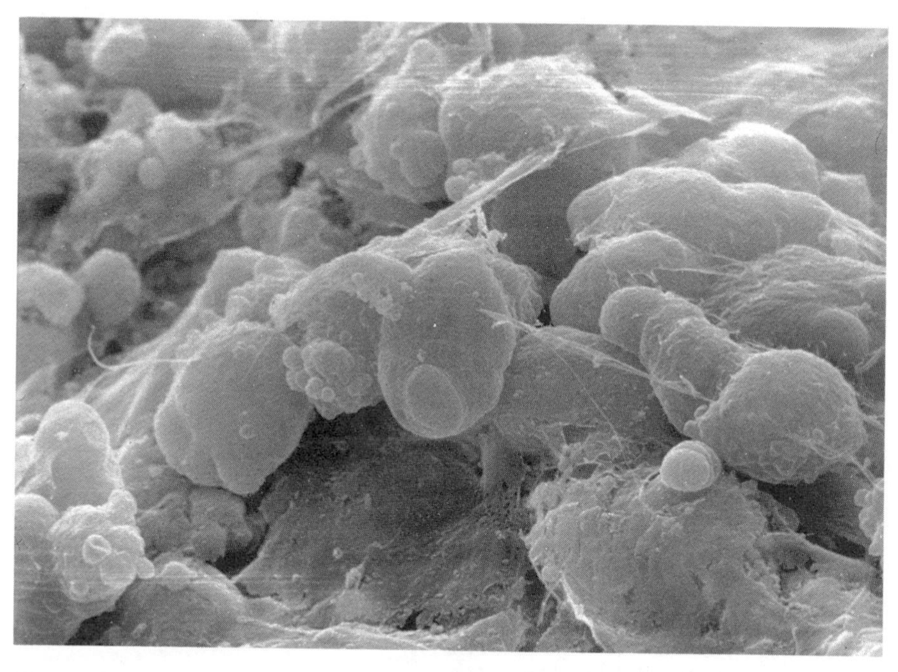

FIGURE 8·5 Scanning electron micrograph of amebic trophozoites adherent to unfixed rat colonic mucosa. Note strands of material, possibly mucus, adherent to amebae. (Reprinted with permission from *Infect. Immun.* 1985, 48:292–297.)

control sera. GalNAc also inhibits adherence of amebae to unfixed rat submucosa. Therefore, the amebic Gal/GalNAc lectin mediates adherence to the colonic mucosa and submucosa; antiamebic antibodies are inhibitory, suggesting that gut antiamebic sIgA could have a protective role.

The isolation of the *E. histolytica* Gal/GalNAc-inhibitable adherence protein was accomplished by producing adherence inhibitory monoclonal antibodies and taking advantage of the protein's carbohydrate-binding properties. When ^{35}S-methionine metabolically labeled amebic proteins from a culture filtrate or detergent-solubilized (octylglucoside) trophozoites are applied to an ASOR affinity column and the column is washed, a peak of ^{35}S activity can be eluted with galactose (Figure 8-6). SDS-PAGE of the eluate under reducing conditions with autoradiography demonstrates 35- and 170-kilodalton (kD) metabolically labeled amebic proteins. The purifed galactose binding protein was confirmed to be the *E. histolytica* Gal/GalNAc adherence lectin by three procedures: (1) application of ^{35}S methionine metabolically labeled amebic proteins to an adherence inhibitory monoclonal antibody–affinity column results in elution of the same amebic proteins, (2) the most adherence-inhibitory monoclonal antibody, F-14, exclusively recognizes by immunoblotting the 170-kD protein purified by ASO affinity chromatography, and lastly, (3) the Gal/GalNAc lectin purified by H8-5

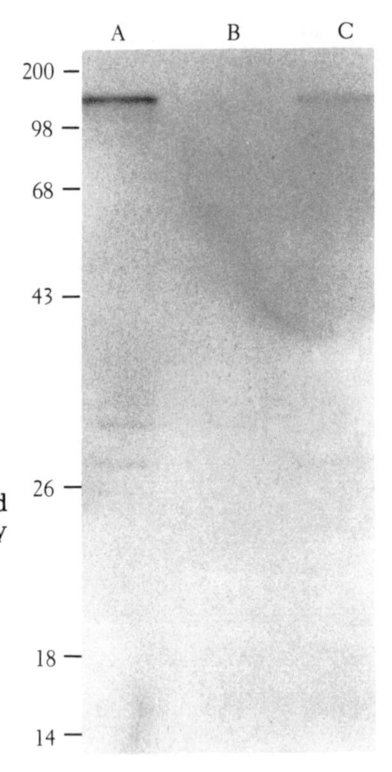

Figure 8·6 PAGE of the isolated amebic adherence lectin. SDS-PAGE autoradiograph of ^{35}S-methionine metabolically labeled amebic proteins isolated by (A) asialoorosomucoid (ASOR) -affinity chromatography of conditioned culture medium, (B) orosomucoid affinity chromatography of conditioned culture medium, and (C) ASOR-affinity chromatography of octylglucoside solubilized amebae. The galactose-eluted fractions were boiled in SDS and β-mercaptoethanol and analyzed by electrophoresis in a Tris/glycine discontinuous SDS-polyacrylamide gel. (Reprinted from the *Journal of Clinical Investigation* 1987, 80:1238–1244 by copyright permission of the American Society of Clinical Investigation.)

immunoaffinity chromatography binds to CHO cells in a galactose-specific manner and competitively inhibits subsequent adherence by viable amebae. In addition, polyclonal antibodies produced by immunization with immunoaffinity-purified lectin specifically recognize the 170-kD lectin subunit on immunoblotting and completely inhibit parasite in vitro adherence to CHO cells. As will be discussed, the Gal/GalNAc lectin is the major antigen recognized by human immune serum, further supporting its central role in pathogenicity.

Surface expression of the Gal/GalNAc adherence lectin depends on parasite protein synthesis rather than on recirculation through an acid pH endocytic vesicle system. The lectin is also the major protein released from viable trophozoites in vitro. The secreted and membrane-associated forms of the Gal/GalNAc lectin account for 26 percent of secreted and 6 percent of total metabolically labeled amebic proteins, as determined by SDS-PAGE, autoradiography, and laser densitometry. A lectin subunit structure has been identified: 35- and 170-kD metabolically labeled proteins are present only when immunoaffinity-purifed lectin is subject to SDS-PAGE under reducing conditions (BME); under nonreducing conditions the lectin has an approximate mobility on SDS-PAGE of 260 kD. The 170-kD heavy subunit mediates attachment, as indicated by its exclusive recognition by adherence inhibitory monoclonal and polyclonal antibodies; the function of the nonimmunogenic 35-kD light subunit is unknown.

The relevant high-affinity receptors for the *E. histolytica* Gal/GalNAc lectin are human colonic mucins. Colonic mucins are rich in galactose and GalNAc; purified human and rat colonic mucins have been found to inhibit amebic adherence to CHO cells and to homologous rat colonic epithelial cells (Figure 8-7). Oxidation and enzymatic cleavage of the mucins' galactose and GalNAc residues abrogate its inhibitory activity. The binding of iodinated rat colonic mucins to axenic *E. histolytica* trophozoites is galactose-specific, saturable, reversible, and pH-dependent. Analysis of ^{125}I-mucins binding to *E. histolytica* trophozoites by the nonlinear least-square technique, assuming a mucin molecular weight of 9×10^5 daltons (D), reveals 2.8×10^3 binding sites per ameba with a dissociation constant (kD) of 8.2×10^{-11}. The low number of apparent binding sites is undoubtedly due to cross-linking of multiple lectin molecules by the large mucin species. A monoclonal antibody (F-14) specific for the 170-kD heavy subunit of the Gal/GalNAc adherence lectin completely inhibits the binding of rat ^{125}I-mucins to amebae. Finally, rat mucins bound to a solid support can affinity-purify the amebic lectin from parasite-conditioned medium. Therefore, colonic mucin

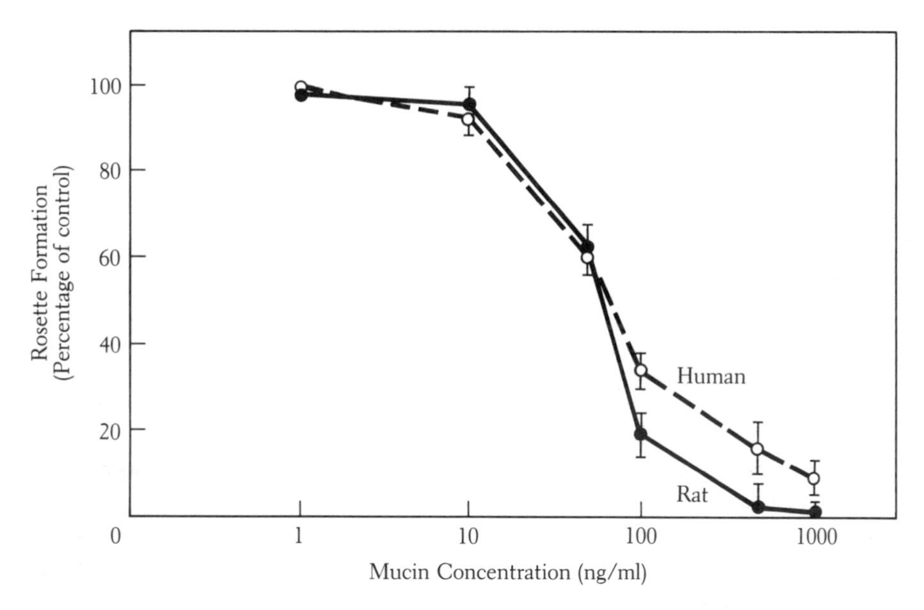

FIGURE 8·7 Inhibition of amebic adherence to CHO cells by purified rat and human colonic mucin glycoproteins. Absolute values for rosette formation in control studies was 77 ± 5.55%. At ≥50 ng/ml, there was inhibition of amebae adherence, $P < .001$ compared with control for each. (Reprinted from the *Journal of Clinical Investigation* 1987, 80:1245–1254 by copyright permission of the American Society of Clinical Investigation.)

glycoproteins act as an important host defense by binding to the *E. histolytica* Gal/GalNAc adherence lectin, thereby preventing amebic attachment to and cytolysis of host epithelial cells. However, adherence to colonic mucins may facilitate intestinal colonization by amebae; inhibition by antilectin secretory IgA antibody could be a potent host defense against *E. histolytica* luminal infection.

An additional amebic lectin has also been described, but its role in pathogenesis is less well defined. Hemagglutinating activity in sonicates of *E. histolytica* is active at acid pH (5.7 and not 7.2) and best inhibited by the trimer or tetramer of N-acetyl-D-glucosamine (chitotriose). This chitotriose-inhibitable lectin was recently isolated by sepharose 4B chromatography and electroelution from SDS-PAGE. The purified 220-kD protein agglutinates human erythrocytes in a chitotriose-inhibitable manner and competes with binding of trophozoites to monolayers of MDCK cells. Monoclonal and polyclonal antibodies raised to the purified chitotriose lectin bind to the surface of viable trophozoites, but only partially inhibit amebic adherence to erythrocytes or MDCK cells.

It is important to note that although galactose or GalNAc can completely inhibit amebic adherence to CHO cells at 4°C, these monosaccharides do not prevent parasite phagocytosis of viable target cells. The possibility that additional parasite surface molecules are involved in phagocytosis is strongly supported by studies of monoclonal antibodies selected by their ability to inhibit *E. histolytica* erythrophagocytosis. These antibodies recognize a 112-kD amebic protein on immunoblotting; this putative *E. histolytica* adherence protein has not been further characterized.

E. histolytica Secretory Enzymes and Toxins

Numerous proteolytic enzymes have been isolated from *E. histolytica*, all of which are inhibited by serum and may require direct parasite contact for release onto host tissue.

Hyaluronidase activity has been demonstrated in multiple strains of *E. histolytica*; however, enzyme activity is not correlated with pathogenicity. Other enzyme activities of both pathogenic and nonpathogenic invasive strains include trypsin, pepsin, gelatinase, and hydrolytic enzymes for casein, fibrin, and hemoglobin, all of which are serum-inhibited. *E. histolytica* has a cathepsin B proteinase of 16 kD, pH 5.0 optimum, which is inhibited by serum β-2 macroglobulin, and has a reversible cell-releasing cytopathic effect on tissue culture cells. As is typical of cysteine proteinases, its activity for specific substrates is inhibited by p-chloromercuribenzoate. Two axenic strains (HM1 and Rahman) were found to have more proteinase activity per milligram of amebic protein than three other less virulent strains (HK-9 and *Entamoeba*-like Laredo and Huff strains).

E. histolytica contains a serum-inhibitable collagenase activity with greater specificity for type I than type III collagen. Collagenase activity can be demonstrated only when amebae make direct contact with collagen substrate. Collagen-lytic activity has correlated with reported virulence in three strains of axenic amebae (HM1, H-200-NIH, and HK-9). A -N-acetyl-glucosaminidase has been purified from two axenic strains of *E. histolytica* (HK-9 and H-200-NIH) with the use of culture supernatant. This enzyme could have a role in the breakdown of phagocytized erythrocytes or disruption of glycoprotein bonds between adjacent intestinal mucosal epithelial cells.

A 56-kD-secreted neutral thiol proteinase has been isolated from *E. histolytica* by FPLC anion exchange and chromatofocusing chromatography. In vitro, this enzyme degrades connective-tissue extracellular matrix and specifically fibronectin, laminin, and type I collagen. The 56-kD proteinase is also cytopathic for culture monolayers, probably by degrading cellular anchoring proteins; it is also a plasmogen activator. Parasite proteinases probably contribute to amebic invasive activities and also complicate attempts by scientists to purify nondegraded *E. histolytica* proteins.

Amebic secretion of a cytotoxin or enterotoxin has been postulated as a mechanism contributing to human disease. Although human pathology and experimental animal studies indicate that colonic mucosal damage occurs only at the site of adherent amebae, such studies do not exclude an enterotoxin effect.

Soluble fractions of *E. histolytica* sonicate have been demonstrated to have enterotoxic activity in indomethacin, 0.1 milligrams per microgram (mg/μg), -treated rats and rabbits. Enterotoxigenic activity is inhibited by fetuin or high concentrations of indomethacin. An *E. histolytica* enterotoxin was recently partially purified and found to be heat labile; to be inhibited by sialoglycoproteins (fetuin and mucin) and by p-chloromercuribenzoate; and to have a molecular weight on SDS-PAGE of 30,000. Lysates of *E. histolytica* applied to the serosal side of rabbit ileum in a Ussing chamber induce a transient increase in short circuit current, a finding suggestive of a secretory effect. This effect can be inhibited with the calcium channel blocker verapamil; amebic lysates contain serotonin, at 4.0 nanograms per milligram (ng/mg), which is at least partly responsible for secretory effects. Whether viable amebae elaborate an enterotoxin or produce an intestinal secretory response upon adherence to or invasion of the submucosa remains to be determined.

E. histolytica Cytolytic Activity

The characteristic pathology of invasive amebiasis consists of trophozoites surrounded by a necrotic eosinophilic debris (Figure 8-8), presumably the result of amebic lysis of surrounding tissues. In vitro studies of parasite cytolytic activity have contributed to our understanding of this event. By cinemicroscopy, kinetic analysis, and suspension of cells in dextran, studies a number of years ago demonstrated that the in vitro cytolethal effect of *E. histolytica* occurs on contact with target cells. The sequence of events in the interaction of axenic trophozoites with target cells has been established: amebic adherence is followed by cytolysis of the adherent target cell, which in turn is followed by parasite phagocytosis of the dead target cell.

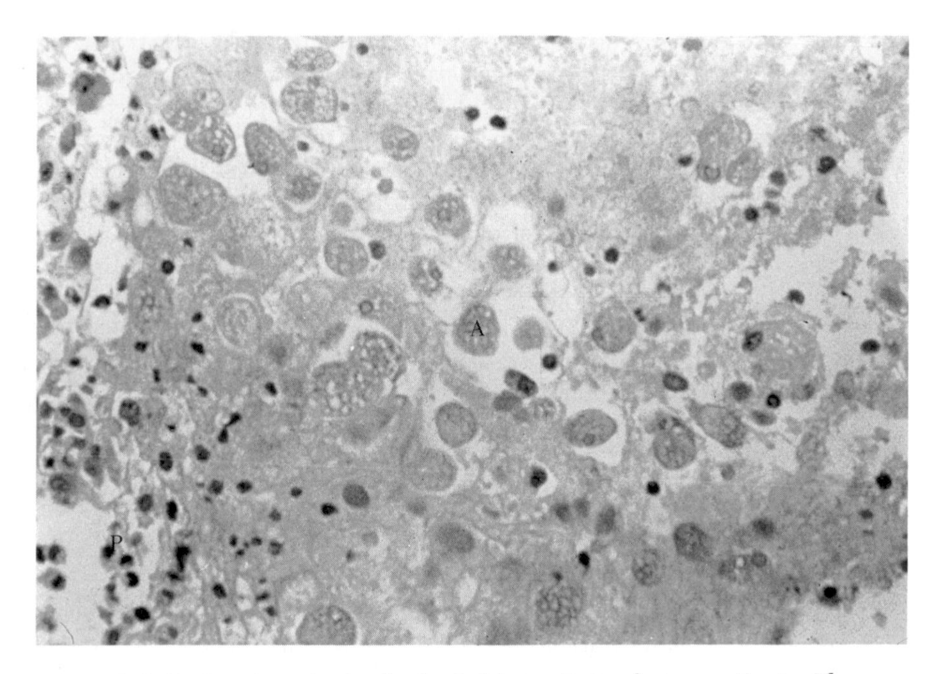

Figure 8·8 Light micrograph of colonic biopsy tissue from a patient with invasive amebic colitis, showing the dissolution of recognizable tissue architecture immediately surrounding numerous *E. histolytica* trophozoites (A) and the acute inflammatory response (P) at the margin of this "microabscess" (hematoxylin and eosin, ×400; courtesy of Dr. Charles Schleupner, Veterans Administration Hospital, Salem, Va.). (Reprinted from R.L. Guerrant et al., *J. Infect. Dis.* 1981, 143:83–93 by permission of the University of Chicago Press.)

In vivo and in vitro studies demonstrate that *E. histolytica* provides a chemoattractant signal for polymorphonuclear neutrophils. Light microscopy studies of the early events in the formation of experimental amebic liver abscess in the hamster suggested that parasite contact-dependent lysis of neutrophils has a deleterious effect on distant hepatocytes. The in vitro interaction of trophozoites with human neutrophils reveals that amebae lyse neutrophils on contact (Figure 8-9) and that release of toxic nonoxidative neutrophil constituents enhances the parasite-mediated destruction of Chang liver cell monolayers. Soluble *E. histolytica* proteins inhibit human neutrophil baseline and stimulated oxidative activities; this finding may account for the parasite's relative immunity to neutrophil oxidative effector activity. Therefore, attraction and contact-dependent cytolysis of neutrophils is an important component in the pathogenesis of invasive amebiasis.

Lysis of target cells in vitro requires the establishment of adherence by the parasite's Gal/GalNAc-inhibitable surface lectin. Galactose or GalNAc (10–40

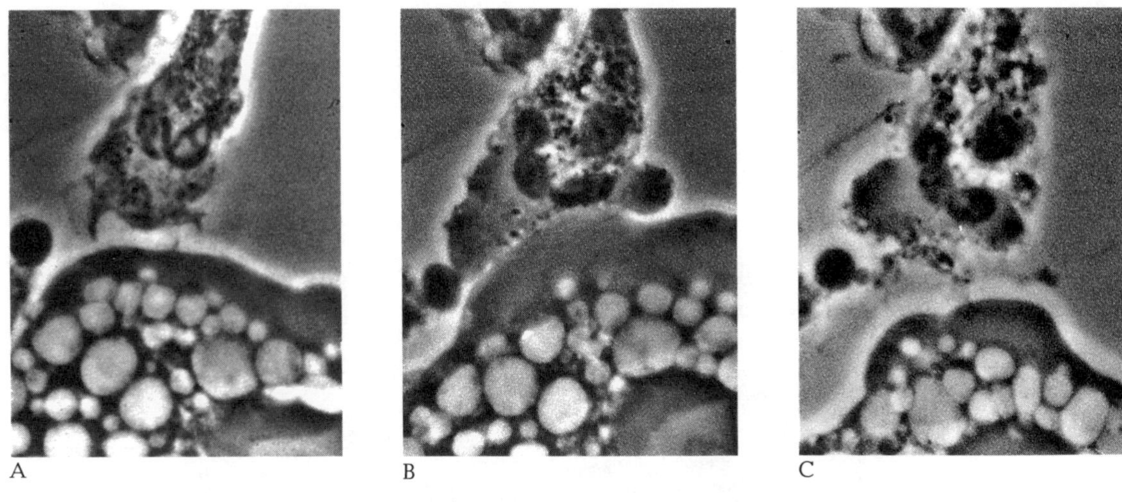

FIGURE 8·9 Time-lapse, phase-contrast cinemicrographs of a polymorphonuclear neutrophil approaching and being killed by an axenic *E. histolytica* trophozoite, strain HM1-IMSS. A PMN approaches the ameba (A), establishes contact (B), and undergoes membrane blebbing and granule disappearance (C) (Zeiss photomicroscope, Sage cinemicrographic apparatus, Bolex 16mm camera, ×2000). (Reprinted from J. I. Ravdin and R. L. Guerrant, *Rev. Infect. Dis.* 1982, 4:1185–1207 by permission of the University of Chicago Press.)

mg/ml) completely inhibits amebic lysis of target CHO cells, as quantified by release of ^{111}Indium oxine or counting in a cell sorter (Figure 8-10). Purified rat colonic mucins, which specifically bind to the Gal/GalNAc lectin, totally inhibit amebic cytolysis of isolated rat colonic epithelial cells. GalNAc, 45 millimolar (m*M*) concentration, inhibits the killing of neutrophils by axenic trophozoites, allowing the neutrophils to turn table and kill the amebae. Amebic killing of activated human monocyte-derived macrophages, T lymphocytes, and BHK cell monolayers has also been found to be galactose-inhibitable.

The Gal/GalNAc lectin molecule may be a cytotoxin and thus directly contribute to target cell death. As discussed later, purified lectin induces a rise in target cell free intracellular calcium concentration. The Gal/GalNAc lectin is also found to be mitogenic for human lymphocytes, as shown by specific inhibition with asialofetuin, coelution of mitogenic and CHO-cell agglutinating activity from a gel filtration column, correlation of mitogenicity with lectin activity in different *E. histolytica* strains, and asialofetuin-inhibitable mitogenicity of immunoaffinity-purified Gal/GalNAc lectin. Further studies on the subunit structure and cytolytic activities of the lectin are necessary.

The relationship of the Gal/GalNAc lectin to virulence was further studied by using axenic strains which express different levels of in vivo virulence and in vitro

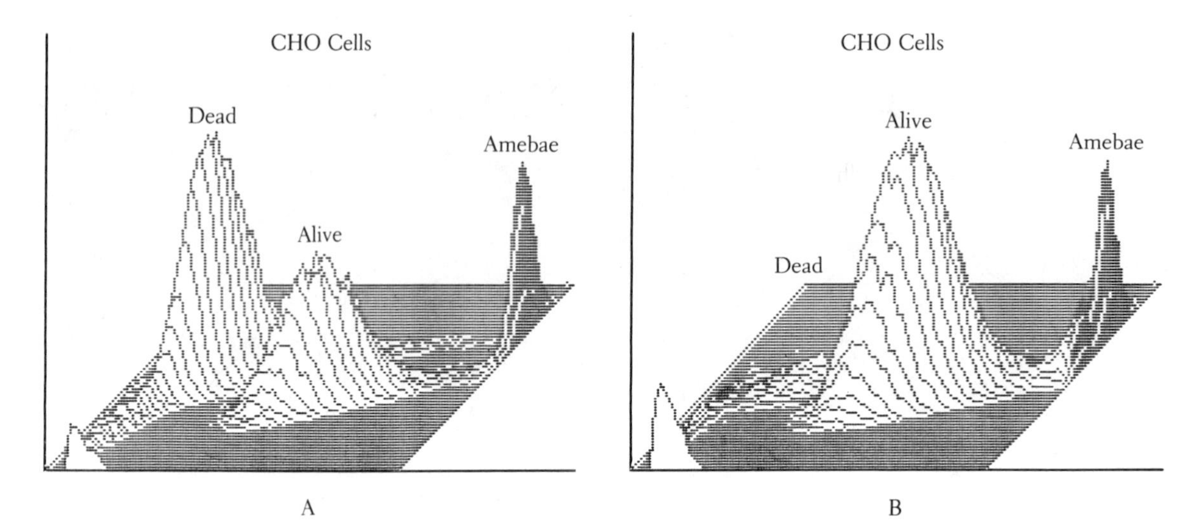

FIGURE 8·10 Galactose inhibitable extracellular cytolysis of target CHO cells by *E. histolytica* trophozoites, as determined by cell sorting. Target CHO cells (2×10^5) with or without amebae (10^4) were centrifuged and incubated for 2 hours at 37°C with (B) or without (A) galactose (20 mg/ml). After incubation, cells were placed on ice, galactose (10 mg/ml) was added to all tubes, and the cells were vortexed to elute adherent CHO cells from amebae. Data are presented as populations of viable CHO cells and dead but still intact CHO cells, as separated by size and density in a cell sorter (ordinate, cell number). With galactose present (B), amebic extracellular cytolysis of target CHO cells is completely inhibited. (Reprinted with permission from *Pathol. Immunopathol. Res.* 1989, 8:179–205.)

cytopathogenicity. In vitro virulence for target CHO cells and human neutrophils (PMNs) in four strains is HM1-IMSS > H303-NIH = 200-NIH > nonvirulent Laredo. Gel filtration chromatography of a soluble fraction of amebic sonicate demonstrates a GalNAc-inhibitable lectin that agglutinates CHO cells, erythrocytes, and PMNs. HM1 amebae contained greater specific lectin activity (agglutinating activity per milligram of protein) and lymphocyte mitogenic activity than did 303, 200, and Laredo amebae. Mutagenized *E. histolytica* clones (L_6, C_{93}, C_{919}), selected by their inability to ingest bacteria, are avirulent in vivo, and adhere to and kill fewer CHO cells than do the parent HM1:IMSS strain. Galactose (10 mg/ml) completely inhibits adherence of all mutants to CHO cells. Avirulent mutants are more susceptible to being killed by human neutrophils; their soluble protein preparations are also markedly less mitogenic for human lymphocytes and have lower CHO cell–agglutinating activity. Indirect immunofluorescence or immunoprecipitation with Gal/GalNAc lectin-specific monoclonal antibodies (F-14 and H8-5) reveals 170-kD Gal/GalNAc lectin heavy subunit in all clones studied. ^{125}I-mucins bind in a saturable and galactose-specific manner to avirulent amebic clones, and the number of mucin binding sites and dissociation constants was comparable to HM1:IMSS. However, all the avirulent clones studied are markedly deficient in internalization and subsequent exocytosis of ^{125}I-mucins. They also exhibit a lower turnover rate of surface lectin molecules, as indicated by studies with cyclohexamide exposure and ^{125}I-mucin binding. Therefore, amebic in vivo virulence and in vitro adherence and cytolytic activities depend not only on the presence of functional surface Gal/GalNAc lectin molecules, but also on their organization on the surface membrane and rate of parasite membrane turnover.

Cytolysis of adherent target cells by *E. histolytica* depends on parasite microfilament function, phospholipase A (PLA) enzyme activity, and maintenance of an acid pH in endocytic vesicles. Parasitic cytolytic activity can be stimulated by phorbol esters and is preceded by a rapid rise in target cell but not amebic free intracellular calcium concentration ($[Ca^{++}]_i$).

E. histolytica trophozoites have been found to contain two PLA enzymes: a Ca^{++}-dependent pH 7.5 optimal and a Ca^{++}-independent pH 4.5 optimal enzyme. The Ca^{++}-dependent enzyme is highly associated with the surface plasma membrane. The molecular weight of both PLA enzymes has been estimated to be 38 kD by gel filtration chromatography and immunoblotting with PLA-specific antibodies. Isoelectric focusing with transfer to nitrocellulose and immunoblotting with anti-PLA antibodies revealed two distinct bands with isoelectric points of 4.5 and 5.5.

Known PLA inhibitors, such as Quinacrine, phosphatidylcholine, and Rosenthal's inhibitor, inhibit parasite cytolysis of CHO cells and the in vitro enzymatic activity of the *E. histolytica* Ca^{++}-dependent PLA. Production of lysophospholipids in amebae (end products of the PLA-mediated hydrolysis of diacylphospholipids), as quantified by thin-layer chromatography of ^{32}P incorporation into diacylphospholipid, is increased during parasite adherence to and lysis of target CHO cells. Addition of anti-PLA serum to amebae and CHO cells at 4°C has no effect on adherence. However, exposure of amebae to heat-inactivated anti-PLA serum for 1 hour at 37°C inhibits subsequent amebic adherence at 4°C; control horse sera has no effect. Anti-PLA inhibited amebic killing of CHO cells. These studies indicate that amebic PLA has a role in promoting parasite adherence and a central role in amebic cytolysis of target cells.

E. histolytica vesicle pH has been measured with fluorescein isothiocyanate linked to dextran; endocytic vesicle pH is 5.1 ± 0.2 by spectrofluorimetry. Concentrations of NH_4Cl sufficient to increase vesicle pH to ≥ 5.7 (Figure 8-11) inhibit amebic killing of target CHO cells. A similar inhibitory effect is observed with the weak bases primaquine and chloroquine. Ammonium chloride affects neither amebic adherence to CHO cells at 4°C nor binding and ingestion of ^{3}H-leucine-labeled bacteria by the parasite. Exposure to NH_4Cl for 48 hours (which has no effect on amebic protein synthesis) or to the protein-synthesis inhibitor cyclohexamide for 3 hours produces persistent inhibition of amebic cytolytic activity. Therefore, an uninterrupted acid pH in amebic intracellular endocytic vesicles is necessary for the delivery and activity of lysosomal proteins that participate in parasite cytolytic activity.

Exocytosis of acid pH vesicles may be required for *E. histolytica* cytolytic activity. Vesicle exocytosis has been studied with the use of release of endocytosed ^{125}Iodine

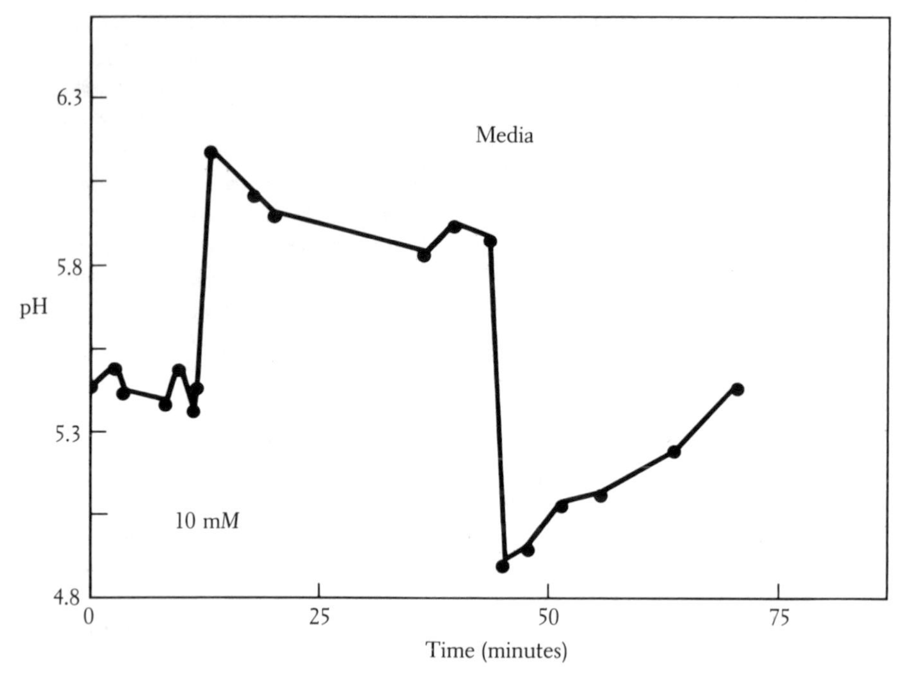

FIGURE 8·11 Effect of NH_4Cl on *E. histolytica* vesicle pH. In this preparation with a baseline vesicle pH of 5.4, NH_4Cl (20 m*M*) produced a rapid rise in vesicle pH of 6.1. When NH_4Cl was removed from the medium by washing, the vesicle rapidly reacidified (hyperacidified) to pH 4.9 and slowly returned to baseline over the next 20–30 minutes. (Reprinted with permission from *J. Protozool.* 1986, 33:478–486.)

(^{125}I)-labeled tyrosine-conjugated dextran. ^{125}I-dextran enters the acid pH vesicles of HM1 amebae and is not degraded. Exocytosis is temperature-dependent and inhibited at 37°C by cytochalasin D, EDTA, or the putative intracellular calcium antagonist TMB-8. Calcium ionophore A23187 and phorbol myristate acetate enhance amebic exocytosis; elevation of vesicle pH with NH_4Cl has no effect. Centrifugation of amebae with target CHO cells results in decreased ^{125}I-dextran release into the supernatant after 30 or 60 minutes at 37°C; interestingly, ^{125}I-dextran release is at control values when galactose or GalNAc is added, but remains decreased in the presence of mannose and N-acetyl-D-glucosamine (which have no effect on amebic adherence and cytolysis). Phagocytosis of inert particles such as serum-exposed latex beads also has no effect on dextran release from acid pH vesicles. In summary, exocytosis of *E. histolytica* acid pH vesicles is temperature-, microfilament-, and calcium-dependent, stimulated by phorbol esters, and apparently down-regulated or directed into target cells during parasite adherence and cytolytic events.

Phorbol esters are specific activators of protein kinase C (PKC) enzymes, which are important in the intracellular transduction of many receptor-mediated calcium-dependent signals. Addition of phorbol 12-myristate, 13-acetate (PMA) at 10^{-6} or 10^{-7} *M* results in a greater than twofold enhancement in amebic killing of target CHO cells over the next 30 minutes. Prior exposure of only the amebae, but not the CHO cells, to PMA for 5 minutes produces a similar effect; the inactive analogue, 4-α-phorbol, has no effect. PMA does not promote amebic adherence but does enhance amebic cytolysis of previously adherent target cells. Sphingosine, a specific inhibitor of PKC, totally blocks PMA stimulated or basal amebic cytolytic activity. These studies indicate that an *E. histolytica* PKC is involved in regulation of parasite cytolytic activities; the mechanisms responsible for the natural in vivo activation of amebic PKC are unknown but should relate to the binding to the target cell by the Gal/GalNAc adherence lectin.

The role of Ca^{++} ions in *E. histolytica* cytolytic activity has been recently investigated. The calcium chelator EDTA and the putative intracellular calcium antagonist TMB-8 inhibit amebic cytolytic activity. In addition, calcium channel blockers verapamil and bepridil inhibit amebic cytolytic activity and increase parasite surface membrane impedance. Hypotheses recently tested were that Ca^{++} functions as a second messenger stimulating amebic cytolytic activity and that entry of Ca^{++} from the extracellular compartment is the final mechanism responsible for target cell death. $[Ca^{++}]_i$ was measured in individual amebae and target cells (CHO cells and human PMN) utilizing the Ca^{++} probe Fura-2 and computer-enhanced digitized microscopy. Motile *E. histolytica* trophozoites demonstrate random cyclic increases in $[Ca^{++}]_i$ at the leading edge or tail regions of cell cytoplasm. There is no increase in regional or total amebic $[Ca^{++}]_i$ upon contact with a target CHO cell. Target CHO cells and PMN exhibit a marked irreversible rise in $[Ca^{++}]_i$ within 30 – 300 seconds after contact by an ameba (Figure 8-12). This increase in $[Ca^{++}]_i$ precedes the development of nonspecific target cell membrane permeability, as determined by leakage of Fura-2. Target CHO cells contiguous on a monolayer to a cell contacted by an ameba demonstrate a rapid but reversible rise in $[Ca^{++}]_i$ and do not die. Galactose totally abrogates the rise in target CHO cell $[Ca^{++}]_i$ that occurs upon contact by amebae. Purified Gal/GalNAc lectin (0.25 μg/ml) induces a galactose sensitive rise in CHO cell $[Ca^{++}]_i$ that is comparable to the rise observed with viable amebae. In summary, elevation of amebic $[Ca^{++}]_i$ does not appear to be required for initiation of parasite adherence and cytolytic activities. Adherence by *E. histolytica* induces a rapid and substantial rise in target cell $[Ca^{++}]_i$, which is mediated at least in part by the Gal/GalNAc lectin. Prolonged elevation of target cell $[Ca^{++}]_i$ directly contributes to, but may not be sufficient for, target cell death.

A putative mediator of target cell ion flux is the *E. histolytica* pore-forming protein described a number of years ago. This 13 – 15 kD parasite protein induces rapid ion flux (Na^+, K^+, and to a lesser extent Ca^{++}) across lipid bilayers or vesicles and depolarizes J774 macrophages, erythrocytes, and mouse spleen lymphocytes. The ionophore is normally not released from viable trophozoites; however, secretion can be induced by calcium ionophore A23187, Concanavalin A, and *E. coli* lipopolysaccharide. The pore-forming protein is packaged in dense intracellular aggregates within

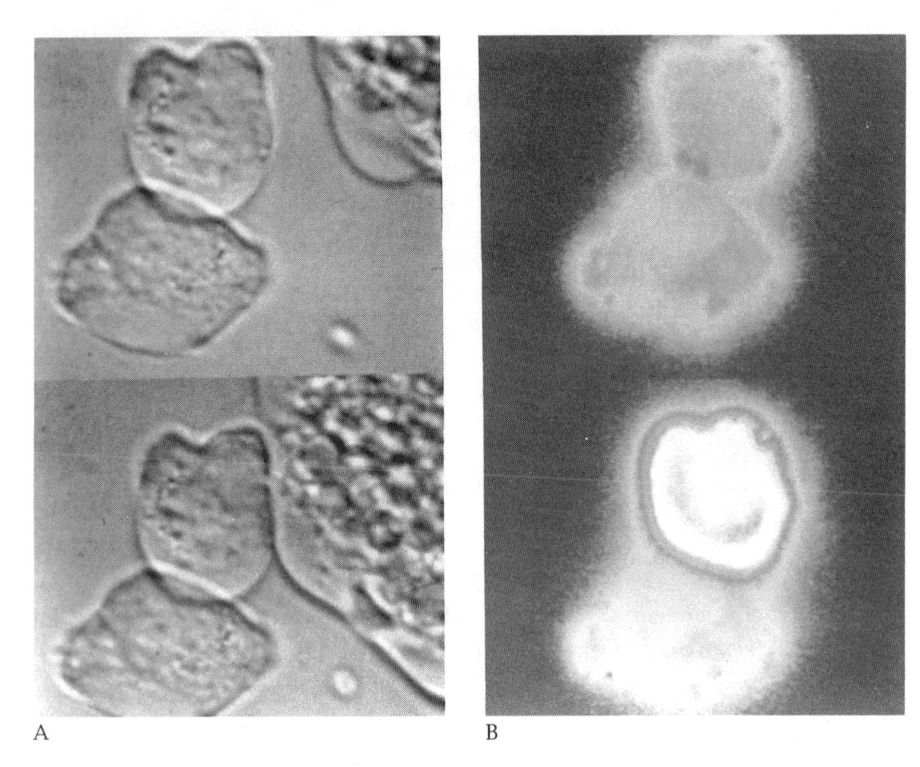

FIGURE 8·12 Photomicrograph of (A) phase and (B) digitized $R_{340/380}$ image of a Fura-2-loaded CHO cell precontact (top) and at 30 seconds a postcontact (bottom) with an *E. histolytica* (Eh) trophozoite. $R_{340/380}$ of Fura-2 fluorescence is directly proportional to the free intracellular Ca^{+} concentration ($[Ca^{2+}]_i$). The two trypsinized CHO cells were in suspension and not directly contiguous; there was a marked rise in $[Ca^{2+}]_i$ only in the CHO cell in direct contact with the ameba, despite the proximity of the other target cell. Bar, 10 μm. (Reprinted with permission from *Infect. Immun.* 1988, 56:1505–1512.)

amebae. Unfortunately, there is no direct evidence as yet linking this cytolytic mediator to the in vitro killing of target cells by *E. histolytica* trophozoites. Potential intracellular functions and the mechanisms by which the parasite avoids harmful ion fluxes at its plasma membrane are unknown.

We do not yet have sufficient information to determine the exact sequence of events and the identification of all the parasite molecules responsible for lysis of target cells by *E. histolytica* trophozoites. A working model is as follows. Adherence initiated by the Gal/GalNAc lectin results in a cellular signal activating parasite protein kinase C, which stimulates amebic cytolytic activities. Parasite phospholipase A activity promotes amebic target cell fusion and is also stimulated by contact with target cells. Disruption of the target cell membrane is preceded by a tremendous influx of extracellular Ca^{++} ions into the target cell, as a result of cytotoxin activities of the Gal/GalNAc lectin or possibly transfer of the amebic pore-forming protein to the target cell membrane. Parasite molecules that are putative final mediators of target cell death include Gal/GalNAc lectin toxic subunits, the pore-forming protein, lysodiacyl-phospholipids that result from amebic PLA_2 activity, and an as yet undefined lysosomal constituent whose activity requires an acid pH.

Human Immune Response to *E. histolytica*

Nonspecific Defense Mechanisms

Barriers to infection by enteric parasites include low gastric pH, digestive enzymes, competition by normal intestinal bacterial flora, and the protective mucus blanket covering the entire length of the gut epithelium. Stomach acid is an effective defense agaisnt trophozoites but not the chitinous cyst form. Bile salts may interact with components of human breast milk to lyse *Giardia lamblia*, but this mechanism is unlikely to be relevant to *E. histolytica* infection. Bacteria promote growth and virulence of *E. histolytica*; however, experiments directed at determining if intestinal species could reduce parasite pathogenicity or interfere with amebic growth have not been performed. Bacterial competition for iron, production of toxic metabolites such as short-chain fatty acids, and effects of bacterial enzymes on trophozoites are potential mechanisms contributing to host resistance.

There is sufficient information to state that colonic mucins serve as a major nonimmune host defense against invasion by *E. histolytica*. As discussed earlier, colonic mucins are the high-affinity receptor for the amebic Gal/GalNAc adherence lectin, and they prevent parasite attachment to and lysis of colonic epithelial cells. Experimental models of invasive colonic amebiasis demonstrate depletion of the mucus blanket before parasite invasion. *E. histolytica* trophozoites show potent mucous secretagogue effects when injected into rat colonic loops, possibly a result of parasite proteolytic enzymes. Immobilization or trapping of trophozoites in colonic mucus may be augmented in immune individuals who experience reinfection, as has been observed in experimental nematode infections.

Complement-Mediated Resistance

E. histolytica trophozoites are susceptible in vitro to complement-mediated lysis through alternative or classical pathway activation. However, invasive amebiasis appears to result from a subpopulation of complement-resistant trophozoites, as shown by study of liver abscess isolates. In addition, one can select complement-resistant clones in vitro by long-term culture with increasing concentrations of serum. The complement membrane attack complex apparently attaches to the ameba's plasma membrane; the mechanism of parasite resistance to complement-mediated lysis is undefined, but it may relate to the ability of trophozoites to rapidly turn over their surface membrane.

Acquired Resistance to Invasive Amebiasis

No adequate prospective controlled studies have yet determined the incidence of invasive amebiasis or intestinal colonization in patients previously cured of amebic colitis or liver abscess. Uncontrolled and anecdotal reports suggest that such individuals have substantial resistance to a recurrence of invasive amebiasis. This is supported by experimental animal models in which cure of invasive infection or immunization with whole parasite proteins results in immunity to subsequent intestinal or intrahepatic challenge with *E. histolytica* trophozoites. There is sufficient data to conclude that both antibody and cell-mediated immune responses contribute to resistance in the immune individuals.

Serum and Intestinal Secretory Antibody Response

Within 1 to 2 weeks from the onset of invasive amebiasis, patients develop a high titer serum antiamebic antibody response, as determined by a variety of quantitative methods. The parasite antigens recognized by human immune sera have been recently defined by immunoblotting (Figures 8-13 and 8-14). The 170-kD heavy subunit of the Gal/GalNAc adherence lectin is the most prominently recognized antigen, as determined by immunoprecipitation (Figure 13); in addition, virtually all patients studied who have been cured of well-documented invasive amebiasis have serum antibody to the lectin. This finding is important because specific antilectin antibody inhibits both in vitro adherence of amebae to colonic mucins or target epithelial cells and parasite destruction of CHO cell monolayers at 37°C. The latter observation is important, as trophozoites have the ability to recurrently aggregate and shed attached antibodies (Figure 8-2). Other major amebic antigens recognized by immune sera (Figure 4) include 37- and 90-kD surface glycoproteins and a 59-kD cytoplasmic glycoprotein. Uncharacterized coproantibodies of the IgA class have been identified in patients with amebic colitis; it is unknown whether an sIgA secretory antibody response occurs in asymptomatic intestinal infection. Development of a serum antibody response does not result in spontaneous resolution of established amebic colitis or liver abscess, but it certainly may contribute to resistance to subsequent invasive disease.

Cell-Mediated Immune Responses

Patients cured of amebic liver abscess demonstrate an antigen-specific T-lymphocyte proliferative response with production of lymphokines, including gamma interferon, capable in vitro of activating monocyte-derived macrophage amebicidal activity. Axenic

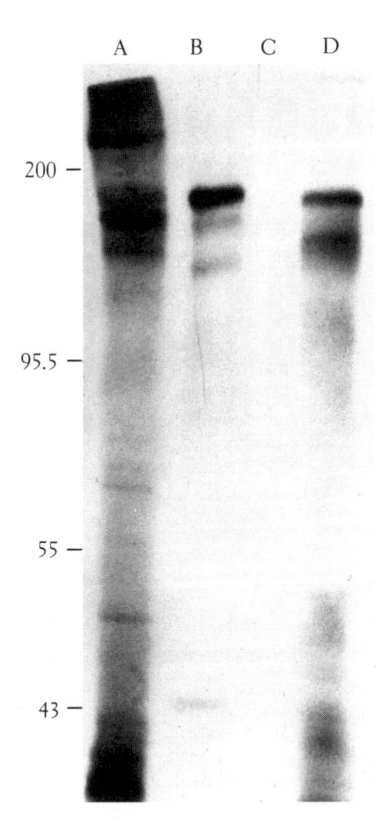

FIGURE 8·13 Immunoprecipitation of [^{35}S] methionine metabolically labeled amebic protein with antilectin monoclonal antibody and human immune serum. Autoradiograph of SDS-PAGE (10% acrylamide separating gel) of [^{35}S] methionine-labeled amebic proteins (A) immunoprecipitated with monoclonal antibody H8-5 (B), normal human serum (C), or pooled human immune serum (PHIS) (D). A 170-kD metabolically labeled amebic protein was immunoprecipitated by lectin-specific H8-5 and PHIS and is the most intensely labeled antigen recognized by PHIS. (Reprinted with permission from *Infect. Immun.* 1987, 55:2327–2331.)

E. histolytica trophozoites kill human neutrophils, mononuclear cells, and monocyte-derived macrophages in vitro without loss of parasite viability. However, activated macrophages are effective in killing trophozoites through both oxidative and nonoxidative effector mechanisms. In addition, cytotoxic T-cell activity can be elicited from patient cells in vitro by incubation with *E. histolytica* antigens. The mechanisms by which activated macrophages and cytotoxic T-cells overcome amebic cytolytic mechanisms, as opposed to polymorphonuclear neutrophils, remains undefined.

Cell-mediated immunity certainly plays an important role in preventing or limiting invasive amebiasis, but it is not relevant to the parasite's initial penetration through the colonic mucosal epithelium. This supposition is supported by an apparently low incidence of invasive amebiasis in individuals with the acquired immune deficiency syndrome (AIDS), despite a substantial prevalence of *E. histolytica* intestinal infection. Some authorities believe this is due to a fortuitous infection by nonpathogenic strains in AIDS patients in developed countries; however, in such areas of the third world as Haiti or Zaire, there has not been an increased occurrence of severe invasive amebiasis in AIDS patients.

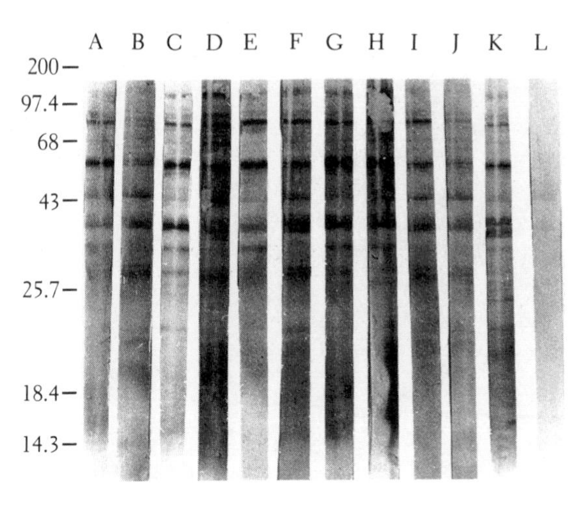

FIGURE 8·14 Immunoblots of a detergent-solubilized amebic protein preparation. Lanes A–K represent the 11 immune sera; lane L is a representative control serum; estimated molecular weights of antigens are indicated, based on the mobility of prestained standards. Note prominent recognition of 37-, 59-, 90-, and 110-kD amebic antigens. (Reprinted with permission from *Am. J. Trop. Med. Hyg.* 1988, 38:74–80.)

As with other parasites, *E. histolytica* may specifically suppress the host's cell-mediated immune response during the pathogenesis of invasive disease. Serum from patients with liver abscess inhibits the in vitro antigen-specific T-cell proliferative response. The mechanism has not been defined; in addition, the parasite could induce suppressive T-cell activities. A candidate for such effects is the released form of the Gal/GalNAc lectin, which is mitogenic in vitro and induces proliferation of a suppressor subset of T-lymphocytes. Further investigation of the effects of *E. histolytica* on host immunity is necessary, especially in regard to future design of an effective vaccine.

Prospects for a Vaccine

There is tremendous need for a vaccine providing effective immunity for *E. histolytica* infection. The highest prevalence of infection and greatest disease burden are found in developing countries that are without the resources or stability to provide sanitary facilities to prevent the spread of amebic infection. Therapy of amebiasis is complex, costly, and not without toxicity, and it certainly is not feasible for the large numbers of asymptomatically infected humans who apparently act as the reservoir for those who suffer invasive disease. Our best hope is that biomedical research will develop a vaccine to prevent the morbidity and mortality of invasive disease or, better yet, provide resistance to intestinal colonization by *E. histolytica*.

Animal models support the concept that immunization with *E. histolytica* proteins will provide protection against invasive amebiasis. However, we lack the epidemiologic studies to be certain that cure of natural human infection provides immunity. Nevertheless, recent biochemical characterization of parasite adherence mechanisms, proteolytic enzymes, cytolytic activities, and antigenic proteins provide a large number of potential approaches for development of a vaccine. We need to determine the genetic code for major parasite virulence factors and antigens in order to optimally identify nontoxic immunogenic parasite proteins and have maximum flexibility for choosing a vaccine delivery system. Work will be hampered by the lack of animal models that truly parallel human intestinal infection. Despite these problems, this author is confident that in the next decade we will surpass the advances that followed development of axenic culture in 1978 and that numerous vaccines will be ready for testing in the field.

Coda

Recent research on the cell biology of *Entamoeba histolytica* has focused on mechanisms of pathogenicity. There is indirect evidence to suggest that ingestion of some intestinal bacteria by ameba may contribute to their capacity for invasiveness. A galactose- or N-acetyl-D-galactosamine-inhibitable lectin has been identified in *E. histolytica* that apparently mediates parasite adherence to colonic mucosa and also binds to human colonic mucins. Such mucins can inhibit parasite binding to mucosal cells and might therefore serve as a first line of non-specific host defense; at the same time, it may enhance colonic colonization. The amebas produce a variety of substances that may promote invasion, including proteases that degrade the extracellular matrix. They also produce enterotoxins that theoretically might contribute to dysentery. Ameba can lyse host neutrophils on contact with the result of enhanced cytotoxicity against Chang liver cells, suggesting one mechanism involved in amebic liver abscess formation. The ameba adherence lectin may also function as a cytotoxin. In addition, a pore-forming protein packaged in dense granules of the ameba may play a role in cytotoxicity; it induces ion fluxes in a variety of cell targets.

The immunology of ameba is in its infancy, lagging behind most of the other clinically important parasites. It is unclear whether protective immunity develops in humans. There is an impression that cell mediated defenses may play some role in limiting (or preventing) invasive disease. Two possibilities are that activated macrophages may be cytotoxic to amebae, even though quiescent macrophages are potential targets of attack by the parasites and that specifically sensitized T lymphocytes are directly amebicidal. Future progress in understanding host defense and the antigenic composition of ameba could provide a basis for vaccine development.

Additional Reading

GITLER, C., and D. MIRELMAN. 1986. Factors contributing to the pathogenic behavior of *Entamoeba histolytica*. *Ann. Rev. Microbiol.* 40:237–262.

MCLAUGHLIN, J., and S. ALEY. 1985. The biochemistry and functional morphology of the *Entamoeba*. *J. Protozool.* 32:221.

MIRELMAN, D., and P. SARGEAUNT. 1987. The pathogenicity of *Entamoeba histolytica:* debate. *Parasitol. Today* 3:37–43.

RAVDIN, J. I. 1989. *Entamoeba histolytica,* from adherence to enteropathy. *J. Infect. Dis.* 159:420–429.

RAVDIN, J. I., ed. 1988. *Amebiasis: Human Infection by Entamoeba histolytica.* John Wiley & Sons.

SALATA, R. A., and J. I. RAVDIN. 1986. Review of the human immune mechanisms directed against *E. histolytica. Rev. Infect. Dis.* 8:261–272.

PART II

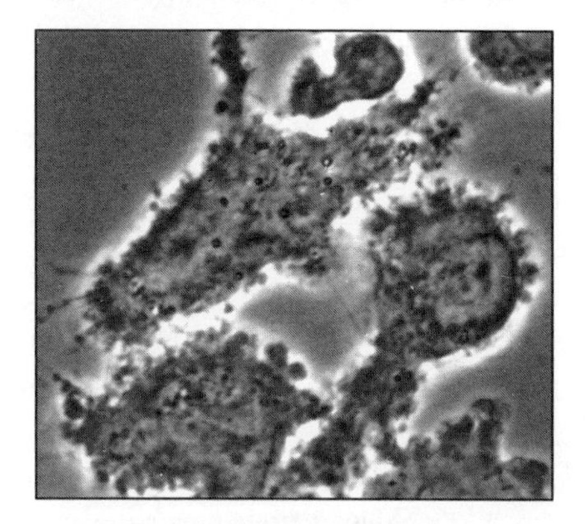

IMMUNOLOGY OF PARASITES

David J. Wyler

Inherent in a parasitic relationship is the ability of the parasite to stave off attack by the host's immune system. The manner whereby this is achieved, on the one hand, and the vulnerabilities of the parasite, on the other, have provided the major focus of interest in parasite immunology. Insights revealed by the application of basic immunological perspectives and techniques to this problem are fascinating and have expanded our appreciation of the complexities of microbial immunity generally. These insights also have defined some of the potential challenges to practical strategies recently defined for vaccine development.

Protozoa and helminths are antigenically more complex than prokaryotic pathogens. Multiple life-cycle stages of the parasite in the host add to this complexity, for in addition to the elaborate array of antigens common to all stages there are also stage-specific antigens. Protozoa and helminths nevertheless provide distinctive challenges to the immune system. For example, intracellular protozoa are potentially susceptible to intracellular antimicrobial defenses such as those imparted by activated macrophages infected with *Toxoplasma* (Chapter 10), *Leishmania* (Chapter 11), or *Trypanasoma cruzi* (Chapter 13); helminths are too large for such defense but are potential targets of extracellular attack (Chapters 14 and 15). Some protozoa, such as *Plasmodium* (Chapter 9), may take up residence inside host cells that lack intrinsic antimicrobial capabilities; these can serve as a haven from immune surveillance. Helminths, most of which reside extracellularly, must use other strategies to avoid being recognized as foreign, such as disguising themselves by acquiring host proteins on their surface; for example, *Schistosoma* adsorb host blood-group and histocompatibility antigens (Chapter 14), which interfere with the host's ability to recognize the parasite as foreign. Some protozoa, namely plasmodia (Chapter 9) and the African trypanosomes (Chapter 12) can rapidly switch their antigenic composition, so that as soon as the host mounts a specific antibody response that might be deleterious to their survival, the parasites alter their identifying antigenic characteristics. This process, which is called antigenic variation and was first defined by parasite immunologists, drew the intense interest of biochemists, who determined the structural basis of the process in African trypanosomes (see Chapter 3); their findings, in turn, motivated molecular biologists to study gene regulation in these parasites (see Chapter 17). A practical consequence of this work is the recognition that achieving a vaccine against African trypanosomiasis is unlikely because the antigenic repetoire played out during the process of variation is so extensive. Antigenic diversity that may derive at least in part from the process of antigenic variation also poses a substantial challenge in the development of malaria vaccines (Chapter 9).

Chronic infectious diseases characteristically trigger a variety of disturbances in immunoregulation. In certain viral, spirochetal, rickettsial, bacterial, and fungal infections, it is possible to detect deficiencies in antigen-specific (and sometimes also in nonspecific) immune responses. For example, delayed hypersensitivity skin test responses and in vitro lymphocyte proliferative responses (both measurements of T-lymphocyte responses) to pathogen-derived antigens may be blunted. Similar observations made in a number of parasitic infections have been interpreted to suggest that one mode of achieving successful parasitism is to induce immunosuppression — to paralyze the host's ability to mount an effective defense. However, the extent to which abrogation of such responses actually contributes to survival of the parasite is less well established. For most parasitic diseases, knowledge about the precise mechanisms that potentially contribute to effective defense is lacking. An exception is leishmaniasis, a disease in which

T-cell-mediated activation of antimicrobial defenses in host macrophages has been shown to underlie defense (Chapter 11). In this case, chronicity and parasite dissemination, in an animal model, can be shown to result from the influences of immunosupressive lymphocytes.

The evolution of knowledge, technology, and theoretical perspectives in the immunology of parasitic diseases has understandably paralleled developments in basic immunology. In the early years of immunology, when antibodies and serologic techniques were the dominant interests of most basic immunologists, parasitologists were most interested in developing serodiagnostic methods and assessing the role of humoral (antibody-mediated) immunity in defense. The adoptive transfer of antibody-containing serum from an immune host to a nonimmune host (passive immunization) has been used to determine whether antibodies might be important in defense against parasites. For example, the ability of investigators in the early 1960s to impart some protection to children with falciparum malaria by this method (Chapter 9) focused attention for the next two decades on the importance of humoral immunity in malaria. The alternative approach, examining the effects on defense to parasites of ablation of B lymphocytes with antibody to the heavy chain of IgM, has also provided insights into the role of humoral immunity.

The recent advent of such new technologies as methods for the production monoclonal antibodies and the development of immunoblotting (Western blot) analysis has permitted investigators to analyze more effectively the antigenic specificity of immune responses. Among other uses, monoclonal antibodies have been exploited in identifying and purifying parasite antigens that might be useful as vaccines. At the most sophisticated level, this approach has provided leads into the fine structure of certain antigens, such as the circumsporozoite protein of *Plasmodia* (Chapter 9), that subsequently stimulated molecular analysis by recombinant DNA technologies. One surprise that emerged from this analysis is that some parasites (such as plasmodia) have antigens comprised of highly repetitive epitopes (that unit of the antigen recognized by the monoclonal antibody) on their surface. Complementing this type of approach, some investigators have used Western blot analysis to compare the antigenic specificity of antibodies produced by hosts resistant and susceptible to specific parasites (for examples, see Chapters 10 and 14). This work is stimulated by the belief that such analysis might permit identification of the protective antigens, ones that could be vaccine candidates.

In addition to antigenic specificity, the qualitative characteristics of antibodies produced during parasitic infection have come under close scrutiny, in investigations motivated in large part by the desire to identify in vitro correlates of in vivo defense. For example, some malariologists have been intrigued by the ability of particular antibodies of immune hosts to block invasion of host cells by sporozoites or merozoites (Chapter 9). Students of Chagas' disease have been interested in the distinctive antibodies that can lyse extracellular *Trypanosoma cruzi* in vitro as potential defenses (Chapter 13). Several immunoparasitologists have focused their attention on antibodies that can mediate binding of eosinophils to the surface of schistosomulae, where these cells may then exert their helminthotoxic effects (Chapter 14). An understanding of the various ways in which antibody might reduce the levels of microfilaria—inhibiting embryogenesis, trapping microfilaria in situ, and removing circulating microfilaria—has emerged from filariasis research (Chapter 15).

A new avenue of investigation in the role of antibodies in parasitic infections has recently followed the recognition that autoantiidiotypic antibodies—made by the host and directed against its own antibodies—may have important immunoregulatory roles that might modify the host's response to specific antigens (for example, see the case of schistosomiasis, Chapter 14).

Research carried out primarily in the early 1960s shed light on the importance of antibody-independent, cell-mediated mechanisms of host defense against intracellular pathogens. *Toxoplasma gondii* and *Leishmania* were found to be susceptible to such cellular defense mechanisms and became important models for cellular immunology research generally (Chapters 10 and 11). An interest in extracellular lymphocyte, macrophage, and natural-killer cell attack on tumor and virus-infected cells (cell-mediated cytotoxicity) was paralleled by studies on the potential role of such defense mechanisms against helminths (Chapters 14 and 15). The unraveling of the complexities of cellular immunoregulation generally—with the emergence of such concepts as helper and suppressor cells and idiotypic regulatory networks—stimulated related work by parasite immunologists (see for example Chapters 11, 12, and 14).

Parasite immunology has evolved beyond simple application of basic immunology to infectious diseases and has developed a unique character of its own. The uniqueness emerged from a heterogeneity of perspectives that emphasize basic aspects of parasite biology: features of the life cycle (see for example, Chapters 9 and 14); host–parasite cellular biology, such as the role of specific receptors in invasion and inhibition of this process by antibody (for example, Chapter 9); clinical medicine, such as the immunopathogenesis of clinical manifestations (for example, Chapters 13 and 14); and even clinical epidemiology, such as the basis for the natural history of the disease (see Chapters 9, 13, 14, and 15).

In some cases, serendipitous observations made in the course of derivative applications of certain immunological methods have generated new lines of interest and new viewpoints. Despite changing emphasis at different times in the history of the discipline, the important questions have remained pretty much constant: What are the mechanisms potentially available to the host for ridding itself of the parasite? By what regimen can these be induced with vaccination? How does the parasite evade host defenses through its intrinsic capabilities or through subverting the host's immunoregulatory apparatus? Which stages of the parasite in the host are most amenable to attack by the immune system, and what are the stage-specific properties that account for these differences?

Presently, an optimism that new technologies may for the first time facilitate the development of vaccines against these scourges of mankind dominates research efforts. If successfully applied, these technologies will provide for the breakthrough that has been hoped for over decades and has been the ultimate goal for parasite immunologists. On the other hand, if the isolation of a candidate vaccine target antigen — and cloning the gene that encodes it (in order to produce sufficient material for a vaccine) — proves insufficient for immunoprophylaxis, we can expect a great deal of attention to return to basic questions of host defense mechanisms and immunoregulation.

This scenario is presently being played out in research aimed at developing an antisporozoite vaccine for malaria immunoprophylaxis (see Chapter 9). After an impressive investment of scientific talent and research funds during the last decade, candidate vaccines emerged and were subjected to preliminary clinical trials. The results were disappointing. Although many pitfalls in approach may be rectified by the continued efforts to enhance the vaccines, some malariologists have come to believe that a better understanding of the basic immunology of malaria may be required before an effective vaccine can be developed. It appears that the pendulum may now be swinging back from emphasis on a practical goal to more basic investigation.

9
Immunology of Malaria

David J. Wyler

A little over a century ago, Laveran discovered that *Plasmodia* organisms cause malaria and Ross defined the parasite's life cycle. This knowledge facilitated the development of successful approaches to disease control directed at both the mosquito vector (control of breeding sites and use of insecticides) and the parasite (chemotherapy and chemoprophylaxis). Since the 1960s, however, insecticide resistance in many vector Anopheles mosquito populations and progressive widespread dissemination of multidrug resistance among *Plasmodia falciparum* strains have contributed to a worrisome resurgence of malaria.

Distressingly, outbreaks of malaria have occurred even in some areas where transmission had been previously reduced or controlled. For example, some of the control modalities that were successful in certain other regions could not be applied or were virtually ineffective in subsaharan Africa. There, malaria remains as one of the greatest threats to childhood survival (over 1 million African children die from malaria each year), creating the need for identifying and instituting novel control strategies. In the 1960s, when the resounding success of smallpox vaccination was being celebrated, the idea that a malaria vaccine might be the necessary new control modality became popular. A major obstacle was the paucity of parasite material from which vaccine antigens could be isolated. This problem was overcome by important technological breakthrough in the 1970s: discovery of methods for long-term in vitro cultivation of *P. falciparum*; perfection of hybridoma methods of preparing monoclonal antibodies; and establishment of techniques for cloning DNA and expressing the products in *Escherichia coli*.

With these new techniques came the possibility of preparing quantities of vaccines needed for field application. It is therefore not surprising that in the last decade major efforts in malaria immunology have been primarily focused on vaccine development. If the empiric approaches to vaccine development prove unsuccessful, one can expect a rejuvenation of research interest into several fundamental unanswered questions: What are the basic mechanisms of antiplasmodial host defense? What factors trigger the striking polyclonal activation and immunosuppression that occur in an apparently paradoxical manner in malaria? Are these immunological perturbations important to a malaria-specific defense mechanism or are they simply epiphenomena? Additionally, what immune responses, if any, play a role in the pathogenesis of acute infections?

The complexity of malaria immunology as we presently understand it and the numerous lacunae in our knowledge make it difficult to synthesize the material into an integrated, coherent, and broadly acceptable scheme. New information is rapidly becoming available and is profoundly influencing the scientific perspectives of host defense and immunoprophylaxis in malaria. This chapter presents a general overview to orient the reader in appreciating the new developments as they are announced. References to important contributions, that regretfully are not cited here, can be found in more specialized reviews such as those listed at the end of this chapter.

Overview of Immunology

Following inoculation of sporozoites by the bites of a mosquito, this infective extracellular form is rapidly cleared from circulation and enters hepatocytes, where exoerythrocytic development proceeds. There exists little convincing evidence that either the sporozoite or the hepatic exoerythrocytic stage in naturally acquired infection is a target of important host defense or that either stage produces clinical illness. Only after the parasites develop into the merozoite stage and are released from the liver to enter erythrocytes is the clinical disease expressed. It is primarily during the parasite's intraerythrocytic cycle that the host immune system responds to the plethora of parasite antigens (and perhaps also polyclonal mitogens). Unfortunately, this exhuberant immune response is not particularly effective against the parasite, and there is some reason to believe that they may even provoke immunosuppressive reactions. To complicate matters, the erythrocytes may provide a haven from host defenses for merozoites. As an additional evasive strategy, the asexual blood-stage parasites are antigenically diverse and can undergo rapid changes in their antigenic composition (antigenic variation). Antigenic diversity represents a major challenge to vaccine development efforts. It therefore appears that the host's relatively ineffective defense and the parasite's evasive strategies contribute to successful parasitism. In areas with endemic malaria, protective mechanisms develop slowly in repeatedly infected individuals. However the best expression of this defense is the conversion of a potentially severe infection with *P. falciparum* (with high parasite density and debilitating illness) to a chronic, more benign infection (with low parasitemia and mild or no symptoms). Sterilizing immunity (eradication of all parasites or prevention of infection) probably occurs rarely in human malaria.

Several hypotheses have been proposed in an attempt to explain how protective immunity works in malaria, but strong support for most of these hypotheses is lacking. The limitations of certain animal models of malaria in attempts to understand human disease have complicated the interpretation of many studies. It is fair to conclude, however, that presently no single mechanism has been identified to account for defense in natural infections; very likely several mechanisms are operative.

In the absence of a clear understanding of defense mechanisms in malaria, present approaches to vaccine development are necessarily empiric, albeit thoughtful and sophisticated. An important concept that underlies much of the present research effort is that immunization with antigens of the extracellular stages (sporozoites and merozoites) elicits immune responses superior to those that accompany natural infections. Special focus on invariant plasmodial antigens is presently considered by some the only hope for circumventing the obstacle of the extensive antigenic diversity characteristic of *Plasmodia*.

Clinical Immunology

It is is useful to begin a consideration of malaria immunology from the perspective its natural clinical history in areas where *P. falciparium* is hyperendemic (Figure 9-1). Clinical epidemiology not only can provide fundamental principles of the disease to direct immunology research efforts, but also defines certain facts against which all experimental data must ultimately be compared. Since nonhuman malaria models are frequently exploited to dissect the immune responses, it is particularly important to consider whether the conclusions derived from such animal studies are likely to be applicable to human malaria.

Natural History of Malaria

Newborns are innately resistant to acute malaria infection and rarely become ill with malaria in the first 3 months of life. Between 3 months and about 2 to 5 years of age, however, children can experience several debilitating and even potentially fatal acute attacks yearly. Thereafter, protective immunity is gradually acquired and is manifested by lower peak parasitemias (blood parasite density) and a relative tolerance to clinical complications of infection. Since mortality in falciparum malaria is directly related to parasite density, the ability to limit peak parasitemia in immune individuals also reduces mortality risk. In such immune individuals, malaria is usually experienced as a brief flulike illness; rarely do they experience the very high fevers, severe debilitation, renal and pulmonary failure, and cerebral dysfunction that can occur in nonimmune patients. Nonetheless, for months parasites may persist in the circulation of these patients. This acquired immunity can be lost during pregnancy (especially in *primipara*) or during extended residence outside of an endemic region.

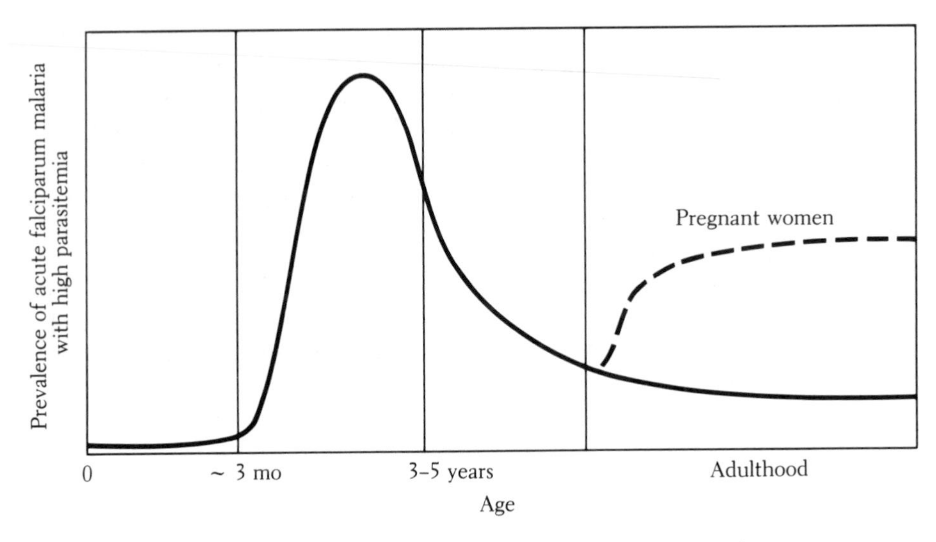

FIGURE 9·1 Natural history of the acute falciparum malaria infections in an area with very high transmission rates (hyperendemic malaria). Newborns are innately resistant to infection, probably as a result of transplacental transfer of protective antibodies and antiplasmodial effects of fetal hemoglobin. After 3 to 6 months of age, individuals become highly susceptible to infection; they can experience high levels of parasitemia. After numerous acute infections over several years, they develop immunity. Thereafter, they generally experience mild disease with low parasitemias. Pregnant women again become highly susceptible to infection.

We can draw fundamental conclusions regarding malaria immunity from these observations. (1) The acquisition of protective immunity is gradual; (2) immunity apparently requires continued boosting through repeated antigenic challenge; (3) immunity is partial in its clinical and parasiticidal expression; and (4) immunity is vulnerable to the physiological or immunological changes that occur in pregnancy. Moreover, protection is also stage-specific. Partial protection against the asexual intraerythrocytic stages does not result in either antisporozoite or transmission-blocking (antigamete) immunity. Confirmation of the stage-specific nature of protection has also been obtained experimentally; vaccine-induced antisporozoite immunity does not influence the course of infection induced by injection of blood-stage parasites.

Innate Resistance

The immunological and physiological factors that contribute to the different stages of susceptibility and partial protection are poorly understood. The relative protection of the newborn may result from its immunological experiences in utero and from nonimmunological determinants as well. Transplacental transfer of protective antibody (IgG isotype) from an immune mother has long been thought to be important in imparting innate immunity to malaria, but compelling evidence remains to be obtained. Theoretically, prenatal induction of antigen-specific or nonspecific defense mechanisms in the fetus might also occur and influence early susceptibility. Indeed, some studies in rodents have provided evidence that maternal infection could subsequently have salutary influences on the course of malaria when induced in the offspring. Epidemiological and immunological studies that address the possibility of prenatal induction of antimalarial defense have yet to be undertaken.

Certain nonimmunological factors have been identified that play important roles in innate resistance, both during infancy and subsequently. For example, erythrocytes containing large amounts of fetal hemoglobin (HbF) provide a relatively nonpermissive environment for the development of *P. falciparum*. When the normal switch from δ to β globin synthesis occurs during infancy and reduces HbF production, this mechanism of innate resistance disappears. There are also indications that diet might influence resistance; folate or iron deficiency in the host may adversely effect parasite development. Erythrocytes that contain certain abnormal hemoglobins (such as sickle hemoglobin) may also impart lifelong protection against *P. falciparum*, because these parasites fail to develop normally inside the variant cells.

Acquisition of Protection

Early childhood, one of the most vulnerable periods for malaria, is also a time when many other pathogens vie for the attentions of the immune system. A widely held belief is that the sluggish development of protection to malaria is largely attributable to the substantial antigenic diversity among populations of *P. falciparum* transmitted in any one geographic area. One possibility is that each acute infection represents a challenge with parasite populations of different antigenic composition. Accordingly, it is reasoned, acquisition of protection must require induction of appropriate immune responses to the repetoire of antigens represented by the local parasite populations. Additional explanations for the sluggish development of protection have also been proposed. One interpretation of the requirement for multiple infections in the acquisition of immunity is that it reflects the need for repeated boosts with poorly immunogenic antigens common to all *P. falciparum* strains (conserved or invariant antigens) that might be the targets of protective mechanisms. According to this interpretation, most of the immune responses to the early infections are against antigens that are irrelevant to protection. Only with time are the immune responses to the relevant antigens sufficiently vigorous to take clinical expression in the form of less severe illness.

Another possibility is that immunosuppressive events—resulting from malaria itself or from other intercurrent infections—adversely modify the pathways that lead to the induction of protection. This hypothesis has as a theoretical corollary the idea that the impact of immunosuppression is attenuated with successive infections, ultimately permitting emergence of effective antiparasitic defense. It must also be considered highly probable that the immune or physiological responses that provide the clinical tolerance that individuals develop with time include ones that have no antimicrobial consequences.

None of these speculations regarding the basis for the sluggish development of antimalarial immunity is supported by entirely compelling evidence. With the present opportunity to define the antigenic heterogeneity of plasmodial populations and to assess humoral and cellular immune responses to specific antigens, it may be possible to determine how antigenic diversity and perturbations in immunoregulation interact in delaying the genesis of antimalarial defense.

Because of its inherent virulence, *P. falciparum* infection might reasonably be expected to be lethal to most of the untreated, nonimmune children. Surprisingly, the majority of children are actually protected from lethal complications during the period of greater vulnerability (although many do succumb). Some genetic and acquired factors that may account for this observation have been identified. The deleterious influence of sickle hemoglobin on intracellular plasmodial development and the fact that falciparum malaria has helped maintain the sickle gene pool through balanced polymorphism are well known to students of biology. On the other hand, the significance of other variant hemoglobins (such as glucose-6-phosphate dehydrogenase deficiency) among residents of areas with long-standing endemic falciparum malaria is controversial. Very little is known about the immunological mechanisms that might serve to rescue young children from potentially lethal malaria. One population study provided a hint that falciparum malaria might select for certain major histocompatibility haplotypes that theoretically regulate important immune responses providing protection to the host. There is also evidence that certain responses to sporozoite antigens in mice are class II major histocompatability complex-restricted. One therefore could imagine that insights into the genetics—including the immunogenetics—of host defense in human malaria might profoundly influence our thinking about the role of immune mechanisms in defense against this disease.

Humoral Immunity

An important clue about how antiplasmodial defense might act in immune individuals was provided in the 1960s by the observations that serum gamma globulins from immune adult residents of The Gambia, West Africa could rapidly reduce the density of asexual-stage parasites when infused into children with acute falciparum malaria. Control serum globulins from residents of Europe had no effect. Furthermore, immune globulins only reduced the asexual parasitemia; the density of circulating gametocytes was unaffected. This now-classical study focused attention on the likely role specific antibodies played in mediating antiplasmodial defense and also established that this humoral defense was parasite stage-specific. Follow-up in vitro studies provided the further suggestion that the protective antibodies might act to prevent the merozoites released at the time of schizont rupture from infecting other erythrocytes, thereby blocking an essential step in the parasite life cycle.

This concept of antibody-dependent defense remains popular and underlies much of the current efforts at identifying target (vaccine) antigens. Theoretically, antibodies (alone or in concert with complement components) could also mediate removal of infected erythrocytes by macrophages in the liver and spleen (immune clearance). This theory has resurfaced in one form or another at various times since its proposal nearly a century ago. Despite the role of immune clearance as a basis of defense in human malaria, the critical studies have yet to be performed. On the basis of the possibility that immune clearance might be a defense mechanism, parasite antigens exposed on the surface of infected erythrocytes that would be involved in such clearance are being structurally analyzed.

Cell-Mediated Immunity

No convincing evidence has been obtained for a role of antibody-independent, cell-mediated immunity in human malaria, although increasingly this possibility is being considered because selected studies in mice suggest such mechanisms. Peripheral blood T lymphocytes have been assessed in patients with acute falciparum malaria. Dramatic reduction in circulating T-cell numbers, abrogation of in vitro proliferative responses to plasmodial antigen, and reduced in vitro lymphokine production have been observed. On the other hand, increased natural-killer-cell activity and circulating interferon levels have been identified in Nigerian children experiencing rising parasitemia. Interestingly, some of these changes very rapidly revert to normal after the institution of antimalarial chemotherapy. Inasmuch as the spleen and perhaps also the liver appear to be a major site of antiplasmodial defense (see below), recruitment of circulating cells to the spleen may account for some of these findings. Down-regulation of T-cell function might also occur but has not been analyzed in detail. As long as we lack a clearer sense of whether and how cell-mediated immune mechanisms function in malarial defense, data from assessment of peripheral blood T-cell numbers and in vitro functions may be difficult to interpret.

Malaria in Pregnancy

That protective antiplasmodial immunity can be dramatically suppressed during pregnancy is a fact that has serious clinical consequences and important public health implications. A few community-based studies have documented that pregnant women who developed immunity before pregnancy can become vulnerable to severe falciparum malaria infections during pregnancy. *Primipara*, for unexplained reasons, are particularly susceptible. The consequences of this regained vulnerability can be devastating: maternal or fetal death or fetal retardation (low birth weight), which is presumably due to placental insufficiency. During the early postpartum period some women may experience spontaneous self-cure while others may experience relapses (emergence of latent liver stage *P. vivax* or *P. ovale* parasites into blood stage infection). The loss of immunity to malaria during pregnancy is unexplained, but could be based on mechanisms that also render pregnant women more susceptible to certain nonparasitic infectious diseases, perhaps including increased levels of steroidal hormones. Because young children and pregnant women are the most vulnerable to fatal malaria, they would be the obvious targets for vaccination programs. Greater attention to understanding relevant immune responses in these populations therefore would seem to warrant a high priority for future research.

Antisporozoite Defense in Humans

Since the intraerythrocytic asexual stage of *Plasmodia* is exclusively responsible for the clinical manifestations of malaria, immune responses to this stage of parasites have traditionally received the greatest attention. More recently, with increasing prospects for a vaccine that could interfere with the invasiveness and development of the infective stage transmitted by mosquitoes (sporozoites), attention has begun to focus on immunity to sporozoites in humans. Older children and adults residing in areas where falciparum malaria is endemic develop antibodies to sporozoites that are not detectable in younger children. These antibodies are specifically directed against the dominant repetitive epitope on the sporozoite that is also being tested as a potential vaccine (see Vaccines). The antibodies can block the invasion of sporozoites into liver cells in vitro. However, there is no direct correlation between the presence of these antibodies and resistance to reinfection in residents of regions with endemic malaria. The lack of clear evidence that antisporozoite immunity develops naturally has not prevented the optimistic belief that a sporozoite vaccine could be effective, a belief based on the idea that even if naturally acquired immunity to sporozoites fails to occur or is only partial, vaccine-induced protection might nonetheless be absolute (prevent development of all sporozoites). Support for this optimism was obtained when it was shown that volunteers repeatedly injected with live irradiated sporozoites that do not induce infection were completely resistant to subsequent challenge with fully virulent nonirradiated sporozoites.

This perspective of the natural clinical history of malaria is applicable to areas with stable and high transmission rates of falciparum malaria. In contrast, in areas with lower levels of transmission, individuals of all ages are potentially susceptible. Disease manifestations (such as cerebral dysfunction) also may vary in different areas. Less is known about the rate of acquisition of protection in such populations and about defense in malaria caused by other species of *Plasmodia*. Apart from its heuristic value, greater understanding of the natural clinical history of nonfalciparum malarias and falciparum malaria in areas with sporadic transmission could expand our understanding of the immunology of malaria. Guidance for design of fruitful immunology research can be expected to derive from the efforts of appropriate clinical epidemiological studies in these areas.

Mechanisms of Host Defense

Because only the asexual intraerythrocytic stages of *Plasmodium* are directly pathogenic (sporozoites, tissue schizonts, merozoites, and gametocytes cause no clinical symptoms), defense against these stages has understandably received the greatest attention. Much of the research has been carried out in animal models (particularly rodents) that permit immunological manipulations not possible in humans. More recently, in vitro parasite cultures have also been exploited. Results of such studies must be cautiously evaluated, however, since neither the existing animal models nor the in vitro culture conditions reliably mimic human malaria. Considerable controversy surrounds interpretation of some of these studies vis à vis their likely relevance to human disease. On the other hand, the one statement with which most malaria immunologists would agree is that host defense in malaria is complex and probably depends on the interplay of several mechanisms.

A number of generalizations about the biology of the asexual intraerythrocytic cycle and disease pathogenesis can help to focus our attention on relevant concepts of protection (for detail, see Chapter 1). Merozoites reside only very briefly (seconds) in the extracellular milieu before they must attach to and enter erythrocytes or lose their invasive capacity and die. Once inside the red cell, they probably depend for their survival and reproduction on selected intracellular constituents as well as on transport of selected substrates from the plasma. The latter presumably must traverse the erythrocyte plasma membrane and cytoplasm as well as the cell membranes of the parasitophorous vacuole and of the parasite itself. The erythrocyte is structurally and functionally altered by infection, and parasite-derived antigens appear on the surface of the infected red cell. Among these are molecules that appear on the surface of *P. falciparum*–infected erythrocytes that mediate binding of these cells to capillary and venular endothelium (a process called cytoadherence). Cytoadherence, which results in sequestration in postcapillary venules of erythrocytes containing the more mature asexual forms (trophozoites and schizonts), presumably facilitates terminal differentiation of the parasite under conditions of low oxygen tension. Antibodies against cytoadherence molecules inhibit this sequestration and can abrogate the infection in monkeys infected with *P. falciparum*.

It is generally believed that most of the life-threatening complications of falciparum malaria result from extensive hemolysis and from the tissue hypoxia caused by the diminished microcirculation initiated by cytoadherence. The additional possibility that there are parasite toxins or soluble products of host cells such as macrophages (e.g., tumor necrosis factor) or toxic oxygen radicals, which could be deleterious to host cell metabolism, has not as yet been convincingly supported but is gaining renewed interest. Host responses that interfere with replication or that accelerate attrition of parasites appear to be primarily directed at merozoites, developing intraerythrocytic trophozoites, or at the infected red cells as a whole. On the other hand, almost nothing is known about how immune individuals are protected from some of the clinical complications of infection, independent of a direct antiparasitic effects. Perhaps such tolerance includes modification of the microvascular pathology or an "antitoxic" response to parasite or host-derived products. Closer scrutiny of these possibilities could lead to new therapeutic approaches.

Invasion Blockade

Interference with the process of erythrocyte invasion by merozoites is generally viewed as a defense mechanism that occurs naturally or at least might be inducible by vaccination (Figure 9-2). Antibodies, including ones present in sera of immune individuals and ones produced in vitro by human or murine B-cell hybridomas can prevent in vitro merozoites from entering susceptible erythrocytes. Complement is not required for this effect. The effect may result, at least partly, from agglutination of merozoites. Unfortunately, the presence of invasion-blocking antibody in circulation does not uniformly correlate with the immune status of the human or experimental host. In some cases sera from immune animals fail to interfere with invasion and parasite development in vitro. On the other hand, the ability of transfused hyperimmune serum to protect rats from rodent malaria *(P. berghei)* has been shown to result primarily from its ability to inhibit merozoite invasion of erythrocytes in vivo. Antibodies may also facilitate merozoite ingestion by macrophages.

Merozoite antigens have been isolated with the use of monoclonal antibodies that effectively block invasion in vitro and have been used with potent adjuvants (complete Freund's adjuvant) to immunize laboratory animals (see Antigens and Vaccines). In some cases, vaccinated animals produce invasion-blocking antibody and are protected from challenge; in other cases, only antibody production is stimulated without imparting protection, or protection develops without production of blocking antibodies. Such inconsistencies indicate that at the very least, defense mechanisms in addition to antibody-mediated blockade of invasion are likely to be important.

Interference with Intracellular Development

Antibody and other soluble factors may interfere with the maturation of plasmodia after they enter erythrocytes. Antibody entering erythrocytes bound to the surface of merozoites might exert a microbistatic effect. It is highly improbable that immunoglobulin enters erythrocytes after infection. Theoretically, antibodies binding to the surface of infected erythrocytes might interfere with the parasite's intracellular development. Since new transport systems appear on the surface of infected erythrocytes, binding of antibody to essential transport molecules might ultimately have a detrimental effect on the growing parasite. This possibility has not been directly tested as yet. On the other hand, antibody-mediated damage to the erythrocyte membrane by a complement "attack complex" or through antibody-dependent cell-mediated cytotoxicity has not been convincingly demonstrated and seems unlikely.

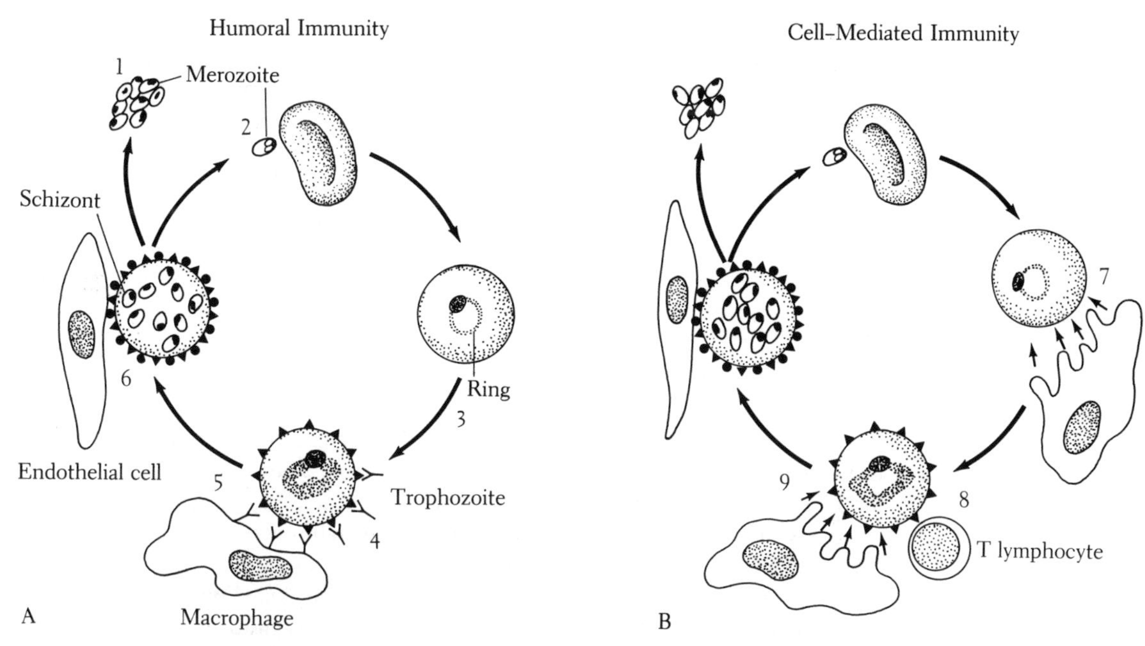

FIGURE 9·2 Potential sites of host defense against asexual (blood) stages of plasmodia. (A) Antibody may interfere with infection by agglutinating extracellular merozoites (1) or blocking ligands involved in attachment and invasion (2) after the release of merozoites from schizonts. If merozoites coated with antibody enter erythrocytes, intracellular development might be inhibited (3), or antibody that binds to the surface of infected erythrocytes might also interfere with development (4). Certain immunoglobulins could promote immune clearance in the spleen and liver either by directly binding to infected erythrocytes or functioning as cytophilic antibodies that are already bound to macrophages (5). Finally, antibody might bind to or near cytoadherence ligands and sterically hinder their interaction with endothelial cells (6).
(B) Cell-mediated immunity theoretically could involve effects of macrophage products that interfere with intracellular parasite maturation (7), or other cytotoxic effects of T lymphocytes (8) or activated macrophages on infected erythrocytes (9). Tumor necrosis factor and oxygen radicals have been suggested as two potential mediators of the antiplasmodial effects. Natural killer (NK) cells (not shown) might theoretically act in a similar cytotoxic manner.

Key: >: parasite antigens on the surface of infected erythrocytes; Y: antibody; O: cytoadherence ligands ("knobs").

Small molecules that can readily traverse the erythrocyte plasma membrane and are toxic to the parasite have been postulated to play some role in defense. For example, a nonimmunoglobulin factor in sera collected from individuals living in an area of low malarial endemicity arrests the intracellular development of *P. falciparum* in vitro. Interestingly, the presence of this factor correlates with apparent resistance to infection (i.e., low titers or absence of the antiplasmodial antibodies that signify recent infection). This factor has not yet been biochemically characterized but appears not to be tumor necrosis factor, as once postulated. In vitro experiments and limited indirect evidence from rodent malaria studies have suggested an antiplasmodial role for toxic oxygen radicals and the presence of scavengers of these radicals in *Plasmodia*. The relative importance of the antigen nonspecific defense that these molecules might mediate in the overall immunity to malaria remains to be established. The fact that antimalarial defense in vivo appears to be remarkably antigen-specific must be reconciled with a role for antigen nonspecific mediators in defense generally.

Erythrocyte Clearance

Antibody binding to infected erythrocytes theoretically could exert a protective effect by facilitating removal of the cells from circulation. Since macrophages possess receptors for the Fc portion of immunoglobulin G (IgG) and for some complement components (such as C3b), IgG or complement-coated infected erythrocytes could be removed by splenic macrophages and Kupffer cells. Under some circumstances, this "immune clearance" might also involve opsonization (antibody-facilitated phagocytosis). Based largely on late nineteenth century – early twentieth century histopathological observations that splenic and hepatic macrophages increase dramatically in malaria and that these cells become laden with parasite pigment (and to a lesser extent with infected erythrocytes), immune clearance was long thought to be an important defense mechanism in malaria. In fact, in vitro studies revealed that antibody could opsonize freed parasites as well as infected erythrocytes for ingestion by macrophages. Direct test of this immune clearance – opsonization hypothesis in *P. berghei* – infected rats failed to verify this long-held hypothesis, however. Specifically, infusion of hyperimmune serum protected rats from challenge with *P. berghei* but resulted in exactly the same rate of removal from circulation of ^{51}Cr-tagged infected erythrocytes as occurred in control recipients of normal rat serum. In a confirmatory study using a murine malaria model, hyperimmune serum that had potent opsonizing properties in vitro lacked the capacity to facilitate accelerated clearance of ^{51}Cr-tagged infected red cells in vivo. Because these studies in rodents may not be representative of events that occur in human malaria, similar clearance studies in malaria patients would be of great interest but may prove difficult to carry out. In primate malaria (in contrast to rodent malaria) immune clearance could yet prove to be an important defense mechanism. In support of this possibility is a recent demonstration that in vaccinated monkeys, resistance to challenge correlated with the ability of their sera to promote opsonization of infected erythrocytes (but not with invasion blockade). In vivo tests of this possibility are now needed.

Interference with Cytoadherence

On the basis of both in vitro and in vivo (simian malaria) studies, it appears that antibody that binds to appropriate antigens on *P. falciparum* – infected erythrocytes can interfere with cytoadherence to endothelial cells. Antibody-mediated interference of cytoadherence excludes the more mature forms from the sites of sequestration deemed essential to their terminal maturation and permits their removal from circulation by the spleen, as explained in Spleen and Malaria. Some investigators who are attempting to identify the relevant target antigens in this defense consider the cytoadherence ligand a potential vaccine candidate. Cytoadherence antigens theoretically could serve as targets not only of immunoprophylaxis but also of immunotherapy, since infusion of antibodies that reverse cytoadherence might rapidly ameliorate some of the microvascular pathology that complicates parasite sequestration. Curiously, it appears that the cytoadherence antigens are diverse. Since it is difficult to imagine that so fundamental a biological process as cytoadherence would depend on ligands that are not highly conserved among parasite strains, it has been suggested that variable as well as conserved domains are present in the cytoadherence antigen "complex."

Complement

The complement system has yet to be ascribed a role in antimalarial defense. Quantification of complement components in American volunteers with vivax malaria revealed that cyclical reduction of C1 and C4 (but not of later components) occurred in association with schizont rupture, but only in patients who had detectable complement-fixing antibodies. Evidence for classical complement pathway activation has also been obtained in studies of *P. falciparum*–infected residents of malaria endemic areas. The significance of these findings in pathogenesis or protection has not been established. Monkeys depleted of some complement components by treatment with cobra venom factor (which activates the alternative pathway) or depleted of C1 and C4 with a shark factor (which activates the classical pathway) experienced the same course of infection (*P. coatneyi*) as untreated controls, indicating that in this model, complement activation was not important in defense. Studies in rodent models confirm this impression. It is not possible to exclude the role of a cell-bound (macrophage) component of the complement system in this defense, however. Nevertheless, the existing data provide little support for an important role of complement in antimalarial defense. Furthermore, in contrast to its role in babesial infection, complement is not involved in facilitating invasion of erythrocytes by plasmodia.

Cell-Mediated Immunity

The notion that cellular immune mechanisms might be important in malarial defense, postulated already in the 1920s, has received renewed interest in recent years, largely as a result of the remarkable progress in basic cellular immunology and in view of the inability to ascribe all of antimalarial defense to one or more antibody-dependent mechanisms. Considering our present understanding of cellular effector mechanisms, we could envision that infected erythrocytes, like tumor cells, might be targets of cytotoxic cells, including natural killer (NK) cells (that might even provide a mechanism of innate resistance) or antibody-dependent or -independent cytotoxic T lymphocytes (providing for antigen-specific, acquired defense). Conceivably, activated macrophages could also participate in antiparasitic defense, perhaps by elaboration of toxic oxygen radicals or soluble factors ("monokines"), as discussed earlier.

Although efforts have been made to assess these possibilities, the data seem generally inconclusive. For example, basal levels of NK cell activity in different strains of inbred mice do not reliably correlate with the susceptibility of these mice to malaria. On the other hand, malaria appears to be a potent stimulus for enhanced NK activity in humans and mice, the significance of which is not clear. Results of efforts to identify cytotoxic T cells directed against infected erythrocytes have been inconsistent. Evidence is slowly accumulating that activated macrophages probably have some role in defense, but in vivo studies will be necessary to assess their importance. One issue that must be reconciled is how macrophages, whose effector mechanisms are generally antigen-nonspecific, can contribute to an antiplasmodial defense that in humans appears to be highly antigen-specific. The cooperation of specific antibody with nonspecific mechanisms might provide for this specificity, as is discussed later. Alternatively, antigen specificity might reside in the T-cell clones that on stimulation activate antigen-nonspecific defenses in macrophages. Very recently it has been demonstrated that interferon-gamma and cytotoxic ($CD8^+$) lymphocytes participate with antibody in defense against sporozoites and hepatic exoerythrocytic stage development. Clearly, well-designed studies are required to further assess how T lymphocytes mediate antimalarial effects.

Malaria is a potent stimulus for hyperplasia of the mononuclear phagocyte (reticuloendothelial) system. Splenic enlargement (splenomegaly) characteristically occurs in patients, especially those experiencing their first acute attacks; it also occurs in malarious animals. The marked increase in the number of splenic macrophages that contributes to organ enlargement has been attributed to secretion of a monocyte chemoattractant by T cells that are nonspecifically stimulated by a substance present in infected red cells. Presumably the chemoattractant recruits monocytes from the circulation. Other T-cell products (lymphokines) such as interferon-gamma can activate the macrophages to exert antiparasitic effects, ones that depend, at least partially, on production of an oxidate burst. The possibility that other lymphokines might also directly interfere with parasite development and replication has not been carefully studied. Thus, while T cells might not have direct cytotoxic effects on infected erythrocytes, they might exert indirect effects through lymphokines.

Spleen in Malaria

Although the cellular basis of malarial immunity remains largely uncertain, it has become increasingly apparent that the spleen is a critical organ of defense. In fact, no other manipulation of an experimental animal so diminishes its resistance to malaria (innate or acquired) as does splenectomy. Since neither autotransplantation of splenic fragments nor infusion of spleen cell suspensions can restore the defect in defense created by splenectomy, it seems that the spleen is not simply a repository of effector cells. Rather, there is evidence that the unique splenic architecture is important for expression of defense.

Studies carried out in rats provides some insight into splenic mechanisms of defense. The spleen is uniquely capable of filtering out of circulation erythrocytes that have decreased deformability (such as old, damaged, or plasmodia-infected cells). Filtration occurs as erythrocytes that have slowly percolated through the splenic cords attempt to traverse the interendothelial slits that separate the cords from the splenic venous sinuses. Normal erythrocytes that possess a high degree of plasticity can undergo the marked deformations needed to traverse these narrow slits, whereas the relatively rigid cells are incapable of deformation and become trapped within the cords. Infected erythrocytes, because of their poor deformability, can be removed from circulation in this manner. Interestingly, in the presence of a rising parasitemia, the filtration capacity of the spleen somehow becomes subnormal. In rats with *P. berghei*, the onset of spontaneous resolution of acute infection (an event that is spleen-dependent) is heralded by establishment of supernormal filtration. Alterations in splenic filtration function in patients with acute falciparum malaria have also been observed, and a rapid salutary effect of antiplasmodial chemotherapy on splenic filtration has been documented in selected patients. The physiological basis for the observed alterations in splenic filtration is uncertain. Alterations in intrasplenic microcirculation and dynamic changes in the width of interendothelial slits that are the filtration sites probably underlie the findings. The clear evidence that the spleen plays an essential role in antimalarial defense should prompt further research on splenic mechanisms.

In summary, in the last two decades research on host defense in malaria has identified the importance of antigen specificity in acquired immunity and determined that antibody probably plays an important role. In contrast, a role for complement in defense has not been established. Antibody-independent T-lymphocyte-mediated or -dependent mechanisms of defense have also been implicated by selected studies in a limited number of rodent malaria models. Lymphokines and monokines are increasingly gaining attention for their potential role in regulatory mechanisms that lead to defense, and they are being considered for their potential in directly exerting antiplasmodial effects. The spleen has been repeatedly rediscovered as an essential element in antimalarial immunity, but its defensive mechanisms demand closer scrutiny. If it were possible to identify a reliable in vitro correlate of protective immunity in human malaria and apply it to studies of the ontogeny and modulation of defense, this correlate would represent a particularly important breakthrough in malaria immunology; presently none has been identified.

Immunoregulation in Malaria

Malaria in humans and animals profoundly perturbs normal cellular and humoral immune functions. This fact has motivated considerable efforts to dissect the pathways that lead to these disturbances. Two clinical observations have also provided a sustained focus of interest in this area. First, as previously discussed, the slow development of protective immunity in residents of malaria-endemic regions could be due at least in part to detrimental immune responses, such as immunodepression, during acute infection. Second, the relatively frequent occurrence of intercurrent viral and bacterial infections in malarious children might also be due to malaria-induced disruption in antimicrobial defense mechanisms through generalized immunodepression. In the absence of a clear definition of defense mechanisms in malaria, it is difficult to be certain how the investigated parameters of immune responsiveness (e.g., lymphocyte proliferation and IL-2 production) ultimately relate to defense in this disease. Systematic documentation of an impact on defense against other infectious diseases by malaria infection in human populations clearly is needed to support the limited existing evidence that malaria enhances susceptibility to other microbial pathogens. In considering the many immunological responses to malaria that have undergone at least some degree of scrutiny, one is struck by two consistent and seemingly paradoxical events.

Polyclonal Activation

The first immunological event that acute malaria infection triggers — in humans and laboratory animals alike — is a striking polyclonal antibody (and seemingly also T-cell) response. That is, malaria infection stimulates proliferation and functional expression of a large number of T- and B-cell clones (polyclonal activation). Interestingly, many of the antibodies produced in response to malaria have specificities for both plasmodial and nonplasmodial antigens, including specificities for autoantigens. In vitro studies have suggested that this veritable explosion of B-cell activity may have its roots in a lymphocyte mitogen present in *P. falciparum*. This nonspecific lymphocyte response may be enhanced by macrophage secretion of cofactors for lymphocyte activation, such as interleukin-1 and perhaps also B-cell growth factors. The clinical consequences of this polyclonal stimulation, other than inducing hypergammaglobulinemia, are not obvious. Certainly, despite the presence of circulating autoantibodies, autoimmune diseases are not recognized complications of malaria. There are, however, conflicting data suggesting that the pathogenesis of anemia in malaria may have an autoimmune component. Evidence for such an autoimmune hemolytic component is not entirely convincing, but conceivably, the products of this polyclonal antibody response, after all, are involved in one facet of disease pathogenesis. Malaria activation of B cells theoretically may somehow facilitate their infection by Epstein-Barr virus and promote malignant transformation, giving rise to Burkitt's lymphoma, a B-lymphocyte malignancy highly prevalent in some malarious areas. It is of interest that malaria significantly enhances the susceptibility of some strains of inbred mice to certain oncoviruses. Speculations that polyconal activation in malaria somehow enhances susceptibility to complications of the acquired immunodeficiency syndrome in Africa have recently been articulated, but they are unsubstantiated and their validity could be very difficult to test.

Immunodepression

The second profound immunological disturbance in malaria, immunodepression, is a generalized reduction in certain humoral and cellular immune responses. Immunodepression has been documented in malaria patients by observing poor specific antibody responses to challenge with certain antigens, as well as by diminished in vitro lymphoproliferative responses to plasmodial antigens. In infected rodents, a number of abnormalities in both humoral and cellular immune responses have been observed, and several mechanisms have been proposed to explain these results. Abnormalities in critical macrophage functions, including effective antigen presentation and secretion of immunoregulatory mediators, have been reported. Generation of suppressor lymphocytes in rodent malaria has also been observed and presumably also occurs in human malaria. Since immunosuppression probably occurs later in the course of infection—following the period characterized by polyclonal activation—down-regulation of immune responsiveness could represent a normal homeostatic immunological control response. That is, the substantial immunological activation accompanying malaria may trigger normal shut-off pathways in the immune response network that occur after any antigenic stimulus. The widespread nature of the immunosuppression therefore theoretically might reflect a normal response to the widespread (polyclonal) nature of the original stimulation.

Explained teleologically, such profound immunological perturbations might actually favor the parasite's survival and represent an evasion mechanism. The effect of parasites showering the immune system with a smoke screen of antigens and mitogens that are not targets of defense might be to delay the host's ability to develop responses to antigens that actually are the appropriate targets. If this scenario is true, then a major challenge for vaccine development against asexual-stage parasites would be to prevent the profound immunological disturbances that possibly interfere with defense. In other words, induction of tolerance to immunodominant parasite antigens that are not targets of defense might augment the efficacy of immunization with target antigens.

Immunosuppression may pose a particularly vexing dilemma for vaccine development. A vaccination program initiated in an endemic area undoubtedly would be targeted to one of the most vulnerable groups, children. However, since many of these children can be expected to harbor parasites at the time of vaccination, poor responses may result. After all, children with symptomatic and asymptomatic malaria already have been shown to mount blunted responses to certain conventional vaccines. Indeed, in preliminary studies, malaria-infected mice and monkeys failed to mount the appropriate responses to sporozoite and gamete vaccines, respectively, that could be elicited in the uninfected vaccine recipients. Accordingly, one might predict that the immunoregulatory disturbances that occur in malaria once more will become of considerable investigative interest if vaccine field trials prove unsuccessful. The relationship—if one exists—of widespread immune perturbations in malaria to oncogenesis associated with Epstein-Barr virus (and perhaps also with retroviruses such as HIV) infection might also be expected to attract greater research interest in the near future.

Vaccine Development

Considering the foregoing discussion, one might reasonably question whether an anti-malaria vaccine could ever be possible. (After all, there is no a priori reason to believe that all infectious diseases will ultimately be amenable to immunoprophylaxis!) Parasites seem particularly shrewd in their strategies to escape host defenses. Malaria—unlike most infections that so far have proved reliably preventable by vaccination—strikingly imparts protection not after one but only after many acute infections. Even then, protection is only partial. Furthermore, protection can be lost rapidly by immune individuals when they reside for extended periods in malaria-free areas, a fact that probably indicates that maintenance of immunity requires constant immunological boosting with live parasite challenge.

The stages of the parasites—sporozoites and merozoites—that are targets for a vaccine rapidly become sequestered in host cells; they have very brief (seconds) extra-cellular existences in their mammalian host. Therefore, if vaccine-induced protective mechanisms are to work against these stages, they presumably would have to do so rapidly. Observations in human malaria and experimental rodent and monkey malaria have largely established that immunity in "natural" infection is certainly species-, stage-, and in some cases also strain-specific. Therefore, even if a vaccination strategy were established that could overcome the problems posed by immunity in natural infection, it would still have to deal with the tremendous antigenic diversity in the parasite challenge. Indeed, this latter difficulty seems to be the most frustrating for present malaria vaccine efforts. Despite these obstacles, the motivations for seeking a vaccine take into consideration the possibilities that (1) vaccination could provide for attack against extracellular parasite stages and for defense mechanisms perhaps not normally induced by natural infection; (2) vaccine-induced protection might be longer lasting than naturally acquired protection; and (3) the use of monoclonal antibodies might make it possible to identify common antigens conserved among the many "strains" of *Plasmodia* species. Furthermore, the decline in efficacy and utilization of traditional malaria control modalities at a time when appropriate research technologies became available (in vitro parasite cultivation, hybridoma methodology, and recombinant DNA technology) have made sophisticated empirical approaches to malaria vaccines an obvious priority.

Identification of Target Antigens

To exploit new technologies developed in the last decade, scientists working in malaria immunology have exerted tremendous efforts to identify antigens that might be potential targets of defense. Two general approaches have been employed.

In the first approach, monoclonal antibodies against *Plasmodia* are produced by hybridization of lymphocytes of immunized mice or by in vitro immortalization of peripheral blood lymphocytes from malaria patients (for example, by EB virus transformation). Antibodies are screened by infusion in vivo (when possible, as in a murine malaria model) followed by parasite challenge, or in vitro, for certain antiparasitic effects. In vitro assays have permitted assessment of the ability of antibodies to interfere with (1) sporozoite invasion of human liver cell cultures; (2) merozoite invasion of erythrocytes and subsequent intraerythrocytic parasite growth; and (3) cytoadherence of schizont-infected cells to endothelial cells or amelanotic melanoma cells (believed to be a correlate of in vivo cytoadherence). To identify targets of transmission-blocking immunity, monoclonal antibodies to sexual and sporogonic stages of the parasite (gametes, zygotes, ookinetes) are incorporated with malarious blood meals on which mosquitoes are then permitted to feed. The subsequent ability of the plasmodia to develop in the mosquito or of the mosquitoes to transmit malaria is then assessed. The monoclonal antibodies found effective in these assays are then employed to biochemically analyze the relevant epitopes and to isolate the target antigens by affinity chromatography. Furthermore, the antibodies can be used in the screening procedures necessary for cloning genes that code for the antigens of interest. Whether purified from the parasites by affinity chromatography or by recombinant DNA technology, the resultant purified antigens can then be tested as vaccines in animal models and eventually humans.

A second approach to identifying potential target antigens has been to examine antigens recognized by antibody in serum of malaria-immune individuals. Since antibody is used in identifying these targets, plasmodial epitopes recognized exclusively by T cells but essential to protection could be missed with this approach. Several investigators have recently adopted the view that T epitopes may be of particular interest. Efforts to identify these epitopes are underway. It can be expected that the methods for identifying T-cell epitopes will improve generally in the next decade and that their application to malaria research specifically could have important practical consequences.

Sporozoite Vaccines

The most advanced area of malaria vaccine research might be called the "all or none" approach and is directed against the sporozoite stage (Figure 9-3). If a vaccine could prevent the entry of these infective stages into the liver or their subsequent development in hepatocytes, malaria might be prevented entirely. On the other hand, if the antisporozoite effects were anything less than 100 percent effective and if even only one parasite evaded the protection, emergence of an asexual stage infection would be expected to occur because protection in malaria is clearly stage-specific. Perhaps the most compelling observation to suggest that an antisporozoite vaccine might be feasible was the finding that some of the human volunteers exposed multiply to live irradiated sporozoite (ones incapable of inducing infection) were resistant to subsequent challenge with virulent sporozoites of the homologous (but not heterologous) species. These vaccinees were fully susceptible to challenge with blood-stage parasites, however. The target antigens of antisporozoite immunity appear to be the major surface protein that surrounds mature sporozoites. In all species studied, this circumsporozoite protein con-

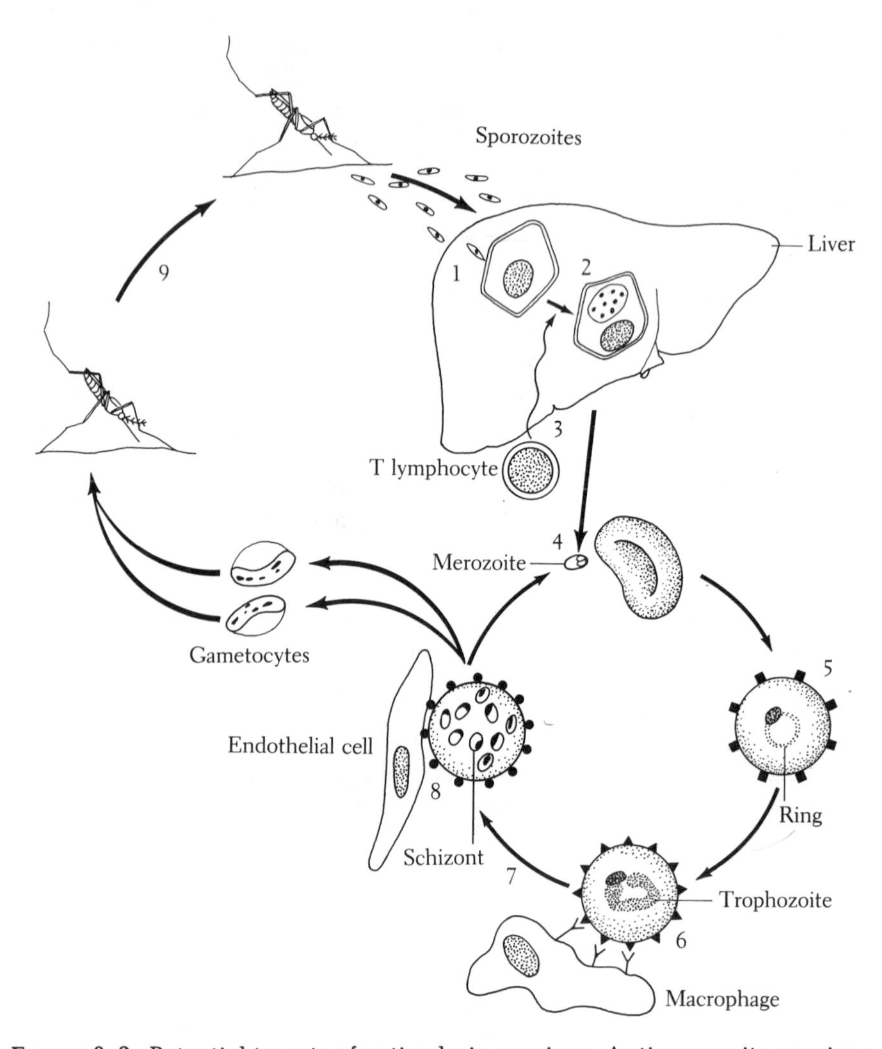

FIGURE 9·3 Potential targets of antimalaria vaccines. Antisporozoite vaccines are directed against the circumsporozoite protein that is thought to be the ligand for attachment to hepatocytes. Antibody produced in response to these vaccines may block sporozoite attachment (1) and also the subsequent development of the hepatic exoerythrocytic schizont (2). Lymphokines (such as gamma interferon) secreted by vaccine-sensitized cells may also interfere with development of the liver stages (3). Antibody elicited by vaccination against merozoite antigens may block invasion of erythrocytes (4). Parasite antigens deposited on the surface of erythrocytes surfaces during invasion (5) may be targets of vaccine-induced protection by unknown mechanisms (possibly immune clearance). Parasite antigens inserted into the surface of erythrocytes containing more mature stages may be susceptible to immune clearance (oponization) (6) or growth inhibition (7) by vaccine-induced antibodies. Vaccine antigens also could give rise to antibodies that block cytoadherence (8), interfering with sequestation of schizonts in capillaries and venules and resulting in their removal by the spleen. Transmission-blocking vaccines have as their targets steps in fertilization and subsequent development of the sporogonic stages in the mosquito that ultimately give rise to infective sporozoites (9)

tains repetitive epitopes corresponding to oligopeptides tandemly repeated and flanked by nonhomologous sequences. The repetitive epitopes have been thought to be the targets of protection. The number of repeats and the composition and size of the oligopeptides that establish the repetitive epitopes vary from species to species. In *P. falciparum*, the sequence asn-ala-asn-pro (NANP) is reiterated approximately 40 times. Since Fab fragments of monoclonal antibodies directed at these epitopes can block sporozoite entry into hepatocytes in vitro, it has been proposed that the repetitive epitopes comprise the ligand by which sporozoites bind to hepatocytes.

Recombinant circumsporozoite antigens of a number of *Plasmodia* species have been produced in *E. coli*. In addition, because of the structural simplicity of the dominant circumsporozoite epitope, it is possible to synthesize peptides of various lengths, couple them to carrier proteins, such as tetanus toxoid and test them as potential vaccines. Both recombinant and peptide vaccine candidates have been subjected to limited clinical trials. Three volunteers vaccinated with the 12-amino-acid synthetic peptide $(NANP)_3$ representing the immunodominant epitope of the *P. falciparum* CSP were challenged with sporozoites. Two had delayed infections; one did not develop infection during a 29-day observation period. Incorporation of genes coding for a circumsporozoite antigen into vaccinia virus has also been achieved and is being tested in animals as a possible vaccine. Whether such a vaccine construction ever will be permissable for human use is debatable.

There are already indications of potential problems with the recombinant and peptide sporozoite vaccines. They apparently elicit low antibody titers that inconsistently are boosted by repeated inoculation. Furthermore, in mice, immune responses to the recombinant vaccine are genetically restricted to a single haplotype of a Class II major histocompatibility locus ($I\text{-}A^b$ in the case of the *P. falciparum* CSP), suggesting that a similar restriction might occur in humans. On the other hand, one fact that is encouraging, in light of the antigenic diversity characteristic of the asexual blood stages, is that the target antigens of sporozoites exhibit little important intraspecies antigenic diversity. Unfortunately, it has been reported that both a recombinant and peptide sporozoite vaccine failed to protect mice against *Plasmodium berghei* (a species that causes rodent malaria), whereas vaccination with irradiated sporozoites was protective. Interestingly, all three vaccines elicited similar antisporozoite antibody responses. The adoptive transfer of lymphocytes but not serum from mice protected by irrradiated sporozoites imparted protection to the recipients. This has been interpreted to suggest that sporozoite vaccine-induced protection is T cell mediated. Accordingly, efforts to identify T-cell epitopes in sporozoite antigens have been initiated, and animal vaccination trials with synthetic constructs combining T-cell epitopes on the CSP outside of the repetitive epitopes and the repetitive epitopes have begun.

---

venules and resulting in their removal by the spleen. Transmission-blocking vaccines have as their targets steps in fertilization and subsequent development of the sporogonic stages in the mosquito that ultimately give rise to infective sporozoites (9).

Asexual-Stage Vaccine

The asexual parasites—merozoites and intraerythrocytic forms—are antigenically more complex and diverse than sporozoites. On the other hand, investigators are motivated to develop vaccines directed at these stages primarily because (1) these stages are the targets of defense in natural infections, and (2) even only partial protection (limiting peak asexual parasitemia) can be expected to dramatically reduce morbidity and mortality. Several classes of antigens are being studied as potential targets of vaccine-inducible protection. These include antigens on the surface of the merozoites, antigens associated with host membranes of infected erythrocytes, and rhoptry proteins (see Chapter 1). The merozoite antigens of greatest interest appear to be a family that are synthesized as large-molecular-weight precursors, 190–220 kilodaltons (kD), late in schizont development and are subsequently processed to smaller fragments, ranging in M_r from 83 to 19 kD, the form in which they are apparently expressed on the merozoite surface. Such antigens were purified from *P. falciparum* culture-derived material and used to successfully immunize squirrel monkeys that were partially protected on subsequent challenge. The antigenic diversity of this group of molecules restricts its utility as a vaccine candidate unless invariant epitopes can be identified. Particularly disturbing but biologically fascinating is the apparent ability of cloned parasites to change their antigenic composition when inoculated into monkeys immunized against them. The great diversity in these merozoite antigens makes it difficult to imagine that they function primarily as ligands for parasite attachment to erythrocytes, since one would expect such ligands to be highly conserved. Two putative glycophorin-binding proteins (155 kD and 130 kD) of merozoites have been identified and proposed as ligands for merozoite-erythrocyte binding (see Chapter 1). Antibodies to the proteins block invasion in vitro. Undoubedtly these proteins will be tested as vaccine candidates, a task that will be simplified by the recent cloning of the genes that code for them.

Proteins that appear on the surface of *P. falciparum* schizont-infected erythrocytes and mediate attachment of the cells to endothelium (cytoadherence) are also being scrutinized. The idea that antibody against these proteins would block cytoadherence and prevent infected cells from being protectively sequestered in the deep vascular beds is supported by in vivo evidence. Surprisingly, the cytoadherence proteins are also apparently antigenically diverse (strain-specific); one might have expected such a fundamental biological process to involve conserved molecules. As with the putative glycophorin-binding proteins, perhaps conserved domains that provide invariant epitopes will be identified as the important ligand of cytoadherence. Without such discovery, this class of proteins can be expected to prove unsuitable as vaccine candidates. Antibody-mediated prevention or reversal of cytoadherence as a defense strategy could also be potentially useful in immunotherapy. Infusion of antibodies directed against cytoadherence ligands might reverse or prevent such complications of falciparum malaria as cerebral dysfunction.

In addition to parasite antigens inserted into the erythrocyte plasma membrane during parasite maturation, there is at least one antigen that is deposited on or below the erythrocyte surface during merozoite invasion. This antigen, termed ring-infected erythrocyte surface antigen (RESA), is a 155-kD protein that apparently is discharged from micronemes of merozoites during invasion and becomes anchored to the erythrocyte cytoskeleton. Some of the monkeys immunized with RESA were partially resistant to challenge with *P. falciparum*.

Numerous antigens have been described, and in most cases their relationship to one another is still uncertain. Cloning genes coding for these antigens will help define the relationships. Perhaps such analysis might also help to identify invariant domains that will deserve particular attention.

Recently successful vaccination against blood-induced *P. falciparum* with polymeric synthetic hybrid peptides derived from merozoite-specific proteins was reported. This exciting finding will no doubt stimulate related approaches. One drawback to such studies, however, is that vaccines may have to experience moderate parasitemias (e.g., 0.5 percent infected erythrocytes) to permit detection of an effect of a vaccine on limiting a further rising parasitemia.

Transmission-Blocking Vaccines

The fact that mosquitoes ingest the host's serum when they take a blood meal suggested to malariologists the possibility of transmission-blocking vaccines. It was reasoned that if a vaccine could induce production of antibody against the gametes, zygotes, or ookinetes (the parasite stages that emerge and develop in the gut of mosquitoes that have ingested gametocytes), the presence of these antibodies in a blood meal might interfere with the sporogonic cycle. Since immunity in malaria is stage-specific, vaccines against these sexual stages would not attenuate the asexual parasitemia that causes illness. Accordingly, transmission-blocking vaccines would be altruistic; they would not directly benefit the vaccinee. The use of such vaccines could therefore be justified only if they were combined with a vaccine against the intraerythrocytic asexual-stage parasites that cause disease. By the same token, a fully effective antisporozoite vaccine that would presumably prevent appearance of all intraerythrocytic-stage parasites, including the gametocytes responsible for transmission, would obviate the need for a specific transmission-blocking vaccine.

Monoclonal antibodies have been produced that, when mixed with gametocyte-containing blood and fed to mosquitoes, prevent parasite development in the vector. These have been used to identify the target antigens. It has been learned that immunodominant target epitopes are expressed on macrogametes and are lost soon after fertilization (fusion of macrogametes and microgametes). These epitopes are thought to be constituents of a receptor-ligand system involved in gamete recognition; masking these ligands prevents fertilization. Other distinct antigens appear on the zygote and ookinete that also are targets of transmission-blocking antibodies. The mechanism of action of these antibodies is uncertain. Fortunately, the target antigens of transmission blockade are much more conserved among different strains than are the asexual-stage antigens.

The populations to which malaria vaccines will be applied when and if available is presently a topic of some debate, albeit an academic one at this time. The number and frequency of booster infections required as well as the duration of protection will be important in determining where malaria vaccines are likely to be applicable. If antisporozoite vaccines can be devised that provide solid but only brief protection, they might prove best suited for tourists and military personnel spending brief periods in areas where malaria is endemic. On the other hand, a vaccine against asexual intraerythrocytic parasites that reduces morbidity and mortality (especially in children and pregnant women) but is not solidly protective could be extremely useful in Africa, where other control modalities have not been effective.

Immunopathogenesis

While it is clear that the clinical manifestations (and death) caused by malaria result directly or indirectly from asexual stage parasites, remarkably little is known about the pathophysiological mechanisms. The hallmark of malaria is fever. Anemia, thrombocytopenia, splenomegaly, and water and electrolyte disturbance are frequent concomitants of malaria from all four species that infect humans. In falciparum malaria, very severe anemia, renal and pulmonary failure, and cerebral dysfunction can occur and presage death. Chronic infection with *P. malariae* is believed to give rise to nephrotic syndrome. Finally, in African Burkitt's lymphoma and tropical splenomegaly syndrome (TSS), malaria is believed to function somehow as a cofactor in pathogenesis. In Burkitt's lymphoma, malaria theoretically might act as a polyclonal stimulus to promote malignant transformation of Epstein-Barr virus–infected B cells, or provide for immunosuppression that interferes with host defense against EB virus–infected cells. In TSS, patients develop massive splenomegaly and very high IgM levels, both without detectable parasitemia; despite the absence of detectable parasitemia they respond to antimalarial chemotherapy. The role of malaria in the etiology of TSS is uncertain. The propensity for this disease in kindreds makes a genetic (or immunogenetic) factor likely to be important. A genetically determined deficiency in restoring the spleen to normal size after induction of splenomegaly by malaria is one (untested) possibility.

Fever

The mechanisms of fever in malaria remain obscure despite long-standing interest in this most dramatic of clinical features. A correlation between schizont rupture and onset of fever has been well established and explains the periodicity pattern of fever in synchronized infections (infections in which all asexual parasites are at the same stage of development at any one time). No exogenous pyrogen released from infected cells has been identified. Reports that some sera from febrile malaria patients can promote gelation of *Limulus* amebocyte lysates ("limulus test" for endotoxin) neither establishes that these patients have endotoxemia (since substances other than endotoxin can cause the test to be positive) nor provides insights into fever mechanisms. One possibility, supported only by an in vitro study, is that ingestion of parasite debris released at the time of schizont rupture may trigger release of an endogenous pyrogen (specifically, interleukin 1) from Kupffer cells and other macrophages. One feature commonly observed in semi-immune patients is an enhanced tolerance for certain levels of parasitemia without the experience of fever. No physiological basis for this observation has been provided, but it is of substantial interest since fever underlies much of the disability and may give rise to some complications (dehydration, hyponatremia) that lead to death. As the role of cytokines (such as interleukin 1 and tumor necrosis factor) in the pathogenesis of fever generally is more precisely elucidated, the pathogenesis of malarial fevers may become increasingly amenable to investigation.

Anemia

Anemia is a common complication in all but the most benign (early, low parasitemia) infections. The red cell destruction caused by parasitization, with resulting hemolysis, does not entirely account for the anemia, since the degree of anemia exceeds the parasite density. In fact, anemia may progress and worsen after chemotherapy. Premature destruction of uninfected erythrocytes is suggested by studies reporting shortened half-lives of ^{51}Cr-labeled autologous uninfected erythrocytes as late as 4 to 5 weeks after parasites have been eradicated by chemotherapy. Efforts by different groups to demonstrate the deposition of immunoglobulin on red cells of such patients have led to conflicting results, so that presently one cannot conclude that the anemia results from immune hemolysis. Other nonimmunologic changes in uninfected erythrocytes (perhaps subtle alterations in lipid content of erythrocyte membranes) that could render these cells more susceptible to splenic removal seems a likely alternative explanation. Finally, evidence that dyserythropoeisis (abnormal generation of erythrocytes and their premature destruction in the bone marrow) may occur adds a potential process. Since splenomegaly and other intrasplenic changes (as well, perhaps, as bone marrow changes) appear to be under T-cell regulation in rodent malaria, and since these changes are likely to contribute to the anemia, a contribution of cellular immune responses in the pathogenesis of the anemia warrants investigation.

Cerebral Malaria

Cerebral malaria, or cerebral dysfunction (usually altered state of consciousness) in a patient with malaria for which no other etiology can be identified, is a serious complication of *P. falciparum* infection in young children or nonimmune older children and adults. The complication is more prevalent in certain geographic regions. Several theories about its pathogenesis have been set forth but none has stood up to recent careful clinical investigation. A striking feature of the condition is that even in its most severe form (with deep coma), it can be entirely reversible, and patients rapidly recover with no detectable residual neurological sequelae. Studies in rodent models of malaria that result in neurological complications suggest that a subset of $Lyl^{+}2^{-}$ T lymphocytes might play a role in pathogenesis; athymic mice or ones treated with specific anti-T-cell or anti-TNF antibodies do not develop the complications. Although these animal models have been criticized for their failure to clinically or pathologically mimic human cerebral malaria, pursuing the findings might provide useful clues that could bear on human disease. One intriguing hypothetical explanation of the findings in rodents is that certain lymphokines might have reversible neurotoxic effects. The striking lack of neurological deficits in successfully treated patients with cerebral malaria makes the concept of a neuropharmacological basis of this complication particularly appealing and deserves investigative consideration.

Glomerulonephritis

Glomerulonephritis has been observed in a few cases of acute falciparum malaria and in some animal models. Since deposits of IgG, IgM, and C3 in the glomeruli have been reported and electron-dense deposits have been observed in the mesangium, this process probably results from immune complex deposition. Antimalarial treatment reverses the acute nephritis. Children with chronic *P. malariae* infection can develop full-blown nephrotic syndrome as a result of a membranoproliferative glomerulonephritis; this syndrome can progress to total glomerulosclerosis. Although immune complexes containing *P. malariae* antigens have been identified in the glomeruli, the fact that antimalarial chemotherapy generally fails to ameliorate the renal disease has shed some doubt on an etiologic role played by malaria in this condition. An autoimmune response initially triggered by malaria has been postulated.

Coda

Malaria remains one of the most important infectious diseases in terms of the number of people afflicted and the severity of illness. Reduced application and efficacy of conventional control measures have resulted in resurgence of transmission in some parts of the world. New technologies have provided the means of identifying parasite antigens and producing them in sufficient amounts to consider the feasibility of a malaria vaccine. Several stages in the life cycle are being investigated as potential targets of such a vaccine. Antigen diversity in the asexual stages poses a particular challenge.

In contrast to the remarkable progress in antigen analysis and the extensive efforts directed at vaccine development, the more fundamental aspects of malaria immunology have received less attention in the last decade. If the present empirical approaches to malaria vaccine development fail, future successes may depend on a greater understanding of the basic aspects of malaria immunology, including host defense and immunopathogenesis. The present state of knowledge and technology does not indicate whether applied or basic research efforts are more likely to lead to the much-needed breakthrough for developing immunological strategies of malaria control.

Additional Reading

BROWN, K. N., K. BERZINS, W. JARRA, and T. SCHETTERS. 1986. Immune responses to erythrocytic malaria. *Clin. Immunol. Allergy* 6:227–249.

CARTER, R., L. H. MILLER, J. RENER, D. C. KAUSHAL, N. KUMAR, P. M. GRAVES, C. A. GROTENDORST, R. W. GWADZ, C. FRENCH, and D. WIRTH. 1984. Target antigens in malaria transmission blocking immunity. *Phil. Trans. R. Soc. Lond. (Biol)* 307:201–213.

HOWARD, R. J. 1984. Antigenic variation of bloodstage malaria parasites. *Phil. Trans. R. Soc. Lond. (Biol)* 307:141–158.

MILLER, L. H., P. H. DAVID, and T. J. HADLEY. 1984. Perspectives for malaria vaccination. *Phil. Trans. R. Soc. Lond. (Biol)* 307:99–115.

MILLER, L. H., R. J. HOWARD, R. CARTER, M. F. GOOD, V. NUSSENZWEIG, and R. S. NUSSENZWEIG. 1986. Research toward malaria vaccines. *Science* 234:1349–1356.

NUSSENZWEIG, R. S., and V. NUSSENZWEIG. 1984. Development of sporozoite vaccines. *Phil. Trans. R. Soc. Lond. (Biol)* 307:117–128.

PLAYFAIR, J. H. L. 1982. Immunity to malaria. *Br. Med. J.* 38:153–159.

WEIDANZ, W. P. 1982. Malaria and alterations in immune reactivity. *Br. Med. J.* 38:167–172.

WYLER, D. J. 1983. The spleen in malaria. In *Malaria and The Red Cell.* Ciba Foundation Symposium 94. D. Evered and J. Whelan, eds. Pitman Books Ltd, pp. 98–116.

WYLER, D. J., and G. PASVOL. 1986. Malaria: perspectives on recent developments. In *Medical Microbiology*, Vol. V, C. S. F. Easmon and J. Jeljaszewicz, eds. Academic Press, pp. 65–109.

10

Immunology of Toxoplasmosis

Somesh D. Sharma

Toxoplasma gondii is a ubiquitous, intracellular, coccidian parasite that was first discovered in 1908 by Nicolle and Manceaux. It infects birds and virtually all mammals. Despite the fact that it is an important pathogen for humans, the first case was not reported until 1920. The development of reliable serodiagnostic tests in the past two decades has led to the recognition that toxoplasmosis is widespread among many human populations. On the basis of these tests, it is estimated that 500 million people are infected. Fortunately, only a minority of those infected develop serious clinical disease. Most acutely infected individuals are asymptomatic or experience a mild, self-limited illness.

Toxoplasma is transmitted to humans by ingestion of oocysts (present in cat feces) or through consumption of infected lamb, pork, or beef that has been insufficiently cooked to destroy the tissue cysts (which contain bradyzoites). Transmission from an infected mother to her fetus can also occur (congenital toxoplasmosis). Rarely, transfusion, organ transplantation, or accidental inoculation of laboratory workers transmits the organism. Infants born to women infected during pregnancy may have at birth or subsequently develop serious disabilities such as blindness, epilepsy, and mental retardation. Congenital toxoplasmosis and infection in immunocompromised individuals give rise to the most serious forms of the disease. Life-threatening toxoplasmosis can occur in individuals with diminished cellular immunity resulting from immunosuppressive agents administered as cancer chemotherapy or in association with organ transplantation. Similarly, individuals suffering from the acquired immune deficiency syndrome (AIDS) frequently experience serious central nervous system toxoplasmosis. After proliferation as tachyzoites during acute infection, toxoplasma transform into nonproliferating bradyzoites that inhabit cysts in various tissues, especially in the central nervous system, where they remain for the lifetime of the individual. Defects in cell-mediated immunity may permit transformation of bradyzoites in the cysts to proliferating tachyzoites, resulting in acute relapses.

Toxoplasmosis is also of considerable economic importance to the livestock industry in England, Japan, New Zealand, and Australia, since it can cause spontaneous abortion in sheep and swine.

This chapter reviews our present understanding of the immunology of toxoplasmosis. Since most of this information has been derived from experimental models, results of animal studies are emphasized.

Genetic Basis of Resistance to *Toxoplasma* Infection

The resistance of certain species of animals to infection with *Toxoplasma* organisms suggests that genetic factors might contribute significantly to host resistance. The size of the challenge dose and the parasite strain can influence the outcome of experimental infection. In mice, the most extensively investigated experimental host, striking differences in susceptibility (mortality following parenteral challenge) have been observed among different inbred strains. For example, mice bearing the $H\text{-}2^d$ major histocompatibility haplotype are more susceptible to the C56 strain of *Toxoplasma* than mice bearing the $H\text{-}2^b$ and $H\text{-}2^k$ haplotypes. Further analysis employing congenic strains of mice and their F_1 and F_2 progenies indicates that for the B10 strain mice, however, those with $H\text{-}2^{a/a}$ and $H\text{-}2^{b/b}$ genotypes are more susceptible than are those with $H\text{-}2^{d/d}$ and $H\text{-}2^{k/k}$ genotypes. Linkage between the $H\text{-}2^a$ allele and greater susceptibility has been demonstrated through breeding studies. These results clearly indicate the importance of the major histocompatibility (H-2) genes in development of host resistance against toxoplasmas in mice. In addition, the existence of a second disease susceptibility gene linked to the H-13 locus has been described.

H-2–linked genes influence not only the mortality of different strains of mice but also the formation of toxoplasma cysts. Whereas most mice form numerous toxoplasma brain cysts following challenge with tachyzoites of certain strains of *Toxoplasma*, five mouse strains and the F_1 generation of a cross between a resistant and a susceptible mouse strain have been identified that form fewer brain cysts. These five strains have in common the d or s haplotype of the H-2 complex. The greater number of cysts have been found in the brains of mice that display a mutation in the β chain of the Ia antigen. It is possible that this mutation results in severe impairment of antigen presentation necessary for induction of immune responses needed to control the dissemination of toxoplasmas to the brain. However, the mechanisms underlying the genetic determinants of susceptibility and cyst formation in these mouse models await elucidation.

Immune Response to *Toxoplasma* Infection

T. gondii infection has potent effects on specific immunity to the homologous parasite as well as on nonspecific immunity to other microorganisms (Figure 10-1). Animals infected with toxoplasma can develop resistance to homologous challenge as well as to

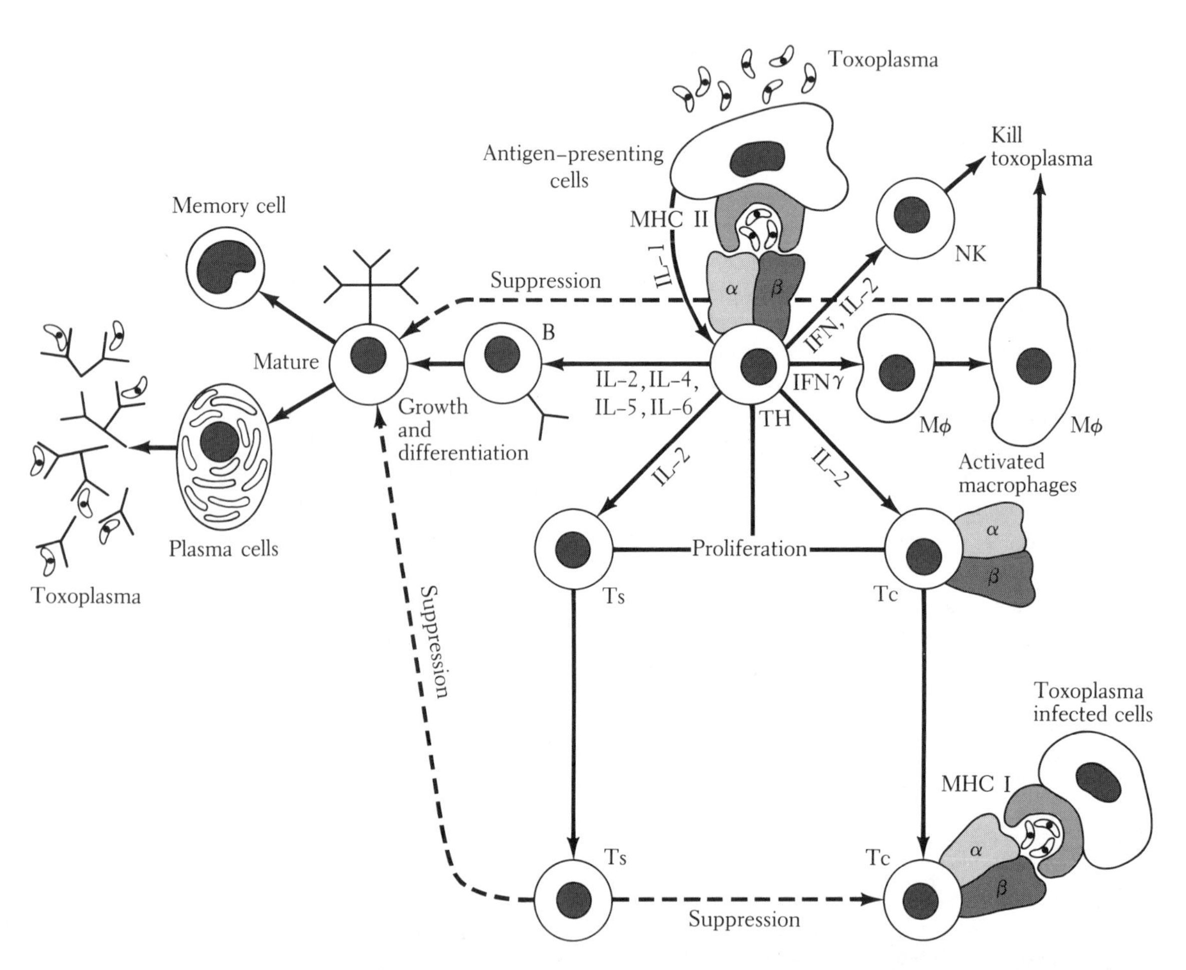

FIGURE 10·1 Induction of cellular and humoral immunity in defense against toxoplasma. An antigen-presenting cell such as a macrophage ingests the organism, digests it, processes the antigens and displays them on the surface in association with a class II major histocompatibility (MHC) molecules, normally present on the cell membrane of antigen-presenting cells. A helper T cell (TH) becomes activated when it binds to the processed antigen and to the MHC protein. Interleukin-1 (IL-1), a protein secreted by the macrophages, is presumed to play an important role in activating helper cells. The activated

challenge with phylogenetically unrelated organisms. Like most protozoal infections, toxoplasmosis is accompanied by the development of both specific humoral and cell-mediated immune responses. The relative contribution of these types of responses to protective immunity in toxoplasmosis is unclear at the present time. Detection of antibodies is the mainstay in the diagnosis of human infection. Cell-mediated immunity elicited following infection is most likely responsible for resolution of acute infections and has also been implicated as the basis of nonspecific immunity induced by experimental toxoplasmosis.

Humoral Response

In humans infected with the toxoplasma, IgM antibodies can be detected by a number of assays (indirect immunofluorescence or ELISA) as early as 7 days after infection, reach maximum titer within a few weeks, and gradually decline over months. Elevated antitoxoplasmal IgM persists for longer periods in some individuals, as determined by ELISA. IgG antibodies appear within 1 to 2 weeks after infection (although in some cases this can be delayed for up to 4 weeks) and rise rapidly to a maximum titer during the next 6 weeks. The IgG titer usually declines gradually over a period of months to years, and low titers commonly persist for life. Detection of antitoxoplasma antibodies in serum is clinically the most useful evidence of prior infection with toxoplasma. The presence of IgM antitoxoplasma antibody is regarded as an indication of recent infection, as is a rising IgG titer. The presence of IgG antibody per se only indicates prior infection and has limited diagnostic value. A very high IgG titer (>1 : 1000) does suggest recent infection.

---

helper cell (TH) secretes interleukin-2 (IL-2) and interferon-gamma (INFγ). Interleukin-2 induces proliferation of a subset of T cells that recognize the antigen in association with class I MHC proteins. Some of these cytotoxic T cells (Tc) could possibly kill toxoplasma-infected cells that display parasite antigen. Another subset of T cells (Ts) which suppress this cytotoxic and other responses (broken arrow) may turn off the defense mechanism. The development of cytotoxic (Tc) and suppressor T cells (Ts) requires assistance from the helper (TH) cells, either through direct contact or through release of soluble factors (lymphokines). Helper cells (TH) also participate in the development of humoral responses against parasite antigens. The activated helper cell binds to B lymphocytes that have independently recognized the antigen on an antigen-presenting cell. Signals between the helper T cell (TH) and B cell leads to a sequence of B-cell stimulation, maturation, and differentiation into a clone of immunological memory and a clone of antibody-secreting plasma cells. Antibody molecules (Y) bind to the parasite (B) and in the presence of complement components destroy it. Lymphokines secreted by helper cells (TH) can also lead to activation of macrophages (MΦ) and natural killer (NK) cells . Activated macrophages and NK cells can then directly kill tachyzoites. Macrophages that are activated can also suppress (broken arrow) antibody responses.

Because congenital toxoplasmosis may be devastating, great efforts have been exerted to develop guidelines for serodiagnosis in the newborn. Clinically apparent defects may only become apparent months or years after birth. Early detection and treatment are desirable because they may prevent late clinical sequelae of infection.

At least four different patterns of antibody responses have been observed in congenitally infected, asymptomatic infants and in infants with clinical features of congenital toxoplasmosis. IgM antibodies can be detected in 80 percent of infected babies at birth or in the first month of life and can either persist for months or may disappear relatively rapidly. All congenitally infected babies have antitoxoplasmal IgG that is maternally acquired but IgM may not be detected in the first days or even for the first year of life of some of these children. Such individuals may show evidence of synthesis of specific IgG at varying times after birth. A number of interpretations have been provided for these variable responses. The time during gestation when the fetus became infected, its immunological competence at the time of infection, and the presence of high titers of maternal antibody might influence the antibody responses of the congenitally infected individual. The relationship of the antibody responses to subsequent disease progression is unclear.

In an effort to improve the serodiagnosis of toxoplasmosis, the fine specificities of antibody responses have been examined (by Western blot analysis). Of interest is that the earliest IgM responses in acute infection are against epitopes on a 6-kilodalton (kD) toxoplasmal protein. Furthermore, the antibodies produced by congenitally infected infants recognize certain epitopes of toxoplasma distinct from those recognized by antibodies of their mothers.

The appearance of toxoplasma antibodies in mice infected with a relatively virulent strain has been extensively analyzed. The first antibody response of mice infected with an avirulent strain occurs within 2 days and is detected by the ability of sera to agglutinate the parasite in vitro. In contrast, such antibodies do not appear until the fifth day of infection in sera of mice infected with virulent strains of *Toxoplasma*. IgG antibodies are first detected 8 days after infection with either strain. The major isotype produced in these infections is IgG2; IgG3 titers are lower and IgG1 antibodies are generally undetectable.

Cell-Mediated Response

Cell-mediated immune response to toxoplasma infection can be detected both in vivo and in vitro. Delayed-type hypersensitivity responses (DTH) to toxoplasma antigen reflect specific in vivo T-cell reactivity in both human and experimental toxoplasmosis and may only appear months to years after infection. DTH to toxoplasma antigens can be elicited in humans. The positive skin test response apparently reflects the presence of chronic infection and largely excludes a recently acquired infection. Population surveys have revealed a close correlation between the positive DTH skin test response and the presence of antitoxoplasma antibodies (detected by a number of methods).

In mice, the initial appearance and intensity of the DTH reaction coincides with the development of high titers of antibody. In this model DTH becomes apparent on day 30 of infection and remains positive during chronic infection. Cellular immune responses to toxoplasma antigens have also been demonstrated in vitro by detection of macrophage migration inhibitory factor (MIF) with use of a guinea pig model. In this model, DTH and in vitro MIF production in response to toxoplasma antigens can be detected as early as 7 days after infection.

In vitro lymphocyte transformation to toxoplasma antigens has been studied as another measurement of cellular immune responsiveness in humans, mice, rabbits, and guinea pigs. Antigen-induced transformation of small lymphocytes of the spleens, lymph nodes, and peripheral blood has been observed in rabbits infected for 7 to 9 weeks. Lymphocytes from patients with suspected acute toxoplasmosis are less likely to proliferate in vitro in response to the antigens than are those from patients with chronic infections. These responses are typically elicited in patients with infection of greater than 1 year duration and has been documented to persist for at least 19 years. This observation implies persistence of long-lived memory lymphocytes and perhaps also chronic stimulation by toxoplasma antigens from tissue cysts.

The helper-inducer T cells ($CD4^+$) have been identified as the subset that proliferate in response to toxoplasma antigens in vitro. Although patients with acute toxoplasmosis have elevated numbers of circulating suppressor cells ($CD8^+$), these cells do not seem to account for the absence of toxoplasma-specific lymphoproliferative responses in the acutely infected patients. Addition of $CD8^+$ cells to cultures of toxoplasma antigen-responsive $CD4^+$ does not abrogate the $CD4^+$ proliferative responses. The possibility that functional suppressor cells that do not bear the classical suppressor T-cell marker ($CD8^+$) are activated during prolonged acute toxoplasma infection is under investigation.

Antigen-specific lymphoproliferation in acute and chronic toxoplasma infection in mice has been systematically examined. Spleen cells from CBA/J mice, but not C3H/He mice (both of which bear the $H\text{-}2^k$ haplotype), respond with marked antigen-specific proliferation when obtained from mice between days 15 and 80 of infection. The T-cell subset that proliferates in response to toxoplasma antigen, as in the case of human studies, is the helper-inducer ($CD4^+$) subset. The lack of proliferative response of C3H/He spleen cells can be restored in part by addition of exogenous recombinant interleukin 2 (IL-2). Culture supernatants from antigen-stimulated C3H/He spleen cells block the growth of an IL-2–dependent T cell line in the presence of exogenous IL-2, indicating the presence of an inhibitory factor and perhaps accounting for the unresponsiveness of spleen cells from this mouse strain. The nature of the inhibitory molecules, their cell source, their mechanisms of action, and the relevance of these findings to host defense remain to be elucidated.

Role of Humoral and Cell-Mediated Immunity in Resistance

The appearance of antibodies and cell-mediated immunity following toxoplasma infection is accompanied by destruction of many of the tachyzoites. Those which escape destruction transform into the bradyzoites which are contained within the cysts. Immunity in toxoplasmosis does not completely terminate infection but rather arrests the multiplication of toxoplasma tachyzoites and the associated host-cell destruction. Since acquired defects in cellular defenses (due to immunosuppressive agents or HIV infection) permit retransformation of bradyzoites within cysts to proliferating tachyzoites, immunity may include maintenance of encystment. Of interest is that chronically infected humans are apparently resistant to reinfection.

Acquired immunity (defined here as resistance to reinfection) has been observed in a number of species of laboratory animals, including mice, rabbits, hamsters, and Rhesus monkeys infected with toxoplasmas. This resistance becomes apparent within 2 to 4 weeks after the initial infection but does not result in elimination of the parasites present in the primary infection. Since resistance is only partial, not all parasites present in the second challenge are eliminated. For example, mice first infected with an avirulent strain and then challenged with another strain rapidly eliminate the challenge strain from the peritoneal fluid, yet the challenge-strain parasites can be isolated from certain tissue. Similarly, female mice chronically infected with an avirulent strain of *Toxoplasma* survive a massive challenge dose of virulent RH strain toxoplasmas, yet subsequently can transmit the avirulent strain to their offspring. The precise reasons for these paradoxical observations—that chronically infected mice that resist challenge with a highly virulent strain of the parasite cannot prevent congenital transmission of the original (avirulent) infecting strain remain unknown. Conceivably the very persistence of the original avirulent infection is critical for expression of resistance to challenge. Such nonsterilizing immunity is also observed with murine malaria and may have analogies with "concommitant immunity" described in schistosomiasis (see Chapter 14). In any case, these provocative observations suggest that this model of toxoplasma immunity could be exploited to gain further insights into the mechanism of protection.

Role of Antibodies

Toxoplasmosis is accompanied by the production of specific IgM, IgG, and IgA antibodies. Apparently these antibodies play a minor role in resistance to toxoplasmas. Passive transfer of serum with high titers of antitoxoplasma antibody fails to protect mice from challenge. Furthermore, inhibiting the antibody response to toxoplasma by treatment of mice with anti-Ia antibodies does not seem to interfere with the subsequent development of resistance to rechallenge. CBA/N mice that have a congenital defect in B-cell function develop very low antibody responses to toxoplasmas. Yet, these mice are protected from infection by rechallenge to the same extent as their normal CBA/J counterparts who mount a vigorous antibody response to infection. Interestingly, the immunodeficient mice develop fewer brain cysts than do CBA/J mice following infection. How antibody might influence cyst formation remains to be elucidated.

In contrast to these results, two studies suggest that antibodies may actually contribute to the development of resistance. Suppression of antibody production in mice by treatment with anti-μ antibody (antibody directed against the heavy chain of

IgM which inhibits antibody formation) results in overwhelming infection unless mice are rescued with sulfadiazine (a drug with antitoxoplasma effect). When serum containing antitoxoplasma antibody is transferred to these mice after sulfadiazine treatment, elimination of parasites is observed in 50 percent of the animals. Furthermore, there are reports that administration of toxoplasma-specific monoclonal antibodies to mice, and immune serum to guinea pigs, successfully transfer immunity against the toxoplasma. Clearly, additional investigation is required to reconcile the apparently conflicting evidence for a protective role of antibody in antitoxoplasma defense.

In vitro studies have suggested at least one way by which antibodies could facilitate resistance. In the absence of antitoxoplasma antibody, tachyzoites are ingested by normal mouse and human monocyte-derived macrophages. The parasites reside in the host cells within parasitophorous vacuoles to which secondary lysosomes do not fuse. Under these conditions the parasites survive and replicate. On the other hand, toxoplasmas first treated with specific antibody are endocytosed into vacuoles to which lysosomes fuse. The parasites are then killed within the resulting phagolysosome. Thus, antibodies may promote the rapid killing of the parasites by normal macrophages. However, once T cell–mediated immunity induces activation of macrophage antitoxoplasma defenses or defenses in infected somatic cells, this defense mechanism may be predominant.

Role of Cell-Mediated Immunity

Adoptive transfer of spleen and lymph node cells from hamsters infected for at least 3 weeks with toxoplasma protect normal recipients from lethal challenge with the highly virulent RH strain. Transfer of lymphocytes from hamsters infected for briefer periods fail to transfer protection but can transfer DTH. It is therefore evident that DTH responses to toxoplasma antigens and protective cellular immune responses are distinct processes (as is also the case in leishmaniasis; see Chapter 11). In contrast to success in the adoptive transfer of protection with lymphocytes in guinea pigs, similar efforts in BALB/c mice have been unsuccessful.

Toxoplasma infection is uniformly fatal in nude mice, even if mice receive sulfadiazine for the first 3 weeks. Sulfadiazine treatment of euthymic littermates, in contrast, permits the development of immunity. Passive transfer of antibody to or bone marrow reconstitution of the athymic mice does not increase survival, whereas thymic reconstitution does permit their development of resistance. In related experiments, the course of toxoplasmosis in MRL/lpr mice has been examined. These mice display ineffective helper T-cell functions which account for certain markedly depressed in vivo and in vitro immune responses. MRL/lpr mice infected with a relatively avirulent strain of *Toxoplasma* die within 2 weeks unless rescued with sulfadiazine treatment, whereas no deaths occur in control mice (including mice with the same H-2 major histocompatibility haplotype). Taken together, these observations suggest that mice with defective helper T-cell function are unable to develop protective immunity and reinforce the strong impression that T-cell–mediated responses are critical for antitoxoplasmic immunity. Helper T cells are important not only for the induction of a protective immune response following infection but also for the maintenance of resistance. Depletion of helper T cells by administration of monoclonal antibody specific for this population into mice chronically infected with toxoplasmas results in reactivation of infection, neurologic deficits, and death. The contribution of other T cells (e.g., cytotoxic T cells) that participate in induction and maintenance of resistance remain unknown. The contribution of the different subsets of T cells that participate in induction of protective immunity also remains to be delineated. Recent studies in mice suggest that both $CD4^+$ and $CD8^+$ lymphocyte populations play an important role in defense.

Lymphokines

Culture supernatants of specifically sensitized lymphocytes from patients or animals stimulated with toxoplasma antigens can activate macrophages to inhibit toxoplasma replication in vitro, implicating a role for lymphokine-mediated macrophage activation in defense. Lymphokines generated in this manner have been analyzed. Activity is detected in fractions (obtained by gel filtration) with apparent molecular weights of 40–80 kD, 30–40 kD, and 3–5 kD. The high-molecular-weight fraction is trypsin sensitive and neuraminidase resistant. Since it was also inactivated by prolonged incubation at pH 2, the possibility has been raised that it might be similar to lymphocyte-produced interferon (gamma interferon; IFNγ), a lymphokine sensitive to low pH. Indeed, recent reports have confirmed that such lymphokines contain IFNγ, as well as IFNα and IFNβ. Normal (unsensitized) human lymphocytes or mouse spleen cells cultured with subcellular components of toxoplasma secrete IFNα and IFNβ but not IFNγ. IFNγ is now recognized as an important lymphokine that can nonspecifically induce antimicrobial defense in macrophages infected with a variety of microorganisms. In addition to affecting cells of the immune system, IFNγ directly inhibits the growth of toxoplasma in somatic cells (fibroblasts) after 1 day of pretreatment. When toxoplasmas and IFNγ are added simultaneously, growth inhibition is observed only after a delay of 2 days. The antimicrobial effect of IFNγ in this case seems to result from a deleterious effect of accelerated tryptophan degradation in fibroblasts induced by the lymphokine. This accelerated degradation results from the induction of host indolamine 2,3-dioxygenase, which converts tryptophan to N-formal kynurenine, which degrades to kynurenine. These metabolites appear not to be toxic to the parasite. Rather, accelerated degradation depletes the host cell stores of tryptophan required by the tachyzoites. In essence, IFNγ induces parasite starvation.

Evidence for the existence of another lymphokine that can prevent multiplication of toxoplasmas within macrophages, kidney cells, and fibroblasts has been reported. In contrast to IFNγ, this lymphokine may induce repsonses more specifically targeted to toxoplasmas than to other microorganisms. The molecular weight of this proposed antigen-specific mediator has been found to be between 3 kD and 5 kD. This lymphokine has unfortunately received relatively little investigative attention since its discovery in the early 1970s. Although recombinant IFNα and IFNβ have no effect on the

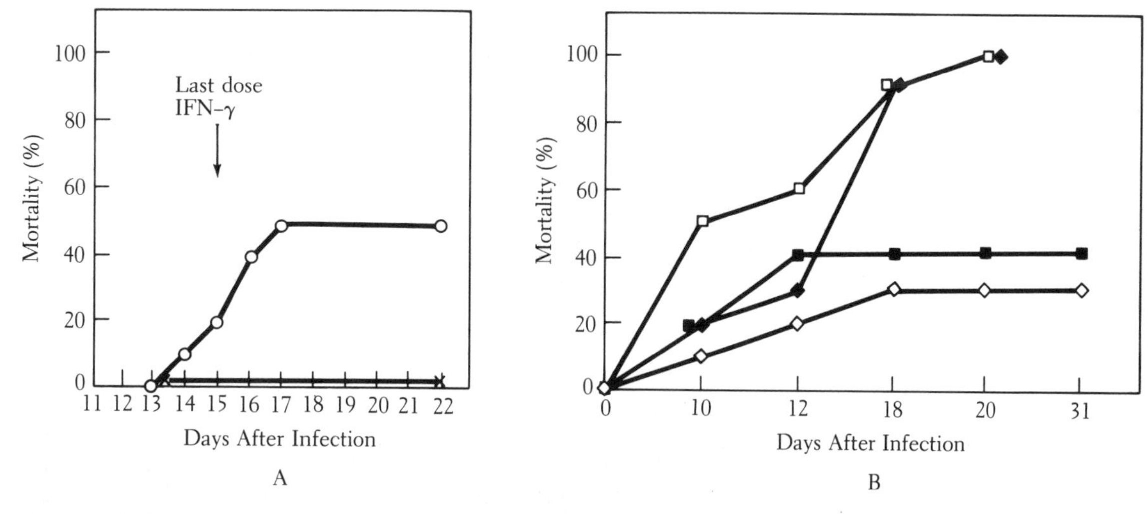

FIGURE 10·2 Effects of lymphokine treatment of mice on the course of *Toxoplasma* infection. A. Mice were treated with 5×10^3 units of recombinant murine interferon-gamma (IFNγ[x]) or with saline (PBS [0]). B. Mice were treated with either 300 units (■), 500 units (◇) of recombinant human interleukin 2 or equal volumes of saline (PBS (□, ◆)). (Adapted from MCCABE, R. E., B. J. LUFT, and J. S. REMINGTON, 1984. Effect of murine interferon gamma on murine toxoplasmosis. *J. Infect. Dis.* 150:961 by permission of the University of Chicago Press; SHARMA, S. D., J. M. HOFFLIN, and J. S. REMINGTON, 1985. In vivo recombinant interleukin 2 administration enhances survival against a lethal challenge with *Toxoplasma gondii. J. Immunol.* 135:4160.)

multiplication of toxoplasmas in fibroblasts, they can activate neonatal and adult human macrophages to kill toxoplasmas, albeit less efficiently than can IFNγ. Tumor necrosis factor, a product of monocytes and macrophages, has been found to have no effect on the growth and multiplication of toxoplasmas within macrophages.

Administration of recombinant murine IFNγ completely protects mice against a lethal challenge of toxoplasmas (Figure 10-2A). Moreover, pretreatment of animals with a monoclonal antibody to IFNγ results in increased mortality following infection with an avirulent strain of *Toxoplasma* and is reversed by administration of IFNγ. Thus, production of endogenous IFNγ appears to be an important event for the development of resistance to acute toxoplasma infection in vivo. Administration of another well-defined lymphokine, interleukin 2 (IL-2), also imparts significant protection against a challenge dose of toxoplasmas that kills all control mice (Figure 10-2B). Recombinant IL-2 recipients who survive develop fewer brain cysts than untreated mice. The exact mechanisms by which these two lymphokines induce a protective response is unknown but may be distinct. IFNγ-treated mice have increased antitoxoplasma antibody and activated macrophages, whereas IL-2 treated mice do not. Toxoplasma-infected mice that receive IL-2 but not infected controls display increased natural killer (NK) cell activity against antigenically unrelated target cells. Since it has been shown that NK cells and cytotoxic T cells can directly kill toxoplasma tachyzoites in vitro, IL-2 may enhance survival by activating NK cells. IFNγ may exert a protective effect by activating macrophages and NK cells and by enhancing antibody production. IFNγ is also known to assist cyst formation within murine astrocytes in vitro cultures, presumably by inhibiting the division of tachyzoites, thereby preventing the destruction of astrocytes and allowing sufficient time for cyst formation. Whether IFNγ plays any role in in-vivo cyst formation needs clarification. It remains to be determined whether other cytokines, such as interleukin 1 (IL-1), granulocyte-macrophage colony stimulating factor, interleukin 4 (IL-4), or interleukin 6 (IL-6) play any role in antitoxoplasmal defense.

Role of Macrophages

Activated macrophages are most likely to play an important role in defense against *Toxoplasma* infection. Activated macrophages rapidly phagocytize toxoplasma tachyzoites in vivo and in vitro. Toxoplasmas continue to multiply within normal mouse and human monocyte-derived macrophages and eventually destroy these cells. Immune and activated mouse and human macrophages, as well as rat peritoneal and alveolar macrophages, contain fewer toxoplasmas after in vitro infection and culture, suggesting that these macrophages are more resistant to infection than normal macrophages. Alveolar macrophages obtained from mice chronically infected with *Toxoplasma* organisms, in contrast to those from uninfected mice, are larger, have a greater tendency to spread on glass, and display an increased rate of infection (all features of "activation"). They also kill or inhibit multiplication of toxoplasmas more readily than do alveolar macrophages from uninfected mice. Alveolar macrophages from normal (uninfected) human subjects and rats kill toxoplasmas without additional stimulation or activation. This finding indicates that there are important differences in the native ability of macrophages obtained from different species and anatomic sites to retard growth of toxoplasmas. Such differences can be exploited in an effort to define the mechanisms of killing and inhibition of multiplication of toxoplasmas.

Biochemical Basis of Defense

The precise mechanisms by which macrophages from a specific anatomical site or host kill toxoplasmas remain unknown at the present time. TLCK, a serine protease inhibitor, can abrogate macrophage antitoxoplasmal activity in a dose-dependent manner. The effect of TLCK is irreversible and is not accompanied by visible morphological damage to macrophages. The molecules affected by TLCK have not been identified. Aminophylline, a drug that increases intracellular cyclic AMP concentrations, also abrogates the inhibitory effect on intracellular growth of toxoplasmas in activated mouse macrophages. It does so without apparent damage to the host cells. TLCK and aminophylline, in contrast to their effects on activated mouse macrophages, have no effect on early elimination of toxoplasmas by human monocytes.

Considerable evidence that provides insight into the relationship between phagocytosis and survival of toxoplasmas has accumulated over the last few years. During phagocytosis of toxoplasmas, monocytes and macrophages that produce reactive oxygen products (toxic oxygen radicals) can inhibit intracellular replication and efficiently kill the parasites. In contrast, macrophage populations that produce little or no reactive oxygen metabolic products during phagocytosis of toxoplasmas are unable to kill or inhibit multiplication. With use of a cell-free system for providing reactive oxygen radicals, it has been observed that toxoplasmas are resistant to high concentrations of reagent H_2O_2 as well as to H_2O_2 generated by the glucose–glucose oxidase system. This lack of susceptibility of toxoplasmas is most likely due to the presence of an endogenous catalase that can scavenge H_2O_2. However, toxoplasmas are susceptible to H_2O_2 in the presence of a peroxidase and halide. In addition, toxoplasmas are efficiently killed by products generated by the chemical reaction of xanthine and xanthine oxidase. The microbicidal activity of this system is inhibited by relatively specific scavangers of hydroxyl and free oxygen radicals, suggesting that hydroxyl radicals and singlet oxygen are involved in killing toxoplasmas.

Intracellular killing of toxoplasmas by monocytes and macrophages may also involve oxygen-independent mechanisms. Monocytes from patients with X-linked chronic granulomatous disease (cells that either lack or are unable to activate the enzymes necessary for the respiratory burst) are not permissive for intracellular replication of toxoplasmas. Furthermore, resident rat alveolar macrophages are able to rapidly kill a large number of intracellular toxoplasmas without a demonstratable respiratory burst or release of toxic oxygen radicals. Similarly, a mouse macrophage-like cell line, incapable of producing toxic oxygen metabolites after appropriate stimulation can inhibit multiplication of toxoplasmas. The precise biochemical basis of intracellular killing by such cells is not known. Acidification of the phagosome (a defense mechanism effective against certain other intracellular microbes) containing the parasites may be responsible for killing toxoplasmas in activated mouse macrophages. However, acidification of the phagosome is probably not required for killing of toxoplasmas by resident rat alveolar macrophages. Finally, toxoplasmas coated with antibody promote the fusion of secondary lysosomes with the parasitophorous vacuole, even in macrophages that are not activated. These coated organisms are killed, presumably by lysosomal contents.

Immunosuppression in Toxoplasmosis

Acute infection with *T. gondii* profoundly affects primary and secondary antibody responses. Mice infected with toxoplasmas have depressed antibody responses to non-toxoplasmic antigens, including both T-dependent and T-independent antigens. The suppressive effect appears to be nonspecific, in that it occurs with several antigens and affects all immunoglobulin isotypes. Immunosuppression is due, at least in part, to interference with initiation of immunological memory. Nonspecific suppression of antibody response also occurs during secondary infection when mice are rechallenged with a large number of toxoplasmas. The suppressive effect of toxoplasma infection on initiation of memory cells is controlled by the induction of radioresistant suppressive macrophages. Such suppression could provide a mechanism for the survival and proliferation of parasites in the host.

Antigenic Structure

The antigenic structure of toxoplasma is complex, as revealed by visualization of proteins of toxoplasma separated by one-dimensional polyacrylamide gels and stained with different dyes. A similar degree of complexity is observed when toxoplasma proteins are separated by one-dimensional polyacrylamide gels, transferred to solid supports (Western, immuno, or protein blotting), and then reacted with serum containing antitoxoplasma antibodies (Figure 10-3a). In contrast, a simpler picture is seen when detergent extracts of intact parasites are immunoprecipitated with immune human or mouse serum. Four cell membrane–associated antigens corresponding to molecular weights of 43 kD, 35 kD, 27 kD, and 14 kD were recognized (Figure 10-3). A 22-kD and a 27-kD surface antigen have also been detected. These findings suggest that there are at least four immunodominant antigens on the cell surface of the toxoplasma. Monoclonal antibodies that precipitate these cell surface antigens have been prepared (Figure 10-3b) and should facilitate their further characterization as well as their role in diagnosis, pathogenesis, and development of resistance against toxoplasma.

It was also revealed by Western blot techniques that during acute infection IgG and IgM directed against a 6-kD antigen is the predominant antibody response, whereas most chronic sera fail to react with this low-molecular-weight material. This antigen appears to be a carbohydrate because reaction with antitoxoplasma antibody is almost completely abrogated by periodate treatment of the antigen, and the antigen binds to concanavalin A yet is resistant to treatment with various proteolytic enzymes. The exact location of this antigen in tachyzoites is unclear. A monoclonal antibody directed against cytoplasmic components and another directed against a membrane component of *T. gondii* both react strongly with the 6-kD antigen.

An internal membrane protein (30 kD) equal to roughly 1 percent of the total cell protein has been characterized in detail. It contains an immunodominant region with repetitive epitopes. Charge shift electrophoresis indicates the presence of extensive hydrophobic regions within the protein. No intrachain disulfide bonds have been detected. Its amino acid sequence remains unknown and gene cloning studies are pending.

In contrast to the results of these studies performed with the tachyzoite stage of the toxoplasma, little is known about the antigenic structure of two other stages of the parasite, the bradyzoite and sporozoite. Recent work performed employing polyclonal and monoclonal antibodies and sera from humans known to be infected by oocysts has lead to the demonstration of antigenic differences between tachyzoites, bradyzoites, and sporozoites. In addition to antigens common to all stages, there are at least two major antigens specifically present in oocysts and sporozoites. These antigens (~67 kD and ~190 kD) react with sera from infected patients. A 22-kD antigen has been found to be present in extracts of bradyzoites but not tachyzoites. The nature of these antigens

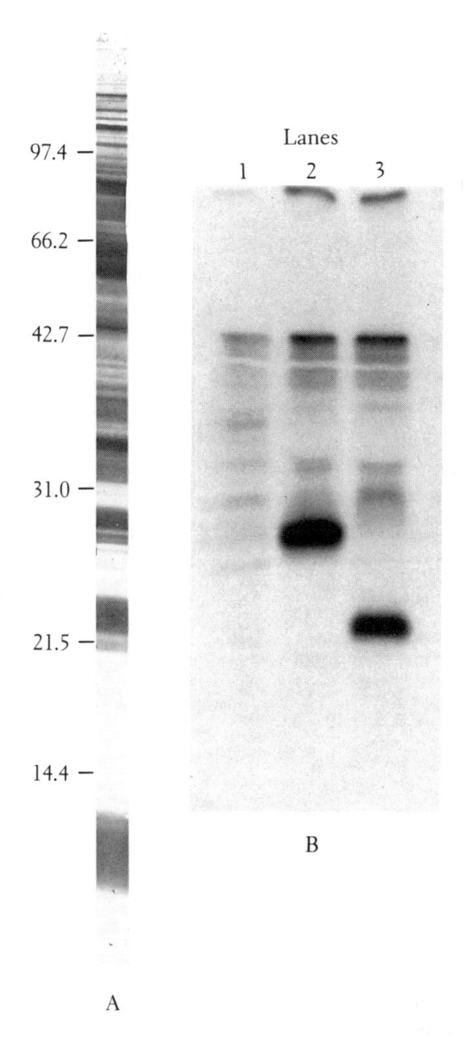

FIGURE 10·3 A. Reactivity of antibodies present in infected human serum to toxoplasma sonicate antigens after electrophoretic transfer of proteins from SDS gels to nitrocellulose membranes. B. Toxoplasma antigens precipitated by monoclonal antibodies directed against parasite surface antigens, left to right: lane 1, antigens precipitated by control supernatant; lane 2, antigens precipitated by a monoclonal antibody directed against the major membrane protein of toxoplasma; lane 3, antigens precipitated by a monoclonal antibody directed against a 22-kD protein.

needs to be established and their relative roles in inducing cell-mediated and humoral immune responses have yet to be assessed.

Toxoplasmic antigens secreted by infected cells have also been described. They have been used in skin tests and have also been found to react with the serum from a patient with acute toxoplasmosis. A circulating antigen of the toxoplasma has also been detected in the serum of mice with acute toxoplasmosis. Circulating antigens can be detected as early as 1 day after infection in experimental animal models and have been detected in a population of patients with recently acquired lymphadenopathic toxoplasmosis. The apparent absence of these antigens from circulation in some patients could represent sequestration in the form of immune complexes or reflect the insensitivity of the present assay methods. Newer, more sensitive methods for detecting circulating antigen are needed.

Vaccines

Attempts have been made to immunize various animal species with crude vaccines prepared from killed toxoplasmas with variable success. Differences in vaccine efficacy in different animal models are most likely related to the natural susceptibility of the host to toxoplasma infection. It is well known that both mice and rabbits are highly susceptible to toxoplasma infection, whereas guinea pigs are naturally more resistant. The use of highly virulent challenge strains in these studies may also have contributed to vaccine failure in the more susceptible hosts; immunization of mice can induce long-term

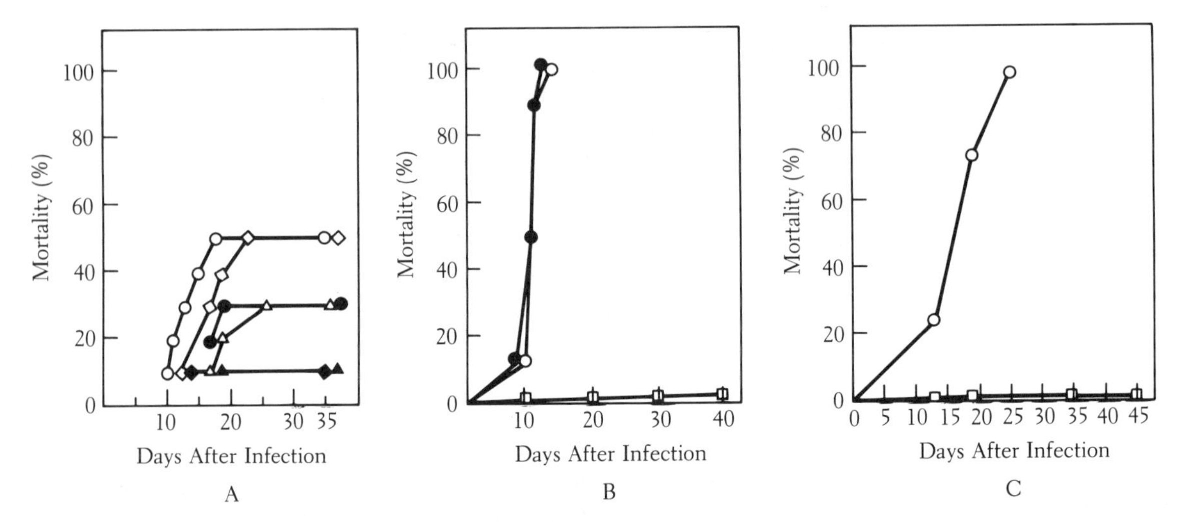

FIGURE 10·4 Protection of mice by adoptive transfer of a monoclonal antibody and by immunization with an antigen purified with this antibody. A. Survival of mice treated with the monoclonal antibody F_3G_3 (closed symbols) or NS 1 (control) supernatant (open symbols) following different challenge doses of Toxoplasma. B. Survival of mice immunized with the affinity-purified antigen recognized by F_3G_3 (□) compared with survival of mice treated with saline alone (O) or incomplete Freund's adjuvant (O). C. Survival of mice immunized with F_3G_3 antigen (□) and challenged with the parasite 6 months later. (From SHARMA, S. D., F. G. ARAUJO, and J. S. REMINGTON, 1984. Toxoplasma antigen isolated by affinity chromatography with monoclonal antibody protects mice against lethal infection with *Toxoplasma gondii*. *J. Immunol.* 133:2818.)

protection against challenge with a relatively avirulent strain. Purified toxoplasma antigens are also protective. An antigen present in the cytoplasmic fraction of the organism induces protective immunity against a relatively avirulent strain of toxoplasma (Figure 10-4). In contrast, the major membrane glycoprotein of the toxoplasma has been found to be suppressive rather than induce protective immunity.

Other approaches have been proposed for the development of a vaccine against toxoplasmosis. Live temperature-sensitive mutants of the parasite are being assessed in experimental vaccine trials. Immunostimulating complexes containing surface membrane proteins of *Toxoplasma* organisms have been prepared and in preliminary studies have been found to protect mice against a challenge dose that killed all unimmunized mice. The availability of an effective vaccine would be particularly useful in the prevention of human congenital toxoplasmosis and spontaneous abortion in sheep. The increased frequency with which potentially fatal cerebral toxoplasmosis is being diagnosed in immunosuppressed patients could argue for an even more widespread use of a toxoplasma vaccine in human populations at risk.

Coda

Toxoplasmosis is widespread in many human populations and has recently emerged as a serious problem in some patients with AIDS. A striking feature of toxoplasmosis is that following immunologically-mediated resolution of the acute infection, the parasite persists for the life of the host in a dormant form (bradycyst). Resolution of acute infection probably involves both humoral and cellular immunity. Reactivation of infection occurs primarily in association with immunosuppression and is most likely due to the loss of lymphocyte-dependent defenses, suggesting that these probably maintain the dormancy of the bradycyst stage. Current research efforts include identifying toxoplasma antigens that are targets of defense and might be utilized in a vaccine, and gaining a better understanding of the cellular and molecular basis of defense. The in vitro and in vivo evidence that IFNγ and other lymphokines may be particularly important in antitoxoplasma defense merits further investigation.

Additional Reading

DUBEY, J. P. 1977. Toxoplasma, hammondia, sarcocystis, and other tissue cyst forming coccidia of man and animals. In *Parasitic Protozoa*, vol. III, J. P. Krier, ed. Academic Press, pp. 450–490.

KRAHENBUHL, J. L., and J. S. REMINGTON. 1982. The immunology of toxoplasma and toxoplasmosis. In *Immunology of Parasitic Infections*, 2nd Edition, S. Cohen and K. S. Warren, eds. Blackwell Scientific Publications, pp. 356–421.

REMINGTON, J. S., and G. DESMONTS. 1983. Toxoplasmosis. In *Infectious Diseases of the Fetus and Newborn Infant*, 2nd Edition, J. S. Remington and J. O. Klein, eds. W. B. Saunders Company, pp. 143–263.

11

Immunology of Leishmaniasis

Mary E. Wilson and Richard D. Pearson

Leishmania species produce a spectrum of disease involving several hundred thousand to several million people on every continent except Australia and Antarctica. The clinical manifestations of the leishmaniases depend on complex interactions between the hosts' immune responses and the parasites' invasiveness, tropism, and pathogenicity. They can be divided into three basic clinical syndromes: cutaneous and mucocutaneous leishmaniasis, which vary from self-healing, localized cutaneous ulcers to disseminated lesions of the skin or mucous membranes, and visceral leishmaniasis, also known as kala-azar, which involves the entire reticuloendothelial system.

Cutaneous leishmaniasis is caused by *Leishmania major*,[1] *L. tropica*, *L. aethiopica*, the *L. mexicana* complex, and the *L. braziliensis* complex. Members of the *L. braziliensis* complex are also responsible for mucocutaneous leishmaniasis in Latin America. The *L. donovani* complex is the major cause of visceral leishmaniasis. It too can cause cutaneous lesions; skin involvement following therapy is termed post-kala-azar dermal leishmaniasis. In a few instances, visceral leishmaniasis has been attributed to *Leishmania* species that usually cause cutaneous disease. Research on *Leishmania*

We wish to thank Mrs. Carol Nelson and Mrs. Terry Hughes for their expert secretarial assistance. This work was supported in part by a grant from the Rockefeller Foundation, a grant from the Thomas F. and Kate Miller Jefress Memorial Trust, and Training Grant T32 AI 07046 from the National Institutes of Health.

1. *Leishmania major* has been widely studied in experimental models. Many older reports refer to it as *Leishmania tropica major* or merely as *Leishmania tropica*. *Leishmania major* should be distinguished from true *Leishmania tropica*, previously termed *Leishmania tropica tropica*. Although both species produce cutaneous leishmaniasis, they are distinguishable by biochemical and kDNA analysis, and produce lesions of different severity in humans and animal models.

species has been fueled by the need for better methods of diagnosis and treatment of infection and by the quest for effective means of immunoprophylaxis.

In addition to their importance as human pathogens, *Leishmania* species have emerged as excellent model systems for the study of the immunology and molecular biology of parasitism. They are prototypic intracellular pathogens that reside only in mononuclear phagocytes. Control of infection is predominantly, if not solely, by cell-mediated immune mechanisms. Furthermore, *Leishmania* species are easily propagated in vitro, and excellent animal models are available.

Studies of leishmaniasis in humans and animals over the past decade have provided important insights into the mechanisms involved in lymphocyte activation of macrophages; the genetic determinants of cell-mediated immunity; the mechanisms responsible for antigen-specific immunosuppression during intracellular parasitic infection; and the immunopathogenesis of disease. The cell biology of host–leishmania interactions is reviewed in Chapter 5. The focus of this chapter is on the immunology of leishmaniasis in humans and animal models.

Clinical Immunology

It has been suggested that in some respects, leishmaniasis, like leprosy, is a spectral disease with polar forms. At one pole lie diffuse cutaneous leishmaniasis (see Figures 11-1, 11-2) and visceral leishmaniasis, clinical forms in which numerous intracellular parasites persist chronically in macrophages, and in which there is little histological evidence of an effective cell-mediated immune response. Delayed type (cutaneous) hypersensitivity reactions (DTH), in vitro lymphoproliferative responses, and production of interferon-gamma (INFγ) by lymphocytes in response to leishmanial antigens are absent in patients with these forms. These findings in leishmaniasis are analogous to observations made in patients with lepromatous leprosy, who have massive bacterial infections of macrophages and little cellular immune response. At the other end of the spectrum lie leishmaniasis recidiva and mucocutaneous leishmaniasis. These syndromes are analogous to tuberculoid leprosy, in which there is an intense mononuclear cell infiltrate with few organisms present. These distinct clinical patterns focus attention

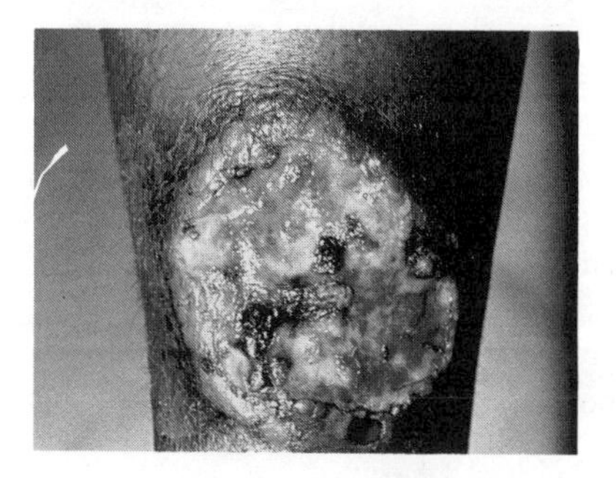

FIGURE 11·1 South American cutaneous leishmaniasis due to *L. braziliensis braziliensis* on the lower extremity. Parasitized macrophages are found at the base of cutaneous lesions. (Reprinted from PEARSON, R. D., and A. Q. SOUSA. Leishmania species (kala-azar, cutaneous, and mucocutaneous leishmaniasis). 1985. In *Principles and Practice of Infectious Diseases, 2nd ed.*, MANDELL, G. L., R. G. DOUGLAS, JR., and J. E. BENNETT, eds., copyright ©1985 by John Wiley & Sons, Inc. Reprinted by permission of John Wiley & Sons, Inc.)

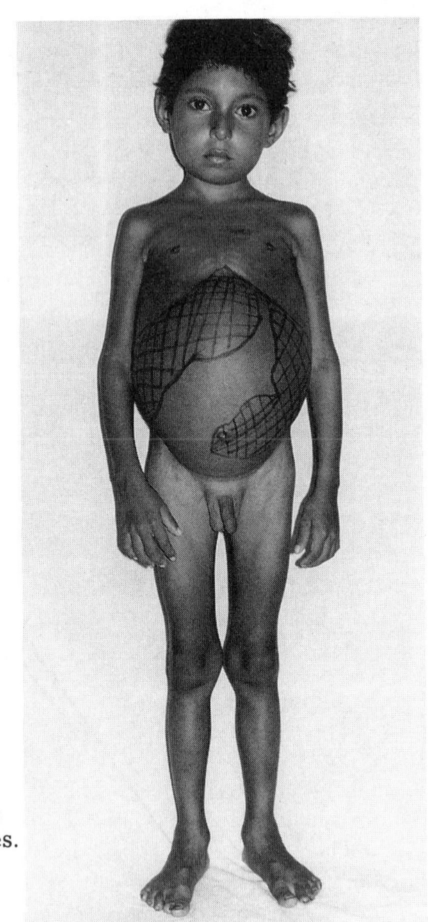

Figure 11·2 Brazilian child with advanced visceral leishmaniasis due to *L. chagasi.* The spleen (markings on left abdomen) and liver (markings on right) are massively enlarged due to extensive parasitization of mononuclear phagocytes. (Photograph kindly provided by Dr. Anastacio de Sousa, Universidade Federal do Ceara, Fortaleza, Brazil.)

to the role of cell-mediated immunity in defense against *Leishmania* organisms and suggest an important contribution of immunoregulation in the outcome of disease.

Infection with *Leishmania* species that can cause visceral leishmaniasis (kala-azar) does not invariably lead to overwhelming disease. In fact, recently reported prospective epidemiologic studies of visceral leishmaniasis in Brazil indicate that asymptomatic and "oligosymptomatic," self-resolving *L. chagasi* infections are common. Risk factors for developing clinical disease include young age and malnutrition before onset of infection.

Of the patients with serological evidence of infection, 23 percent remained asymptomatic and 17 percent had classical kala-azar. The remaining patients manifested oligosymptomatic disease. Of these, 25 percent progressed to classical kala-azar, while in the other 75 percent, illness resolved after a mean duration of 35 months. It therefore appears that protective immune responses arise in the majority of humans infected with *L. chagasi*: a minority, however, develop progressive kala-azar and have no demonstrable DTH or lymphoproliferative responses to leishmanial antigens. The factors that govern the outcome of *L. donovani* infection are currently under study. Genetically determined host susceptibility is one consideration, as discussed in the next section.

Cellular immune responses detected in patients with cutaneous leishmaniasis may indicate elements involved in successful host defense. Peripheral blood lymphocytes from patients with active American cutaneous lesions proliferate and produce INFγ in vitro in response to parasite antigens. DTH responses also develop over the course of infection, but their emergence does not seem to correlate well with healing. Despite histopathologic and immunologic studies of relatively large groups of patients with cutaneous leishmaniasis, only two patients have been followed longitudinally: a laboratory worker was accidentally infected with *L. tropica* amastigotes; the other patient acquired natural *L. mexicana* infection while in Belize. In the laboratory worker, sequential in vitro assays of cell mediated responses were performed during the development of a cutaneous lesion. Peripheral blood lymphocyte proliferation and interleukin-2 (IL-2) production in response to soluble *L. tropica* antigens appeared within 5 weeks of infection and reached maximum levels coincident with ulceration of the lesion. Thereafter the magnitude of both in vitro responses diminished as healing occurred, but lymphocyte proliferative responses declined more slowly than IL-2 production. Healing was complete in 20 weeks. The patient with naturally acquired *L. mexicana* infections was studied before and at intervals for 6 months after therapy with stibogluconate sodium. Before therapy, the patient's peripheral blood lymphocytes proliferated normally in response to concanavalin A and secreted MAF-containing lymphokines, but showed no response to *L. mexicana* antigens. No suppressive humoral factors were identified. The patient's monocyte-derived macrophages readily killed his own infecting *L. mexicana* strain when stimulated with lymphokines in vitro. One month after treatment, the patient's T cells showed variable but measurable proliferative responses to *L. mexicana* antigens, and by 6 months, responses were fully developed in the proliferative and lymphokine-generating assays. In this case the development of antigen-specific lymphoproliferative and INFγ-generating responses of peripheral lymphocytes coincided with clinical improvement. The relative delay in lymphocyte responsiveness in this case compared with the case of *L. tropica* infection is noteworthy.

These findings raise several important questions. Why is the emergence of leishmania-specific T helper cells so slow? Why is healing delayed after helper/inducer T cells appear in the blood? Is cutaneous ulceration a consequence of the immune response? Finally, which T cell subsets mediate protective immunity after ulcers have spontaneously healed?

In a subset of patients with *L. braziliensis braziliensis* infection the primary cutaneous ulcer heals spontaneously, but there is the later appearance of mutilating mucosal lesions of the face and head. The time between healing of primary lesions and mucosal involvement is generally several years, but has varied from a month to 24 years. The leishmanial skin test is usually positive in mucocutaneous leishmaniasis. A granulomatous response is observed in biopsy specimens, and parasites are sparse. Peripheral blood lymphocytes proliferate in response to leishmanial antigens and produce INFγ, which is capable of inducing macrophages to inhibit intracellular amastigote growth in vitro. The immune response is apparently effective in limiting (but not eradicating) mucosal infection. Whether the persistence of a small number of parasites is sufficient to perpetuate a locally destructive, hyperergic response or whether tissue destruction results from autoimmune responses is as yet unresolved.

Genetic Determinants of Susceptibility and Resistance to Leishmaniasis

The degree to which the clinical findings result from an expression of the parasite's genetic constitution (virulence) and the host's innate susceptibility or capacity to respond (immunity) have been the focus of recent investigative efforts.

Humans

Little is known about the genetic determinants of susceptibility or resistance to leishmanial infection in humans. Clinical observations suggest that there are important differences among individuals in terms of severity of disease. For example, as noted earlier, only a minority of those infected with *L. donovani* develop progressive visceral leishmaniasis; the majority appear to have self-healing, asymptomatic or oligosymptomatic, infection. Moreover, in regions where *L. aethiopica* or *L. mexicana* species are present, a subset of patients develop anergic diffuse cutaneous leishmaniasis while others have localized cutaneous ulcers that spontaneously heal.

In residents of the eastern Andes of Bolivia, extreme facial mutilation from mucocutaneous leishmaniasis has been observed almost exclusively among persons of African ancestry, even though many more cases of cutaneous leishmaniasis were identified among indigenous Amerinds (Indians). Blacks also had more vigorous leishmanial skin-test responses. These observations suggest that the severe mutilation observed in blacks may have resulted from exaggerated (but ineffective) hyperergic immune responses and also that genetic factors may be important determinants of disease. This hypothesis awaits testing in a more rigorous manner.

The possibility of a relationship between susceptibility to leishmaniasis and ABO blood group types has also been raised. It has been postulated that *Leishmania* organisms may utilize externally disposed blood group antigens as a form of molecular mimicry to evade host defenses. However, no correlation between ABO blood group type and the development of clinically apparent visceral leishmaniasis has been found in studies in Brazil or India.

Leishmania donovani Infection in Mice

The past decade has witnessed dramatic progress in defining the immunogenetics of leishmanial infections in mice. At least two characteristics affect the course of murine infection; the ability of the parasite to bind, enter, survive, and multiply within host macrophages, and subsequent T cell–dependent immune responses. Inbred strains of mice fall into two categories: either resistant to *L. donovani* (Lsh^r) or susceptible (Lsh^s), with Lsh^r behaving as an incomplete dominant allele and Lsh^s as a recessive one. Within a given mouse strain, there is little variation in the spectrum of disease. The course of visceral leishmaniasis in susceptible mice such as BALB/c or C57BL/10 differs from that in humans with kala-azar or experimentally infected Syrian hamsters in that susceptible mice do not usually die of *L. donovani* infection despite high parasite burdens. Lsh^r strains such as C3H/HeJ spontaneously clear their infections.

Studies with recombinant inbred mouse strains facilitated mapping of the *Lsh* alleles to a single locus on chromosome 1, and not to the major histocompatibility complex. The regulator of resistance to two other intracellular pathogens, *Salmonella typhimurium (Ity)* and *Mycobacterium bovis (Bcg)*, appears to be under control of the same locus on chromosome 1. Some Lsh^s mouse strains spontaneously reduce their parasite burdens (cure), whereas others retain high parasite loads indefinitely (noncure). With the use of congeneic mouse strains on a B10 background (which differ genetically only at the small piece of chromosome that comprises the H-2 complex), it has been shown that acquired immunity among susceptible mice is H-2 linked, although it is modulated by non-H-2 genes. Mice homozygous for the H-2^b haplotype spontaneously reduce their parasite burdens or "cure" late in infection, while mice homozygous for H-2^d or heterozygous mice behave as "noncure." Study of other H-2 haplotypes has revealed different responses, with H-2^s and H-2^r mice clearing their infections more rapidly than H-2^b (early vs. late cure). Moreover, studies using recombinant mice that carry alleles of different haplotypes at either end of the H-2 complex have localized control of the response to the K end of H-2. The difference between cure and noncure for different H-2 types has been verified with congeneic mice on different backgrounds (e.g., BALB or CXB). Although the rate of recovery varies considerably among strains, the same pattern of H-2–associated recovery has been observed. Subsequent studies have revealed that the H-2 differences are dose dependent and that the influence of alleles at the H-11 locus can override those at the H-2^b–associated determinant.

The *Lsh* gene product appears to be expressed at the level of the liver macrophage (Kupffer's cell). Infection with *L. donovani* in Lsh^s mice leads to proliferation of amastigotes in vivo in Kupffer's cells, whereas Lsh^r strains limit the intracellular growth of amastigotes after a brief period of proliferation. Explanted liver macrophages mirror this response in vitro. Macrophages from Lsh^s mice support the growth and division of amastigotes in vitro, whereas Lsh^r cells limit amastigote proliferation after 2 days in culture. This lag period may be the time that it takes for *Lsh* gene product expression to occur. A similar lag time occurs in vitro and is unaltered in lethally irradiated mice, suggesting that the effect is independent of T cells.

The ability of different mononuclear phagocyte populations to support the intracellular growth of leishmanias also varies according to the site from which they are obtained (e.g., peritoneum, Kupffer's cells, skin), their age (extent of differentiation), and their state of activation. Not all mononuclear phagocytes are equally susceptible to lymphokine activation of antileishmanial effects. For example, elicited peritoneal cells are relatively resistant to such activation in comparison to resident peritoneal macrophages.

Leishmania major Infection in Mice

Infection with *L. major* is also under genetic control, although the strain susceptibilities do not parallel those of *L. donovani* (Table 11-1). BALB/c and other mice that are susceptible to *L. major* develop progressive local infection after subcutaneous inoculation that disseminates widely and eventually leads to death. In contrast, B10 and other resistant strains display localized, self-curing ulcers. Susceptibility to *L. major* and *L. mexicana* is mediated by one or more genes which are distinct from *Lsh*. Attempts to map them have given variable results. It is noteworthy that individual mouse strains may be sensitive to one *Leishmania* species but quite resistant to another, even though the species have the same tissue tropism (e.g., *L. major* and *L. mexicana*). Among the inbred strains, BALB/c mice have proved to be unique in their universal susceptibility to all *Leishmania* species studied.

Early studies using radiation chimeras indicated that susceptibility to *L. major* was determined by descendants of donor hematopoietic bone marrow cells. Susceptibility to disease appears to depend on both a "permissive" macrophage and T-cell responses. Cultured macrophages from susceptible strains of mice allow extensive intracellular amastigote growth. Macrophages from some resistant mouse strains permit replication

TABLE 11·1
Susceptibility of inbred strains of mice to infection with *L. donovani* or *L. major*

Strain	*L. donovani*	*L. major*
A/Jax	R[a]	S/R[b]
CBA/Jax	R	R
C3H/HeJ	R	R
DBA/2	R	S
C57BL/6	S	R
C57BL/10	S	R
$B10.D_2$	S	R
BALB/c	S	S

[a] R, resistant; S, susceptible.
[b] Reports are conflicting.

of the parasite in vitro, whereas macrophages from other strains do not. Recent studies (discussed in the next section) indicate that progressive *L. major* disease is due not only to a failure in the elicitation of protective T-cell clones but also to the induction of disease-enhancing T-cell populations.

The H-2 genes, which have a profound effect on the course of *L. donovani* infection in susceptible mice, have little effect on the development of *L. major* or *L. mexicana* skin lesions. A major determinant for *L. major* infection is another gene, *Scl-1*, which on current evidence maps to mouse chromosome 8. The *Scl-1* gene exerts some influence over *L. mexicana* infection as well. H-11–linked differences also have an effect on cutaneous disease caused by *L. major* and upon visceralization and metastatic lesion development in *L. mexicana* infection. Thus, the H-11–linked gene controls elements of resistance in both visceral and cutaneous leishmaniasis. Neither the H-11– nor *Scl-1*–linked gene products have yet been identified.

Histopathology of Leishmaniasis

Studies of the histopathology of leishmaniasis in experimentally infected humans and animals have provided insights into the sequence of cellular immune responses in vivo. When cultured *L. donovani* promastigotes were inoculated into the skin of human volunteers, nodules appeared at the site of inoculation. A granulomatous response was observed in biopsies of these nodules.

A hamster model has been used to characterize the initial events in infection. When Syrian hamsters, which are highly susceptible to *L. donovani*, were infected intradermally with approximately 100,000 stationary-phase promastigotes, a local mixed polymorphonuclear (PMN) and mononuclear phagocyte inflammatory response was observed within 1 hour. Although intracellular parasites were distributed equally between the two types of phagocytes, they assumed amastigote morphology only in the latter. Leishmania were killed within PMN, but amastigotes survived in mononuclear phagocytes and proliferated during the ensuing 2 weeks. Granulomas (composed of epithelioid cells, Langhan's type giant cells, and eosinophils) formed between 4 and 6 weeks, and the parasites were eliminated from the skin without ulceration. Despite the reduction of parasites at the site of inoculation, visceral dissemination occurred in some of the animals.

Experimental cutaneous leishmaniasis of guinea pigs caused by *L. enriettii* infection also results in a heavy local infiltration of macrophages containing amastigotes. Ulceration of the overlying skin occurs. Eventually, a surrounding ring of lymphocytes and fibroblasts develops, followed by a decrease in the number of infected macrophages and restoration of the epidermis. Only fibrosis and cellular debris remains. The influx of lymphocytes is accompanied by development of cutaneous DTH to leishmanial antigens, suggesting that T helper cells play a role in the clearance of amastigotes.

Immunocytochemical and electron microscopic studies have supported the role of T cells as important local effector cells in the healing of murine cutaneous leishmaniasis. C57BL/6 mice infected with *L. mexicana amazonensis* display progressively enlarging cutaneous lesions over the first 8 weeks with epidermal thickening, ulceration, and accumulation of eosinophils and Ia^+ infected macrophages. Healing after 12 weeks of infection is associated with a local influx of T helper ($L3T4^+$)[2] and T cytotoxic/suppressor (Lyt-2^+)[2] cells into the dermis, and Ia antigen expression on epidermal keratinocytes. Intracellular parasites become scarce as T-lymphocyte infiltration increases. In contrast, lesions in genetically susceptible BALB/c mice continue to enlarge and never heal. There is minimal T-lymphocyte influx. Keratinocyte Ia expression remains absent in BALB/c lesions. However, there is no deficiency in circulating T cells in either mouse strain; both $L3T4^+$ and Lyt-2^+ cells are present in normal numbers in the peripheral circulation of both strains during infection.

Cell-Mediated Immune Responses in the Resolution and Prevention of Leishmaniasis

Clinical observations and histopathologic findings point to the central role of cellular immunity in control of leishmanial infection. This role has been born out by studies of experimental leishmaniasis in animals and in vitro observations of lymphocyte–macrophage interactions. Studies with *L. enriettii*, which produces cutaneous ulcers in guinea pigs, indicated that cell-mediated immune mechanisms are responsible for resolution of cutaneous leishmaniasis. Spontaneous healing of *L. enriettii* ulcers is preceded by the development of DTH and the appearance of lymphocytes capable of proliferating in vitro in response to leishmanial antigens. Although antileishmanial antibodies are produced during infection, their presence does not correlate with resolution of disease. Suppression of cell-mediated responses by administration of antithymocyte antisera increases the severity of disease, providing further proof that T lymphocytes play a critical role in controlling infection.

Studies of mice infected with *L. major* further demonstrate the central role of cell-mediated immunity in the spontaneous resolution of cutaneous leishmaniasis and in protection against reinfection. In contrast to normal mice, ones rendered T cell–deficient by thymectomy and irradiation fail to spontaneously heal cutaneous ulcers. Moreover, T cells obtained from spleen, peritoneal exudates, or lymph nodes of healed mice confer protection to syngeneic recipient mice. Serum from healed mice itself is not protective although it can enhance the effects of lymphoid cells administered concurrently. Immunity against *L. donovani* is also conferred to naive mice by transfer of T

2. $Ly1^+2^-$ and $L3T4^+$, refer to antigens on the T-cell surface that are found to be either present (+) or absent (−) by monoclonal antibody binding. Lymphocyte surface antigens sometimes correlate with the function of the lymphocyte subset.

lymphocytes from healed, syngeneic donors; transfer of immune serum does not provide protection.

The late "cure" response in BALB/cJ mice chronically infected with *L. donovani* is determined in part by genes at the H-2 locus. Spleens from chronically infected BALB/cJ mice contain T cells that respond to leishmanial antigen by proliferating and producing INFγ. These cells appear to be present at low frequency. An $Lyl^{+}2^{-}$, $L3T4^{+}$ T-cell line was isolated from a late curing mouse that was also boosted with *L. donovani* antigen. This cell line produced INFγ in response to *L. donovani* antigens. When cocultured with *L. donovani*–infected BALB/cJ macrophages, it was able to activate macrophages to kill intracellular amastigotes. Furthermore, after adoptive transfer of this cell line, naive BALB/cJ mice challenged with amastigotes demonstrated a marked reduction in parasite burden, compared with control mice. In contrast, another *Leishmania*-dependent T-cell line, also $Lyl^{+}2^{-}$, $L3T4^{+}$, which was derived from a draining lymph node of a subcutaneously immunized mouse, produced no INFγ. It failed to protect macrophages against *L. donovani* in vitro or in vivo. In summary, the spontaneous reduction in *L. donovani* burden observed late during chronic infection of BALB/cJ mice may be mediated by $Lyl^{+}2^{-}$, $L3T4^{+}$ T cells that proliferate and produce INFγ in response to leishmanial antigens. Recent studies indicate that $Ly2^{+}$ (CD8) cells also contribute to protective immune responses.

The resolution of cutaneous leishmaniasis due to *L. major* and development of protective immunity also appear to depend on complex interactions between protective and disease-enhancing T-cell populations. The ratio of $Lyt2^{-}$, $L3T4^{+}$ leishmania-specific T cells to $Ly2^{+}$, $L3T4^{-}$ T cells may be critical in determining the outcome of infection. When the ratio is low, as in CBA mice or in BALB/c mice treated with anti-$L3T4^{+}$ antibodies to lyse this cell population, disease resolves. When the ratio is relatively high, as in infected BALB/c mice, the disease progresses. However, when near complete eradication of $L3T4^{+}$ cells is achieved, progressive *L. major* infection again occurs. In summary, the spontaneous resolution of cutaneous leishmaniasis is mediated by one or more subsets of T helper cells, but their proliferation and expression seem to be affected by multiple factors, such as their preinfection frequency and interaction with other mononuclear cells. The protective T cells appear to belong to the subpopulation of $L3T4^{+}$ lymphocytes known as Th1 cells and the disease-enhancing cells, also $L3T4^{+}$, may be members of the Th2 subpopulation. The former are characterized by their secretion of IFNγ and IL-2; Th2 cells secrete neither of these lymphokines but secrete interleukin 4 (IL-4) and interleukin 5. Support for this notion derives from at least two observations. First, the spleens and draining lymph nodes of C57BL/6 mice infected with *L. major* (self-healing model) contained significantly greater steady-state levels of mRNA for IFNγ than did the organs from infected BALB/c mice (disseminating, chronic model) for all but weeks 4 and 6 of infection. In contrast, mRNA for IL-4 was detected only in the infected BALB/c mice. Second, two T cell lines produced with specificity to two *L. major* antigens imparted either protection or disease enhancement on adoptive transfer to BALB/c mice. The protective line had the characteristics of Th1 cells; the disease-enhancing line had Th2 characteristics. Interestingly, the former line was specific for antigens that could impart protection when used in a vaccine; the latter was specific for antigens that caused a more severe disease when used in a vaccine. Additional evidence for the interaction of different T-lymphocyte subpopulations in regulating the outcome of disease is discussed in the section on immunoregulation.

T-Lymphocyte–Dependent Mechanisms of Defense: The Role of Lymphokines

Important insights into the immunobiology of leishmaniasis and basic principles of cellular immunology have emerged from studies of lymphocyte–macrophage interactions in leishmaniasis. Human monocyte-derived macrophages or murine macrophages, which are exposed to cell-free supernatants from lymphocytes cultured with antigens or lectin mitogens, under the proper in vitro conditions, can kill or inhibit the growth of intracellular amastigotes of several *Leishmania* species. A basic principle of cellular immunology is that T helper cells produce soluble lymphokines in response to specific antigens, but the effects of these lymphokines on target cells such as macrophages are generally nonspecific. Macrophage "activation" by soluble lymphocyte factors is a complex and intricately coordinated event. There are alterations of macrophage metabolism, morphology, and constitutive proteins, which culminate in an enhanced ability to kill microorganisms. Macrophage activation is not an all-or-none phenomenon: both priming and triggering steps are involved. Under certain conditions, macrophages have been activated to kill tumor targets or intracellular pathogens such as *Listeria* or *Toxoplasma* without killing *Leishmania*. The mechanism for differential states of activation has not been elucidated, but the requirements for killing intracellular *Leishmania* amastigotes appear to be more stringent than for killing tumor targets or other intracellular microbes.

Studies of leishmaniasis have led to the identification of INFγ as a principal macrophage activating factor (MAF). Initially, supernatants from lymphocytes stimulated with plant lectins (concanavalin A) or nonleishmanial antigen (PPD) were found to have MAF activity that resulted in the killing of *L. major* and *L. donovani* amastigotes by murine and human macrophages. Antibodies against INFγ abrogated this effect. Subsequently, recombinant INFγ was found to activate macrophages to inhibit and kill *L. major* and *L. donovani* amastigotes to the same degree as lymphocyte supernatants. Peripheral blood mononuclear cells from patients with kala-azar when cultured in vitro with *L. donovani* antigens neither proliferate nor make INFγ. It therefore has been proposed that the lack of INFγ production in response to parasite antigens represents a key defect in the immune response during kala-azar. INFγ is not the only MAF; there are other factors that are capable of priming and triggering macrophages to kill intracellular leishmanial amastigotes. Recent evidence suggests that GM-CSF and interleukin-4 may have this capacity; these factors have not yet been assessed in kala-azar patients.

Unexpectedly, murine macrophages exposed to lymphokines before incubation with *L. major* ingest fewer amastigotes, in addition to inhibiting replication of ingested parasites (see Figure 11-3). Their ability to ingest other particles, such as IgG-sensitized erythrocytes, is unimpaired. One possible explanation for the decrease in parasite

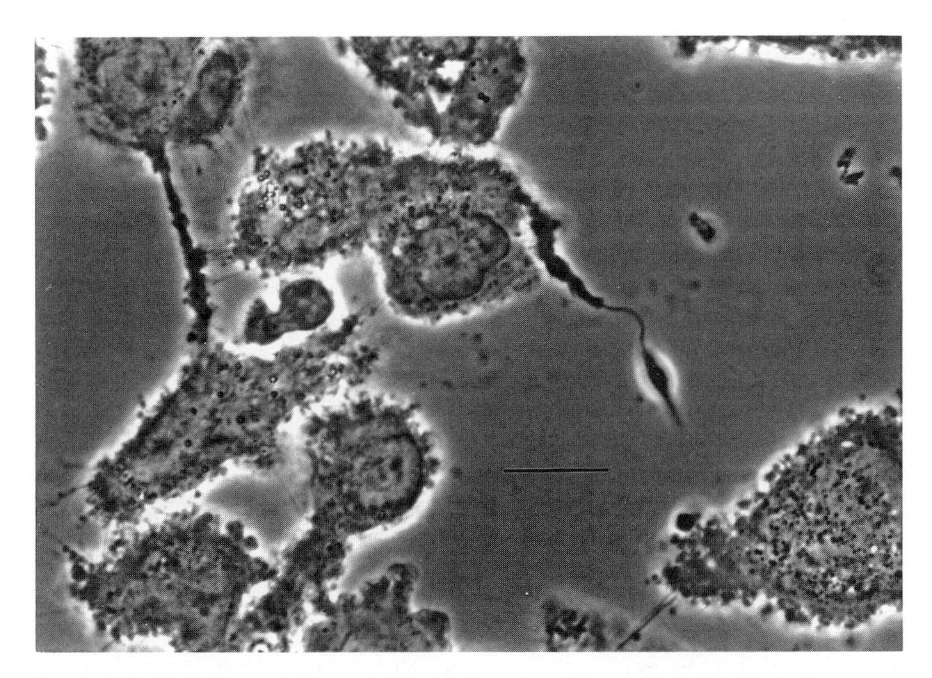

FIGURE 11·3 Human monocyte-derived macrophage which has bound an *L. donovani* promastigote by its flagellum. Note the macrophage pseudopod enveloping the flagellum. Bar, 10 μm. (Reprinted from PEARSON, R. D., J. A. SULLIVAN, D. ROBERTS, R. ROMITO, and G. L. MANDELL, 1983. Interaction of *Leishmania donovani* promastigotes with human phagocytes. *Infect. Immun.* 40:411–416.)

ingestion is that lymphokines down-regulate macrophage surface receptors that mediate ingestion (Figure 11-4). For example, the number of mannose/fucose receptors on macrophages decrease after lymphokine exposure; mannose/fucose receptors, as well as other receptors on macrophages, are involved in the attachment and ingestion of leishmania. Peritoneal macrophages from *L. major*–infected mice are specifically deficient in their ability to ingest amastigotes, suggesting that this effect of lymphokines also occur in vivo.

Patients with active cutaneous lesions have circulating lymphocytes that produce INFγ in response to leishmanial antigens. Their peripheral blood monocytes inhibit the growth of intracellular amastigotes after exposure to INFγ in vitro. From these observations, one must conclude that the protracted course of cutaneous leishmaniasis is not due to a failure in the development of *Leishmania*-specific cell-mediated defense mechanisms; the presence of circulating lymphokine-producing T cells is not sufficient for disease resolution. As previously noted, adoptive transfer of T-cell lines that produce MAF(s) may, under certain circumstances, actually exacerbate infection with *L. major*.

Lymphocytes can also produce factors that inhibit macrophage activation. Studies of supernatants obtained from a T cell line (phorbol myristate acetate–stimulated EL-4 thymoma cells) indicate that certain lymphocyte subsets have the capacity to

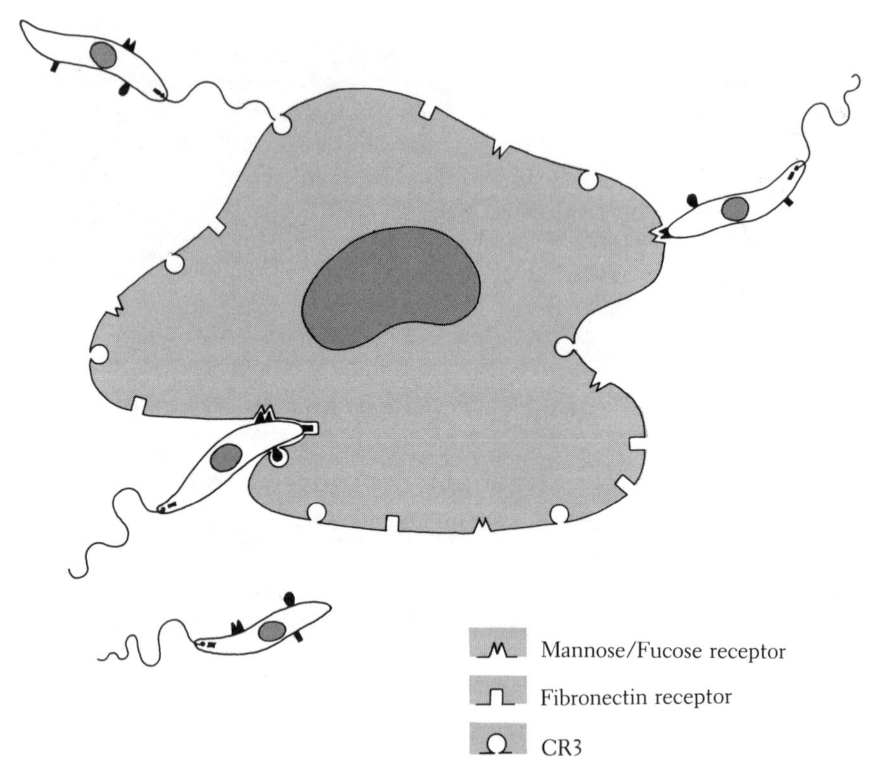

FIGURE 11·4 Suggested mechanism for receptor-mediated phagocytosis of *Leishmania* sp. promastigotes. Several macrophage receptor systems may act in concert, like sequential teeth on a zipper, to facilitate parasite ingestion. Reduced ingestion of parasites by lymphokine-treated macrophages may be due to reduced receptor expression or receptor mobility.

produce suppressive factors that can prevent macrophage priming by other lymphokines. The effects of EL-4 factors are selective; lymphokine-induced macrophage killing of *L. major* is inhibited without affecting macrophage killing of extracellular tumor targets.

The mechanisms by which lymphokine-activated macrophages kill intracellular amastigotes remain controversial. Macrophages exposed to lymphocyte supernatants or recombinant INFγ in vitro develop enhanced oxidative microbicidal potency, which correlates with acquisition of leishmanicidal capacity. Oxidative microbicidal mechanisms no doubt contribute to the killing of amastigotes by macrophages, but nonoxidative microbicidal mechanisms are also operative. Monocytes from patients with chronic granulomatous disease do not mount an oxidative burst but can be activated by lymphocyte supernatants or INFγ to kill *L. donovani* amastigotes. It is not clear whether nonoxidative killing is due to an active process or to changes in macrophage metabolism that result in starvation of parasites. Oxidative and nonoxidative microbicidal systems need not be exclusive, and both probably contribute to the destruction of amastigotes by activated macrophages. Finally, different *Leishmania* isolates have been observed to vary in their susceptibility to killing by lymphokine-activated macrophages. This may be important in the pathogenesis of some forms of leishmaniasis.

Activation of Macrophages by Contact with T Cells

Recent studies indicate that macrophages can be activated to inhibit growth and kill *Leishmania* by direct contact with T cells. Activation by this mechanism does not involve secretion of soluble macrophage factors nor is there evidence of cytotoxicity to host cells. For example, lymphocytes from popliteal lymph nodes draining infected foot pads of C57BL/6 mice induce *L. major* – infected syngeneic peritoneal macrophages to kill intracellular amastigotes when the cells are cocultured. Lymphocytes from popliteal lymph nodes of uninfected control mice or from mice injected with complete Freund's adjuvant exert little antileishmanial effect. Amastigote death cannot be attributed to cytotoxic effects on infected macrophages. The maximum effects of leishmanial sensitized lymphocytes are observed with lymph node cells obtained at 4 to 5 weeks on infection, a time that coincides with spontaneous resolution of lesions. This mechanism is specific to leishmanial antigens and is genetically restricted.

Several lines of evidence indicate that macrophage activation in these studies is truly independent of soluble lymphokines. First, lymphocytes physically separated from amastigote-infected macrophages by a semipermeable membrane show substantially reduced antileishmanial effect. Furthermore, there is no detectable MAF activity in supernatants collected from the macrophage-lymphocyte cocultures, and monoclonal antibodies against murine INFγ does not abrogate the contact-mediated effects. Since contact-mediated activation is antigen-specific, and Ia restricted, it is distinct from the nonspecific and genetically unrestricted effects of soluble MAF.

Circumstances in which lymphokine production or lymphokine effects are minimal do not appear to affect contact-mediated macrophage activation. Cyclosporin A blocks in vitro secretion of lymphokines that mediate antileishmanial effects but does not inhibit the capacity of T cells to exert contact-mediated activation, even under stringent conditions. In addition, inflammatory peritoneal macrophages (starch-elicited) are relatively refractory to lymphokine activation in vitro, but contact-mediated activation of these macrophages results in killing even of an *L. mexicana* isolate that is resistant to killing by lymphokine-activated macrophages (see Figure 11-5).

Humoral Factors in Leishmaniasis

Promastigotes of the many *Leishmania* species are susceptible to killing by the complement membrane attack complex (C5b-C9) generated in human serum. The mecha-

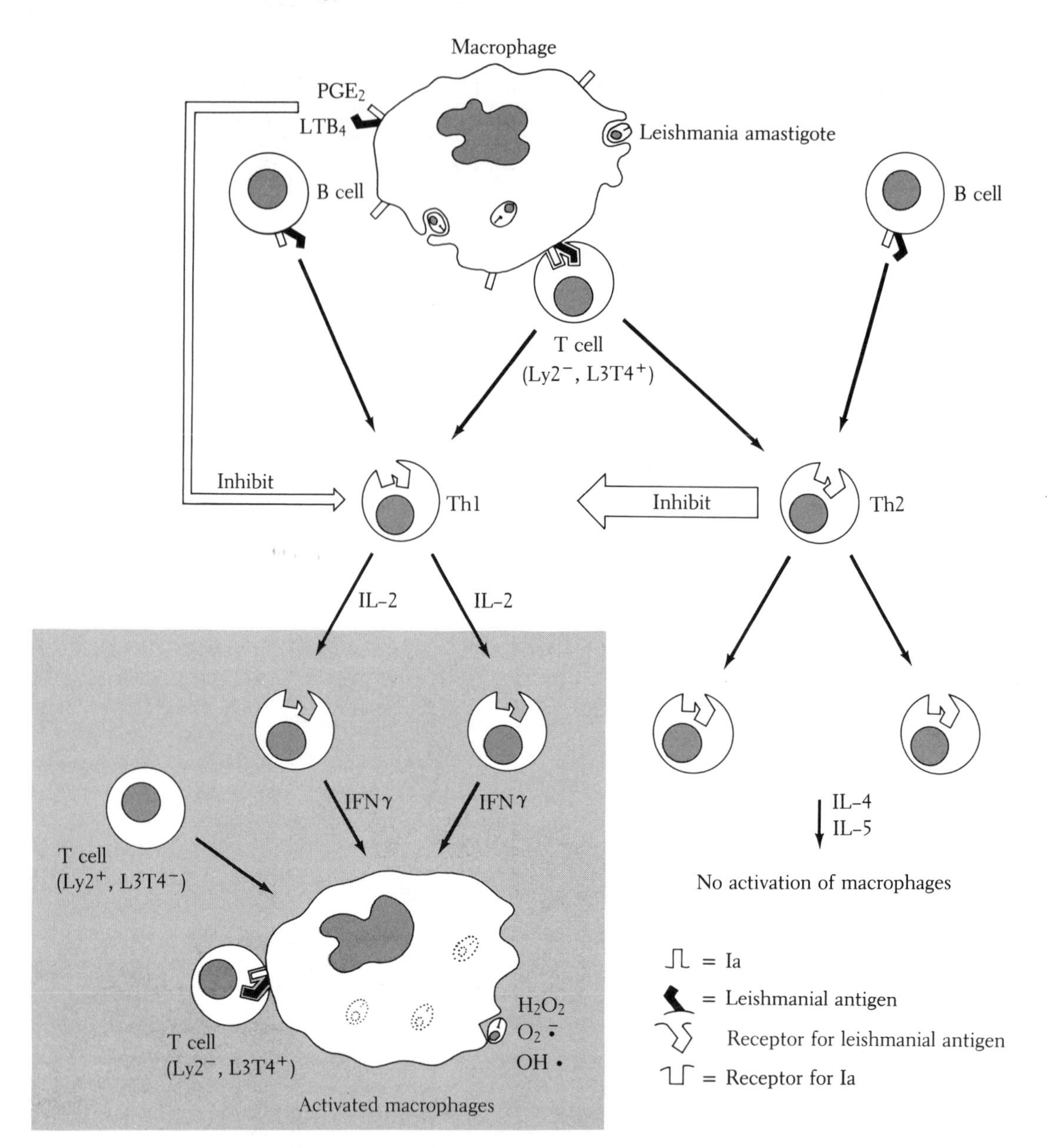

FIGURE 11·5 Proposed scheme of events leading to resolution of murine leishmaniasis. The outcome of murine infection with *Leishmania* species depends on cellular immune responses. The initial event is processing and presentation of leishmanial antigens to T cells by macrophages or B lymphocytes. Th1 T cells ($Ly2^-$, $L3T4^+$; T helper/inducer phenotype) produce INFγ and IL-2. Th1 T cells activate macrophages to kill intracellular amastigotes by secretion of INFγ or by direct contact, which is Ia restricted. Th2 T cells (also

nisms of complement activation varies: some *Leishmania* species activate complement through the alternative pathway. Stationary growth-phase promastigotes are more resistant than logarithmic-stage promastigotes to complement-mediated killing. Fixation of complement on the parasite surface results in death if the membrane attack complex is functional or in opsonization with C3b and C3bi if it is not. If polymorphonuclear leukocytes are present, ingestion of complement-opsonized promastigotes results in intracellular parasite killing. Conversely, complement opsonization facilitates entry of the promastigote into its preferred host cell, the macrophage, in which it converts to the amastigote stage. Activation of complement by promastigotes also results in macrophage chemotaxis toward the parasite.

The effects of complement on amastigotes varies with the *Leishmania* species. *Leishmania donovani* amastigotes are relatively resistant to killing by complement. Conversely, *L. major* amastigotes are killed after alternative pathway activation of the complement cascade. The entire membrane attack complex need not be intact for death of *L. major* amastigotes to occur. The relative resistance of *L. donovani* amastigotes may contribute to their ability to spread throughout the reticuloendothelial system. In contrast, the susceptibility of *L. major* amastigotes may prevent or limit their dissemination.

Markedly elevated immunoglobulins, resulting from polyclonal B-cell activation, and high antileishmanial antibody titers are present in humans and animals with visceral leishmaniasis. Antileishmanial antibodies are also present in cutaneous leishmaniasis in its various forms. In general, the presence and magnitude of antibody responses are inversely proportional to the extent of disease. Attempts to provide protective immunity by transfer of immune serum have failed in multiple animal systems. However, several investigators have observed that sera from convalescent or healed animals can augment the level of adoptive immunity expressed in recipient animals that concurrently receive lymphocytes from immune donors. The mechanism is unknown. In addition, treatment of *L. mexicana*, *L. major*, or *L. braziliensis* promastigotes with monoclonal antibodies directed against various surface antigens before inoculation into experimental animals has a protective effect. However, passive transfer of monoclonal antibodies to naive animals has not yet been shown to be protective.

Immune complexes and rheumatoid factors are found in the sera of a high percentage of persons with visceral leishmaniasis and severe mucocutaneous leishmaniasis. The circulating immune complexes contain immunoglobulins and parasite antigen(s). It has been postulated that they may play a role in modulating immune re-

---

$Ly2^-$, $L3T4^+$) produce lymphokines other than INFγ and IL-2. Th2 cells do not activate macrophage leishmanicidal activity and appear to inhibit expansion of leishmania-specific, potentially protective Th1 cells. $Ly2^+$, $L3T4^-$ cells (suppressor/cytotoxic phenotype) contribute to the killing of amastigotes by undetermined mechanisms. The outcome of infection depends on which T-cell population comes to predominate. Macrophages may also contribute to the outcome by secreting immunosuppressive PGE_2 and LTB_4. Leishmania do not elicit macrophage secretion of IL-1. The killing of amastigotes in activated macrophages occurs by oxidative and nonoxidative mechanisms.

sponses. Sera from patients with visceral leishmaniasis have also been shown to inhibit lymphocyte proliferative responses to lectins such as concanavalin A, and in one instance, to leishmanial antigens. The role that circulating immune complexes, rheumatoid factors and other serum components play in modulating immune responses during progressive visceral and mucocutaneous leishmaniasis warrants further study.

Immunoregulation

Studies in Mice

The presence of cell populations that can interfere with protective immune responses has been most extensively studied in a murine model of leishmaniasis. BALB/c mice are susceptible to progressive, ultimately fatal infection with *L. major*. One of the first indications that development of protective immune responses can be inhibited by T cells came from studies in which BALB/c mice were given sublethal irradiation at levels known to deplete precursors of functionally suppressive T lymphocytes. Infection resolved in these irradiated BALB/c mice. However, when the irradiated mice were reconstituted with T cells from normal BALB/c mice, the disease once again progressed to death. The findings suggested that a subset of T cells could prevent development of protective immune responses in *L. major*–infected BALB/c mice.

Further evidence came from studies of congenitally athymic (nude) BALB/c mice, which, like normal BALB/c mice, develop progressive *L. major* infection. When nude BALB/c mice are given a limited number of Ly-1^+2^- T cells from naive BALB/c donors, they spontaneously resolve *L. major* infection. However, reconstitution with large numbers of T cells results in progressive disease. One explanation for these observations is that potentially protective T cells (Th1) are present in normal BALB/c mice in higher frequency than disease-enhancing T cells (Th2), but the latter come to dominate during infection. Alternatively, excess numbers of T cells potentially capable of imparting protection, paradoxically, might promote disease progression. There is also evidence that $Ly2^+$ (CD8) T cells are involved in protection against *L. major* and *L. donovani*.

B lymphocytes, but not antibodies, appear to play an important role in immunoregulation as well. When the highly susceptible BALB/c mice are treated from birth with antibodies against the heavy chain of IgM (anti-μ antibodies) to suppress B-cell development, they experience self-resolving disease when challenged with *L. major*. Transfer of T cells from these animals to normal BALB/c recipients protects them against amastigote challenge. Conversely, anti-IgM treatment of genetically resistant CBA mice renders them susceptible to progressive, uncontrolled infection. Since antibody production is not the critical determinant in these instances, it appears that antigen-presenting B cells (targets of the anti-μ antibody treatment) may be involved in the expansion of either disease-enhancing or protective T-cell subsets.

Macrophages have been implicated as important immunoregulating cells as well. Unlike most microbial pathogens, *L. major* and *L. donovani* do not stimulate interleukin-1 (IL-1) secretion during ingestion by monocytes or macrophages in vitro. Furthermore, intracellular *L. major* inhibits IL-1 secretion by monocytes stimulated with substances (such as killed staphylococci) that are potent inducers of IL-1 secretion in uninfected cells. IL-1 is known to play an important role in T-cell activation and in a variety of inflammatory processes. The lack of an IL-1 response to intracellular *L. major* may contribute to the delay in development of protective immune responses during infection. In addition, infection of macrophages by *L. donovani* apparently reduces the expression of class I and II histocompatibility antigens.

Products from infected macrophages may depress potentially important T-cell responses. Spleen cells from mice obtained within the first 2 months of *L. donovani* infection show depressed lymphoproliferative responses and lymphokine secretion in response to the lectins phytohemagglutinin and concanavalin A, and no response to leishmanial antigens. These decreased responses to mitogens are reversible by indomethacin treatment, which inhibits prostaglandin production, and are due to an adherent spleen cell population(s) (probably macrophages). Similar observations have been made with spleen cells from *L. major*–infected BALB/c mice. In vitro infection of macrophages with *L. donovani* amastigotes results in increased macrophage production of both cyclooxygenase and lipoxygenase metabolites of arachidonic acid. Prostaglandin-E_2 and leukotriene-C_4, which are produced in excess by *Leishmania*-infected macrophages, are known to reduce the proliferative responses of spleen cells to phytohemagglutinin. Nonspecific immune suppression by macrophages may play a role in the evolution of leishmaniasis.

Studies in Humans

Leishmania-specific T-cell responses are absent in patients with classical kala-azar. Peripheral blood mononuclear cells neither proliferate nor produce INFγ in response to leishmanial antigens. The in vivo T-cell unresponsiveness is antigen-specific in most patients, although depressed responses to heterogenous antigens and mitogens are observed in some patients, possibly secondary to severe malnutrition. To date, no circulating suppressor T-cell population has been identified, but a circulating inhibitor has been detected in some cases in sera. Furthermore, depletion of Leu 2^+ ($OKT8^+$) T cells and the addition of exogenous IL-2 to in vitro cultures have not reversed the leishmanial antigen-specific unresponsiveness of peripheral T cells in patients with Indian kala-azar.

Parasite-specific T-cell responsiveness develops in a majority of kala-azar patients after successful chemotherapy and is present in a subset of asymptomatic household and family members, who presumably have experienced self-resolving, asymptomatic *L. donovani* infection. It is believed that these patients have protective immunity against *L. donovani*.

Patients with diffuse cutaneous leishmaniasis from the Dominican Republic lack DTH and lymphocyte proliferative responses to leishmanial antigens. Reducing the number of glass-adherent cells (presumably monocytes) or adding indomethacin to mononuclear cell cultures permits expression of lymphocyte proliferation to leishmanial antigens. The data suggest that mononuclear phagocytes, rather than lymphocytes, are responsible for the failure of protective immune responses to develop in this population. In contrast to murine models, no antigen-specific, disease-enhancing T-cell populations have been identified in humans with cutaneous forms of leishmaniasis.

Immunopathogenesis

An important issue that has not yet been thoroughly studied is the degree to which the clinical manifestations of leishmaniasis are a by-product of the host's immune response. Ulceration of cutaneous lesions, in both naturally infected humans and experimental animals, is in all likelihood a consequence of immune responses. First, ulceration is not observed in the anergic syndrome, diffuse cutaneous leishmaniasis, even though the parasite burden is massive. Second, necrosis of infected macrophages and adjacent cells is observed both in humans and animals at a time when DTH and lymphocyte proliferative responses to leishmanial antigen are evident. In one human case followed prospectively, induction of *Leishmania*-specific lymphocyte proliferation and IL-2 production in response to *L. tropica* antigen reached maximal levels coincident with ulceration of a skin lesion. Third, the destructive, granulomatous response that characterizes mucosal leishmaniasis is most likely a hyperergic response. The parasite density in mucocutaneous lesions is low, and peripheral blood lymphocytes from these patients respond vigorously to leishmanial antigens.

Direct evidence for the role of immune responses in tissue destruction comes from ingenious studies in which mice were exposed to local, suberythematous levels of ultraviolet-B (UV-B) irradiation. Inoculation of *L. major* into the irradiated skin was followed by parasite replication in macrophages but no ulceration. Nonirradiated control skin inoculated with parasites ulcerated. UV-B radiation had no effect on parasite viability and did not kill host cells. However, UV-B irradiation did abrogate the induction of hypersensitivity to dinitrofluorobenzene and the induction of DTH responses to leishmanial antigens. The cost to the host of an effective immune response in cutaneous leishmaniasis seems to be local tissue destruction, ulceration, and resultant scar formation.

Even in visceral leishmaniasis, immune factors probably have detrimental consequences. Humans and hamsters infected with *L. donovani* become anorectic and profoundly cachectic. Parasite metabolism may partially account for the wasting, but not for the anorexia. Patients with visceral leishmaniasis are reminiscent of those with the "consumption" of miliary tuberculosis or the cachexia of AIDS or neoplasm. It has been hypothesized that the wasting observed during visceral leishmaniasis is due to the production of macrophage mediators such as IL-1 or tumor necrosis factor (TNF)/

cachectin. The presence of fever provides circumstantial evidence that IL-1 or TNF/cachectin, which is also a pyrogen and can trigger release of IL-1, or other yet to be identified endogenous pyrogens are produced. High levels of cachectin recently have been found in the peripheral circulation of patients with visceral leishmaniasis. Although neither *L. major* nor *L. donovani* elicits IL-1 secretion by monocytes or macrophages infected in vitro, as discussed earlier, it has been proposed that IL-1 secretion may be stimulated by circulating immune complexes or other factors in visceral leishmaniasis.

Vaccines and Antigens

Studies in Humans

Spontaneous cure of cutaneous leishmaniasis is associated with the apparent development of life-long immunity against the infecting *Leishmania* strain. For centuries, residents of the Middle East exposed the bare bottoms of their infants to *Leishmania*-infected sandflies or inoculated material taken from active human lesions. Infants developed local lesions, and after healing, were immune to reinfection, thereby preventing cosmetically embarrassing lesions.

Infection with a live promastigote "vaccine" has been used successfully in Israel and Russia. Inoculation of cultured promastigotes of *L. major* gives rise to typical cutaneous lesions and ultimately imparts a high degree of protection against reinfection. As of 1980 it was estimated that more than 20,000 people had been immunized in this way. The practice has lost favor, however, because the parasites lose virulence in culture, and intolerable untoward effects have been encountered in some patients. A vaccine composed of killed promastigotes provided partial protection in a recent study carried out in Brazil. BCG in combination with promastigotes apparently imparts defense to anergic patients with diffuse cutaneous leishmaniasis in Venezuela.

Studies in Mice

BALB/c mice immunized with live-irradiated, heat-killed, or sonicated *L. major* promastigotes by intravenous or intraperitoneal route with adjuvant are protected against an *L. major* challenge. The cell population(s) responsible for protection can be identified in cell transfer experiments. Protection is bestowed on naive BALB/c recipients by transfer of T cells $Ly2^-$, $L3T4^+$; Th1) from intravenously immunized animals. Transfer of antileishmanial antibodies has no protective effect.

Curiously, immunization with the same antigen preparations subcutaneously does not protect BALB/c mice and actually inhibits spontaneous resolution of lesions in genetically resistant strains of mice. This disease-enhancing activity is mediated by $Ly2^-$, $L3T4^+$; Th2 lymphocytes. Thus, depending on the route of immunization, it is possible to elicit either protective or disease-enhancing T cells of the helper-inducer phenotype. Both the $Ly2^-$, $L3T4^+$ disease-enhancing T cells that emerge after subcutaneous immunization and the protective T cells that result from intravenous immunization appear to "help" *Leishmania*-specific antibody production in vitro.

A defined vaccine has many potential advantages over a living promastigote vaccine or a crude antigen preparation. Several candidate leishmanial molecules have already been identified. Studies of an amphipathic lipophosphoglycan of *L. major* (thought to mediate attachment of parasites to macrophages) indicate that it can induce protective immunity in mice when the molecule is intact and is administered with BCG or *Corynebacterium parvum* as adjuvants. However, after exposure to phospholipase CIII, a carbohydrate is released that has disease-enhancing properties. For this reason, it is unlikely that the glycolipid in its native state will be used in a human vaccine, but a chemically stabilized form might be safe and effective. Another lipophosphoglycan has been purified from *L. donovani*. Alternatively, there may be protein antigens that can elicit protective immunity. Antibodies against the major leishmanial membrane glycoprotein, gp63, have been shown to decrease the infectivity of *L. braziliensis* and *L. mexicana* promastigotes for macrophages in vitro. Gp63 incorporated into liposomes has also been shown to be protective in vivo. A protein vaccine might have an advantage over a glycolipid vaccine because current technology should permit production of a recombinant protein antigen. Recent advances in the areas of immunology and molecular biology have brought us closer to the development of defined vaccines.

Coda

Leishmania species are prototypic intracellular parasites of macrophages. Human leishmaniasis varies in its manifestations from localized, self-healing cutaneous ulcers to disseminated infection of the entire reticuloendothelial system. Not only are *Leishmania* species important pathogens in widely dispersed areas of the world, but they also have emerged as excellent model systems for the study of cell-mediated immune responses. Most *Leishmania* species can be propagated as promastigotes in culture. Parasite–phagocyte interactions can be easily studied in vitro, and excellent animal models of infection are available. Recent studies of the immunology and molecular biology of *Leishmania* species have provided important insight into general principles of biology and brought us closer to developing safe and effective vaccines.

Additional Reading

CHANG, K.-P, C. A. NACY, and R. D. PEARSON. 1986. Intracellular parasitism of macrophages in leishmaniasis: in vitro systems and their applications. *Methods Enzymol.* 132:603–626.

HOWARD, J. G. 1986. Immunological regulation and control of experimental leishmaniasis. *Int. Rev. Exp. Pathol.* 28:79–116.

MITCHELL, G. F., and E. HANDMAN. 1985. T-lymphocytes recognize *Leishmania* glycoconjugates. *Parasitol. Today* 1:61–63.

PEARSON, R. D., and A. Q. SOUSA. 1989. *Leishmania* species: visceral (kala-azar), cutaneous, and mucosal leishmaniasis. In *Principles and Practice of Infectious Diseases*, 3d ed, G. L. Mandell, R. G. Douglas, Jr., and J. E. Bennett, eds. Churchill Livingstone, pp. 2066–2077.

PEARSON, R. D., D. A. WHEELER, L. H. HARRISON, and H. D. KAY. 1983. The immunobiology of leishmaniasis. *Rev. Inf. Dis.* 5:907–927.

SYPEK, J. P., and D. J. WYLER. 1988. Host defense in leishmaniasis. In *Contemporary Issues in Infectious Diseases, Parasitic Infections*, vol. 7, J. H. Leech, M. A. Sande, and R. K. Root, eds. Churchill Livingstone, pp. 221–242.

12

Immunology of African Trypanosomiasis

John M. Mansfield

African trypanosomiasis is a parasitic disease of medical and veterinary importance that has adversely influenced the economic development of sub-Saharan Africa. Trypanosomiasis is geographically restricted to an area of West, Central, and East Africa that is larger than the continental United States. This is the area defined by the ecological range of the trypanosome intermediate host and vector, the tsetse fly, and is an area in which sporadic epidemics of human trypanosomiasis occur and in which European breeds of cattle cannot be raised with impunity. Although the disease has been recognized for more than 100 years, scientific advances have not resulted in the control or cure of trypanosomiasis. This lack of progress, compared with successes for many other infectious diseases, can be attributed to several key factors: trypanosomiasis is a Third World problem; experimental and clinical studies are difficult to perform outside of the areas affected; and research funding for this, and other, tropical diseases historically has been inadequate.

In microbial diseases the immune system usually provides a protective interface between a pathogen and host tissue integrity. In African trypanosomiasis, host immune responses to parasite antigens play a central role in controlling fulminating parasitemias. Yet, as discussed in this chapter, the biological relevance of host immune responses to defined trypanosomes antigens is not clear and is being questioned for the first time. It is fair to say that very little is *understood* about the immunological relationship between the trypanosomes and their hosts, despite a fairly descriptive and somewhat paradoxical literature on trypanosome immunology that has stemmed primarily from experimental studies. Thus, the present chapter will bring the reader up to date with respect to our current information, thoughts, and understanding of the immunology of African trypanosomiasis. This chapter focuses on parasite-specific immune responses and resistance mechanisms, on the regulation of these mechanisms in the infected host, and on trypanosome gene expression as a modifier of host resistance. Information for this chapter is drawn not only from research experience in our laboratory but also from the results of our colleagues worldwide who are involved in this intriguing area of medical and veterinary research.

The Disease Process

Trypanosomes are protozoan hemoflagellates that infect humans and other animals. Infections are initiated when metacyclic trypomastigote forms of *Trypanosoma* spp. are transmitted to the mammalian host by the bite of tsetse flies *(Glossina* spp.*)*, which are the intermediate host and vector for trypanosomes. The inoculated organisms transform into larger trypomastigote forms (those typically seen in infected blood; Figure 12-1) and replicate extracellularly at the site of inoculation. A chancre usually develops at this site which is characterized histologically by mononuclear cell infiltration. Trypanosomes subsequently spread from the chancre through the lymphatic vessels to the blood and other tissues of the body, where they continue to replicate extracellularly. When trypanosomes appear in the blood during infection, their numbers fluctuate in a cyclical manner. This intermittent parasitemia is a characteristic of mammals infected with the African trypanosomes and is a consequence of two main events: (1) antigenic variation by the African trypanosomes within the host tissue and (2) ability of the host to mount

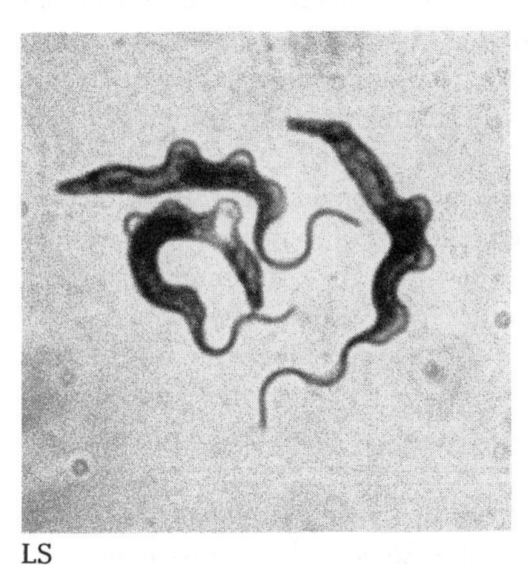

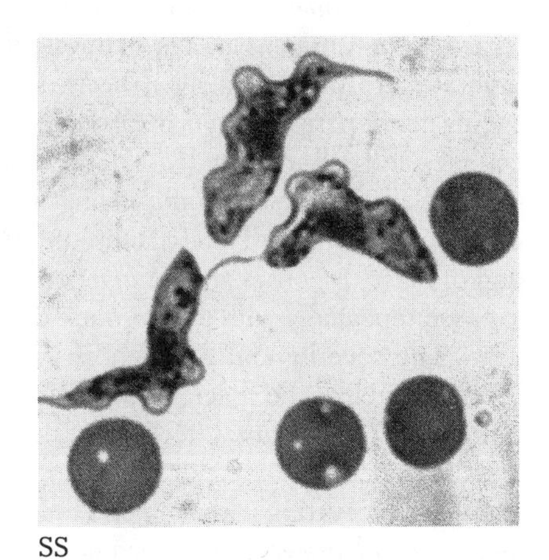

FIGURE 12·1 Trypomastigote forms of *Trypanosoma brucei rhodesiense* present in infected blood. In terms of the pleomorphism evident, it is suspected that the long, slender (LS) forms are those that actively divide, whereas the short, stumpy (SS) forms have differentiated to a nondividing metabolic state suitable for further development when taken up by the tsetse fly.

antibody (Ab) responses to the variant antigenic types (VATs) present, which results in a temporal reduction in organisms in the blood and other tissues. However, sterile immunity rarely occurs, and as the cycle continues, trypanosomes eventually enter the cerebrospinal fluid and brain tissue. These events signal the impending death of the host.

The symptoms of infection in humans and cattle are not specific for trypanosomiasis and may be variable from one individual or animal to another. The earliest symptoms in humans are a recurring fever and headache, which are closely associated with the appearance of trypanosomes in the blood. Another early development is that superficial lymph nodes and the nodes draining the chancre become markedly enlarged. Concurrently, facial edema, large, circular erythematous skin rashes, pruritus, arthralgia, tachycardia, emaciation, and deep hyperesthesia (Kerandel's sign) develop. Marked lassitude and other symptoms of central nervous system involvement may appear subsequently: ataxia, hand and limb tremors, mental changes, and somnolence progressing to coma. As in infected humans, infected cattle exhibit febrile changes associated with the appearance (or elimination) of trypanosomes in blood, and superficial lymph nodes are enlarged. However, a chronic wasting process occurs, leading to extreme emaciation as well as a terminal weakness and recumbency.

The pathology of trypanosomiasis is not pathognomonic of the disease, as there are no histological changes that are unique to trypanosome infections. Several hallmarks of infection are a marked elevation in serum Ig levels, immunomodulation of T- and B-cell responses, hepato- and splenomegaly, lymphadenopathy, thymic atrophy, anemia, and the appearance of lesions in numerous tissues, including the brain. The lesions are inflammatory foci caused by the presence of immune complexes in the tissues and by T cell–mediated responses (initiated primarily in the early stage of disease) to trypanosome antigens in the extravascular spaces. For example, mononuclear cell infiltrates are present in liver, heart, skeletal muscle, and brain tissue as interstitial aggregates or perivascular cellular accumulations. That many of these lesions (immune complex and cell-mediated) are T-dependent has been revealed from studies in which infected athymic mice do not exhibit the pathological changes observed in infected euthymic littermates. The trypanosome antigens predominantly involved in the generation of these responses are the *in*variant antigens to which the host immune system is exposed repeatedly with the immune destruction of each variant population.

The severity and duration of the disease process depend largely on the host species as well as the species (and strain) of the infecting trypanosomes. For example, untreated *T. gambiense* infections of humans are chronic in nature, with death occurring after a period of a year or more; *T. rhodesiense* infections of humans are more acute, with death occurring within a period of weeks to months. Different strains of *T. brucei* and *T. congolense* vary in their pathogenicity for catte. Some of the strains can produce infections that spontaneously cure, have favorable long-term survival rates, or are associated with low parasitemia. Other, more virulent strains cause infections that result in less favorable short-term survival and are associated with high parasitemia. The severity of trypanosomiasis is influenced by a number of factors, including host genetics and parasite gene expression. An understanding of the contributions of both host and parasite to the disease process may be necessary for developing effective strategies to control the disease as it exists in Africa. This will be discussed in more detail in the sections that follow.

The Variant Surface Glycoprotein (VSG)

The trypanosome plasma membrane is covered by a monomolecular surface "coat" composed of approximately 10^7 glycoprotein molecules, each with a molecular weight of approximately 60 kD and containing 7 percent to 17 percent carbohydrate (Figure 12-2). This molecule is the variant surface glycoprotein, or VSG molecule, to which the infected host mounts protective B-cell (antibody) responses, and which is the molecular basis for antigenic variation in African trypanosomiasis.

The VSG molecule is composed of approximately 450 amino acids and is exposed on the plasma membrane with the carboxy-terminus portion of the molecule anchored in the membrane by a phosphoglycolipid residue and the amino-terminus portion extended away from the membrane. The VSG molecule is displayed in a dimer configuration in some species and conforms to a highly symmetrical internal alpha-helix construction. The VSG molecules are expressed in such a highly dense array on the membrane that only epitopes present in the N-terminal one-third of the molecule normally are exposed to Ab binding (Figure 12-3).

When the primary structure of VSG molecules from different VATs are examined, remarkable differences and similarities in the structure are revealed: the N-termi-

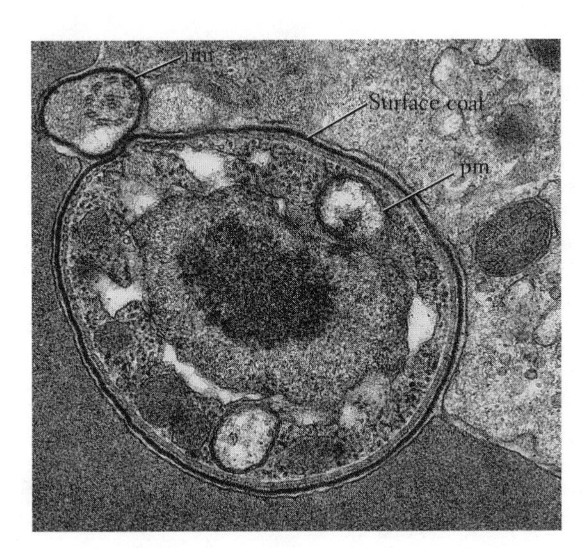

FIGURE 12·2 Electron photomicrograph of an intact trypanosome *(T. b. rhodesiense)* present in a section of infected host tissue. The VSG surface coat is evident as an electron-dense layer covering the plasma membrane (pm) of the cell, including the undulating membrane (um) which contains the flagellum. (From Smith and Mansfield, unpublished.)

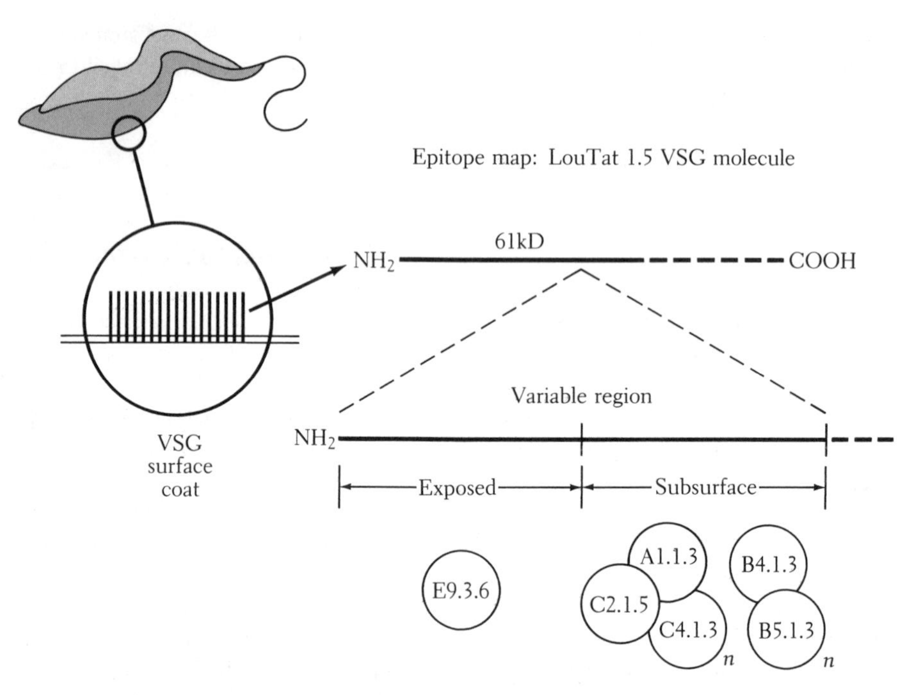

FIGURE 12·3 A. Topological display of VSG determinants on the trypanosome surface. Exposed N-terminal epitopes are shown diagrammatically, as is the dense, rodlike display of VSG dimers on the intact trypanosome plasma membrane. B. A VSG molecule (monomer) showing variable and constant regions and epitopes detectable in these regions by monoclonal Ab mapping. (From Theodos and Mansfield, unpublished.)

nal two-thirds of the VSG molecule (the variable region) differs almost completely when compared to VSG molecules from other variants, yet the C-terminal one-third (the constant region) exhibits extensive amino acid homology and appears to be conserved among different VSG molecule families. Despite the amino acid heterogeneity existing among the variable regions (N-terminal two-thirds) of different VSG molecules, close examination reveals several cysteine residues present at discrete positions within the molecules; these residues may serve to provide intrachain (and perhaps interchain) disulfide bridges that result in the molecule being expressed as a densely packed array of rigid monomers or dimers on the plasma membrane. The result of this "lawnlike" VSG alignment is that only a subregion of the molecules is exposed to antibodies in the external environment. These observations are important from the viewpoint of immune recognition, in that the structure, alignment, and display of VSG molecules, and in particular the N-terminal epitopes, on the plasma membrane predict that the exposed epitopes will be presented to the immune system in a *topological* array mimicking epitopes present on T-independent antigens (which exhibit an array of identical determinants repeated on a single molecule).

Several observations support this hypothesis. For example, VSG-epitope mapping studies performed with monoclonal antibody probes have revealed that there are relatively few exposed N-terminal epitopes and that many of these exposed epitopes are conformationally labile. Furthermore, immunization of animals with soluble VSG results in Ab that recognizes predominantly the subsurface epitopes of the molecule, in contrast to the results obtained after infection (or immunization with intact but nonviable trypanosomes). Also, athymic *nu/nu* mice make significant surface-specific IgM responses to the VSG molecule after infection and are able to control parasitemia during the course of infection. It is clear from the latter studies that the trypanosome also must provide a set of antigen-nonspecific second signals to B cells in order to generate a mature clonal B-cell response to the T-independent surface epitopes. Indeed, the polyclonal B-cell activating effect of trypanosomiasis has been recognized for some time, and it may be that trypanosomes release factors that directly or indirectly produce second signals. Thus, although T-dependent (e.g., IgG) B-cell responses occur to VSG determinants in thymus intact *nu/+* mice, T-independent B-cell responses to surface epitopes also occur that are sufficient to control parasitemia.

Carbohydrate residues associated with the variant surface glycoprotein are present within the subsurface variable region and within the constant region, near the plasma membrane; by virtue of their location, they are not exposed on the viable organism. Associated with the C-termini of VSG molecules in the plasma membrane is a phospholipase C-like enzyme that cleaves the phosphoglyceride tail from the VSG molecule, resulting in the release of VSG from the membrane. This cleavage may be responsible for VSG liability and the appearance of soluble VSG in the blood of infected animals, as well as the release of VSG that precedes the expression of a new surface coat. Disruption of VSG surface coat integrity or loss of parasite viability result in the exposure not only of the carbohydrate residues but also of other buried VSG epitopes, the cross-reactive determinants associated with the C-terminus, and invariant antigens present on the plasma membrane and within the cell.

The genetic bases of antigenic variation, which lead to the switch in the VSG surface coat phenotype, are the subject of intensive study (see Chapter 17). Trypanosomes spontaneously express new VSG genes and, hence, display new VSG epitopes to the host immune system at the approximate rate of one in 10^6 trypanosomes. The switching event is not *initiated* by an antibody response to existing surface epitopes (but may be *enhanced* by such responses which will select for low numbers of new variant trypanosomes present in a population), since antigenic variation occurs in vitro and in immunosuppressed infected animals. Expression of a new VSG gene in some VATs is preceded by gene rearrangements in which formation of an expression-linked copy (ELC) from an existing basic copy of the VSG gene occurs. The ELC is subsequently transposed to a telomeric expression site on one of the trypanosome's chromosomes, which contains an upstream promotor site. If a VSG gene already is in residence at the new telomeric transcription site, it is lost, degraded, or inactivated when the new VSG gene is inserted into the telomeric expression site. Subsequently the newly inserted VSG gene is transcribed, and the translational events ultimately result in expression of a new VSG surface coat. In some variants, activation of a VSG gene that already is within a telomeric expression site occurs to produce a new surface coat. Alternatively, a new VSG gene may be constructed not from a basic copy of the gene but through a recombinational event involving two different telomere-associated VSG genes. The events responsible for regulating VSG genes within an expression site are not yet known. However, it is clear that host immune responses do not directly influence molecular events regulating VSG expression.

Given that from 100 to 1000 copies of different VSG genes may exist in a genetically homogenous trypanosome population (clone), the parasites are ensured of extended survival in an immunologically hostile host environment. Certainly in nature, survival time of the parasite within the host is sufficient to result in transmission of the disease by the tsetse fly vector before death of the host occurs. Despite earlier hopes that a multivalent VSG vaccine might be effective, the sequence in VSG gene expression is virtually random, so that potential target antigen VSGs cannot be predicted. Some VATs do appear with a greater frequency than predicted by chance during experimental infections initiated with a single VAT. VAT sequence predominance probably depends on numerous factors, including VSG immunogenicity and efficiency of host immune elimination of the existing VAT population(s), VSG gene switch frequency of a particular VAT population, differences in the growth rates of different VATs, and biological competition among VATs. The heterogeneity of VATs and unpredictable VSG expression sequences in infected hosts therefore preclude development of multivalent VSG vaccines. Vaccination with the metacyclic (insect-borne) trypomastigote VATs that are injected by infected tsetse flies is also problematic because similar metacyclic VSG gene heterogeneity and instability occur. Genetic exchange in the tsetse fly between developmental forms of trypanosome populations present may further expand the genetic and phenotypic diversity of trypanosomes (new VSG gene sequences and surface coats).

Immune Responses to Trypanosome Antigens

Trypanosome-infected hosts are exposed to a multitude of parasite antigens during the course of infection. These antigens include a series of distinct variant VSG epitopes as well as invariant VSG, membrane, cytoplasmic, and nuclear antigens. While T-cell-dependent immune responses to trypanosome invariant antigens may be associated with immunopathology (as discussed earlier in The Disease Process), B-cell responses to exposed variant VSG epitopes are associated with the cycle of antigenic variation and immunoselection of VAT populations (Figure 12-4). Variant-specific B-cell responses are the central focus of this section.

As noted earlier, N-terminal epitopes of the VSG molecule comprise the "protective" epitopes of the surface antigen (that is, they are the targets of protective antibodies). Antibodies to these epitopes ultimately result in the elimination of the variant population expressing them. Evidence for the role of VSG-specific B cells in controlling parasitemia is extensive and includes the following types of observations:

(1) VSG-specific antibody appears during infection with trypanosomes, and this appearance is temporally associated with VAT elimination; (2) immunization of animals with purified VSG will protect the animals from a challenge infection with trypanosomes expressing the homologous VSG but will not protect against challenge with VATs expressing heterologous VSG molecules; (3) passive transfer of VSG-specific antibody or primed B cells to immunocompromised recipient animals will result in the passive transfer of variant-specific immunity, whereas passive transfer of T cells alone has not such effect; and (4) administration of VSG-specific antibody to an animal infected with the homologous VAT will result in elimination of those trypanosomes.

During an infection, trypanosome replication occurs extracellularly, and parasite numbers increase both within the blood and extravascular spaces. During this period trypanosome VSG determinants stimulate B cells through T-cell-independent and T-cell-dependent pathways to produce antibodies to the surface coat. When sufficient antibody is produced, trypanosome elimination occurs rapidly, and only heterologous antigenic variants remain to repopulate the blood and other tissues. Each wave of parasitemia is controlled by a subsequent wave of VSG-specific Ab, resulting in cycles of rising and falling parasitemia.

The kinetics of VSG-specific immune responses have been studied in responder animals (see Figure 12-5). A strong IgM response occurs after 3 to 4 days of infection, at a time when parasite numbers in blood are quite high (e.g., $10^7 - 10^8$ trypanosomes per milliliter (ml) of blood). Subsequently, trypanosomes may disappear within hours of the appearance of VSG-specific IgM. VSG-specific IgG usually appears in the serum after the trypanosomes have been eliminated and therefore is not involved in trypanosome

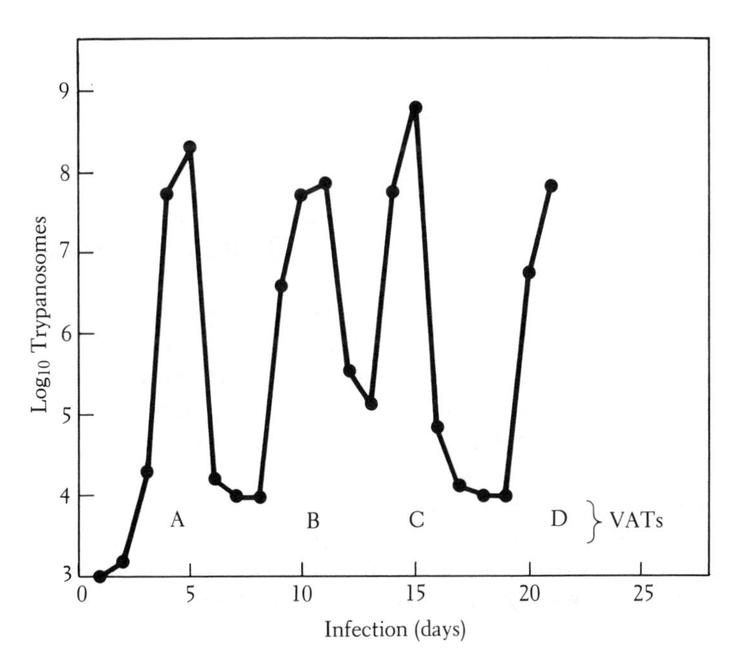

FIGURE 12·4 Host immune responses to trypanosomes. The association between the cycle of host Ab responses to the VSG surface coat of VAT populations and of antigenic variation during the course of infection is illustrated. These events are responsible for the fluctuating parasitemias.

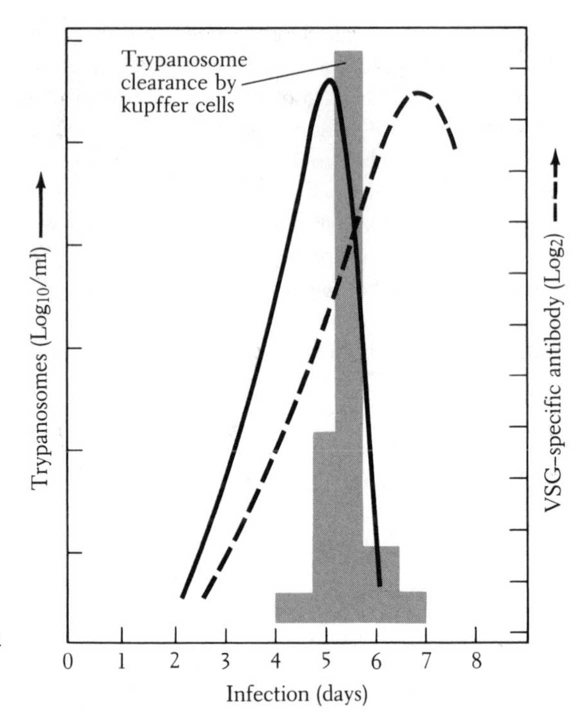

FIGURE 12·5 Kinetics of antibody response to VSG surface epitopes, trypanosome elimination, and Kupffer cell uptake following the growth of a defined VAT population in the infected host. (From Smith and Mansfield, unpublished.)

clearance but is expressed as a normal consequence of B-cell clonal expansion. The total IgM response frequently is greater than IgG response, which perhaps reflects the largely T-independent configuration of exposed VSG epitopes. Over a period of several weeks, after elimination of a variant population, VAT-specific IgM and IgG concentrations decline to nearly undetectable levels in blood, but IgM and IgG responses to invariant trypanosome antigens remain elevated, reflecting continual stimulation by such invariant antigens. Although antibodies to VSG epitopes are involved in elimination of parasites that comprise defined VAT populations, there is also evidence that production of these antibodies does not entirely account for resistance to disease (as discussed later in Genetics of Resistance).

Trypanosome elimination is an antibody-mediated event that depends largely on uptake and destruction of the organisms (see Figure 12-6) by Kupffer cells (KC) of the liver. Although the mechanism of trypanosome uptake is associated with antibody binding to VSG N-terminal epitopes, the exact mode of attachment to and uptake by KC is not known. IgM binds to exposed VSG determinants and results in activation of complement, mediating trypanosome attachment to KC through their C3b and C3bi receptors. Subsequently, parasites are engulfed by the KC.

It is not clear that this is the sole mechanism of trypanosome destruction. It may be that antibody and/or complement binding to the VSG also results in perturbations ("tufting") of the surface coat in a manner that results in exposure of normally buried ligands (e.g., carbohydrate residues) that are then exposed to KC receptors for such

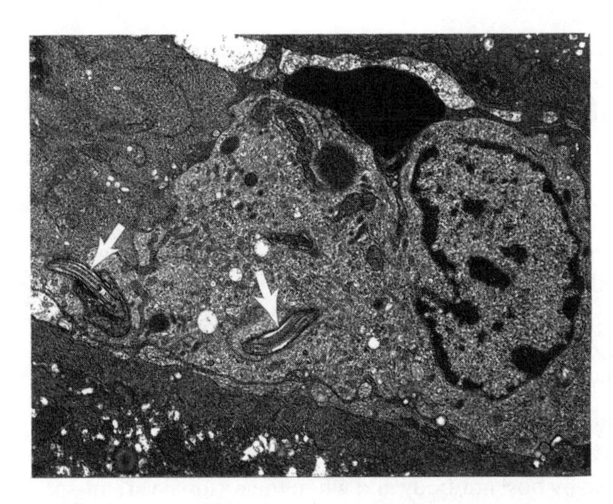

FIGURE 12·6 Trypanosomes within the phagolysosomes of a Kupffer cell in the liver. Extensive degeneration of the parasites is evident at 1 minute post phagocytosis. (From Smith and Mansfield, unpublished.)

residues. Alternatively, Ab binding may result in VSG perturbations such that C3 binding to trypanosome acceptor sites may occur; in fact, several studies have reported that trypanosome membranes or VSG molecules may bind complement proteins through C1-dependent or C1-independent pathways, but the biological relevance of this binding is not yet clear. Trypanosome cytolysis as the result of antibody and complement binding does not seem to play a major role in parasite destruction in vivo, because of the trypanosome membrane-associated complement decay accelerating factor present.

Regardless of the mechanism of trypanosome attachment to the KC membrane, subsequent phagocytosis of the parasite occurs and is rapidly followed by parasite destruction within phagolysosomes. Electron microscopy reveals that Ab-coated trypanosomes are partially degraded within KC phagolysosomes already 1 minute after their uptake from the blood (Figure 12-6). Although radiolabeled trypanosome proteins may remain associated with the KC for hours after uptake, structurally recognizable trypanosomes cannot be found several minutes after in situ phagocytosis by the KC. It is not known to what extent other tissue macrophages play a similar role in destroying extravascular trypanosomes. Most trypanosomes in the circulation end up within the liver KC after immune clearance, and not within phagocytic cells of other tissues such as the spleen (in contrast to clearance of plasmodia-infected erythrocytes; see Chapter 9). This is a surprising finding because the spleen becomes markedly enlarged during infection primarily as a result of increased macrophage populations.

Immunoregulation

How VSG-specific antibody responses are regulated is unknown. It appears that surface VSG epitopes are presented to the host immune system as type I (T-cell-independent) antigens (Figure 12-3). Type II (T-cell-dependent) B-cell responses (IgG production) to the VSG molecule also occur, but it is not known whether these responses are directed to the same epitopes as recognized in the T-cell-independent response. Other than production of IgG to VSG, there is little direct evidence that significant VSG-specific helper T-cell stimulation occurs or that other regulatory T cells are primed by VSG epitopes during infection. Some degree of regulation of antibody responses to exposed VSG epitopes occurs since VSG antibody responses are sharply up- and down-regulated during infection and since B-cell responses to exposed VSG epitopes are regulated independently of responses to buried epitopes. Furthermore, mice that are infected with a defined VAT and then are rechallenged with the same VAT at different times after the primary infection do not always respond with the same magnitude of the secondary antibody production. Indeed, at two periods during the infection, it is impossible to induce a secondary B-cell response to surface epitopes of the infecting VAT. One period is at 2 weeks after the primary infection, at a time when the primary infecting VAT has been eliminated by host antibody and when the residual variant-specific antibody levels are declining. Challenge with the homologous VAT fails to induce an increase in the antibody titer, although B-cell responses to other VAT populations (e.g., those to which the infected animals had not yet been exposed) are unimpaired. On the other hand, if mice are first drug-cured of their active infections before the challenge infection, a secondary response occurs. Another period of host unresponsiveness to the challenge VAT occurs during the final week of infection just prior to host death. At this time, the infected host is unresponsive not only to the infecting VAT but also to the VSG molecules displayed by unrelated variants. Failure of B-cell function before death may be an agonal event that may account for the failure of the infected host to respond to any VAT population, since all T-cell-independent B-cell responses seem to degenerate at this time, as discussed later. Thus, an active infection can cause modification of host B-cell responses to the VSGs. An early event is the down regulation of variant-specific antibody responses and occurs shortly after immune elimination of a VAT, and a preterminal event in which nonspecific depression of B-cell responses occurs prior to death of the host.

The molecular and cellular basis for regulation of parasite-specific immunity is ill-defined and is under investigation. One possibility is that regulatory B and T cells that express relevant antiidiotypic receptors may regulate the production of variant-specific antibodies. One reason for believing that such idiotope-specific regulatory cells might be involved is that whereas MHC-restricted T-dependent immune responses, including antigen-specific helper-, cytotoxic- and suppressor-T-cell responses are depressed after 1 week of infections, T-cell-independent B-cell responses remain vigorous throughout most of the infection, as discussed later. Recent studies have shown, in fact, that rapid down regulation of VSG epitope-specific Ab production occurs at a time when lymphocytes that express receptors against the VSG antibody idiotypes (i.e., antiidiotypic lymphocytes) are detectable in trypanosome-infected mice.

Another potential mechanism of regulation of parasite-specific immune responses involves macrophages. For example, a radioresistant cell population has been identified in susceptible animals (semiallogeneic radiation chimeras) that can interfere with genetically determined antitrypanosomal immune responses as well as the VSG-specific B-cell responses. Furthermore, infected macrophages have been found to be altered in their ability to present trypanosomal antigens to T lymphocytes in vitro, an observation that may have bearing on parasite-specific immunity, as discussed later.

Observations on the antigen *non*specific immunodulation that occurs in infected animals may have broader significance in terms of our understanding the immunological basis of the disease process. From a consideration of results obtained in several laboratories, a pattern of events emerges. As parasitemia first becomes established after an infection, a number of lymphocyte populations appear to be nonspecifically stimulated. These include helper and suppressor T cells as well as B cells. Macrophages also become stimulated and in situ proliferation of T, B, and macrophage cells results. Increased seeding of the lymphoid tissues by bone marrow–derived lymphocyte and monocyte populations also occurs. Associated with the nonspecific proliferation of these cell populations is spontaneous increase in B- and T-cell immune responsiveness to a variety of non-parasite-related antigens. Consequences of such polyclonal B- and T-cell activation include elevated "background" responses to many antigens, as well as idiotypic and antiidiotypic antibody responses to bacterial antigens. Paradoxically, the ability to induce T-cell-dependent B-cell responses by specific immunization is decreased during this period of polyclonal B- and T-cell activity. Subsequently, the ability to induce all sorts of T-cell responses declines (including helper, suppressor, and cytotoxic functions), whereas the ability to induce T-cell-independent B-cell responses becomes enhanced. After the first week of infection until just prior to death, only T-independent B-cell functions remain relatively intact.

It is not clear how trypanosome infection triggers these paradoxical immunological changes (polyclonal stimulation and immunosuppression). Macrophages appear to play a central role in these events. Suppressor macrophages appear during trypanosome infection that have been demonstrated to down-regulate T-cell responses to antigenic and mitogenic stimuli. The interaction of trypanosomes with macrophage target cells (perhaps through Ab-mediated phagocytosis) apparently serves to subvert macrophages to the role of suppressor cell. The subcellular events that account for this change in macrophage function are largely unknown. Some macrophage populations from infected mice display enhanced Ia expression, H_2O_2 production, IL-1 secretion, and PGE_2 release. In addition to these alterations in phenotype, the infected macrophages fail to present antigen to appropriate responder helper T cells after approximately 1 week of infection, as discussed later.

Can these observations on the macrophage explain the spectrum of antigen nonspecific immune dysfunctions in trypanosomiasis, and to what extent do these dysfunctions extend to the parasite-specific immune response? It may be that the initial uptake of trypanosomes by macrophages, in an Ab-dependent manner, results directly or indirectly in macrophage activation (i.e., expression of the new phenotypic characteristics noted). This early activation may lead to stimulation of trypanosome antigen-specific helper T cells with trypanosome antigen, as well as nonspecific stimulation of helper T cells as a result of elevated IL-1 production or other stimulatory macrophage-derived factors. Helper T cell subpopulations which have been specifically or nonspecifically stimulated subsequently produce a variety of factors that help to stimulate other

T cells (e.g., through IL-2 production and the induction of IL-2 receptor expression) as well as factors that affect B cells (such as IL-4 and IL-5), macrophage functions (IFNγ and IL-4), and cells of the bone marrow (IL-3). Continual uptake of trypanosomes by macrophages leads ultimately to an inability to properly present antigen to T cells, however.

This apparent macrophage defect alters the processing and presentation of variant and invariant trypanosome antigens since infected macrophages do not present trypanosome antigens in a normal manner to infected or immune helper T-cell populations. Theoretically, the inability to present antigen could be due to diminished IL-1 production, decreased or inappropriate Ia expression (e.g., I-E versus I-A), or increased catabolism of antigen. Studies reveal that the trypanosome antigen-presenting cell functions of infected splenic macrophages decline within a few days of infection. Yet, this occurs at a time when infected macrophages actually express elevated levels of Ia and continue to do so throughout infection. IL-1 is also produced normally by infected macrophages. Thus, the altered trypanosome antigen-presenting cell capacity of infected macrophages, which declines early in infection, cannot be ascribed to any decline in Ia expression or loss of IL-1 production. By a process of elimination, it may be that infected macrophages either do not properly process trypanosome antigens or do not properly express the immunogenic peptides in the context of Ia on the macrophage membrane.

What causes these alterations in macrophage function? Continued stimulation of infected macrophages by T-cell-derived IFNγ or by autoregulatory factors (TNF) might cause excessive prostaglandin release (e.g., PGE_2), which could interfere with further T-cell stimulation. Alternatively, the ingested trypanosome constituents might cause inhibitory alterations in macrophage functions directly. Macrophage activation and down regulation in trypanosomiasis may occur in a T-cell-independent manner since these are observed in infected congenitally T-cell-deficient athymic mice. Thus, trypanosome constituents may serve as a new class of macrophage modulatory agents. While the nature of the parasite constituents responsible for modulation is not known, preliminary evidence implicates membranes and membrane lipid components.

Does this hypothetical sequence of events account for the well-documented suppressor-cell activity previously associated with infected macrophages? Perhaps, since the in vitro cultures routinely employed to demonstrate macrophage suppressor cell activity depend on culture conditions in which numbers of infected macrophages are greater than numbers of uninfected macrophages present that normally serve as functional antigen-presenting cells.

The picture of trypanosome immunology that emerges is one of a host that is immunologically modulated by the parasite, primarily through an effect on macrophages. The ultimate result is that host T-cell-dependent immune responses to trypanosome antigens are depressed but that T-independent B-cell responses to VSG surface epitopes remain intact. What potential evolutionary advantage (if any) may have been afforded the trypanosome by these effects on the host's immune system? One potential benefit might be that T-dependent pathology (T-cell-mediated monocyte accumulation in tissues, IgG immune complex deposition) in response to trypanosome invariant antigens (to which the host is exposed repeatedly in successive waves of parasitemia) is depressed, permitting the host (and the parasite) to survive for a prolonged period of time. This depression of T-cell-dependent histopathology is coupled with survival of an intact T-independent B-cell response to the surface epitopes of VSG molecules, so that parasitemias are still controlled by the host. In evolutionally more recent hosts for trypanosomes (e.g., humans, domestic breeds of cattle, experimental animals), it appears that the depression of T-cell-dependent responses extends to all antigens requiring macrophage–helper T cell interaction since not only trypanosome-specific responses but also heterologous T-cell-dependent responses are affected. One might predict, therefore, that evolution-adapted hosts (wild animals, trypanotolerant cattle) exhibit a more finely tuned immunomodulation in which only T-cell-dependent responses to parasite antigens are affected. The hypothesis regarding both the mechanism and biological significance of immunoregulation in trypanosomiasis requires substantially more investigation.

Genetics of Resistance

Trypanosome infection results in a spectrum of disease in humans and animals. Clinically and experimentally, it has been known for some time that certain hosts are more susceptible than others to the pathological consequences of infection. For example, individuals have been identified in geographically defined regions who appear to suffer less from the disease than others, and rare cases of spontaneous cure have been reported in humans. There also are anecdotal reports of tribes in Africa that have emigrated from trypanosome nonendemic areas to endemic areas. These individuals suffer more acutely fatal forms of trypanosomiasis than local residents with similar exposure to and incidence of the disease. Cattle bred in Africa (e.g., N'dama) experience trypanosome infection without severe pathology (and are therefore referred to as trypanotolerant), whereas European breeds of cattle suffer debilitating, rapidly fatal infection.

No studies have been performed that suggest a genetic basis for susceptibility to trypanosomiasis in infected human populations; studies of experimental infection in domestic and laboratory animals do indicate a genetic basis for resistance, however. Most studies have examined infection with *T. brucei* (or *T. brucei* subspecies) and *T. congolense*, and *relative* resistance (time to death of inbred mouse strains after infection) has been assessed.

In all studies, a spectrum of relative resistance to trypanosomes has been identified; some inbred strains are more resistant (e.g., live longer) than other strains (Figure 12-7). When H-2 congenic mice were examined for resistance, it became apparent that resistance was not linked to the MHC H-2 locus of the mice. For example, C57BL/10 mice (H-2^b), B10.D2 (H-2^d), B10.A mice (H-2^a), and B10.BR mice (H-2^k) all share the same resistance characteristics and non-MHC "background" genes but have different H-2 haplotypes. Conversely, B10.BR, CBA, and C3H mice all share the H-2^k haplotype but are relatively resistant (R), intermediately resistant (I), or susceptible (S),

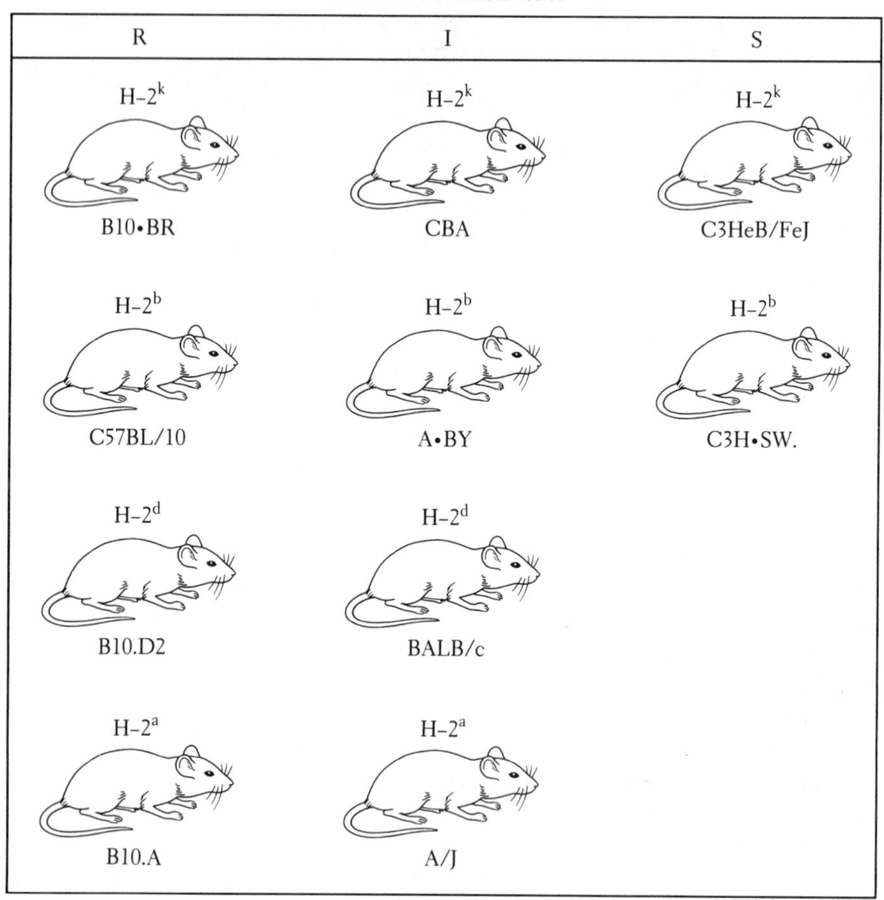

FIGURE 12·7 Resistance is not linked to the MHC of infected mice. H-2 compatible animals, which share the same H-2 haplotype but which differ in non-MHC background genes, exhibit a broad spectrum of resistance, whereas animals sharing the same background genes but exhibiting different H-2 haplotypes express only one general level of resistance. (Adapted from LEVINE, R. F., and J. M. MANSFIELD, 1981. Genetics of resistance to the African trypanosomes. I. Role of the H-2 locus in determining resistance to infection with *Trypanosoma rhodesiense*. *Infect. Immun.* 34:513; and from LEVINE, R. F., and J. M. MANSFIELD, 1984. Genetics of resistance to the African trypanosomes. III. Variant specific antibody responses of H-2 compatible resistant and susceptible mice. *J. Immunol.* 133:1564.)

respectively. Thus, genes linked to the MHC do not seem to be directly involved in the resistance patterns of mice to trypanosome infection.

Susceptibility (e.g., short survival times) is a dominant trait in mice infected with *T. rhodesiense* (see Table 12-1). F_1 hybrids derived from crosses of R and S mice show

TABLE 12·1
Inheritance of immunity to trypanosomes

Mouse strain	Survival (%)[1]	% Variant-specific immunity[2]
B10.BR/SgSnJ	R (100)	100
C3H3B/FeJ	S (100)	0
(B10.BR × C3H)F1	R (0) I (5) S (95)	100
(B10.BR × C3H)F2	R (7) I (57) S (36)	93
B10.BR × F1	R(41) I (46) S(13)	100
C3H × F1	R (3) I (36) S (61)	38

[1] Percentage of infected mice exhibiting R, I, or S survival times.
[2] Percentage of all mice exhibiting antibody-mediated control of infecting VAT.
Source: Cell. Immunol., 1984, 87:85 by permission of Academic Press.

F_2 generations and backcrosses to R and S parents show that there is a multigenic basis to the inheritance of resistance traits in mice and that the genes of interest are autosomal in nature. Different stocks of *T. brucei* or *T. congolense* produce different resistance patterns in mice employed in genetic studies. The contribution of specific parasite features and the broad genetic backgrounds of apparently R and S parental mouse strains used in such studies to the outcome is unclear. In any case, the results of most of the studies have indicated a multigenic basis for resistance in mice.

When examinations of R, I, and S mice are made with respect to parasitemia profiles, distinguishing differences are observed in these patterns (Figure 12-8): R mice control the infecting VAT population, as well as multiple relapse populations, before death occurs; I mice control the infecting VAT but in a delayed manner, and do not control subsequent parasitemias; and S mice do not control the infecting VAT population. Examination of a possible linkage of VSG-specific Ab responses to resistance patterns of mice has led to some interesting and unexpected results.

An association was made between the resistance patterns and the kinetics of the VSG-specific Ab response to the infecting VAT population (Figure 12-9): infected R mice have early and predictable VSG-specific responses that are associated with clear-

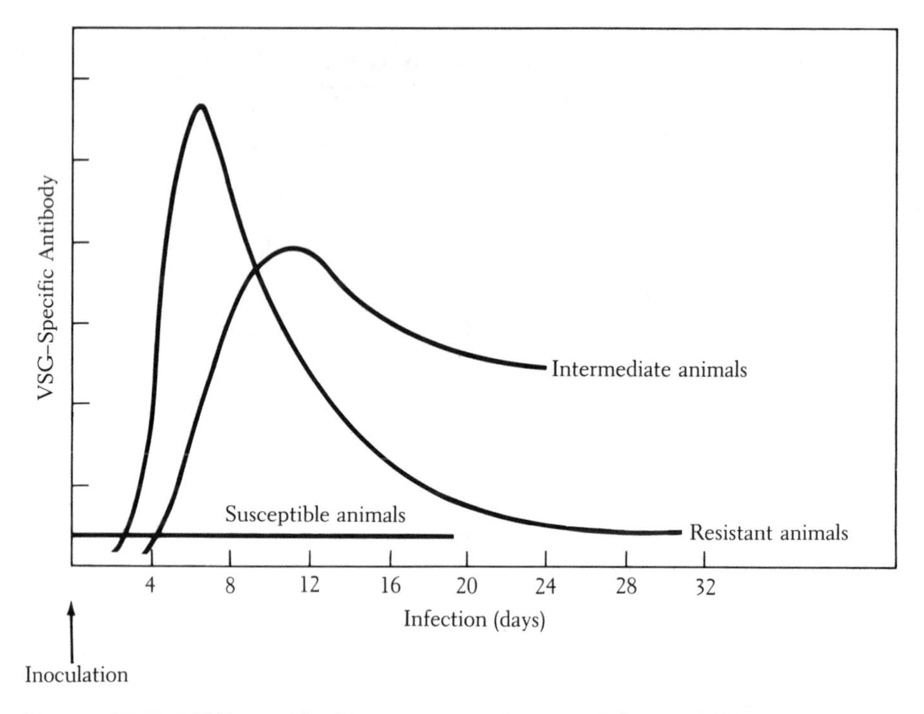

FIGURE 12·8 VSG-specific Ab responses of mice exhibiting different resistance characteristics. (Adapted from LEVINE, R. F., and J. M. MANSFIELD, 1984. Genetics of resistance to the African trypanosomes. III. Variant specific antibody responses of H-2 compatible resistant and susceptible mice. *J. Immunol.* 133:1564.)

ance of the infecting VATs; infected I mice also make VSG-specific Ab responses, but the responses occur later and peak titers are lower, with delayed clearance of the infecting VAT; the infected S mice, however, fail to make detectable VSG-specific immune responses or control parasitemia. While these results indicate a *correlation* between resistance status and parasite-specific immune responses, they do not provide a rigid test of functional or genetic *linkage* of such responses to resistance.

Evidence indicating that such linkages do not occur came from a study of infected semiallogeneic radiation chimera mice (Figure 12-10). These and subsequent studies are detailed here because they have redirected our perspective of the significance of VSG-specific Ab responses of infected hosts. A radiation chimera study was designed to examine the cellular bases of resistance and immunity. It revealed that R + S chimeras (i.e., lethally irradiated S mice reconstituted with H-2-compatible R donor bone marrow cells) make a B-cell response to the VSG of the infecting VAT and can control parasitemia. In contrast, intact S mice (as discussed earlier) and S + S chimeras did not make VSG-specific antibody or control parasitemias. However, despite the presence of VSG-specific antibody and the ability to control parasitemia, R + S chimeras did not

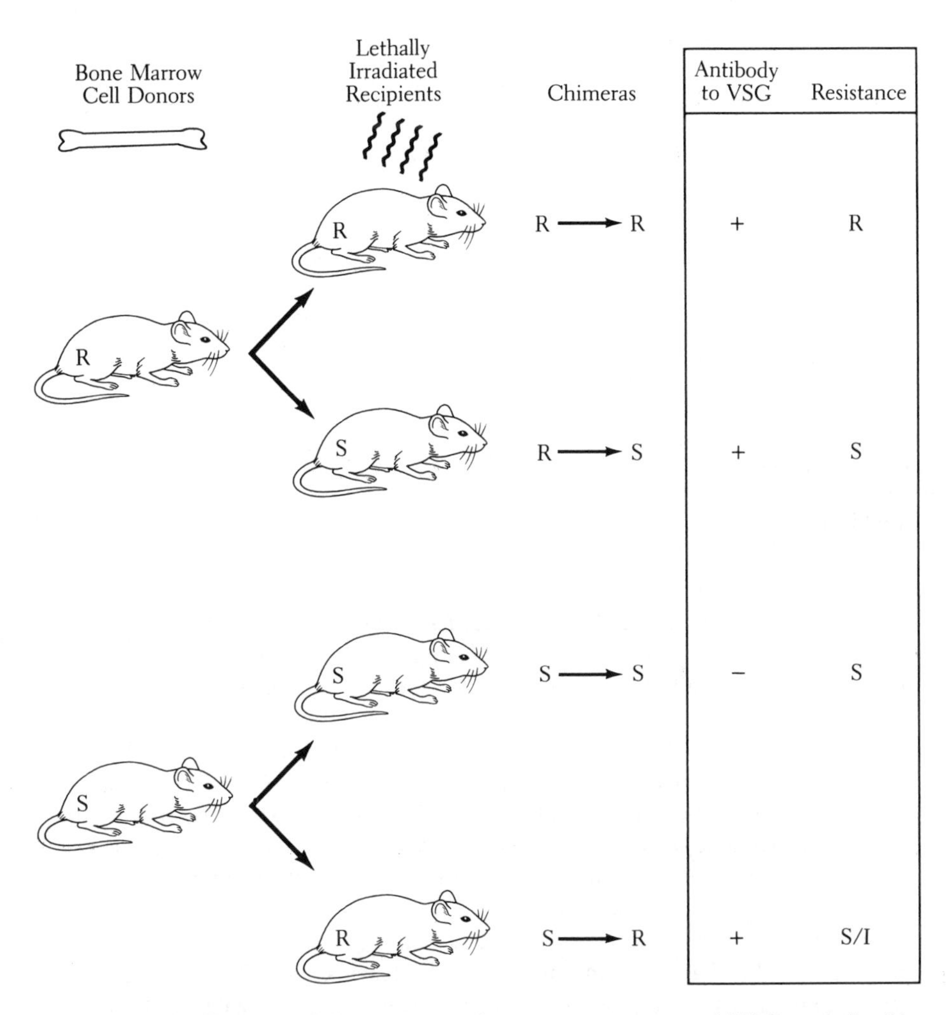

FIGURE 12·9 Functional dissociation of resistance status and VSG-specific Ab responses in semiallogeneic radiation chimeras infected with the African trypanosomes. (Adapted from DEGEE, A. L. W., and J. M. MANSFIELD, 1984. Genetics of resistance to the African trypanosomes. IV. Resistance of radiation chimeras to Trypanosoma rhodesiense infection. *Cell. Immunol.* 87:85 by permission of Academic Press.)

live as long as intact R or R + R mice, and their resistance characteristics were like those of the S or S + S mice. These results indicated that there was no link between the antibody responses and the overall resistance of the host. Resistance requires undefined radiation-resistant host cells, as well as cell(s) of bone marrow origin.

Results of studies examining the inheritance of VSG-specific immunity and resistance provided specific genetic evidence supporting the conclusion regarding a lack of linkage between these traits (Table 12-1). F_1 hybrid offspring of R × S crosses

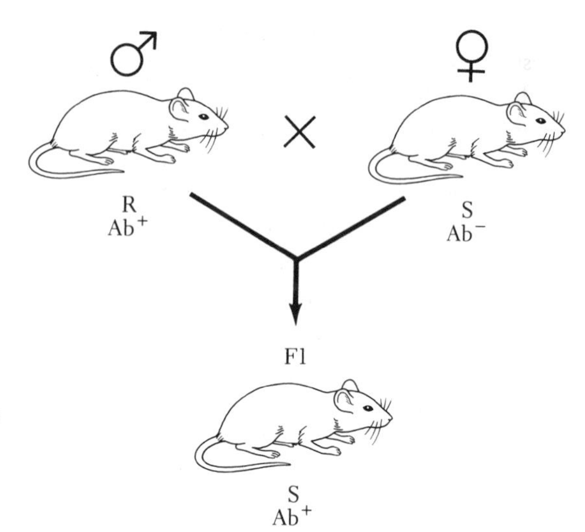

FIGURE 12·10 Genes regulating resistance segregate independently from genes regulating VSG-specific antibody responses. (Adapted from DEGEE, A. L. W., R. F. LEVINE, and J. M. MANSFIELD, 1988. Genetics of resistance to the African trypanosomes. VI. Heredity of resistance and variable surface glycoprotein-specific immune responses. *J. Immunol.* 140:283.)

produce VSG-specific Ab and also control parasitemia, but they are as susceptible (defined by short-term survival) as the S parent. Such susceptibility, after all, is a dominant trait. Thus, the ability to mount VSG-specific Ab responses and the ability to survive for a prolonged period of time after infection are distinct, genetically unlinked capacities of the host. Additional support for this hypothesis is seen in additional studies. Recombinant inbred mice derived from R and S parental strains exhibit a wide range of resistance characteristics regardless of the control of parasitemia by VSG-specific antibodies. Infected cattle also display different resistance traits, but no correlation has been established between the magnitude of the parasite-specific antibody response and resistance.

If parasite-specific Ab responses are not directly associated with resistance in infected animals, what determines how long an infected animal lives? Interferon has been measured as another possible mediator of defense. Paradoxically, trypanosome-infected mice express high serum levels of interferon (IFN) when infected, and differences in the quantity and quality of the IFN response exist among R, I, and S mice. Infected R mice experience an early peak in their serum IFN levels, IFN that is primarily IFNα/β. This peak is followed by a second peak of IFN that is predominantly IFNγ. S mice make no IFN whereas I mice make only IFNα/β. The IFN responses do not seem to be directly linked to resistance, however. Administration of IFN inducers or recombinant IFNγ to S mice does not render them resistant. Nor does administration of monoclonal Ab to IFNγ result in susceptibility in R mice. Thus, the biological relevance of the IFN response to trypanosomal defense is not clear. Perhaps it reflects an early helper-T-cell or NK-cell stimulation by trypanosomes.

Radiation chimera studies (discussed earlier) demonstrated that resistance is dictated by both donor bone marrow–derived cells and recipient radiation-resistant cells (or factors they produce). Possibly VSG-specific B-cell responses coupled with as yet unknown aspects of macrophage function are the key elements in determining resistance, and these two elements might segregate independently, providing for the data suggesting no apparent linkage of VSG-specific antibody responses to overall resistance. While macrophages may be the relevant radioresistant cells involved with the radiation chimera studies, perhaps the assumption is warranted that other recipient's cells produce factors that directly or indirectly control infection.

In summary, determinants of resistance to African trypanosomes are multigenic and multifactorial. The aggregate functions of B cells, macrophages and, perhaps, other host cells or their factors may be the key elements in the patterns of resistance observed. Clearly, new initiatives are needed in order to determine the molecular and cellular bases of resistance to trypanosomes. Interestingly, host resistance traits (whatever their bases) are modified dramatically by alterations in parasite gene expression. This is the focus of the following section.

Trypanosome Virulence

The disease spectrum caused by the African trypanosomes has its basis not only in host genetic makeup but also in trypanosome genetic constitution. It has been recognized for some time that certain subspecies and strains of African trypanosomes cause differences in the disease progression. Some organisms produce acute, fulminating infections resulting in early death of the host while others may produce a more chronic disease characterized by minimal parasitemias and prolonged survival times for the same host species. The molecular and genetic bases for these differences in the expression of parasite virulence are unknown.

It is of central importance to determine whether trypanosome gene expression within a genetically homogeneous population may influence host resistance or immunity. One observation relative to trypanosome virulence was made in which a resistant mouse strain was infected with a genetically homogeneous trypanosome population, as discussed later. At specific times after infection, different antigenic variants were isolated from the blood of individual mice, and these populations were further cloned and expanded. Each of these trypanosome clones, which represented daughter cell populations derived from the parent population but which expressed different surface antigens, were used to infect mice for comparisons with the parent clone.

The results of this study showed that daughter cells arising in individual animals from a genetically homogeneous parent population expressed marked differences in virulence for the host (Figure 12-11). For example, infection with the parent clone resulted in host survival times of approximately 60 days, whereas some daughter cell populations caused death as early as day 25. Extensive studies subsequently showed that trypanosome growth rates, immunogenicity of the individual VSGs, host clearance of the infecting VATs by KC, and induction of antigen nonspecific immunosuppression were equivalent for all clones. A difference was seen, however, in the ability of each population to induce trypanosome-specific immunosuppression: the more virulent

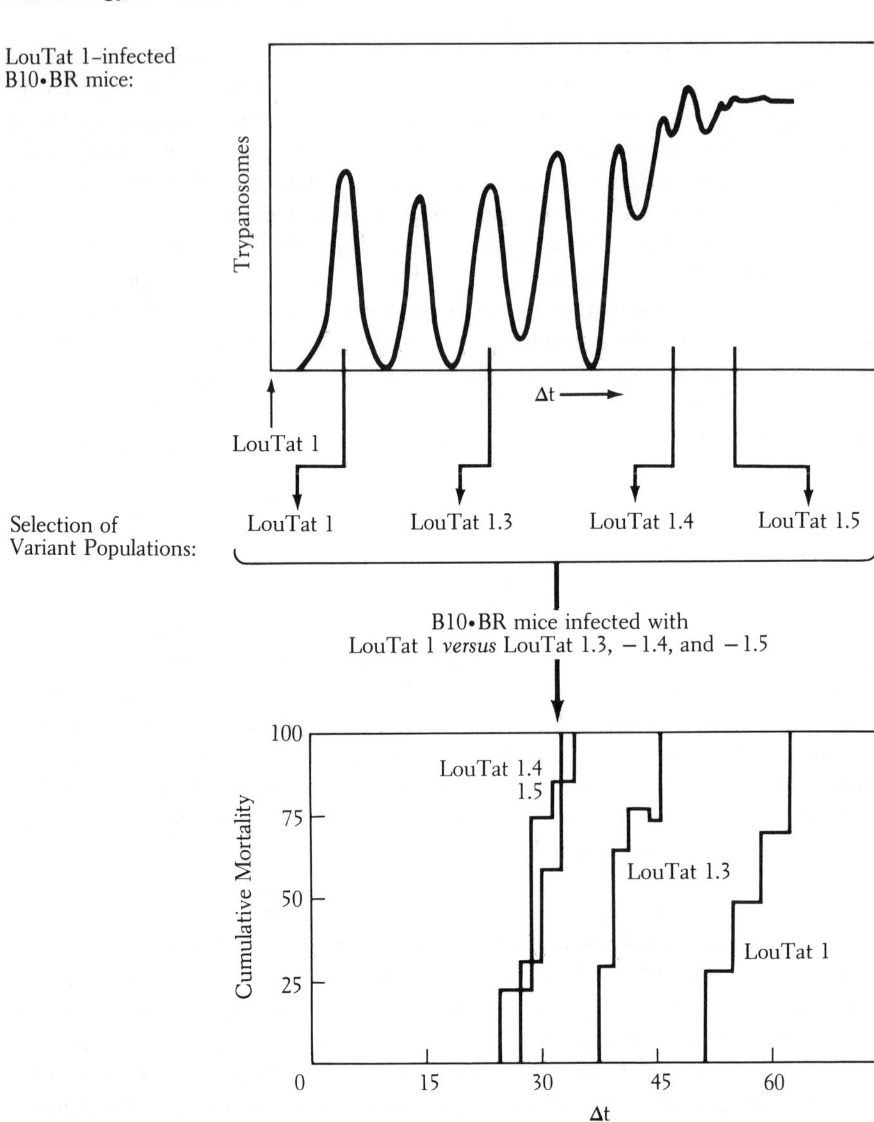

FIGURE 12·11 Daughter cells arising from a genetically homogeneous parent population (clone) show differences in virulence for the same host-mouse strain. (Adapted from INVERSO, J. A., and J. M. MANSFIELD, 1983. Genetics of resistance to the African trypanosomes. II. Differences in virulence associated with VSSA expression among clones of *Trypanosoma rhodesiense*. *J. Immunol.* 130:412; and from INVERSO, J. A., A. L. W. DEGEE, and J. M. MANSFIELD, 1988. Genetics of resistance to the African trypanosomes. VII. Trypanosome virulence is not linked to variable surface glycoprotein expression. *J. Immunol.* 140:289.)

clones induced immunosuppression that prevented the host from responding to the antigenic variants arising after growth of the infecting clone.

In this representative study, all of the more virulent trypanosome populations expressed different VSG phenotypes than the parent population. It was proposed, therefore, that a specific VSG molecule might exhibit the virulence traits associated with a particular VAT population, as a variety of biological functions previously have been ascribed to the VSG molecules. Alternatively, it also seemed plausible that the genomic rearrangements leading to expression of certain VSG genes within their telomeric sites were associated with the expression of hypothetical virulence genes. However, in two separate approaches it was demonstrated that virulence neither was linked to a specific VSG molecule or VAT nor varied with the expression of new surface coats. In one approach, trypanosome sublines and subclones that expressed the same VSG as a less virulent parent population were derived. By rapid subpassage and subcloning, several trypanosome subpopulations were isolated that expressed the same VSG gene and surface coat as the parent population. Despite VSG homology, one virulent subclone killed mice at 3 days of infection, compared to death at 60 days post infection with the parent clone.

In another approach, the VSG identical but highly virulent subclone was used to infect a heterologous host species (rabbits), and the virulence characteristics for mice of VAT populations arising in these animals were compared with the virulence characteristics of VATS arising in animals infected with the less virulent parent clone. While the parent clone gave rise to several VATs with virulence for mice that differed from the infecting population, all VATs arising from the virulent subclone expressed the virulence characteristics of the infecting population. Overall, these studies show that virulence expression may occur independently from and is not linked to VSG expression in the African trypanosomes and that virulence, once expressed for a host species, is a stable phenotype that does not change with subsequent VSG gene rearrangements. Although no virulence genes or their products have yet been identified, there is indirect evidence for the existence of a trypanosome virulence gene or genes. First, as noted before, trypanosome virulence seems to be a stable phenotype once expressed, perhaps indicating that a virulence gene (or genes) not normally expressed in a population becomes constitutively expressed regardless of other gene rearrangements, such as VSG genes. Second, these subclones are virulent for all mouse strains tested, in that infection of an R mouse strain with a virulent subclone results in severe S-like infection. Third, virulence is associated with major stable genetic changes. For example, virulent organisms derived from a parent clone display different chromosomal profiles compared with the parent clone or other less virulent VATs that arise from the parent clone. Also, there are numerous differences evident at the mRNA and expressed protein levels when less-virulent and more-virulent clones are compared, and genes preferentially expressed by virulent clones have been isolated in this laboratory.

Differences in virulence clearly exist among the African trypanosomes and these differences are independent of the VSG molecule itself or VSG expression. Virulence expression by trypanosomes affects both parasite-specific immunity and host resistance. The molecular and genetic basis for the expression of virulence remains to be uncovered as the next major hurdle to our understanding of the immunobiology of trypanosomiasis.

Coda

Evidence has been presented in this chapter that there is an early immunological event involving the interaction of trypanosomes (or their products) and macrophages (Figure 12-12). This parasite-host cell association initially results in stimulation of both T and B cells at an early stage of infection (stage 1). Subsequently this stimulatory influence converts into a suppressive one that affects predominantly MHC-restricted antigen presentation to T cells (stage 2).

A possible sequence of events is postulated. The *central* event in the trypanosome-macrophage association involves the uptake of trypanosomes by macrophage cells

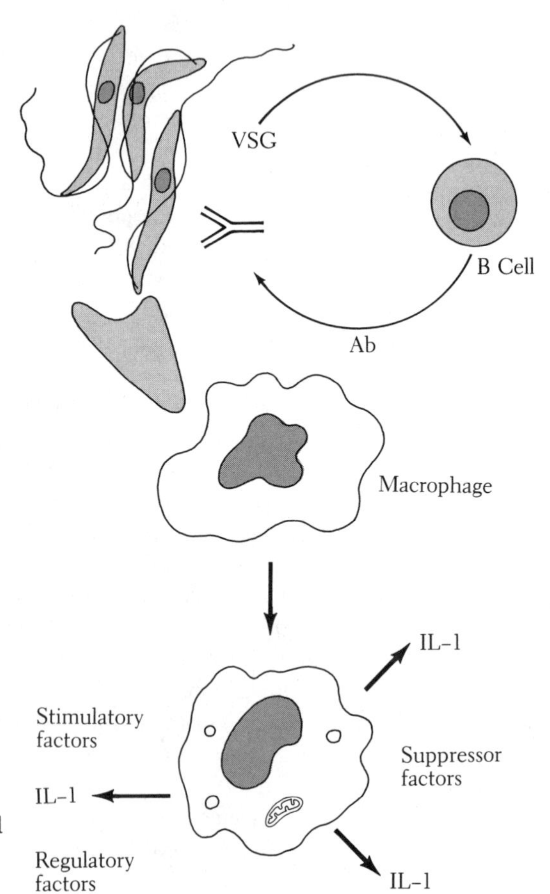

FIGURE 12·12 Overview of parasite-specific and antigen-nonspecific immune responses and immunoregulation in African trypanosomiasis.

in an antibody-dependent but predominantly T-independent manner. In a classical T-independent interaction, B cells are stimulated by an array of surface VSG epitopes plus one or more maturation signals from the trypanosome, or from macrophages, to undergo clonal expansion and differentiation, resulting in production of IgM to the surface VSG determinants. The resulting IgM response promotes uptake of trypanosomes by macrophages. During this cycle, the regulation of parasite-specific antibody responses (especially the sequential VSG-specific antibody responses) involves idiotype-specific (antiidiotypic) B and/or T cells.

A suggested description of peripheral events follows. Early in stage 1 of infection, it is believed that macrophages are activated directly by stimulatory components of the ingested trypanosomes, by trypanosome products present in the tissues, or as a result of antigen-specific stimulation of helper T cells, with the results that the macrophages express high levels of Ia antigens and spontaneously secrete IL-1. T cells, in response to enhanced IL-1 production and to trypanosome antigens plus Ia, secrete IL-2, which induces IL-2 receptor expression in the T helper cells, with resultant stimulation and clonal expansion of these IL-2 – stimulated cells. The subsequent production by T-cell subpopulations of lymphokines presumably influences the activity of a variety of cells. INFγ activates or produces additional activation signals for macrophages, which then further amplify T-cell stimulation. Other T-cell-derived factors (IL-4, IL-5) also cause stimulated B cells to undergo clonal stimulation, expansion, and differentiation in response to numerous variant and invariant trypanosome antigens. The anticipated end result would be not only increased trypanosome uptake by macrophages, as the result of increased parasite-specific antibody production, but also polyclonal B- and T-cell activation. The production of IL-3 by stimulated T cells presumably is a basis for increased cellular output from the bone marrow and seeding of lymphoid tissues with cells of the granulocyte-monocyte series. All these proposed events occur within the first week of infection, at a time when macrophages are maximally exposed to trypanosome constituents after initial clearance of the organisms from the blood.

A second stage of infection follows soon after the initial burst of macrophage activation and subsequent T-cell stimulation. It may be initiated by the production of PGE_2 by activated macrophages or by an inhibitory trypanosome component. The result of stage 2 is a profound cessation of certain macrophage-dependent activities, principally those leading to T-cell stimulation. Antigen processing or presentation is down-regulated or altered significantly. Suppressor factors that have T cells as their target are released. This suppressive stage results in depression of all parasite-specific T-cell responses and persists until the animal dies.

Infected animals that are VSG nonresponders possess B cells that are not stimulated appropriately by trypanosomes (or are blocked in their maturation), with the result that parasite-specific IgM responses are depressed or delayed. Also, resistant animals that are infected with virulent trypanosome clones appear to resemble the susceptible and VSG nonresponder hosts, in that trypanosome-specific B-cell responses are depressed. Depending on the species or subspecies of *Trypanosoma*, the early events may be accelerated or prolonged, so that the suppressive events of stage 2 occur earlier or later during infection. There is also evidence that macrophages from different anatomical sites may be differentially influenced by trypanosome infection. Thus, the sequence of events proposed in Figure 12-12 may be skewed significantly, depending on host and parasite gene expression, as well as on the lymphoid organs examined in the infected host.

The proposed events occurring in trypanosomiasis are unique in several respects. For example, the initial activation step for macrophages may be trypanosome- and B-cell-dependent but is T-cell independent. Subsequent T-cell stimulation is followed by a suppressive effect of the macrophages, which clearly is directed towards the T-cell compartment of the host immune system. The hypothetical result of this event is that T-dependent immunopathology may be depressed yet T-independent B-cell responses, which control parasitemia, may remain intact. Despite the intellectual allure of this overall picture, it must be realized that comprehensive experiments have not been performed to prove or disprove the sequence of events hypothesized. Rather, this scheme is derived from numerous experimental observations performed in a number of different laboratories. This collective hypothesis must yet withstand the rigors of contemporary molecular and cellular immunological testing.

The picture that this chapter has painted of trypanosome immunology is one in which differences in host or parasite gene expression may influence both the resistance traits of the host and the molecular and cellular components of host VSG-specific immunity. The genetic bases of these differences remain to be elucidated, as do the mechanistic bases of the differences. It is hoped that the results of such studies will elucidate, in a clearer manner, the nature of the host–parasite immunobiological relationship and reveal the specific manner in which differences in host or parasite gene expression influence this relationship.

Additional Reading

ASKONAS, B. A. 1985. Macrophages as mediators of immunosuppression in murine African trypanosomiasis. *Curr. Top. Microbiol. Immunol.* 117:119.

BLACK, S. J., C. N. SENDASHONGA, C. O'BRIEN, N. K. BOROWY, M. NAESSENS, P. WEBSTER, and M. MURRAY. 1985. Regulation of parasitemia in mice infected with *Trypanosoma brucei*. *Curr. Top. Microbiol. Immunol.* 117:93.

MANSFIELD, J. M. 1981. Immunology and immunopathology of African trypanosomiasis. In *Parasitic Diseases*, vol. 1, *The Immunology*, J. M. Mansfield, ed. Marcel Dekker, Inc., p. 167.

MANSFIELD, J. M. 1985. Genetics of resistance to the African trypanosomes. In *Genetic Control of Host Resistance to Infection and Malignancy*, E. Skamene, ed. Alan R. Liss, p. 483.

MULLIGAN, H. W. 1970. *The African Trypanosomiases*. Wiley-Interscience, 950 pp.

PEARSON, T. W. ed. 1986. *Parasite Antigens*. Marcel Dekker, Inc., 424 pp.

ROELANTS, G. E., and M. PINDER. 1984. Immunobiology of African trypanosomiasis. *Contemp. Top. Immunobiol.* 12:225.

13

Immunology of Chagas' Disease

Zigman Brener and Antoniana U. Krettli

Chagas' disease begins in humans as an acute illness of short duration (1 to 2 months) during which certain lesions at the portal of entry may be present either in the ocular conjunctiva (Romaña's sign) or the skin ("chagoma" of inoculation). Systemic manifestations may include fever, edema, generalized lymphadenopathy, various electrocardiographic changes, and heart enlargement. Parasites can be detected in blood and tissue. A subpopulation of patients are only mildly symptomatic and have low levels of parasitemia ("inapparent" cases). Diagnosis during the acute phase is established by detecting circulating bloodstream forms of the trypansomes (BTry) or by serological methods.

Manifestations of the acute phase, including those related to the heart, gradually subside and symptoms disappear within a few months. Mortality rates of 2 percent to 8 percent have been reported in the past in untreated patients. The recent availability of nitrofuran and nitroimidazole derivatives for chemotherapy have improved prognosis dramatically, however. This early stage of the disease is succeeded by a chronic infection, in which parasites are barely detectable by parasitological methods such as xenodiagnosis and blood cultures. Since the acute phase may go undetected in endemic areas, Chagas' disease is most commonly diagnosed during the chronic phase. Chronic infection is a life-long disease; spontaneous cure rarely, if ever, occurs. Recrudescence of acute illness in chronically infected patients is not a common event in patients receiving immunosuppressant chemotherapy. A challenge infection of chronically in-

We thank the Brazilian agencies National Research Council (CNPq) and Financiadora de Estudos e Projetos (FINEP) and the International Development Research Centre of Canada for financial support and to Dr. Seymour H. Hutner for review of the manuscript.

fected animals with virulent parasites is usually not followed by a new acute phase. However, the original and the superimposed *Trypanosoma cruzi* populations can both be isolated from the challenged animal. Some patients have also been found to harbor more than one *T. cruzi* strain, defined as zymodemes (parasites that contain enzymes reacting with the same substrate but differing in electrophoretic mobility) or schizodemes (parasites that contain kinetoplast DNA that differs in the electrophoretic mobility of fragments after treatment with restriction endonucleases). Chronic infection by *T. cruzi* elicits an immune response that suffices to maintain extremely low parasite levels and to suppress parasitemia induced by reinoculation of homologous or heterologous parasite strains. This immunity, however, is ineffective in eradicating an established infection.

During the chronic phase, Chagas' disease may manifest in different forms. Clinical manifestations of the cardiac form, which affects 30 percent to 40 percent of patients, can range from sudden death or heart failure to minor electrocardiographic changes. In the gastrointestinal form, found in 8 percent to 10 percent of patients, abnormalities can range from mild functional changes in bowel motility to marked dilatation of the esophagus and colon. In the indeterminate form, the category to which most chronically infected patients belong, clinical manifestations are by definition absent when patients are given routine physical examination, xray, and EKG. However, more refined clinical methods may detect abnormalities.

Natural *T. cruzi* infection of sylvatic reservoir hosts has been reported in a large area of the American continent and includes over 100 mammalian species belonging to different orders (Marsupialia, Endentata, Chiroptera, Carnivora, Lagomorpha, Rodentia, and Primates). Domestic reservoirs, such as dogs, cats, rats, and mice, are often infected with *T. cruzi* in the endemic area. Marsupials with domestic and peridomestic habitats may be important in introducing *T. cruzi* into the domestic cycle. Several laboratory animals are used as experimental hosts for *T. cruzi* — mice, rats, guinea pigs, primates, dogs, and rabbits. The course of infection is greatly influenced by the host species (and in the case of inbred mice and rats, by strain differences) and by parasite strain. Infective stages can be derived from infected vertebrates, vectors, and axenic or tissue culture.

The usual course of *T. cruzi* infection (i.e., an acute phase followed by a life-long chronic phase) is observable in most experimental hosts. However, for a chronic experimental model to adequately resemble the chronic forms of the disease in humans and be suitable for immunological and immunopathological studies, the following requirements ideally should be fulfilled: (1) permitting isolation of the parasite; (2) eliciting antibodies detectable by available serological methods; (3) developing relevant cardiac and gastrointestinal lesions; and (4) inducing antitrypanosomal antibodies that cross-react with host cell components (see the later section Immunopathology). Unfortunately none of the available experimental hosts satisfies these requisites for an ideal chronic model. Nevertheless, mice, dogs, and rabbits are being used in experimental work with *T. cruzi*. The infection of certain inbred strains of mice with different *T. cruzi* strains that exhibit unique biological characteristics has provided murine models attractive for immunological studies of Chagas' disease. Puppies are extremely susceptible to

T. cruzi infection that can progress to the chronic stage, rendering them suitable for studying the mechanisms of heart damage. The appropriateness of rabbits as models for immunopathological studies is controversial, but this model has proven useful for certain immunological and parasitological studies.

Host Defense

Natural Resistance

Amphibians and birds are completely resistant to infection with *T. cruzi* under natural conditions. The mechanism of resistance in birds is antibody-independent and related to complement (C) activation by the BTry (bloodstream forms) surface membrane. In vitro activation by BTry of the alternative complement pathway in chicken serum results in parasite lysis. BTry injected intravenously into normal chickens are destroyed within 1 minute. If chickens are first treated with cobra venom factor to deplete complement components, the parasites remain viable in the circulation. In other animal models for *T. cruzi* (e.g., rats) in which the blood parasites may remain undetectable, the role of complement in natural resistance has not been determined. Fresh human sera and other mammalian sera do not destroy BTry unless these forms are first coated with specific antibodies, as will be discussed later.

Natural resistance to *T. cruzi* in humans has not been demonstrated; existence of uninfected individuals in highly endemic areas has been ascribed to epidemiological circumstances rather than to inherent individual refractoriness to infection. Some patients during the acute phase of illness experience only very mild symptoms that contrast with those experienced during the more typical severe acute illness. These "inapparent" acute infections are usually detected only by documenting that seroconversion has occurred. It is thought that the attenuated course of acute infection is related to a depression of thymus-derived (T) lymphocyte function induced by *T. cruzi*. Other distinguishing features of these patients have not been investigated.

A significant and early increase in natural killer (NK) cell activity in mice experimentally infected with *T. cruzi* has been reported. This early enhancement of NK activity might suggest a basis for some "natural" resistance against *T. cruzi*, but this interpretation remains speculative.

The genetics of natural resistance to *T. cruzi* has been studied in inbred mice. Different inbred strains of mice, when infected with *T. cruzi*, express a spectrum of resistance. The C3H strain is highly susceptible, experiencing high parasitemia and mortality, whereas C57BL/10 mice are strikingly resistant. Irradiation of resistant mice results in higher parasitemia after *T. cruzi* challenge, suggesting that a radiosensitive immune response might participate in control of the infection. The contribution of the major histocompatibility complex of mice (H-2) in the regulation of anti-*T. cruzi* defense has been investigated. It has been clearly established in studies with congenic mice inoculated with *T. cruzi* that the H-2 haplotype influences disease outcome. The

H-2^k haplotype (susceptible C3H mouse strain) is associated with a significantly higher mortality than is the H-2^b haplotype (resistant C57BL/10 strain). Interestingly, experiments with the highly resistant B.10.S (H-2^s) strain strongly suggests that low parasitemia is a dominant trait influenced by multiple genes outside the H-2 region, whereas survival is a trait strictly associated with the H-2^s haplotype. Non-H-2 genetic loci that influence defense have not been precisely mapped. A mutation at a single locus in recombinant inbred BXH-2 strain renders mice unable to limit the early multiplication of *T. cruzi* in the acute phase of infection. Early proliferation of parasites takes place despite normal antibody responses. The antitrypanosomal defense mechanisms controlled by H-2–linked or non-H-2 genes has not been elucidated. Certain antibody isotypes are higher in resistant than in susceptible mice. In addition, resident peritoneal macrophages from susceptible, chronically infected mice become more heavily infected with *T. cruzi* than do macrophages collected from resistant strains. Differences in proliferative response to T-cell mitogens of lymphocytes from resistant and susceptible mice strains inoculated with *T. cruzi* have been reported. Possibly, immunoregulatory perturbations occurring during infection are under genetic control and influence disease outcome.

Mononuclear Phagocytes

Cells of the mononuclear phagocytic system (monocytes and macrophages) play a dual role in relation to *T. cruzi*, serving as host cells that support the parasite's intracellular multiplication and differentiation into infective stages on the one hand, and serving as effector cells of the immune response on the other. In vivo parasitism of macrophages by *T. cruzi* has been demonstrated in various hosts. Evidence that certain *T. cruzi* strains selectively infect mononuclear phagocytes has been obtained by observing the preferential parasitism of macrophages from the spleen, liver, and bone marrow in inoculated mice. Whether this selective parasitism occurs in humans or other mammals is not known. Certainly, *T. cruzi* can infect other cells of mesenchymal origin. Fc receptors of macrophages are not essential for uptake of *T. cruzi*. BTry collected from immunosuppressed mice, and consequently deprived of membrane-bound immunoglobulins, are readily ingested by intact macrophages. Opsonization of BTry significantly enhances phagocytosis, however. Sera from chronically infected hosts contain antibodies that can enhance phagocytosis of (that is, opsonize) BTry by mouse peritoneal macrophages. In contrast, sera from mice or rabbits immunized with fixed parasites and sera from treated and cured patients do not opsonize BTry.

Normal resident peritoneal mouse macrophages and thioglycolate-induced macrophages are permissive to the intracellular growth of *T. cruzi*. Macrophages from mice immunized with BCG or infected with pathogenic protozoa unrelated to *T. cruzi* (*Toxoplasma gondii* and *Besnoitia jellisoni*) inhibit replication of BTry and cultured trypomastigotes. The trypanocidal activity of macrophages can also be induced by incubating macrophages with lymphokines released upon stimulation of specifically sensitized T lymphocytes with heat-killed *T. cruzi* trypomastigotes or lymphokines secreted by lymphocytes treated with concanavalin-A. Peripheral blood mononuclear cells (including lymphocytes) obtained from patients with chronic Chagas' disease (but not from uninfected controls), when stimulated with heat-killed *T. cruzi*, BCG, or concanavalin-A (Con A), elaborate a macrophage-activating cytokine that can induce trypanocidal effects in infected macrophages. There are also in vivo observations that reinforce the concept of a role for macrophages in defense. Administration of silica particles to mice (particles selectively toxic to macrophages) markedly enhances parasitemia and mortality associated with *T. cruzi* infection. On the other hand, administration to animals of the macrophage activators BCG and formalin-fixed *Corynebacterium parvum* decreases the severity of infection.

Destruction of intracellular *T. cruzi* stages in activated macrophages is mediated at least partly by toxic oxygen metabolites (superoxide, hydrogen peroxide, and hydroxyl radicals). This sensitivity to an oxygen-related mechanism is shared by other intracellular protozoa, notably *T. gondii*, *Leishmania*, and *Plasmodia*. The antimicrobial effects exerted on *T. cruzi* by activated macrophages can be abrogated by catalase, a scavenger of H_2O_2. Furthermore, release of H_2O_2 parallels the microbicidal effectiveness of activated macrophages.

Acquired Resistance: The Acute Phase

As previously mentioned, the acute phase in Chagas' disease, in which parasites are usually detected by examination of fresh blood, is followed by a long-lasting chronic phase with extremely low parasitemia that is only detectable by more elaborate parasitological methods (blood culture and xenodiagnosis). What controls *T. cruzi* proliferation in acute and chronic phases is unknown; the mechanisms may be different for each stage. That specific antibodies play a central role in the chronic disease is reasonably well established, but such a role in the acute phase has not been defined. Defense mechanisms in the acute phase have been investigated primarily with laboratory-adapted *T. cruzi* strains that induce in mice intense parasitemia, severe immunosuppression related to polyclonal activation, and high mortality. *T. cruzi* strains freshly isolated from humans or animal reservoirs are much less virulent in mice and are not being used for such immunological studies.

Intravenous challenge with virulent BTry in mice infected with *T. cruzi*, followed by subcurative treatment with an antitrypanosomal drug to abolish the first parasitemia wave, has been used as a model to study acquired resistance. On the basis of this model, it is concluded that acquired resistance is mounted soon after the initial inoculation, since mortality and parasitemia levels are reduced or even absent after challenge. Interestingly, such resistance in the first 3 weeks of infection is unassociated with the presence of detectable anti-*T. cruzi* protective antibodies. Such antibodies only appear after week 4. Although specific IgM and IgG antibodies are synthesized at the first weeks of infection, they do not kill the parasite. These in vivo findings may imply that early control of *T. cruzi* infection is antibody-independent. However, from 4 to 5 weeks onward after the initial infection, protective antibodies can be detected, and after week 5 over 90 percent of parasites in an intravenous challenge are cleared from the circulation within 3 to 10 minutes.

The role of cell-mediated immunity in early control of *T. cruzi* has not been clearly established. Antithymocyte serum administration and neonatal thymectomy result in enhanced parasitemia in mice. Adoptive transfer of T cells from immune mice early in infection also can transfer resistance. Despite the in vitro evidence for participation of lymphokine-activated macrophages in resistance against *T. cruzi*, it remains to be established if T cells are protective in these in vivo studies by virtue of secreting macrophage-activating lymphokines, by acting as helper cells in antibody production, or by serving some other function.

Antibody-Mediated Resistance: The Chronic Phase

There is sound evidence that resistance in the chronic phase is largely humorally mediated. Passive transfer of immune sera or purified IgG obtained from chronically infected hosts protect mice against an otherwise lethal challenge. Moreover, the in vitro incubation of viable virulent trypomastigotes with sera from chronically infected hosts, but not from those with acute infections, greatly reduces parasite infectivity. Moreover, parasitemia and mortality are greater in Biozzi low-responder (Ab/L) mice that mount poor antibody responses than in Biozzi high-responder (Ab/H) mice, which mount normal antibody responses. Also consistent with the conclusion that antibody is critical in defense during the chronic phase is the observation that adoptive transfer of spleen cells from mice that recovered from acute infection induces protection against a lethal challenge with *T. cruzi*. This protection may be prevented by removing B cells, whereas T-cell depletion only reduces the degree of resistance transferred. Mouse IgG antibody, but not IgM, has been demonstrated to be involved in the acquired resistance against *T. cruzi*. Specifically, transfer of IgG1 and IgG2 antibodies imparts protection.

Strains of *T. cruzi* differ in their susceptibility to antibody-mediated defense. Trypomastigotes of the Y strain are agglutinated by both anti-Y and anti-CL mouse serum, as well as immune human sera from chronically infected patients, whereas the CL strain of trypanosomes are not agglutinated by the homologous or heterologous antisera. Furthermore, in vitro incubation of trypomastigotes with immune sera renders Y strain but not CL strain parasites avirulent on subsequent injection into mice. Similarly, injection of immune sera from chronically infected mice into naive animals significantly decreases the parasitemia and mortality rates induced by a challenge infection with Y-strain blood forms, but not with those of the CL strain. Furthermore, blood forms isolated from immunocompetent mice at the acute phase of illness also vary in their susceptibility to complement-mediated lysis. Y trypomastigotes are readily lysed by addition of complement in vitro, whereas CL parasites are not affected, despite the fact that both strains bind antibodies. This finding has raised the possibility that CL blood forms might be capable of evading C activation and lysis by a unique mechanism.

The CL and Y strains apparently represent two "polar" forms with respect to susceptibility to humoral defense. This polarity is taken to reflect a substantial diversity among parasite strains that can be isolated from naturally infected hosts. BTry of several other *T. cruzi* strains have been assessed by in vivo and in vitro methods similar to those discussed. Most exhibit characteristics that are CL-like. The Tulahuen and the Berenice strains, however, resemble the Y strain and are readily destroyed in vivo or in vitro by antibodies plus complement.

Protective versus Nonprotective Antibodies

Immunization of the vertebrate host with attenuated *T. cruzi* or with parasite homogenates or fractions ellicits a strong humoral immune response with high titers of specific antibodies. Nevertheless, contrary to the pronounced acquired resistance induced by chronic infections, the resistance achieved after immunization with fixed parasites or their fractions is at best partial or is absent. Protective antibodies (ones so far found only in chronic infections) recognize epitopes on the surface of living trypomastigotes that are targets of complement-mediated lysis. The protective antibodies are therefore referred to as lytic antibodies (LA). Antibodies present in patients with chronic infection that do not recognize epitopes on living trypomastigotes, but do react with epitopes on the surface of fixed or air-dried *T. cruzi* stages as well as with parasite fractions, serve as the basis for serodiagnosis. They are called conventional serological antibodies (CSA). CSA but not LA are also detected in immunized, uninfected hosts and do not correlate with acquisition of anti-*T. cruzi* defense (Table 13-1).

CSA can be detected by several methods that employ as antigen either parasite fractions (i.e., for ELISA, complement-fixation, indirect hemagglutination, or radioimmune assays) or fixed intact parasites (for indirect immunofluorescence or direct agglutination of trypsin-treated epimastigotes). The following methods have been found suitable for detection of protective antibodies: (a) complement-mediated lysis (CoML), an assay in which LA antibodies bind to live trypomastigotes and converts them into activators of the complement cascade; (b) indirect immunofluorescence with living trypomastigotes; (c) serum neutralization, in which trypomastigotes are treated in vitro, injected into normal mice, and the reduced parasitemia and mortality imparted by these treatments are assessed as evidence of the presence of protective antibodies; (d) passive transfer of specific sera from chronically infected hosts, which decreases parasitemia and mortality in mice challenged with virulent *T. cruzi* BTry, is used as an assessment; (e) assessment of the rate of clearance of trypomastigotes in vivo after their in vitro pretreatment; (f) assessment of the rate of macrophage phagocytosis of trypomastigotes and of antibody-dependent cell-mediated cytotoxicity against trypomastigotes.

Effective treatment of Chagas' disease halts the production of protective antibodies within a few months, whereas the nonprotective antibodies remain detectable for several years. Chemotherapy of *T. cruzi*–infected mice that result in parasitological cures is followed by the gradual decline of acquired resistance, as gauged by the

TABLE 13·1
Protective or lytic antibodies (LA) and conventional serology antibodies (CSA) in Chagas' disease

Vertebrate hosts		Serum antibodies present	Target cell to detect antibodies	Methods for antibodies detection[1]
Rabbits and mice	Chronically infected	LA, CSA	LA Living tyrypomastigotes	LA CoML IFV ADCC SNA Agglutination Phagocytosis
	Immunized with *T. cruzi* antigens	CSA		
Humans and mice	Chronically infected	LA, CSA	CSA Fixed parasites or purified *T. cruzi* antigens	CSA IIF ELISA CFR Agglutination
	Cured after specific treatment	CSA		

[1] (CoML): Complement-mediated lysis; (IFV): Immunofluorescence with viable trypomastigotes; (ADCC): Antibody-dependent cytotoxicity; (SNA): Serum neutralizing antibodies; (IIF): Indirect immunofluorescence with fixed parasites; (CFR): Complement-fixation reaction.

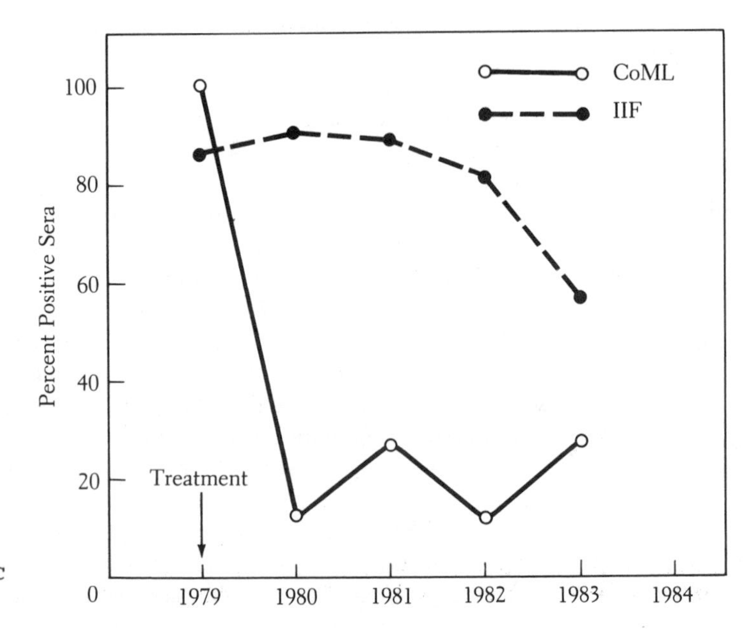

FIGURE 13·1 Evolution of the complement mediated lysis (CoML) and indirect immunofluorescence-(IIF) tests in a chagasic patient followed up for 4 years after specific treatment.

progressive decreased ability of the mice to resist challenge with virulent blood forms. Within 5 to 7 months after successful treatment, such reinoculations evoke high parasitemias similar to those that develop in naive mice. As expected, disappearance of protective antibodies (=LA) parallels this loss of acquired resistance. On the assumption that disappearance of anti-*T. cruzi* protective antibodies might be a good index of parasitological cure, antibodies were assessed in 156 chagasic patients treated with benznidazole or nifurtimox and followed for 2 to 5 years. In only 12 percent, treatment resulted in disappearance of LA and CSA; 69 percent had positive LA and CSA; and the remaining 19 percent had persistently negative LA and positive CSA. The results have been interpreted to indicate that the first group was cured and that the second (and largest) group was not. In the third group (Figure 13-1), disappearance of LA (an antibody correlate of protective immunity) is taken as evidence that these patients are cured. A small number of patients in this group of LA-negative individuals experienced disappearance of CSA after several years. Blood cultures for trypanosomes were negative in several patients from the first and third groups, supporting the interpretation of the serological results. Why CSA remains circulating for many years despite absence of protective antibodies (LA) is unknown.

Complement Activation by *T. cruzi*

Different *T. cruzi* stages vary in their ability to activate complement (C) and in their susceptibility to C-mediated lysis in vitro. The epimastigotes forms (EF), found in the insect vector and cultivatable in cell-free media, can directly activate the alternative pathway in the absence of antibodies. The intracellular amastigote forms resist C-mediated lysis whether or not they are first exposed to protective antibodies. Trypomastigotes isolated from blood or from cultures also resist direct C-mediated lysis unless they are first treated with trypsin, heat (43°C), or incubation with specific anti-*T. cruzi* antibodies present in acutely or chronically infected patients. Fragments of antibody—$F(ab')_2$ and monovalent Fab—also render these trypomastigotes capable of activating the alternative complement pathway.

Conversion of trypomastigotes into activators of the complement alternative pathway by univalent Fab fragments from different immunoglobulin isotypes suggests that LA interfere with C cascade regulation, which normally takes place at the surface of the trypomastigotes. It is surprising that antibody-mediated lysis of trypomastigotes does not require the presence of the Fc portion of antibody. Proteins regulating complement activation are present in plasma and on the membrane of blood cells, including erythrocytes, and also on trypomastigotes. Released from these parasites at 43°C (or by enzyme treatment), such membrane components, of molecular weight (MW) 60–65 kilodaltons (kD), affect the formation of C3 convertase and have been considered to be similar to the decay-accelerating factor that is present on erythrocytes. Understanding how LA destroy the blood trypomastigotes might enable isolation of protective antigens that could be employed in vaccines against Chagas' disease as well as clarify the puzzling ability of these parasites to circulate for decades in a vertebrate chronically infected yet resistant to rechallenge.

Evasion of Host Defense

The intracellular localization of the *T. cruzi* amastigotes is probably an important factor in parasite survival in acute and perhaps also in chronic infection. Collections of intracellular amastigotes are readily detected and numerous during acute infection. During the chronic phase, they are rare but nonetheless detectable, even at postmortem examination of patients dying many years after the initial infection. The mechanisms whereby amastigotes evade host defenses have not been specifically elucidated. On the other hand, several evasion mechanisms have been proposed for BTrys. Among them is the severe immunosuppression that is induced by parasite antigens. Paradoxically, during the acute phase polyclonal activation of B cells is conspicuously evident. Many antibodies produced by the host early in infection are irrelevant for protection and are ones perhaps involved in disease pathogenesis. This polyclonal activation is followed by a period of immunodepression. This pattern of polyclonal activation followed by immunosuppression has also been observed with malaria and African trypanosomiasis and may be linked causally (i.e., polyclonal activation may trigger immunosuppressive responses).

Eventually protective antibodies against the surface of the bloodstream parasites are formed and may be detectable at high titers. It is unclear how the trypomastigotes survive in this potentially hostile, immunoglobulin-rich environment, evading destruction by antibody-mediated effector mechanisms. Antigenic variation, a strategy of evasion employed by African trypanosomes (see Chapter 3) in which parasites successively change the antigenic composition of their surface coat, is unlikely to occur with *T. cruzi*. Whereas antigenic variation provides for several waves of parasitemia in *T. brucei* infection, *T. cruzi* gives rise to only one acute wave of parasitemia, succeeded by a long-lived subpatent infection undetectable on blood smears.

Incorporation of various host plasma proteins, including immunoglobulins and complement components on the surface membrane of trypomastigotes, has been demonstrated, and may provide a way whereby BTry escape immune recognition. *T. cruzi* trypomastigotes can bind the Fc portion of IgG immunoglobulins. Since some mechanisms of antibody-mediated resistance, such as antibody-dependent cell-mediated cytotoxicity and antibody-dependent complement lysis, require exposure of the Fc domain of IgG on the target cell, the binding of Fc on the parasite surface might interfere with such mechanisms.

BTry also can cleave membrane-bound IgG molecules by elaborating a specific protease. This process results in release of the Fc portion. Fab fragments of specific anti-*T. cruzi* antibody that remain bound to the parasite surface presumably permit parasites to circulate without being susceptible to defense mechanisms that depend on host components (cells or proteins) interacting with the Fc domain. This cleavage process is termed "FABulation." Its importance in vivo remains to be elucidated.

Escape might also occur by capping and shedding of immune complexes (parasite membrane components plus specific antibody) from the *T. cruzi* surface.

Immunopathology

The inflammatory process found in the heart muscle cells (myocarditis) and other tissues occurs only when *T. cruzi* intracellular stages (amastigotes) are released by rupture of the host cells they parasitize. Mononuclear cell (lymphocyte and macrophage) infiltrates in the heart are initially focal, but gradually become diffuse and disassociated with tissue parasites. This inflammatory process subsides within a few months, coincident with remission of clinical manifestations. Neuronal degeneration of the intracardiac ganglia and ganglia of the digestive tract (Auerbach's plexus) occurs in the acute phase and may progress during the chronic phase.

During chronic Chagas' disease, most patients (representing the indeterminate form) have small, scattered foci of inflammatory cells in the heart, usually without clinical or electrocardiographic expression. More-severe cardiac lesions occur in 20 percent to 30 percent of the patients and are apparent only after the first or second decades of infection, when diffuse myonecrosis (muscle cell destruction), mononuclear cell infiltrates, and interstitial fibrosis may become evident. Damage to the cardiac conduction system also occurs and leads to conduction abnormalities, such as heart blocks. Patients with gastrointestinal involvement experience neuronal destruction and pathological dilatations of the esophagus and colon. Inflammation and hypertrophy of the muscle layers of the bowel may also be appreciated histopathologically. The pathogenesis of these lesions, which develop late in the course, at a stage when parasites are scarce, is unclear. Experiments suggest that neuronal destruction in the acute phase might result from a sequence of events: release of parasite antigens; their binding to noninfected neuronal cells; destruction of these cells by the host-antibody-mediated responses; and, finally, emergence of autoimmune antibodies that attack the neurons. This hypothesis has been challenged on the grounds that significant reduction of cardiac neuronal cells in infected mice occurs even before immune response against neurons can be detected. In vitro experiments have shown that host cells, passively sensitized by *T. cruzi* antigens, can become targets that are readily lysed by cytotoxic T lymphocytes present in chagasic mice. Even more controversial are suggestions that chronic cardiomyopathy and the digestive tract pathology are caused by an autoimmune mechanism elicited by *T. cruzi* antigens that mimic host-cell components and evoke cross-reactive immune responses. The demonstration that lymphocytes from chronic chagasic rabbits can bind and destroy both normal and *T. cruzi*–infected rabbit heart cells provided the first evidence in support of a cell-mediated autoimmune mechanism in the pathogenesis of cardiomyopathy. Unfortunately, allogeneic cells were used in these experiments, making interpretation of their results problematic. In related studies, injection of *T. cruzi* subcellular fractions into normal rabbits evoked cardiac lesions resembling ones that develop in patients. The interpretation of these experiments is disputed because the investigators failed to control for the possibility that the immunogen contained host antigens in the extract preparation of *T. cruzi* that had been cultured in rabbit heart cells.

Autoantibodies have been detected in patients with Chagas' disease and are postulated to play a pathogenic role. Presently, it is uncertain whether their presence represents an immunological epiphenomenon or contributes to damage of myocardial cells and neuroenteric (Auerbach's) plexus. Antibody that binds to heart endothelial, blood vessels, and interstitium (designated EVI, for endothelium, vessel, and interstitium) is detected in a high percentage of sera from chagasic patients. It is more frequently detected in patients with cardiac involvement than in those with the indeterminate form of the disease. The potential role of EVI antibodies in the pathogenesis of cardiomyopathy has been questioned since these antibodies are also detected in a small percentage of patients with other cardiac diseases. Perhaps these antibodies result from rather than cause myocardial damage. Deposits of IgG, IgM, and complement have been detected in heart biopsy specimens obtained from patients with chronic Chagas' cardiomyopathy. This finding has been presented in support of the hypothesis that myocardial damage might result from antibody-dependent cell-mediated cytotoxicity.

Sera from humans and Rhesus monkeys infected with *T. cruzi* contain high titers of an antibody that reacts with laminin (a basement-membrane glycoprotein) but not with other connective tissue components. Antibodies from the infected monkeys, purified by laminin-Sepharose affinity chromatography, and antilaminin antibodies raised in uninfected rabbits by immunization, recognize antigens in the endocardium, vessels, and interstitium of the heart, where laminin is known to exist. Furthermore, EVI antibodies are absorbed by laminin-Sepharose. Together, these findings suggest that laminin is the major antigenic component inducing EVI. Since purified antilaminin antibodies (IgM and IgG) react with trypomastigotes and amastigotes derived from tissue culture cells unable to synthetise laminin, either the parasite stages are able to synthesis laminin or a lamininlike substance or they possess constituents that antigenically mimic host laminin.

More recently, it has been reported that antilaminin antibodies from chagasic sera (and also from sera of leishmaniasis patients) recognize a carbohydrate epitope galactosyl-α-(1 – 3) galactose (galα1 – 3 gal) which is present in mouse laminin but not human laminin. Since antilaminin antibodies do not react with human laminin, they are likely to be of heterophile nature and harmless to human cardiac tissue. Natural antibodies are found in normal individuals and are apparently induced by gastrointestinal and pulmonary bacteria harboring terminal α1 – 3 gal epitopes. It is still uncertain whether the high levels of such antibodies found in sera from acute chagasic cases and, in lower concentration, from chronic chagasic patients are boosted by galα1 – 3 gal epitopes known to be present in *T. cruzi* trypomastigotes' surface membrane.

Antibodies against peripheral nerves are also detected in a high percentage of patients with chronic Chagas' disease. As previously mentioned, there is evidence for the existence of antibodies against certain *T. cruzi* antigens that cross-react with neuronal antigens. A monoclonal antibody raised against rat dorsal root ganglia recognizes epitopes synthesized by different *T. cruzi* developmental stages. In addition, two monoclonal antibodies raised against *T. cruzi* were found to react with murine brain and spinal cord, providing further evidence of epitopes common to *T. cruzi* and mammalian neurons.

Vaccines and Antigens

Control of Chagas' disease is readily achieved by insecticide spraying—a control measure that dramatically reduces the density of intradomiciliary vectors and prevents *T. cruzi* transmission. Accordingly, what would be the role of a vaccine in the disease control? An effective vaccine might be an alternative in countries reluctant to take on vector control programs.

At least four requirements are thought essential for development of a vaccine against *T. cruzi* infections: (1) it should not induce active infections, as may happen with living vaccines; (2) it must confer total protection; (3) it should not induce an autoimmune response; and (4) it should not be immunosuppressive. Since it has not been demonstrated that attenuation or suppression of the acute phase prevents late lesions, only vaccines that meet the second requirement—conferring a sterilizing immunity—would be acceptable. Because of the lively controversy over autoimmune responses in Chagas' disease, *T. cruzi* vaccines are expected to be free of host antigens as well as parasite antigens that might evoke autoimmune responses.

Several vaccines have been tried in experimental Chagas' disease employing live attenuated parasites, killed intact organisms, cell homogenates and subcellular fractions, and purified antigens. Live parasites attenuated by prolonged serial passage in acellular culture medium afforded only partial protection, because vaccinated mice experienced subpatent infections. Vaccines composed of live trypanosomes, rendered incapable of replication by treatment with agents that inhibit DNA replication (actinomycin D, ethidium bromide) or by irradiation, fail to induce infection but only confer partial protection to challenge. Vaccinations with organisms killed chemically, parasite homogenates, and viable monogenetic trypanosomatids (phylogenetically related to trypanosomes but not pathogenic to mammals) also confer negligible protection. On the other hand, heat-inactivated *T. cruzi* metacyclic trypomastigotes have been reported to confer protection in mice.

Since there exists the concern that *T. cruzi* antigens present in intact or crude preparations of disrupted parasites might elicit undesireable autoimmune responses, such preparations are not considered likely vaccine candidates. The search for defined subcellular vaccines is therefore underway. Subcellular fractions obtained by differential centrifugation of disrupted culture forms have been used to vaccinate mice. A flagellar fraction imparted protection, as demonstrated by a lower rate of positive xenodiagnosis in vaccinated mice than in controls. This fraction did not produce cardiac lesions when injected into normal animals, whereas two other fractions induced myocarditis and myositis, validating the concern about the immunopathogenic potential of some *T. cruzi* components.

T. cruzi cell-surface carbohydrates or glycoproteins participate in cell recognition and interiorization, and might serve as appropriate targets of a vaccine. Stage-specific or ubiquitous surface glycoproteins have been purified from blood-, tissue-, or axenic-culture-derived forms and tried as vaccines. A cell membrane surface glycoprotein (gp) of 90 kD (gp90) present in all *T. cruzi* developmental stages has been extensively investigated. The gp90 appears not to elicit autoimmune responses, and when injected into mice and marmosets, it affords partial protection. A 72-kD (gp72) glycoprotein has been isolated from epimastigotes and trypomastigotes derived from axenic cultivation. Immunization of mice with gp72 results in reduced parasitemia after challenge with vector-derived metacyclic trypomastigotes but not with blood forms. Another surface glycoprotein, 25 kD (gp25), has been purified from culture-derived epimastigotes; it was also detected in trypomastigotes. Sera from chagasic patients contain antibodies against gp25, but immunization of mice and rabbits with this antigen failed to protect them. An 85-kD glycoprotein has been identified that seemingly is involved in host cell recognition by *T. cruzi*. Anti-gp85 immune sera inhibit in vitro binding to and penetration of *T. cruzi* trypomastigotes in various nonphagocytic cells, suggesting that this glycoprotein might also be worth testing in experimental vaccination.

The limited efficacy of these chemically defined vaccines apparently is due to their failure to induce protective (LA) antibodies; they induce only the nonprotective CSA antibodies, which are unable to bind to *viable* BTry. It was recently demonstrated that a 160-kD membrane surface glycoprotein of BTry and tissue culture trypomastigotes is recognized by antibodies in sera from chronically infected patients or animals displaying LA but not by sera from immunized animals producing only CSA. Thus, gp160 could be a critical protective antigen; its isolation could open avenues for development of immunoprophylactic measures for Chagas' disease.

Coda

Although existing data point to a central role of humoral immunity in defense to *T. cruzi*, additional roles for cellular immunity may be defined in the future. The precise mechanisms involved in pathogenesis remain to be elucidated. Present evidence suggests a role for autoimmune mechanisms in the damage of myocardial and neural (neurenteric plexus) cells. Studies are needed to substantiate this hypothesis. The prospects for a vaccine against *T. cruzi* are uncertain, although candidate antigens are being subjected to testing in animal models. The concern that some antigens might enhance pathogenic responses by inciting autoimmunity has tempered enthusiasm for the concept of disease control through vaccination. While strides have been made in understanding host defense, pathogenesis, and antigen analysis, many aspects of the immunology of Chagas' disease remain to be explored.

Additional Reading

BRENER, Z. 1980. Immunity to *Trypanosoma cruzi*. *Adv. Parasitol.* 18:247–292.

CAMARGO, M. E., A. W. FERREIRA, B. A. PERES, L. MENDONÇA-PREVIATO, and J. SCHARFSTEIN. 1986 *Trypanosoma cruzi* antibodies. *Methods Enzym. Anal.* 11:362–82.

HUDSON, L. Immunological consequences of infection and vaccination in South American Trypanosomiasis. *Phil. Trans. R. Soc. Lond.* B307:51–61, 1984

HUDSON, L. 1985. Immune response to South American Trypanosomiasis and its relationship to Chagas' disease. *Br. Med. Bull.* 41:175–180.

KIERSZENBAUM, F. 1986. Autoimmunity in Chagas' disease. *J. Parasitol.* 72:201–211.

NOGUEIRA, N. 1983. Host and parasite factors affecting the invasion of mononuclear phagocytes by *Trypanosoma cruzi*. In *Cytopathology of Parasitic Disease*, Ciba Foundation Symposium. 99:52–73.

SNARY, D. 1985. Biochemistry of surface antigens of *Trypanosoma cruzi*. *Br. Med. Bull.* 41:144–148.

TRISCHMANN, T. M. 1983. Natural and acquired resistance to *Trypanosoma cruzi*. In *Host Defenses to Intracellular Pathogens*, T. K. Eisenstein, P. Actor, and H. Friedman, eds. Plenum, pp. 365–382.

14

Immunology of Schistosomiasis

Anthony E. Butterworth

The three main species of schistosome that are pathogenic for man, *Schistosoma mansoni*, *S. haematobium*, and *S. japonicum*, are estimated to affect some 200 million people, causing extensive morbidity and some mortality, especially in children and young adults. Although considerable success in controlling transmission and disease has been achieved in some countries, notably Brazil and China, schistosomiasis remains an increasing problem in many others, especially in Africa. Increasing population density there has put heavy pressure on effective land usage, leading to a progressively widespread development of large- and small-scale irrigation schemes that increase the potential for transmission.

The biology of the host–parasite interaction in schistosomiasis presents several unusual features that we need to bear in mind in order to put the role of the immune response in proper perspective. First, the adult worms have little or no pathogenic effect. Chronic disease is largely attributable to the deposition of eggs in the tissues, with granuloma formation and subsequent fibrosis; transmission is maintained through the release of eggs into the feces. Second, the adult worms do not replicate within the mammalian host. The extent of disease depends on the number of eggs that are deposited in the tissues, which in turn depends on the number of adult worms that are present, their egg-laying capacity, and their longevity within the host. Intensities of infection in human communities, as determined by measuring egg output in the feces, classically show a negatively truncated binomial distribution, with a few individuals

The author gratefully acknowledges the participation of Dr. Daniel Colley and Dr. Dov Boros in the preparation of the section on immunopathology and Dr. David Wyler in the preparation of the section on hepatic fibrosis.

harboring most of the parasites (Figure 14-1); these are the patients who will go on to develop severe disease. Under these conditions, a partial degree of immunity, incompletely effective in reducing worm burdens within individuals, may be of great value in preventing disease, by markedly reducing the numbers of individuals who fall within the high-intensity category. Such partial immunity also might be of value in limiting transmission, by reducing the overall output of eggs into the environment. Third, adult worms have a long life span in man: the mean life span for different species has been estimated at 2 to 5 years, but individual infections may persist for decades after a patient has moved from an endemic area. The implication is that protective immunity, if it occurs in man, is unlikely to have a major effect on the mature adult: the developing larva is a more likely target. Finally, the parasite passes through a series of major differentiation and maturation steps within the mammalian host, from its time of entry as a cercaria, through its development as a schistosomulum, to its differentiation into mature adults and the deposition of eggs. Although some antigens or epitopes are shared between the life-cycle stages, others are stage-specific and elicit separate, independent responses. A response that is important with respect to one stage may be inactive against another. Each stage must therefore be considered separately; in particular, the protective immunity that is directed against the young schistosomulum and the immunopathology of the egg granuloma appear to be mediated by quite different mechanisms.

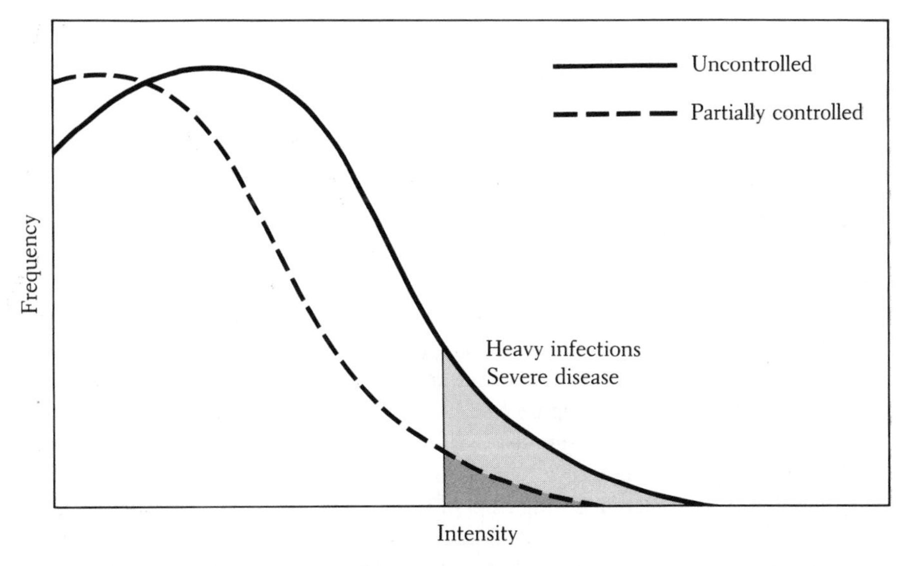

FIGURE 14·1 The effect of partial immunity, acting throughout an infected population, would be to reduce markedly the proportion of patients with high-intensity infections. Prevalence of disease therefore would be reduced more dramatically than prevalence of infection.

This chapter deliberately emphasizes the mechanisms of protective immunity and the antigens against which protective responses are directed. During the last few years, the application of recombinant DNA techniques has led to a reasonable optimism that a schistosome vaccine is now a practical proposition. Although the advent of a new generation of safe antischistosomal drugs, especially praziquantel, has markedly improved the possibility of widescale control of schistosomiasis, chemotherapy cannot be regarded as a final solution. The high costs of the available drugs, as well as of diagnosis and drug delivery, may well be beyond the means of many national governments; this problem is compounded by a need for repeated surveillance and retreatment of reinfected individuals, especially younger children. As with any other infectious disease, drug resistance (which has already developed under field conditions for oxamniquine, a widely used antischistosomal agent) may well become a general problem. In contrast, the vaccine strategy for control not only has the intrinsic advantage of preventing infection, but may be of longer duration and, most important, can be integrated within existing programs of immunization, thus obviating the requirement for establishing a separate hierarchical structure for the control of this particular disease.

This chapter also addresses the pathogenesis of the egg granuloma and the consequent fibrosis — in terms of disease, rather than infection, one of the most important aspects of the host–parasite interaction. Other areas, such as immunodiagnosis, although of considerable practical importance, are of less theoretical interest and are given less consideration. Finally, although due weight is given to experimental animal models of schistosome infection, it should be remembered throughout that these are only models, and may not reflect the situation in man. Emphasis is therefore placed on the problems and recent results of studies of immunity and immunopathology in man.

Protective Immunity

The Idea of Concomitant Immunity

Age-specific intensity curves for communities in areas endemic for schistosomiasis characteristically show a sharp rise in mean intensity of infection, as estimated by the output of eggs into the feces, from the time of a person's first exposure during early childhood until the early teens. Thereafter, mean intensities of infection progressively decline throughout adult life (Figure 14-2): neither at the individual nor at the population level is there evidence for a dramatic self-cure, accompanied by a solid resistance to reinfection, of the type that is seen in many other infectious diseases. The slow decline in mean intensity of infection during adult life might be attributable to a slow spontaneous death of adult worms, together with a slowly acquired immunity to superinfection. However, the decline might be equally well attributable to the same slow death of adult worms, together with a reduced exposure to contaminated water in the older age groups; or to alterations in the fecundity of adult worms; or to the capacity of eggs to escape into the feces. Simple examination of age-intensity curves does not distinguish between

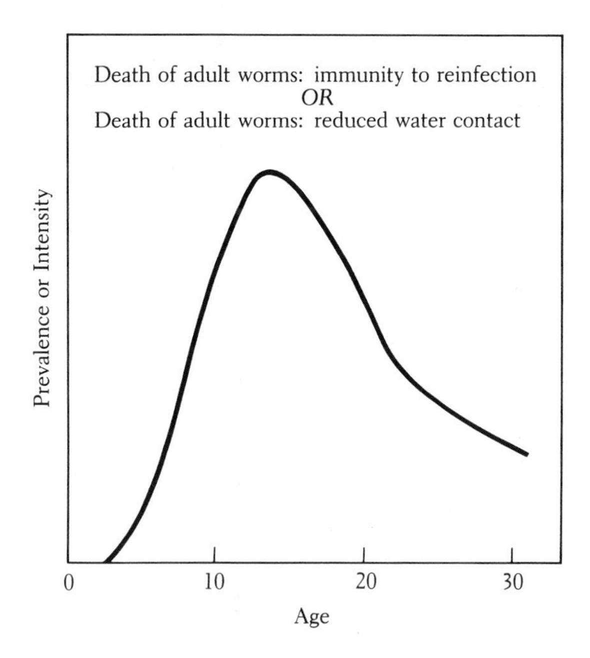

FIGURE 14·2 The decline in prevalence and intensity of infection among older members of communities living in endemic areas may be attributable to a slow spontaneous death of adult worms, together with either a slowly acquired resistance to superinfection, or a reduced level of exposure in the older age groups, or both.

these possibilities. Until recent years, during which studies have been carried out in which simultaneous measurements of exposure and immune responses have been attempted (described below), there has been much controversy over the existence of acquired immunity in man, and no hard evidence concerning its relative importance in limiting schistosome infection has emerged.

Instead, much of our knowledge about the nature and mechanisms of acquired immunity has come from studies in a range of experimental animal models. During the 1960s, techniques were established for quantifying infection in various experimental hosts, and it was observed that following small primary infections of mice or rhesus monkeys, adult worms of the first infection continued to survive, while the host was capable of rejecting challenge organisms of a secondary infection. This phenomenon of acquired resistance to superinfection in the presence of a continued primary infection has been called concomitant immunity, by analogy with a similar situation seen with some experimental tumors.

Evasion of the Immune Response by Adult Worms

Concomitant immunity is a subtle adaptation toward the establishment of a stable host–parasite relationship. The adult worms of a primary infection are capable of surviving for long periods, yet the host develops an immunity to superinfection and therefore rarely dies as a result of overwhelming infection. The presence of infection does not prevent survival of the host: both host and parasite can live for many years,

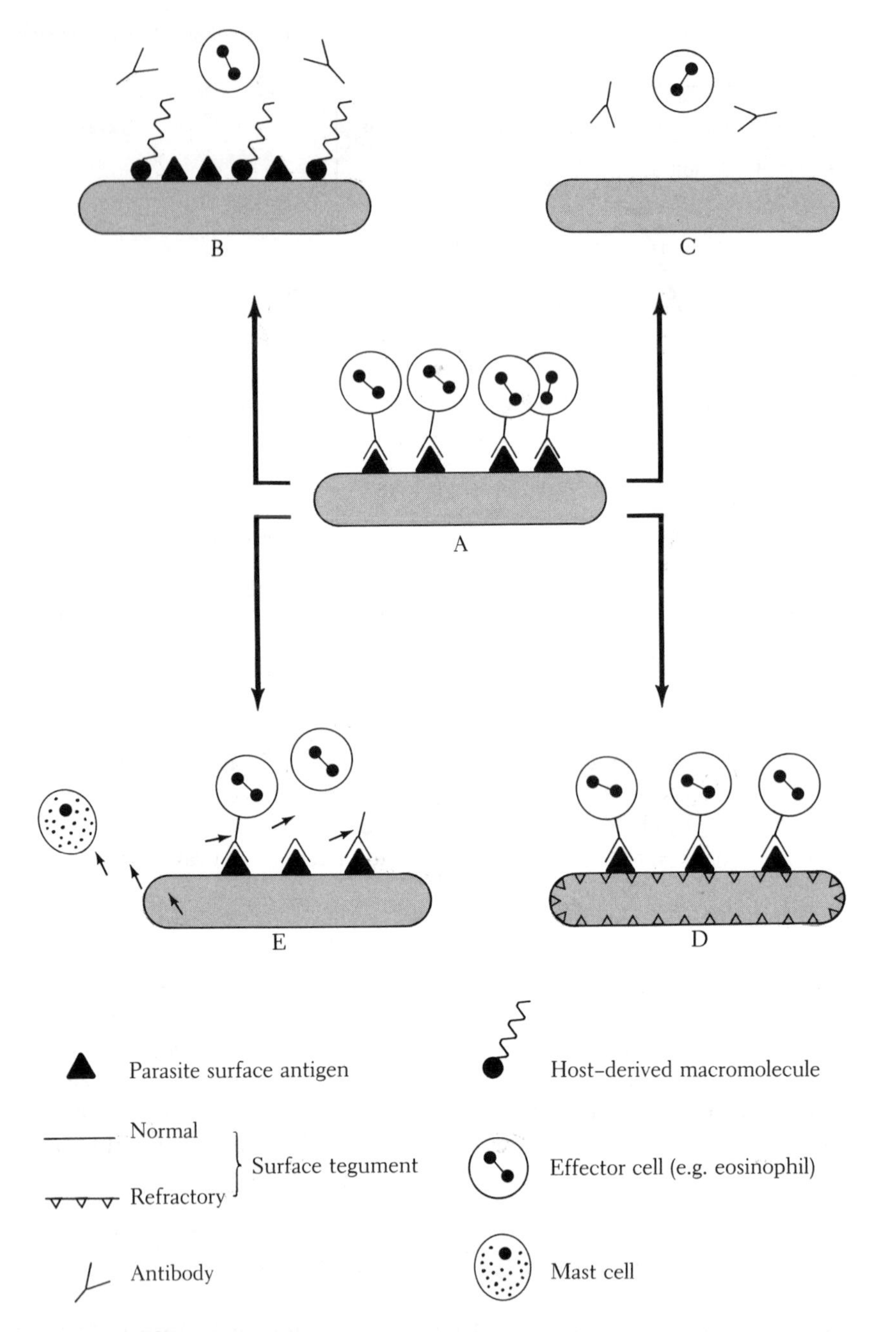

FIGURE 14·3 Postulated mechanisms of evasion of the immune response by developing schistosomula and adult worms. A. The young schistosomulum, expressing surface antigens, is susceptible to attack by a variety of antibody-mediated, cell- or complement-dependent effector mechanisms. B. Masking of parasite antigens by host-derived molecules prevents antibody binding.

thereby allowing the opportunity for prolonged transmission of the parasite. This is in contrast to more acute and lethal infectious diseases, which may rise and decline in epidemic form, and in which the parasite may eventually vanish from the host population.

The mechanisms whereby the adult worm evades attack in otherwise immune individuals is still a matter of some controversy, in spite of much useful work over the last fifteen years (Figure 14-3). One of the earliest explanations to emerge was that the adult worm passively acquires a coating of host molecules ("host antigens"), which mask the antigenic determinants on the worm's surface, thereby preventing immune recognition. Uptake by the developing schistosomulum of a range of host molecules has now been demonstrated, both in vivo and in vitro, and such uptake parallels the loss of susceptibility of the worm to a variety of immune effector mechanisms. Host molecules that may be demonstrated on the worm's surface include blood group substances and products of the major histocompatibility complex. Of the blood group substances, there is preferential uptake of those that are expressed on the red cell surface as glycolipids. The mechanism of uptake is unknown, but one possibility is that it occurs through fusion of the outer of the two schistosomulum tegumental membranes with closely adjacent host cells or cell fragments (see Chapter 9 for a detailed discussion). Adult worms bearing host-derived molecules are susceptible to immune attack, both in vivo and in vitro, by antibody-mediated effector mechanisms directed against those molecules, even though they are not susceptible to attack by antiparasite effector mechanisms.

These findings suggest that the presence of host molecules may effectively mask parasite antigens and prevent immune attack, even though the organism is still intrinsically susceptible to immune damage. It is somewhat difficult to understand, however, how a complete masking of parasite determinants might be achieved. There is no evidence, for example, of a thick layer of host material covering the parasite surface, of a density comparable to the coating of the trypanosome surface with the variant surface glycoprotein (VSG): instead, such evidence as there is suggests that host molecules are freely mobile in the outer of the two lipid bilayers of the tegumental membrane. There is no evidence from other systems that the presence of one molecule within a lipid bilayer will prevent the binding of antibody to a second, independent molecule, nor is it easy to see why this should be the case in the schistosome. However, various lines of evidence indicate that absorbed host molecules can effectively mask at least some parasite determinants. Schistosomula recovered from the lungs of mice fail to bind most monoclonal antibodies: yet, if such schistosomula are subsequently cultured in vitro, binding of at least one monoclonal antibody can be demonstrated, associated with a loss of host-de-

---

C. Loss of parasite antigens from the outer of the two tegumental membranes also prevents antibody binding. D. Antibody may bind, but the tegument of the worm becomes refractory to immune damage. E. The organism releases a variety of locally active immunoregulatory mediators, including (1) proteolytic enzymes that cleave bound immunoglobulin; (2) lymphocyte inhibitory factors (SDIF); (3) mast-cell inhibitory factors.

rived molecules. In addition, treatment of older organisms with formaldehyde results in a loss of expression of host molecules and a reappearance of detectable surface parasite determinants.

More recently, other suggestions have been put forward to explain the progressive loss of susceptibility to immune attack of developing schistosomula and adult worms. For example, during culture of young schistosomula in vitro, it is found that organisms maintained in serum-free medium, in the absence of host macromolecules, lose susceptibility to attack at almost the same rate as organisms cultured in the presence of a source of such macromolecules and that this loss of susceptibility is associated with a concomitant loss of the organism to bind antiparasite antibodies. These findings suggest that there is an actual loss of expression of parasite antigens from the outer of the two bilayers, so that the membrane becomes immunologically "inert." Labeling studies have shown no difference between the parasite stages in the overall composition of the isolated tegumental membrane, indicating that, although antigens are lost from the outer membrane, they are still present within the whole tegumental structure. Experiments with monoclonal antibodies further suggest that different antigens are lost from the outer membrane at different rates: for some major antigens, a rapid loss of expression occurs within 24 to 48 hours after transformation of the cercaria into the schistosomulum; other antigens show a more prolonged expression, being detectable on the surface, with the appropriate monoclonal antibody, as late as five days after skin penetration. New antigens, not previously detected on skin-derived schistosomula, may also be expressed at this late stage.

In addition to showing a reduced expression of some surface antigens, the tegument of the developing schistosomulum may become progressively refractory to immune attack. Covalent coupling of the trinitrophenyl group (TNP) as an artificial hapten onto surface molecules allows experiments on the susceptibility to antibody-dependent damage of those stages of the parasite that, under normal circumstances, fail to bind antibody. Under these conditions, organisms recovered from the lungs of mice, 5 days after infection, are less susceptible to eosinophil- or complement-mediated, antibody-dependent damage than freshly prepared skin-stage schistosomula, even when both stages can be demonstrated to bear comparable amounts of TNP and to bind similar amounts of anti-TNP antibody, indicating that it is not solely the amount of antibody bound that limits susceptibility. This phase of reduced intrinsic susceptibility to attack appears to be limited to the lung-stage organism.

Alternative mechanisms of evasion include the production by the parasite of substances that actively inhibit host effector mechanisms. Developing schistosomula produce and release at least two proteases, one with trypsinlike activity. This protease production results in a cleavage of surface-bound immunoglobulins, with a consequent failure of antibody-dependent cell-mediated effector mechanisms that depend on an intact Fc region, and also results in the release of small peptides, such as the tripeptide Thr-Lys-Pro, that actively inhibit macrophage cell function. In addition, adult worms can be demonstrated to release soluble factors (such as schistosome-derived inhibitory factor, SDIF) that inhibit lymphocyte function and mast cell degranulation.

This list of evasion mechanisms is likely to grow. There is no reason to suppose that a parasite as complex as the schistosome will have evolved only one, or even a few, mechanisms that ensure its continued survival within the immune host. The picture that emerges so far is of a progressive loss of susceptibility to surface immune attack, starting soon after the cercaria penetrates the skin. Even this picture, however, must be treated with caution: recent work in both the mouse and the rat models, described later, have suggested a later phase of susceptibility to immune attack in vivo, after the developing schistosomula have left the lungs but before they reach the hepatic portal system. The reasons for the reappearance of susceptibility, and the possible role of parasite surface antigens in this susceptibility, are still obscure.

Studies on Acquired Immunity in Experimental Hosts

Two problems have hindered the interpretation of experimental animal models of acquired immunity to schistosome infection. First, each species of experimental host differs markedly in both the nature and the extent of its immune response to schistosome infection, and in the mouse such differences are further complicated by a considerable strain variation. There is no a priori reason for considering one host to be a better model of human immunity than any other. It is sometimes assumed, for example, that as a permissive host, the mouse is more valuable than the rat, in which worms fail to reach sexual maturity or to produce eggs. However, a failure of worms of a primary infection to reach maturity need bear no causal relationship to the events that lead to the immune rejection of the young larvae of a secondary infection, and it may well be that events that occur in the rat are more closely related to those that occur in man. Such controversies will be resolved only by the direct demonstration of the mechanisms of immunity in man: in the meantime, all experimental models may be regarded equally and impartially as useful indicators of the types of immune mechanism that should be sought in man.

Second, recent studies in at least one species of permissive host, the mouse, have indicated that a major part of the resistance to reinfection that is observed during a chronic primary infection is attributable not to a specific antiparasite response, but rather to an abnormal migration of schistosomula caused by previous egg-induced pathology. Although interesting in its own right, the relevance of this phenomenon to human disease is questionable. The extent of such pathology depends on the numbers of eggs deposited in the tissues and therefore on the adult worm burden; on a weight-for-weight basis, a worm burden of one worm pair in a mouse (the lowest that can be achieved for egg-laying to occur) would be equivalent to 2000 to 3000 in an adult human, a very high, and possibly rarely attained, level. In the mouse model, attention therefore has switched in recent years to the undoubtedly specific acquired immunity that is achieved by immunization with irradiated larvae.

Bearing these cautionary points in mind, we can gain useful and interesting information from experimental animal models. Immunity is assayed by comparing the recovery of worms at different developmental stages from animals that have been immunized and challenged with appropriate control groups that have received the immunization only or the challenge only. Although it may be convenient and informative to determine worm recoveries at the lung stage, this approach will fail to detect any immune killing that may occur after migration through the lungs; recovery of adult worms, after their arrival in the hepatic portal system, is therefore a preferable although more laborious assay. It is generally agreed that permissive hosts can be rendered immune to reinfection by immunization with irradiated cercariae or schistosomula, whereas nonpermissive hosts can be immunized either in the same way or by a primary infection, in which worms fail to mature to an egg-laying stage. Immunity is usually species-specific, but there is no evidence for strain or isolate specificity. Furthermore, immunity is rarely complete, a proportion of challenge organisms surviving attack and completing maturation. The mechanisms of such immunity are probably multifactorial, with different mechanisms predominating in different hosts or experimental situations, and acting at different stages during differentiation of the schistosomulum.

In the rat, for example, there is clear evidence for a role for antibody-dependent, cell-mediated effector mechanisms involving both the IgE and IgG2a isotypes. The development of immunity after a primary infection is a T-dependent event, and T cells recovered from recently infected animals can transfer protection. This ability subsequently declines, and at a later stage immunity can be transferred only with serum. The effect of transferred serum has not been shown clearly to depend on an intact complement system in the recipient animal. Depletion of serum from infected rats, showing a strong resistance to reinfection, of either IgE or IgG2a results in a marked diminution of the capacity of such sera to transfer immunity, and immunity can be transferred with an IgG2a monoclonal antibody against a schistosomulum membrane antigen. Immunity following primary infection of the rat is a biphasic event; during the early phase, up to 6 weeks after infection, IgG2a is the most effective isotype in mediating transfer, while at a later stage, from 8 weeks after infection, IgE antibodies are more effective. During the intervening period of lack of immunity, high levels of IgG2c antibodies can be demonstrated; as discussed later, these may block the IgG2a-dependent cytotoxic effects of eosinophils, by competition both for schistosomulum surface epitopes and for eosinophil Fc receptors.

In the mouse model, much of the early work involved the use of animals bearing a chronic primary infection. In this system, much of the acquired resistance that is observed may be the consequence of trapping of schistosomula as a result of previous egg-induced pathology. In the irradiated vaccine model, however, there is also evidence that antibody-mediated mechanisms can be effective. Transfer of immune serum confers protection only inconsistently, but a range of monoclonal antibodies have now been prepared that consistently mediate moderate levels of protection, with IgM, IgG1, and IgG2b isotypes being effective in different situations. Early experiments in the chronic infection model showed that acquired resistance was a T-dependent event and suggested a role for eosinophils in mediating antibody-dependent immunity, with no absolute requirement for complement, but these experiments now need repeating in the irradiated vaccine model.

More recently, considerable evidence has also been obtained that supports a role for macrophage activation in mediating immunity in the mouse. Early experiments showed that mice immunized by procedures that lead to macrophage activation, for example, by intraperitoneal inoculation of BCG or *Corynebacterium parvum*, show a nonspecific resistance to schistosome challenge, whereas macrophages recovered from such animals show the capacity to kill schistosomula in vitro. In later studies, it has been found that strains of mice deficient in the capacity of their macrophages to undergo activation cannot be immunized with irradiated cercariae and that immunization of normal mice, either with irradiated organisms or with antigenic extracts, by procedures that enhance macrophage activation leads to an enhancement of immunity. In contrast to the situation with antibody-dependent mechanisms, such immunity is not necessarily elicited by or directed against surface antigens.

In both the mouse and the rat models, there is still some uncertainty about the stage of the parasite against which immunity is directed. The finding that some monoclonal antibodies confer maximum protection when transferred just before or at the time of cercarial challenge, and progressively less when administered at later periods, indicates that at least some immunity can be directed at the skin-stage organism. However, experiments involving the tracking of selenomethionine-labeled organisms in immune and nonimmune animals, the transfer of lung-stage schistosomula directly into immune animals, or the quantitative recovery of organisms from different sites at different times after cercarial challenge of immune animals also indicate that a major part of the attrition that is seen in immune animals may occur at a later stage, after their passage through the lungs. Little is known about the physiological changes that occur in developing schistosomula during this phase of their maturation, and this is a major area for future research.

A third small-animal host that has recently been used for studies on immunity is the guinea pig. Like the mouse, the guinea pig is permissive of *Schistosoma mansoni* infection and develops a strong immunity to reinfection following either chronic primary infection or immunization with irradiated cercariae.

Primate models are less susceptible to experimental manipulation, and little is known about the mechanisms of immunity in demonstrably resistant animals. However, at least one study has provided evidence for a death of schistosomula in the skin of immune baboons, in association with a marked eosinophil infiltrate. Since the baboon is a natural host of *S. mansoni*, resembling man in the slow development of immunity, this finding deserves further investigation.

Most of the experimental work outlined so far has been carried out with *S. mansoni*, largely because this species is the easiest one to maintain in the laboratory. However, the levels of immunity induced in baboons by immunization with irradiated cercariae are considerably greater for *S. haematobium* than for *S. mansoni*, while immunization of baboons with irradiated cercariae of *S. japonicum* has been reported to be ineffective. Although there is no evidence yet that the mechanisms of immunity differ substantially between the three species, the extent of immunity that can be achieved is clearly of relevance to the development of effective vaccines. Immunity is frequently species-specific, especially after immunization with irradiated cercariae, an event that may be related to the recognition of species-specific surface antigens, treated in a later section.

Studies In Vitro on Immune Effector Mechanisms

Experiments on immune effector mechanisms active against schistosomes in vitro have the obvious disadvantagereby offering an insight into immune effector mechanisms that may be operative in man; second, they allow a detailed analysis of the mode of action of each effector mechanism and of its interaction with others.

Most such studies have involved the use of the young schistosomulum as target, prepared either by allowing the cercariae to penetrate an isolated skin preparation in vitro or by mechanical disruption and incubation of cercariae. Partly because of their relative unavailability, less work has been done on lung-stage schistosomula, the evidence so far indicating that they are resistant to most forms of immune attack in vitro. There have been virtually no studies on the susceptibility to attack of the potentially important postlung schistosomulum.

A plethora of immune effector mechanisms active against young schistosomula in vitro have now been described, with the use of materials derived both from infected humans and from a range of experimental animals (Table 14-1). The relative effectiveness of each mechanism differs from species to species and, unfortunately, from laboratory to laboratory; this is still a controversial field. In some cases, the mechanism of damage appears to be entirely comparable to that inflicted on more conventional targets. Examples might include the membrane damage induced by classical pathway fixation of

TABLE 14·1
Effector mechanisms active against schistosomula in vitro

Mechanism	Species	Comment
Antibody + complement	Humans, rhesus monkeys, rats	Large amounts of antibody and complement required
Neutrophils + antibody ± complement	Humans, baboons, guinea pigs, rats,	Possibly by oxidative killing mechanisms
Eosinophils + antibody ± complement	Humans, baboons, guinea pigs, rats	Primarily by release of toxic granule contents; also oxidative mechanisms; antibody may be IgG or IgE
Macrophages + IgE antibody	Humans, rats	Preincubation of macrophages with IgE required
Activated macrophages	Humans, mice	Macrophages activated in response either to infection or immunization or to irrelevant antigens

complement, in the absence of added effector cells, and the antibody-dependent damage mediated by neutrophils, described by some groups, which has been reported to involve the generation of active oxygen species. In addition, however, there are unusual mechanisms whose action appears to be selectively directed against helminth parasites (including schistosomes) and that may represent an adaptation by the mammalian host to cope with large, tissue-invasive, noningestable pathogens. Of particular interest are the role of IgE and other anaphylactic antibodies in mediating or enhancing cell-dependent damage to schistosomula, and the striking capacity of eosinophils to mediate antibody-dependent damage.

Evidence for a direct role for IgE in mediating cell-dependent damage to schistosomula came first from studies on rat macrophages. It was found that macrophages from normal rats, preincubated for several hours in unheated sera from immune animals, showed a marked capacity to kill schistosomula. A period of preincubation of the macrophages in the immune serum was required for this effect to be demonstrated, and there was a continued requirement for immune serum during the phase of interaction with the schistosomula. The activity in the immune serum was heat-labile, and could be absorbed on anti-IgE immunoabsorbent columns. The activity mediating macrophage-dependent damage could also be absorbed on columns bearing antischistosome antibodies, indicating that the IgE mediating the effect was present in the serum in the form of immune complexes. However, later work showed that monoclonal IgE antibodies would mediate the same effect. These findings were unexpected, since at that time there was no indication that macrophages bore IgE receptors, a necessary prediction that was subsequently confirmed. The requirement for a period of preincubation suggested that a change was occurring in the macrophage that was necessary for killing to occur, other than simple binding through IgE, and subsequent studies have confirmed that a range of metabolic changes occur in such cells, comparable to those that occur during conventional macrophage activation. However, the identity of the change that is responsible for the enhanced capacity of such cells to mediate killing has not yet been clearly established. Subsequently, similar IgE-dependent, macrophage-mediated effects have been demonstrated in both baboons and man.

In later studies, in both rats and man, a direct role for IgE in mediating damage was demonstrated with other effector cells. In the rat, eosinophils can mediate damage through both classes of anaphylactic antibody, IgE and IgG2a; in man, the "hypodense" eosinophils that can be detected in patients with elevated eosinophil counts can be demonstrated to bear low-affinity IgE receptors and to mediate IgE-dependent damage. Human and rat platelets also can mediate IgE-dependent damage to schistosomula, and the transfer of rat platelets bearing surface-bound antischistomulum IgE to naive animals confers protection against challenge.

Apart from its direct effect in mediating damage to schistosomula, by interaction with effector cells bearing low-affinity IgE receptors, IgE also may act indirectly in enhancing protection. Release of mediators from mast cells sensitized with IgE leads to an enhancement of effector cell activity, especially of eosinophils; these mediators include particularly the chemotactic tetrapeptides of eosinophil chemotactic factor of anaphylaxis, Ala-Gly-Ser-Glu and Val-Gly-Ser-Glu. Release of such mediators may also serve to localize eosinophils and other effector cells to the immediate hypersensitivity reaction that occurs at the site of parasite invasion.

Thus, studies in vitro indicate that the marked IgE response associated with schistosomiasis and other helminth infections maybe of value in protecting the host against challenge, both by localizing and stimulating effector cells at the site of challenge and by mediating directly a variety of cell-dependent effects on the challenge organisms. Evidence from experiments in vivo in the rat, mentioned earlier, supports this hypothesis: in particular, the observations that immune sera depleted of IgE show a reduced capacity to transfer protection, which can be restored by simultaneous administration of IgE monoclonals, and that rats depleted of IgE by neonatal treatment with antiepsilon chain sera show a reduced capacity to mount a protective response to secondary cercarial challenge.

Among the various cell types that have been reported to mediate antibody-dependent damage to schistosomula in vitro, eosinophils are particularly effective, especially in the rat and in man. Such damage may be mediated through either IgE or IgG antibodies, with a role for complement being still somewhat controversial. Antibody-dependent eosinophil-mediated damage is associated with an initial tight attachment of the cell to the surface of the antibody-coated organism (Figure 14-4), followed by degranulation and the release of granule contents onto the surface. These granules contain, apart from eosinophil peroxidase, large amounts of toxic and highly cationic proteins, especially the eosinophil major basic protein (MBP) and the eosinophil cationic protein (ECP), which contribute directly to membrane damage by an unknown mechanism. In addition, some groups have proposed a role for oxidative damage, associated with hydrogen peroxide release and peroxidase-mediated halogenation of surface molecules.

In man eosinophils recovered from the blood of normal individuals are capable of mediating such damage (Figure 14-4). However, helminth infections, including schistosomiasis, are classically associated with an increase in circulating eosinophil levels. Eosinophils recovered from the blood of such eosinophilic individuals show a markedly enhanced capacity to kill schistosomula. This capacity is associated with the capacity of monocytes from eosinophilic individuals to produce during culture without added stimuli one or more mediators that enhance a range of eosinophil functional properties. These mediators may include eosinophil-activating factor, eosinophil colony-stimulating factor, and tumour necrosis factor, the exact relationship between them remaining to be determined. Supernatants containing such mediators, or the partially purified mediators themselves, will not only enhance antibody-dependent eosinophil-mediated killing of schistosomula, but also enhance degranulation, spontaneously or after interaction with antibody-coated surfaces; the production of active oxygen species; and various aspects of lipid metabolism, including the production of leukotriene C4 and of platelet-activating factor. Such cells also show fine structural changes compatible with their designation as "activated" cells.

IgE responses and eosinophilia are closely associated in helminth infections, and a reasonable working hypothesis, although by no means proven, is that the two together represent an adaptation to protection against tissue-invasive, multicellular parasites. A

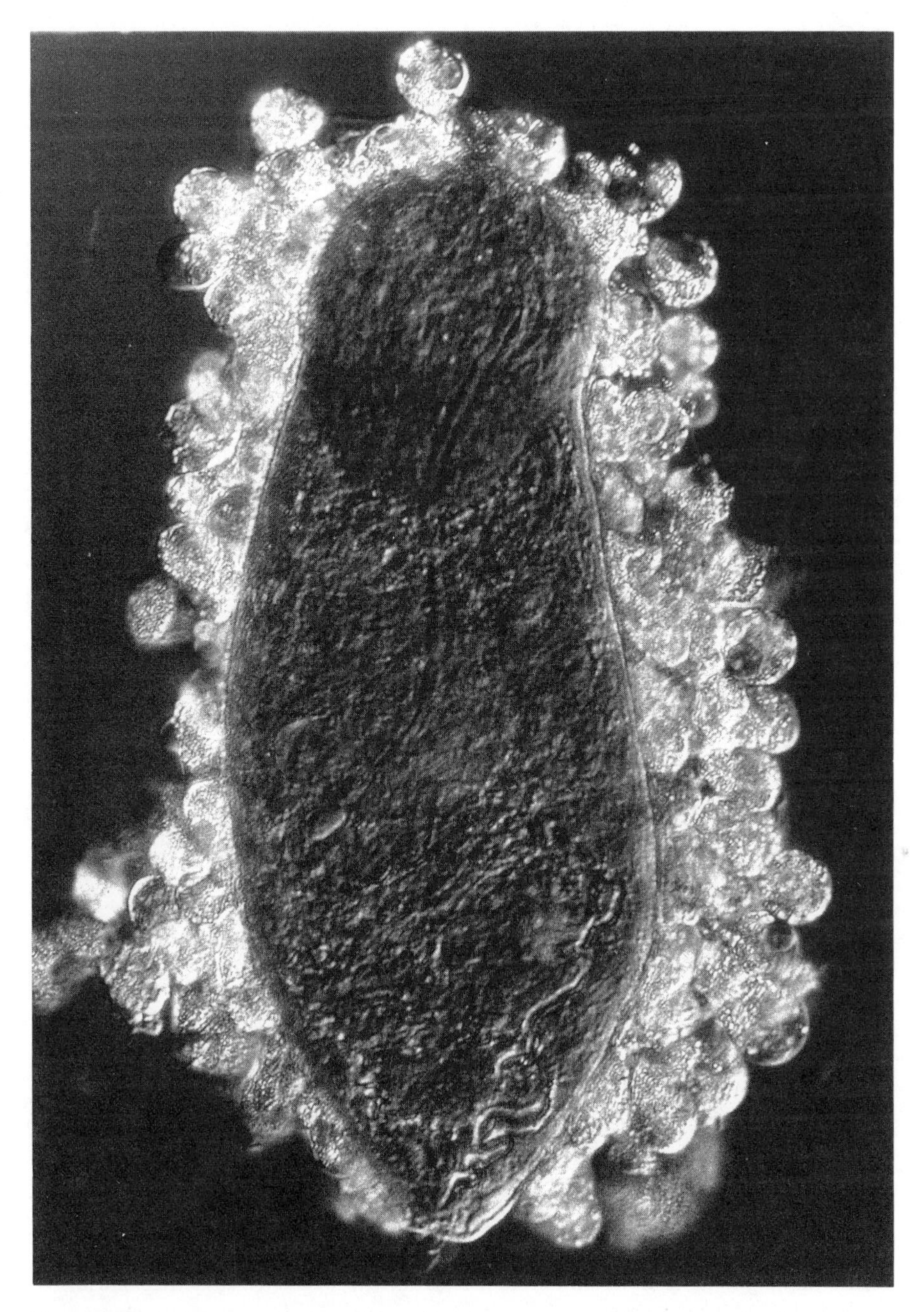

FIGURE 14·4 Human eosinophils adhering to an antibody-coated schistosomulum: the reaction leads to death of the organism over 24 to 48 hours. (Nomarski optics.)

thymus-dependent response to particulate antigens in peripheral tissue sites may lead to increased production of eosinophils, increased functional activity of the mature cells, and IgE antiparasite antibodies. These antibodies may then be involved in both the localization of eosinophils to the site of challenge organisms, and to mediating eosinophil-dependent damage, associated particularly with the capacity of that cell for degranulation and the release of a range of toxic moieties.

Most of the effector mechanisms listed in Table 1 depend on the presence of antibodies with specificity for schistosomulum membrane antigens, and the isotypes of such antibodies are of key importance. In the rat, a period of depressed immunity between 5 and 7 weeks after a primary infection is associated with a predominant production of IgG2c antibodies: and a monoclonal IgG2c antibody will block the eosinophil-dependent killing mediated by an IgG2a antibody with specificity for the same 38-kilodalton (kD) schistosomulum surface antigen. This blocking effect is mediated both by a competition for epitopes on the schistosomulum surface and by a competition of the two isotypes for a single Fc receptor on the eosinophil, distinct from the receptor for IgE. A similar situation has now been found in man, with IgM antibodies blocking the eosinophil-dependent effects of IgG antibodies from the same sera. Such IgM antibodies also react with miracidial antigens, and the interesting possibility is now being tested that such blocking antibodies, whether of IgM or of ineffective IgG isotypes, are elicited in response to carbohydrate epitopes on polysaccharide or glycoprotein egg antigens presented during early infection. Their possible role in delaying the appearance of resistance is discussed later.

In the light of recent observations from the mouse model, a possible role for antibody-independent damage mediated by activated macrophages must also be considered. Activated macrophages from mice immunized with BCG or other agents that conventionally lead to macrophage activation can kill schistosomula in vitro, and part of this effect has been ascribed to the release of arginase. Similarly, macrophages from mice immunized with irradiated cercariae or with crude schistosome antigens administered intradermally with BCG can also damage schistosomula, an effect that is not seen in the P/J mouse, which fails to develop resistance after such procedures. Such activated macrophages can kill schistosomula in the absence of added antibody, although the effect is more marked if immune serum is also present. This mechanism is potentially important since it might be elicited in response to antigens other than surface membrane antigens and might be expressed on stages of the parasite, such as the postlung schistosomulum, that do not express antigens on their surface.

Immunity in Man

Three main problems have hindered studies on the existence and mechanisms of protective immunity in man. First, the continued presence of adult worms from early infections, dying at an unknown and possibly variable rate, has made it difficult to estimate the extent of superinfection that may be occurring in an individual. Second, it has proved difficult to distinguish between immunity and lack of exposure as possible reasons for a lack of superinfection or of reinfection after treatment. Third, immunity —if it occurs in man—may well be incomplete, and yet still of considerable value in reducing both the transmission of infection within a population and the extent of disease, since this is proportional to the adult worm burden.

The characteristic age-specific prevalence or intensity curve (Figure 14-2), with a peak of infection during the early teens and a slow decline thereafter, could well be attributable to a slow spontaneous death of adult worms from early infections, together with a slowly acquired resistance to superinfection. However, it could equally well be attributable to the same slow death of adult worms together with a reduced exposure in the older age groups, and some workers have argued strongly in favor of this second alternative. Various early observations, however, suggested that acquired resistance may play a major part in limiting infection. For example, studies on adult immigrants to endemic areas have revealed the same pattern of a slow rise and decline in prevalence or intensity of infection, independent of age but dependent on the duration of residence within the area. Since water contact patterns, and hence exposure, are strongly age-dependent, this dissociation from age makes it unlikely that the observed decline after prolonged exposure was attributable to a reduced exposure. Similarly, observations among groups who, by virtue of their work, are heavily exposed—such as fishermen and canal cleaners—have revealed the same pattern of rising and declining infections. More recent studies on *S. haematobium* in The Gambia, in which changes in infection were compared among communities in which transmission had or had not been partially controlled by the application of molluscicides, indicated that maintenance of a stable egg output within age groups depended on an equilibrium between a rapid loss of established infections and a rapid acquisition of new infections, with the relative failure of adults to acquire new infections being only very partially attributable to their relative lack of exposure.

In an attempt to overcome some of these problems, various recent studies have been designed to follow the intensities of reinfection after treatment among selected cohorts of intensively studied individuals, whose levels of exposure are monitored by determining the duration and nature of contact of each individual with water known to contain infected snails or cercariae, at the same time that measurements of various immune responses detectable in vitro are carried out (Figure 14-5). Such studies obviously suffer from the marked heterogeneity and variability that is observed among water contact patterns, as among any other forms of human behavior; however, the differences in reinfection intensities observed among different groups are sufficiently great as to overcome this intrinsic behavioral variation.

Several studies, involving both *S. haematobium* and *S. mansoni*, are still in progress, but some useful information has already emerged.

1. There is clear evidence that some patients show a relative lack of reinfection after treatment in spite of high levels of observed contact with water known to contain infected snails. In such patients, the relative lack of reinfection cannot reasonably be attributed to a lack of exposure; they must, instead, be resisting reinfection. Such resistance is an acquired, age-depen-

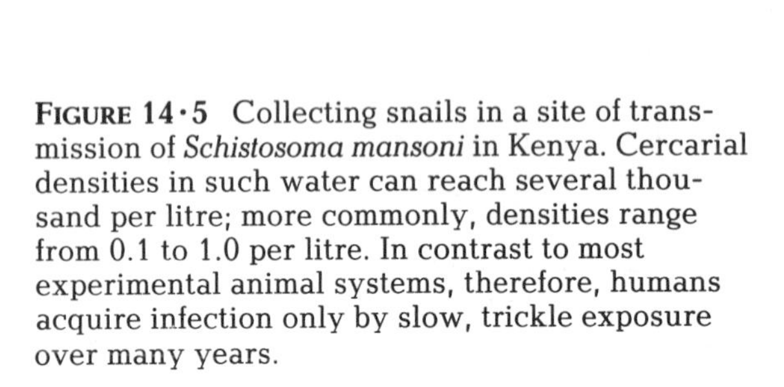

FIGURE 14·5 Collecting snails in a site of transmission of *Schistosoma mansoni* in Kenya. Cercarial densities in such water can reach several thousand per litre; more commonly, densities range from 0.1 to 1.0 per litre. In contrast to most experimental animal systems, therefore, humans acquire infection only by slow, trickle exposure over many years.

dent event, not obviously related to previous egg-induced pathology. Resistance is not demonstrable in young children under 10 years old, but it increases progressively thereafter.

2. Two studies have demonstrated an association between high eosinophil counts, before treatment or immediately after treatment but before the next transmission season, and low subsequent intensities of reinfection, which supports the hypothesis that eosinophils — especially "activated" eosinophils — may be involved in mediating resistance in man. However, this indication has not been confirmed in a third study.
3. Heavily exposed young children mount a wide range of potentially protective responses directed against the young schistosomulum, but they fail to resist reinfection. One possible explanation is that such children also have demonstrable IgM or other "blocking" antibodies. These antibodies recognize schistosomulum surface antigens, and inhibit the eosinophil-dependent killing of schistosomula mediated by IgG antibodies from the same sera. They also recognize epitopes present in egg or miracidial antigens; a working hypothesis compatible with current available data is that polysaccharide antigens released from eggs during early infections elicit predominantly blocking antibodies (of IgM or ineffective IgG isotypes) that can also bind to carbohydrate epitopes on schistosomulum surface glycoproteins, thereby blocking the binding of effector IgG antibodies elicited simultaneously in response to those glycoproteins and preventing the expression of immunity. As the child ages, these blocking responses may decline, while the effector responses persist, and the child then expresses resistance. This hypothesis would explain the slow development of resistance in the face of an early development of a range of potentially protective responses and would have implications both for

treatment and for vaccination as control measures, but the details have yet to be confirmed.

This sort of work is still sparse. Relatively few studies are completed, and the results are still open to more than one interpretation. Such studies are extremely difficult and time-consuming to carry out, but they are essential; they offer insights into the nature and importance of immune responses in the *natural* host.

Antigens Associated with Protective Immunity

Because of the apparent importance in immunity of antibody-dependent, cell-mediated effector mechanisms active against the developing schistosomulum, much effort has been directed toward the identification, isolation, and genetic cloning of molecules expressed on the surface of the young larva. Early experiments, in which surface proteins and glycoproteins were radiolabeled and precipitated with a range of polyclonal sera from different hosts immunized in various ways, revealed immediately that a wide range of antigens are expressed at the surface. In contrast to some protozoal infections, therefore, one does not have to think of a single molecule as the only possible target for immune attack, although some antigens appear to predominate both in their level of expression at the surface and in their capacity to elicit immune responses in most hosts. Subsequently, the preparation of monoclonal antibodies with specificity for individual surface antigens and the demonstration that some of these monoclonals can confer protection in vivo have allowed the definition of at least some molecules that are good candidates for future work. Finally, at least one antigen, affinity purified by the use of monoclonal antibodies, has been successfully used to elicit protection in vivo.

A major group of *S. mansoni* schistosomulum surface antigens is a series of glycoproteins of 32 to 38 kD. These antigens, demonstrated to be interrelated by partial protease digestion, form a major component of the material that can be radiolabeled on the schistosomulum surface by peroxidase-catalysed iodination, and are ubiquitously recognized by a range of infection or immunization sera from humans as well as mice and rats. A rat monoclonal, of the IgG2a isotype, with specificity for the 38-kD molecule mediates eosinophil-dependent killing of schistosomula in vitro and confers high levels of resistance to challenge after passive transfer in vivo. In contrast, an IgG2c monoclonal against the same antigen fails to confer protection and inhibits the eosinophil-dependent effect of the IgG2a monoclonal. The same antigen is also recognized by at least one protective mouse monoclonal.

Other moieties are also good candidates as protective antigens, because (1) they are strongly and ubiquitously recognized by human infection sera; (2) they are selectively recognized by sera raised in animals rendered highly resistant to infection (for example, by immunization with irradiated cercariae); or (3) they are recognized by monoclonals that confer protection in vivo. These include several low-molecular-weight products, especially of 20 kD, but also of 17 and 15 kD, as well as some higher-molecular-weight antigens, of 160, 130, and 92 kD, respectively. Immunization of mice with either of two affinity-purified antigens, of 155 and 53 kD, respectively, in a manner designed selectively to elicit an IgE response, has led to demonstrable protection, as has immunization of rats with a 28-kD molecule eluted from gels. Antigens not expressed at the schistosomulum surface may also be protective in some circumstances. After the demonstration that activated macrophages have a role in mediating protection in the mouse model, it was shown that a whole adult worm extract, administered intradermally with BCG, can elicit protection, the relevant component being a 97-kD molecule recently identified as paramyosin. The importance of these findings is that although there is a multiplicity of potentially protective antigens, *individual* antigens are clearly expressed at a sufficiently high density that, by themselves, they can elicit a protective response. The levels of protection obtained, however, have not been great, and more work is needed on the most effective means of administration of the purified antigens to elicit the maximum protective response, either by the use of appropriate antigens or by the development of live cloning vectors.

One of the reasons for attempting to test the protective effect of individual antigens is that although live irradiated cercariae can induce protection in many circumstances, the constraints of the parasite life cycle make it impractical to generate these organisms on a scale or at a cost adequate for use in man. Long-term approaches to vaccination in man must depend on cloning of individual protective peptides. There is considerable activity in this area at the moment, with a range of schistosome peptides now cloned and expressed in a variety of phage and plasmid vectors. One recombinant peptide (p28, a molecule transiently expressed on the schistosomulum surface) has been found to elicit high levels of protection in rats and hamsters and is currently being tested in baboons; other candidate vaccine antigens are being actively developed by several groups.

In other parasitic infections, a major problem with the gene-cloning approach has been the existence of antigenic variation and of antigenic diversity between different strains. At present, although schistosome immunity is species-specific, there is no evidence for strain diversity or for antigenic variation within a given strain: this does not appear to be a major parasite escape mechanism of the schistosome. In one study, an extensive effort was made to select for the appearance of strains insusceptible to immune attack, by repeated passage through partially immune animals; there was no evidence that such insusceptibility developed.

A potential problem with the recombinant DNA approach to the production of antigens for use in vaccines may be that the epitopes that can be recognized on the native molecule may be carbohydrate in nature, and therefore not available on the unprocessed molecule. At least some of the schistosomulum surface antigens are glycoproteins, and at least some protective monoclonals may recognize carbohydrate epitopes. A theoretical approach to this problem is to prepare antiidiotypes to protective monoclonals and to use them for active immunization. In the rat model, antiidiotypes against a protective IgG2a monoclonal have now been raised: these elicit protection in vivo, and the corresponding antibodies against the antiidiotype mediate damage to schistosomula in vitro. This represents an exciting and viable alternative for an approach to vaccination against nonpeptide epitopes.

Suppression of Egg Output as a Mechanism of Protection

Most studies on protective immunity have concentrated on the young schistosomulum, at some stage from its transformation from the cercaria to its development into the adult worm, and as we have seen, the adult worm itself is resistant to rejection by immune hosts. However, the pathology of disease is attributable entirely to the effects of egg deposition in the tissues, and immune events that act to reduce egg deposition by adult worms may be markedly beneficial to the host, even though the worms themselves are not damaged in the process.

There is little evidence for an immunologically mediated suppression of egg output in most rodent models of schistosome infection, but there is some evidence from the baboon. Individual animals bearing primary infections of either *S. mansoni* or *S. haematobium* may show a sudden and unexpected fall in egg output, which may later recover spontaneously. In the case of *S. haematobium*, adult worms can be recovered during the period of suppression of egg output; such worms appear normal, and after direct transfer to the hepatic portal system of naive recipients, they rapidly resume egg production. Similar striking reductions in egg output can occur in man, and the possibility of an immunologically mediated suppression should be considered among the many possible explanations. The maturation of the female worm, as well as the subsequent deposition of eggs, depends on the close proximity of the adult male in a complicated and poorly understood interaction. Subsequently, egg production by the female is associated with the enhanced expression of a range of antigens, whose regulation at the gene level is now under study. Such peptides could form suitable targets for immune attack, and this process is a fruitful area for further study.

Immunopathology of Infection: The Egg Granuloma

Primary infections with *S. japonicum* (and less frequently with *S. mansoni*) may lead to a severe acute condition referred to as Katayama fever. This complication generally occurs between 2 and 11 weeks after infection and is thought to be associated with deposition of eggs in tissue. The syndrome is characterized by fever, abdominal pain, diarrhea, lymphadenopathy, and marked eosinophilia. However, since the manifestations of acute schistosomiasis have occasionally been observed before oviposition, factors other than egg deposition may also be involved in their pathogenesis. The clinical features are reminiscent of those that can be induced by immune complexes, and indeed circulating immune complexes are detectable in some schistosome patients. Despite these similarities, we lack compelling evidence that these complexes are causative in the acute schistosomiasis syndrome.

Patients who undergo repeated exposure to cercariae of schistosomes (including species that do not mature to adults in humans) may show a transient cercarial dermatitis, or swimmer's itch. An additional, less common complication of infection results from deposition of eggs in ectopic sites such as the central nervous system, where resulting granulomatous inflammation or microvascular obstruction and tissue infarction can cause neurological dysfunction.

By far the most important cause of pathology in all forms of schistosomiasis is the deposition of eggs in the tissues and the subsequent host responses leading to granuloma formation and fibrosis. In the case of *S. haematobium* infection, fibrosis of the bladder (especially around the ureteral orifices) may lead to ureteric obstruction, and finally pyelonephritis, hydronephrosis, and renal failure. In contrast, in *S. mansoni* infections, fibrosis in the liver leads to portal hypertension, hepatosplenomegaly, ascites formation, and the development of esophageal varices that can rupture, leading to fatal hematemesis (vomiting blood).

Studies in the mouse have demonstrated that the pathological consequences of infection are largely the result of the host's immune responses rather than a virulence factor of the parasite. Deposition of eggs in the livers of acutely infected animals or injection of eggs into the lungs of naive (uninfected) recipients leads to a granulomatous reaction to the egg. Over a 16-day period, the egg becomes surrounded by a dense cellular infiltrate composed mainly of lymphocytes, macrophages, and variable numbers of eosinophils held together by excess extracellular matrix. Such lesions do not occur in mice depleted of T cells or in congenitally athymic (nude) mice. Reconstitution of nude mice with a thymic graft permits development of the type of granulomatous responses observed in euthymic littermates. In contrast, granulomatous reactions occur normally both in chickens rendered B-cell-deficient by bursectomy and in mice comparably impaired by treatment with anti-μ antibody. Clearly, then, granuloma formation is both T-cell-dependent and T-cell-mediated.

Intravenous inoculation of eggs into mice recently infected with *S. mansoni* results in the formation of pulmonary granulomata that develop more quickly and to a larger size than those seen in normal animals, indicating an anamnestic reaction. This enhanced reaction can be transferred to normal uninfected recipients by the administration of Lytl^{+}2^{-} splenic lymphocytes from acutely infected donors. Interestingly, eggs inoculated into chronically infected mice elicit reactions that are smaller than those observed in acutely infected animals. The reduction in the granulomatous reaction is associated with concomitant decline in various in vitro lymphocyte responses to egg antigens, including lymphocyte proliferation and the production of the lymphokine macrophage migration inhibitory factor and eosinophil stimulation promoter (a lymphokine that promotes eosinophil locomotion). In mice bearing infections 16 weeks or longer the granulomatous response is reduced, or modulated. Indirect evidence (such as reduced in vitro lymphoproliferation responses to egg antigens) suggests that a similar modulation occurs in chronically infected humans. The extent of tissue damage occurring in schistosomiasis is thought to depend in large part on the magnitude of granulomatous inflammation. Accordingly, efforts to understand the immunoregulatory features that control modulation continue to attract intensive investigation.

Transfer of splenic lymphocytes from chronically infected mice to normal recipients results in a reduced granulomatous reaction to a subsequent inoculation of eggs.

This indicates a role for suppressor cells in modulation. Fractionation of the spleen cells before adoptive transfer has shown that the cell capable of transferring modulation bears the $Lyl^{-}2^{+}$ phenotype characteristic of a lymphocyte subpopulation capable of suppressor function in immune responses. $Lyl^{-}2^{+}$ regulatory T cells suppress MIF production by $Lyl^{+}2^{-}$ cells. Recently, an in vitro model of granuloma formation and modulation has been reported. Reminiscent of the in vivo findings, granuloma formation in vitro is enhanced by the presence of $Lytl^{+}2^{-}$ cells from acutely infected mice. A mixture of $Lytl^{+}2^{-}$ cells abrogates their enhancing influence on the granuloma formation. Similar observations have been made with peripheral blood lymphocytes from chronically infected but asymptomatic humans. These lymphocytes show reduced proliferative responses to egg antigens in vitro and can exert suppressor influence on in vitro granuloma formation. The majority of patients with chronic hepatosplenic infections show strong responsiveness to the same egg antigen preparation.

The mechanisms of modulation of granuloma formation are almost certainly multifactorial. Recent studies have revealed the presence in sera of chronically infected mice of both the soluble immune responses suppressor (SIRS) and of autoantiidiotypic antibodies, both of which may contribute to modulation. SIRS can also be detected in culture supernatants of isolated granulomas from mice late in infection (modulated granuloma) but not from mice early in infection (vigorous granuloma). A promising approach to the further analysis of modulation has been the development of T-cell clones with relevant functional activity. Several such clones that bear the $L3T4^{+}$, $Lyt2^{-}$ phenotype enhance granuloma formation in vitro and mediate delayed hypersensitivity reactions in vivo. They possess a variety of functional properties, including the capacity to produce a range of lymphokines that includes gamma interferon, macrophage migration inhibitory factor, interleukin 2, and eosinophil stimulation promoter. Comparable immunoregulatory (suppressor) clones have not yet been developed. As more such clones are developed, it will be of interest to examine their antigenic specificity.

Efforts to assess antigen specificity of granuloma regulation are beginning. At present, fractionation of soluble egg homogenates has demonstrated a series of glycoproteins of high molecular weight that can sensitize mice for enhanced granuloma formation. However, these antigens are not yet available in a purified form. It is not known whether the suppressor cell populations recognize the same epitopes, different epitopes on the same antigen, or entirely different antigens. Autoantiidiotypic cells have been detected in S. *mansoni* – infected patients that, on exposure to antiegg antibodies, respond by proliferation and subsequently also suppress in vitro granuloma formation. The involvement of regulatory idiotypic – antiidiotypic circuits is easily seen in both human and murine S. *mansoni* and has been well documented in murine S. *japonicum* infections.

At first sight, it would appear that the host response to egg antigens, by leading to granuloma formation, is solely deleterious to the host. However, some workers have reported that mice depleted of T cells show an increased mortality from low-dose schistosome infections. In these mice, granulomatous reactions around eggs are absent; instead there is evidence of hepatic cell necrosis not observed in euthymic mice. The component in soluble extracts of eggs that can induce hepatic cell damage has been isolated and partially characterized. Antibodies elicited against this component protects athymic mice against the hepatotoxic effects. Granuloma formation therefore can both contribute to the pathology and protect the host against toxic parasite products. Curiously, T-depleted animals also exhibit reduced excretion of eggs into the feces. Transfer of lymphocytes or serum from infected animals to these animals restores the normal progression of eggs through the intestinal walls into the bowel lumen. As well as contributing to pathology and protecting hepatocytes, the granulomatous response may also actually aid in the continued transmission of infection.

Hepatic Fibrosis

Tissue fibrosis, the most serious consequence of schistosomiasis, is the basis for important morbidity and mortality. In *S. mansoni* and *S. japonicum* periportal fibrosis can develop leading to life-threatening portal hypertension. Scar formation in the bowel wall can lead to stricture formation and obstruction. In *S. haematobium*, fibrosis of the urinary tract substantially contributes to obstruction of urine flow, which ultimately can lead to renal parenchymal failure. Despite the obvious importance of fibrosis to disease pathogenesis, its regulation in schistosomiasis has received relatively little investigative attention. With our present understanding, two compelling questions arise. Why do some, but not all, infected individuals develop the serious fibrotic sequelae? What are the important factors that initiate and maintain the fibrotic process? Answers to these questions could have practical significance. For example, if the population at risk for developing hepatic fibrosis could be prospectively identified, immunopreventive or therapeutic interventions might be selectively directed at them rather than at the whole community. Moreover, if the molecular basis of fibrosis in schistosomiasis were clearly delineated, it might be possible to devise pharmacological means of preventing or treating the process. The rapid progress now being made in studies of basic aspects of regulation of fibroblast function generally lends optimism to the second possibility.

Autopsy data have provided the strong impression that heavy worm burden may be a prerequisite to development of the characteristic form of hepatic fibrosis (so-called Symmer's clay-pipestem fibrosis) unique to schistosomiasis. Clinical studies support this impression, although the premortem data are open to some criticism. Until recently, reliable clinical criteria have not been available for establishing a specific diagnosis of Symmer's fibrosis without resorting to liver biopsy. The recent introduction of hepatic ultrasonography to assess hepatic fibrosis can be expected to overcome this shortcoming. Two additional features are noteworthy. First, not all heavily infected individuals progress to clinically important liver fibrosis. Epidemiological data indicate that the prevalence of this complication (adjusted for egg burden) can vary by as much as less than 3 percent to 30 percent in different geographic regions. Disproportionately higher prevalence in some kindreds argues for the possibility of genetic predisposition. Despite some efforts, the genetic markers have not been identified. The new technology of restriction fragment length polymorphism (RFLP) for genetic analysis has yet to be applied to the problem. Second, hepatic fibrosis can develop relatively rapidly, since even preteen-age children can develop endstage portal hypertension. There are suggestions that antihelminthic chemotherapy can reverse the process in the early stages. To confirm this impression with greater certainty, improved noninvasive diagnostic criteria of hepatic fibrosis (such as ultrasonography) will need to be employed. Advanced fibrosis apparently is not spontaneously reversible after the elimination of the adult worms.

The observations that granulomas isolated from livers of mice spontaneously elaborate various biologically active mediators stimulated the hypothesis that secretion of fibrogenic factors by the granuloma might play a critical role in hepatic fibrogenesis. Indeed, culture supernatants from isolated granulomas have been shown to contain fibroblast chemoattractants, growth factors, and stimulators of fibroblast collagen and fibronectin synthesis. It appears that at least in the mouse hepatic fibrosis is a T-cell-dependent process. Congenitally athymic (nude) mice experience minimally enhanced collagen deposition in the liver when infected with schistosomes, whereas this deposition is substantial in infected euthymic mice. The fact that the isolated egg granulomas from nude mice secrete an inhibitor of fibroblast growth and fail to secrete a collagen synthesis stimulating factor (in contrast to euthymic mice) presumably can account for the in vivo findings. Recognizing that one aspect (size) of the granuloma is "modulated" during the course of infection, it is of interest that production of certain fibroblast-stimulating factors by modulated granulomas is reduced, compared with production by vigorous granulomas. We do not yet know if this down-regulation of production of fibrogenic factors is under similar immunological control as the granulomatous inflammatory response itself or if abnormalities in down-regulation select patients for developing Symmer's fibrosis. The possibility that it might be is suggested by the finding that adoptive transfer of spleen cells from mice chronically infected with S. *mansoni* into recently infected mice interferes with the ability of granulomas of the recipient to produce fibrogenic factors.

Immunodiagnosis

Immunodiagnosis is not covered in detail here; however, some principles associated with immunodiagnosis of parasitic infections in general, and of schistosomiasis in particular, are worth mentioning.

First, the quality of an immunodiagnostic test for the detection of circulating antibody, in terms of sensitivity and specificity, usually depends on the quality of the antigen used in the test. Ideally, the antigen should be one that is recognized by antibody from all infected patients and that shows no cross-reactions with antigens of other parasites to which the patient may be exposed. The intensity of the antibody response measured ideally should reflect the intensity of infection and should decline rapidly after treatment. These criteria are unlikely to be fulfilled by assays that employ crude preparations containing a mixture of different antigens: a purified preparation is preferable. At present, the most promising antigen for serodiagnosis is a cationic fraction of egg antigens. Assays using this antigen have proved both sensitive and specific in a limited trial that is now being extended. In the long run, recombinant peptides may be yet more valuable, as these may avoid the problems of cross-reactive carbohydrate epitopes that are present in certain other preparations.

A promising approach to the detection of active infection is the detection of circulating antigens, rather than antibody. Circulating antigens, either free or in the form of immune complexes, can be detected in patients with schistosomiasis. Tests based on the detection of such antigens, involving the use of either polyvalent or monoclonal antibodies, are being developed and may eventually prove useful. Since preparation and examination of fecal preparations in diagnosis are skilled, time-consuming, and uncongenial activities, and serological methods to detect active infection would be highly desirable.

Coda

As with many other chronic parasitic infections, the main interest in schistosomiasis is in the balance established between the host and the parasite that allows both to survive (at least at the population level) for long periods of time. In schistosomiasis, several features contribute to the establishment of this balance.

The organism is of relatively low pathogenicity. Symptoms of acute schistosomiasis are seen only in the unusual circumstance of a massive initial exposure to infection. More commonly, over a span of years, repeated exposure to very small numbers of cercariae occurs and is associated with a slow rise in worm burdens and the development of a persistent, chronic infection. During the chronic stage, disease attributable to the presence of adult worms is negligible, and most pathology is associated with the deposition of eggs and the development of fibrosis. However, the inflammatory response to the egg becomes less severe during chronic disease, by a down-regulatory process called modulation. Typically, infection is associated with relatively little morbidity and mortality; by contrast, disease occurs more commonly under conditions of abnormally high transmission, such as occurs, for example, in association with irrigation schemes.

The high prevalence of infection in rural areas may be associated with relatively low mean intensities of infection. High intensities are most commonly seen among older children and decline in adult life. Several explanations for this observation are possible, but at least one major factor that restricts adult worm burdens in older individuals may be the development of acquired immunity to superinfection in the presence of a continued primary infection, an effect referred to as concomitant immunity. There is now good evidence that such immunity occurs in man, and there are suggestions that the slow development of immunity is attributable to a slow loss of "blocking" immune responses rather than to a slow acquisition of effector responses. There also is evidence from a range of experimental animal models that acquired immunity develops. One of the most useful models involves immunization with irradiated larvae that fail to mature to adult worms. In several models, there is evidence that antibody-dependent, cell-mediated effector mechanisms directed against the young schistosomula are important; however, there is also good evidence in some systems for a role for activated macrophages and for immune events directed against older schistosomula, after their migration through the lungs. There is now a considerable body of information on the antigens on the young schistosomulum against which antibody-dependent mechanisms may be directed: these are multiple; there is no evidence for antigenic diversity or variation. Much less is known about the important antigens involved in events directed against older schistosomula.

One of the implications of the idea of concomitant immunity is that the adult worm succeeds in evading immune responses that are capable of destroying the young organisms of a challenge infection. The reasons for this evasion are difficult to determine, and there are conflicting views on the relative importance of the acquisition of host-derived molecules, the loss of expression of parasite surface antigens, the release of inhibitory molecules, and the development of a tegument that is refractory to damage. It is likely that evasion is a multifactorial event and that other mechanisms remain to be discovered: this is a field in need of additional study. Of particular interest is why the schistosomulum, after leaving the lungs, appears transiently to regain its susceptibility to immune attack.

Another area of considerable importance is the development of the pathogenetic egg granuloma and the associated fibrosis. Much is known about the role of cell-mediated responses in the initiation of the granuloma during acute infections and of the modulation of granuloma formation by suppressor cells and other immunoregulatory mechanisms, but the precise relationship between granuloma formation and subsequent fibrosis requires closer scrutiny.

During the 20 years that have elapsed since the development of adequate animal models of schistosome infection and of in vitro techniques for assaying human immune responses, there has been an explosion in our understanding of the immunological events that may contribute to the development of a stable host–parasite interaction. Much more remains to be done, both in furthering our understanding of the basic mechanisms involved and in transforming this understanding into information that may be of value in controlling the disease in man.

Additional Reading

BUTTERWORTH, A. E. 1984. Cell-mediated killing of helminths. *Adv. Parasitol.* 23:143–235.

CAPRON, M., and A. CAPRON. 1986. Rats, mice and men—models for immune effector mechanisms against schistosomiasis. *Parasitology Today* 2:69–75.

COLLEY, D. G. 1987. Dynamics of the immune response to schistosomes. *Balliere's Clin. Trop. Med. Commun. Dis.* A. A. F. Mahmoud, ed. 2:315–332.

PHILLIPS, S. M., and D. G. COLLEY. 1978. Immunologic aspects of host responses to schistosomiasis: Resistance, immunopathology and eosinophil involvement. *Prog. Allergy* 24:49–182.

SMITHERS, S. R., and M. J. DOENHOFF. 1982. Schistosomiasis. In *Immunology of Parasitic Infections*, 2nd edition, S. Cohen and K. S. Warren, eds. Blackwell Scientific Publications, pp. 527–607.

WARREN, K. S. 1982. Immunology of schistosomes. In *Comprehensive Immunology*, Vol. 9. Immunology of human infection. Part II: viruses and parasites; immunodiagnosis and prevention of infectious diseases, A. J. Nahmias and P. J. O'Reilly, eds. Plenum Medical Book Co., pp. 445–458.

15

Immunology of Lymphatic Filariasis

Willy F. Piessens, Juliet A. Fuhrman, and Ann C. Vickery

Filarial parasites have complex life cycles that include an obligatory maturation period in a hematophagous arthropod vector and a stage of further differentiation followed by reproductive activity in a vertebrate host (Figure 15-1). Like most species of filarial nematodes, those that cause onchocerciasis *(Onchocerca volvulus)* or lymphatic filariasis in humans *(Wuchereria bancrofti, Brugia malayi,* and *B. timori)* have a limited host range in nature. However, some filarids develop normally in experimentally infected laboratory animals that are not their natural hosts. These experimental models of lymphatic filariasis differ from natural infections in three variables that are known to affect immune responses to defined antigens: the frequency of exposure, the initial route of presentation, and the amount of parasite antigen presented to the host. Most experimental infections are induced by a single parenteral injection of a large number of infective larvae. In contrast, natural infections are acquired over a period of weeks to months through "trickle infections," or repeated penetrations of the skin by small numbers of larvae deposited by the vector. Because of these differences, some immunologic phenomena observed in animal models may not be relevant to host–parasite interactions in naturally acquired lymphatic filariasis.

In natural as well as in laboratory infections, several developmental stages of filarial worms can be present simultaneously in the final host, and these are likely to

The authors' research is supported by the Filariasis Component of the United Nations Development Programme–World Bank–World Health Organization Special Programme for Research and Training in Tropical Diseases, the Rockefeller Foundation, the Thrasher Foundation, and the United States Public Health Service through NIH NIAID Grants AI 16479, AI 20052, AI 20102, and AI 22418.

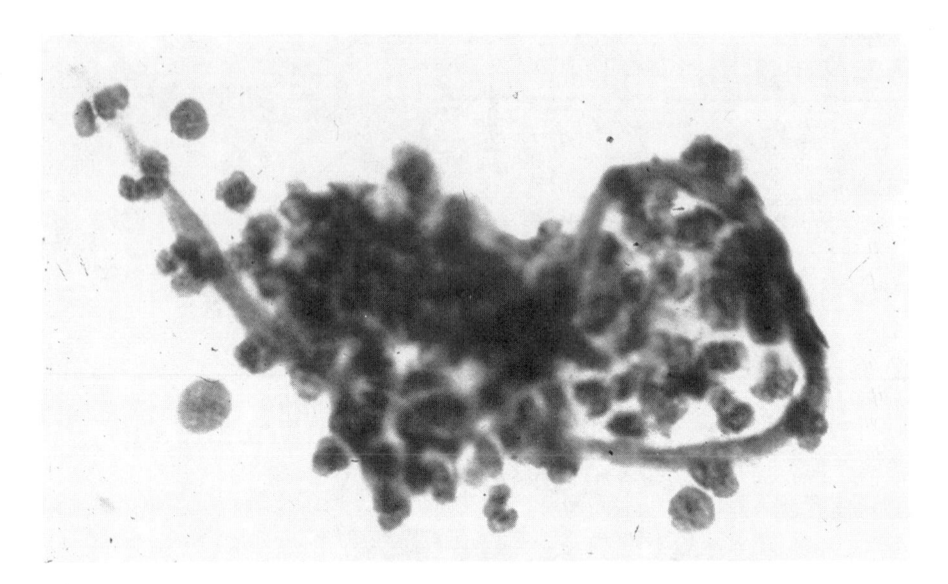

FIGURE 15·1 Killing of microfilariae by antibody-dependent cell-mediated cytotoxicity reactions (ADCC). In this process, worms are killed by cells attached to parasites; attachment results from the interaction of the cells' Fc receptors to the Fc portion of antibodies bound to microfilarial antigens through the $F(ab')_2$ portion of the antibody molecule. The isotype of the adherence-promoting (opsonizing) antibody and the nature of the effector cell vary among parasite species.

interact with the immune system in different ways. For example, infective larvae, adult worms, and microfilariae reside in the skin, lymphatic vessels, and blood, respectively; their being in different tissues influences the way the host is sensitized to stage-specific parasite antigens and where immune effector mechanisms must operate (Table 15-1). Different filarial stages also present distinct target surfaces to the immune system: larval and adult worms are bounded by a typical nematode cuticle, whereas microfilariae of some species are enclosed in a multilayered sheath derived from the original eggshell and separated from the underlying cuticle by an electron-lucent space (Table 15-1). These distinct physical and chemical properties of the worms' surface are likely to influence the susceptibility of various developmental stages to different immune effector mechanisms.

The simultaneous presence of multiple worm stages that share many common antigens also complicates analysis of host reactions to stage-specific antigens. To circumvent some of these problems, increasing use is being made of the "proxy" method of infection. In this approach, a single developmental stage isolated from a suitable host is surgically transferred to a new host of the same or different animal species. In this way, one can study reactions to one worm stage without interference from host responses

TABLE 15·1
Parasitological characteristics of filarial worms

Parasite species	Natural hosts	Location		Sheathed microfilariae
		Adult worms	Microfilariae	
Brugia malayi, B. timori	Humans, primates	Lymphatics	Blood	Yes
B. pahangi	Primates, cats, other	Lymphatics	Blood	Yes
Wucheria bancrofti	Humans	Lymphatics	Blood	Yes
Dipetalonema viteae	Jirds	Subcutaneous	Blood	No
Litomosoides carinii	Cotton rats	Pleura	Blood	No
Dirofilaria immitis	Dogs	Heart, lungs	Blood	No
Onchocerca volvulus	Humans	Subcutaneous	Skin, eyes	No

elicited by earlier developmental stages. Studies of this nature indicate that adult worms or microfilariae can survive in some species of animals that are innately resistant to infection with filarial larvae. Such studies reinforce the concept of the stage specificity of protective immune responses in filariasis.

This chapter focuses on four major topics of the immunobiology of lymphatic filariasis: mechanisms of resistance to infection, immune reactions that control the number of microfilariae circulating in blood, mechanisms of immune evasion, and the pathogenesis of filarial disease syndromes. It is not intended to be an exhaustive review of the subject; our aim is to illustrate general principles with selected examples. Where appropriate, nonimmunologic variables that affect the ultimate outcome of an individual's exposure to filarial helminths are briefly mentioned. The immunology of onchocerciasis is not considered here, because several excellent recent reviews of this topic are available. Other aspects of the general biology of filarial helminths are the subject of recent reviews listed at the end of this chapter.

Finally, some terms used in this chapter require definition. Prepatency is the interval between the onset of infection and the appearance of microfilariae in blood, patency is the period of microfilaremia, and latency is the period of postpatent infection. Latent infections often are referred to as "occult" filariasis, a term we will not use, because prepatency can also be considered a period of occult filariasis, that is, one during which microfilariae are absent from the circulation. Filarial infections in animals often progress through these three stages in a timely fashion, and lymphatic filariasis in human beings probably follows a similar course.

Immune Effector Mechanisms

Innate and Acquired Resistance to Filarial Infections

Microfilaremia or clinical symptoms of lymphatic filariasis are absent in a variable but sizable proportion of long-term residents of endemic areas, but whether immune resistance develops during natural infections is a question that defies simple answers. Filarial helminths complete several steps of their complex life cycle in the vertebrate host. Each one of these maturational stages constitutes a potential target of protective immune responses. However, not all stages can be readily quantified, and therein lies the difficulty of assessing immune resistance in filariasis. Quantitative recovery of infective larvae, the most direct measurement of immune resistance, is difficult to achieve even under optimal experimental conditions because of the rapid migration of larvae from the infection site. It is relatively simple to monitor blood levels of microfilariae, but this measurement of host resistance is inadequate because the number of microfilariae in peripheral blood does not correlate with adult worm burdens, and, as is discussed later in this chapter, microfilaremia may be absent when fertile adult worms are present in host tissues. Thus, quantitative recovery of adult worms is the most commonly used index of host resistance in filariasis. This indirect measurement should be interpreted with caution, however, because the number of adult worms that can be recovered from individual animals infected with the same number of infective larvae varies tremendously even among inbred animal strains.

Despite our inability to measure the phenomenon directly, several lines of evidence suggest that host immune responses to antigens on early larval stages confer some degree of resistance to primary infections with filarial helminths. Several species of filarids develop normally in naturally resistant hosts, provided the infection is initiated with larvae that have undergone partial development in a proxy host rather than with vector-derived larvae. For example, increasing numbers of adult *Brugia pahangi* are recovered when progressively more mature larval stages are transplanted from jirds to naturally resistant mice. If such patterns of resistance are truly related to immune phenomena, rather than to differences in metabolic requirements of distinct parasite stages, then they would be an example of remarkable stage specificity of immune effector mechanisms in filariasis.

Direct evidence for a role of immune reactions in innate resistance to filariasis comes from studies in animals with congenital or acquired immune deficiencies. Such animals are more susceptible than their immunocompetent counterparts to filarial infections. Although adult mice resist *B. pahangi*, a proportion of mice inoculated shortly after birth with larvae of this parasite species go on to harbor patent infections. This has been attributed to the relative immune incompetence of mammals during the neonatal period. Normally resistant rats that are immune suppressed before inoculation with *Litomosoides carinii* or *Dipetalonema viteae* infective larvae likewise develop patent infections and yield large numbers of adult worms.

Models of selective immunodeficiencies suggest that a thymus-dependent immune response contributes to innate resistance to filarial parasites. Congenitally athymic rats are susceptible to infection with *D. viteae*, a parasite to which their euthymic littermates are resistant, and *B. pahangi* readily infects athymic animals of strains of rats that are innately resistant to this parasite. Similarly, athymic (nude) mice are fully susceptible to larval-initiated infection with *B. pahangi* and *B. malayi*. However, implantation of a thymic graft several weeks before inoculation of infective larvae not only confers complete protection against infection but also enables the mice to produce cellular and humoral immune responses to larval antigens.

It should be emphasized that host immunity most likely is only one of the factors that determine primary resistance to filarial infections. Some of the variability in adult worm recovery from inbred animals can be explained by such nonimmunological factors as the age and sex of the experimental host. Genetic differences between strains, mostly unrelated to major histocompatibility loci, also influence to some extent the susceptibility of experimental animals to filarial infections. Furthermore, innate or immune-mediated resistance appears highly target-specific: nude mice that are susceptible to *Brugia* spp. remain resistant to *Wuchereria bancrofti*.

These examples of natural immune-mediated resistance to larval infection suggest that susceptible populations might be rendered resistant through vaccination. This idea prompts two major questions: (1) What is the precise mechanism of larval killing in naturally resistant animals, and (2) how can the components of the larvicidal mechanisms be elicited in otherwise susceptible hosts?

In vitro studies reveal that infective larvae can be killed by antibody-dependent cell-mediated cytotoxic (ADCC) immune reactions similar to those that act on microfilariae; they are described below. These studies led to the widespread belief that ADCC reactions mediate resistance to filarial worms, as has been postulated for several other helminth infections (see Chapter 14). However, the presence of antibodies that mediate this reaction correlates poorly with resistance to infection. Sera from virtually all residents of areas endemic for brugian or bancroftian filariasis contain such antibodies, even though many of these individuals remain susceptible to infection. Some cats repeatedly infected with third stage larvae of *B. pahangi* become amicrofilaremic and develop partial resistance to reinfection. Yet antibodies to larvae and adult worms are present in both "resistant" as well as in susceptible (microfilaremic) cats. Similarly, mice, jirds, and hamsters differ widely in their susceptibility to *D. viteae*, but resistance is not correlated with the development of antibodies to the cuticle of infective third-stage larvae of this parasite species. Thus, it is clear that the in vitro ADCC assay does not adequately reflect the process that kills larvae in vivo.

Recent studies in two animal models nevertheless indicate that cooperation between humoral and cellular components of the immune system can lead to some degree of resistance to filarial infections. Adoptive transfer of cells and sera from jirds infected with *L. carinii* retards the development of larvae injected into the recipient; neither immune cells nor immune serum alone is effective. Likewise, nude mice given immune serum or B cells from donors immunized with *B. pahangi* larvae are not protected against a larval challenge. However, transfer of primed B and T lymphocytes from similarly immunized donors confers complete protection on the recipients, which promptly produce high titers of antibodies and delayed hypersensitivity reactions to larval antigens.

The success of such adoptive transfers in conferring resistance indicates that protective immunity might be elicited by direct immunization. Conventional mice, which kill larvae of *B. pahangi* within 3 to 4 weeks after inoculation and are therefore considered resistant to infection, eliminate subsequent larval inocula within 10 days and exhibit greatly potentiated humoral and cellular immune responses to larval antigens after rechallenge. This suggests that reexposure to parasite antigens augments immune responses that may contribute to the "natural resistance" of refractory hosts.

Unfortunately, the opposite may be true in "naturally susceptible" hosts of filariasis. For example, jirds infected with *B. pahangi* manifest increased rather than decreased susceptibility to reinfection. In already infected jirds, recovery rates of adult worms resulting from challenge infections are twice those in previously uninfected jirds, although worms developing in the previously parasitized animals are smaller than in the naive recipients. As will be discussed later, this intriguing result has been attributed to immune suppression induced by the primary infection.

Thus, a primary infection in a susceptible host may not elicit immunologic resistance to reinfection. On the other hand, while hyporesponsiveness to filarial antigens is a characteristic of filariasis in humans and in animal models, repeatedly infected animals, like residents of highly filariasis-endemic areas, seldom suffer from overwhelming parasitemias. The number of adult worms increases with each administration of larvae during most experimental trickle infections, but the proportional increment in parasite load decreases with time, and the development of larvae in later challenges appears to be delayed. This result has been observed even in natural host–parasite combinations such as *L. carinii* infections in cotton rats and *D. viteae* infections in jirds. In most cases, direct evidence that these phenomena are immune-mediated is lacking.

Although it is uncertain whether natural hosts develop immune resistance through previous infections, it is clear that resistance to infection can be induced by several manipulations that enhance host immune responses to filarial antigens. Soluble parasite materials, whether they are somatic extracts or excretory-secretory products, are generally inefficient in conferring protection against larval challenge, but they protect when combined with an adjuvant. For example, homogenates of *L. carinii* larvae produce a high degree of resistance to infection in albino rats when administered as an emulsion in Freund's adjuvant, but not when used alone.

Irradiated larvae have been more successful than parasite extracts to vaccinate animals against several filarial species. For example, dogs can be protected against *Dirofilaria immitis*, albino rats against *L. carinii*, and cats and jirds against *B. pahangi*. The rationale for using such vaccines is that by inhibiting the development of the larvae, the host will be exposed to antigens on early larval stages for a longer period than during natural infections, but experimental evidence to support this assumption is not available. In at least one model, irradiated larvae induce a moderate degree of lymphatic pathology that may not be acceptable in human beings.

Termination of filarial infections with drugs before the parasite matures and causes lymphatic pathology is a promising variant of the attenuated larval vaccine approach. This protocol induces resistance to *D. immitis* in ferrets, and to *B. pahangi* in jirds. However, resistance of monkeys to reinfection with *B. malayi* apparently does not occur after successful drug treatment of patent infections. Nor is it clear whether elimination of microfilaremia with diethylcarbamazine, which reverses the state of immune unresponsiveness to parasite antigens associated with patent brugian filariasis in humans, leads to increased resistance to reinfection. Epidemiologic studies of previously treated populations suggest it does not. In contrast, immune reconstitution of nude mice parasitized by *B. pahangi* or *B. malayi* induces resistance to reinfection and kills adult worms but levels of circulating microfilariae are not affected. Thus, immune effector mechanisms that mediate protection against reinfection appear to be highly stage-specific, being both elicited by and directed against antigens on larval stages. Much remains to be learned about the nature of protective immune responses and the antigens that elicit these reactions.

Control of Microfilaremia

Microfilaremia is essential for the transmission of lymph-dwelling filarial helminths. Their survival as a species depends on the presence of microfilariae in peripheral blood of the vertebrate host in numbers adequate to infect the appropriate vector. In most natural hosts, the duration of patent filariasis far exceeds the estimated life span of individual microfilariae, even when reinfection is known not to take place. In contrast, patency lasts only for a relatively brief period in many experimental models of filariasis. Infections with this characteristic, such as *L. carinii* in albino rats and *D. viteae* in hamsters, have been used extensively to study the nature of host immune responses that lead to latent filariasis. Some observations made in these models undoubtedly reflect the uniqueness of the "unnatural" host–parasite combination, but studies of natural hosts that develop latent filariasis, while differing in detail, generally support the broad concepts derived from these model infections.

Foremost is the notion that microfilariae themselves are the primary target of immune reactions that lead to latency. This idea is based on the observation that fecund adult worms can be recovered from animals with latent infections; these worms continue to produce microfilariae in vitro and upon transfer to naive recipients in vivo. Three lines of evidence indicate that host immune responses control microfilaremia in animals with latent filariasis. First, immune suppression of primary hosts with postpatent infections results in the reappearance of microfilariae in blood. Second, adoptive transfer of sera from animals with latent filariasis reduces microfilaremia in the recipients. Third, immunization with microfilarial antigens prevents the development or reduces the level and duration of microfilaremia but usually has no effect on adult worm recovery. These observations indicate a remarkable stage specificity of host immune reactions that control microfilaremia. Three mechanisms by which immune sera can reduce levels of circulating microfilariae have been identified.

Inhibition of embryogenesis Sera from hamsters with latent *D. viteae* infections inhibit embryogenesis and prevent the release of microfilariae in vitro by affecting the fecundity, but not the fertility, of female worms. Whether this phenomenon occurs in vivo is not easy to ascertain, because it is technically difficult to distinguish between inhibition of embryogenesis and subsequent destruction of microfilariae in intact infected animals.

In situ trapping of microfilariae Immune sera also participate in host reactions that prevent microfilariae from entering the bloodstream. After 3 to 4 months of infection, latency develops in albino rats infected with *L. carinii*. The disappearance of microfilariae from peripheral blood is caused by a localized immune process that traps the worms in the pleural cavity. When adult worms are transplanted into the peritoneal cavity of rats with latent infections, microfilariae appear in recipients' blood. This does not happen when adult worms or microfilariae are transferred into the pleural cavity of such animals (Table 15-2). Microfilariae released in the pleural cavity of rats with latent *L. carinii* infections become covered with leukocytes and, one presumes, are killed in situ before they are able to enter the circulation. Similar selective trapping of microfilariae also occurs in the pleural cavity of cotton rats (the natural host of *L. carinii*) immunized with microfilarial antigens before infection.

This adherence phenomenon can be reproduced in vitro. Antiparasite IgE present in sera from rats with latent infections promotes cellular adherence to *L. carinii* microfilariae; another heat-labile serum component, probably complement, augments this reaction. Neutrophils, eosinophils, and macrophages adhere to microfilariae in vitro, but only neutrophils and macrophages kill the parasite, whereas peripheral blood lymphocytes or peritoneal mast cells neither adhere to nor kill *L. carinii* microfilariae. In this in vitro model, the worms appear to be killed by an antibody-dependent cell-mediated cytotoxic (ADCC) reaction.

These in vitro observations fail to represent the in vivo situation in two respects. Examination of the pleural content of rats with latent *L. carinii* infections often reveals cells, mostly lymphocytes and macrophages, attached to living microfilariae. Hence, the in vitro model evidently does not identify the type of cell that kills the worms in vivo. Second, sera from rats with latent infections readily promote cellular adherence to *L.*

TABLE 15·2
Compartmentalization of immune responses to *Litomosoides carinii* microfilariae[1]

Worm stage transplanted	Recipient status	Transplant site	Microfilaremia
Adult	Normal	Pleura	Yes
	Latent	Pleura	No
Adult	Normal	Peritoneum	Yes
	Latent	Peritoneum	Yes
Microfilariae	Normal	Pleura	Yes
	Latent	Pleura	No

[1] Based on studies by Bagai et al.

carinii in vitro, but adoptive transfer of such sera often fails to reduce microfilaremia. Furthermore, microfilariae are not retained and killed in the peritoneal cavity of rats with latent infections, even though the required constituents of the in vitro killing system should be present at this site as they are in the pleural cavity. Thus, anatomical compartmentalization of host immune mechanisms may influence the in vivo killing of microfilariae in ways which cannot be adequately reproduced in vitro.

Such compartmentalization of immune effector mechanisms could explain several clinical features of lymphatic filariasis in natural hosts. For example, the absence of microfilaremia in patients with elephantiasis could be due to the development of local immunity at sites that harbor adult worms, thereby preventing microfilariae from reaching the peripheral circulation. The appearance of microfilariae in the blood of amicrofilaremic patients with elephantiasis treated with corticosteroids, and the induction of localized acute adenitis with microfilaricidal drugs such as diethylcarbamazine support this hypothesis. A similar phenomenon of immune compartmentalization can be invoked to explain the pathogenesis of the tropical pulmonary eosinophilia (TPE) syndrome, discussed in a later section.

Removal of microfilariae from the circulation A third antibody-dependent mechanism that leads to latent filariasis is the clearance of microfilariae from blood. In many animal models, adoptive transfer of immune sera or monoclonal antibodies induces a rapid decline in the number of circulating microfilariae due to sequestration of worms in host tissues. This result has also been observed in natural infections such as those caused by *W. bancrofti* in humans and *D. immitis* in dogs. Other particulate antigens such as antibody-coated or malaria-infected red cells are cleared mostly by the reticuloendothelial system of the spleen and liver, but microfilariae are retained predominantly in the lungs of dogs infected with *D. immitis* or mice infected with *B. malayi*. The same process probably occurs in humans with lymphatic filariasis, although in some patients with brugian filariasis the spleen appears to be the major organ that traps microfilariae. Microfilariae cleared from blood are believed to be killed by an ADCC reaction similar to that observed in the pleural cavity of *L. carinii*-infected rats, as described earlier. That such a reaction can kill microfilariae in vivo is supported by studies on the fate of *D. viteae* microfilariae contained within diffusion chambers implanted into hamsters with latent infections (Table 15-3).

The relative ease with which ADCC reactions can be produced in vitro has prompted numerous investigators to study which antibody-cell combinations can "kill" filarial worms (Figure 15-2). In most of these studies, loss of microfilarial motility is equated with parasite death. It should be emphasized that the true measure of microfilarial damage or death should be the parasite's ability to infect and develop in the appropriate vector. Nonmotile parasites conceivably could recover function when suspended in the blood ingested by mosquitoes. Conversely, parasites that remain motile may in fact be sufficiently damaged to prevent their subsequent development in the vector. Nevertheless, such studies have yielded interesting results.

The isotype of antibodies and the nature of cells that participate in ADCC reactions to microfilariae greatly differ among various host–parasite combinations

TABLE 15·3
ADCC reactions kill microfilariae in vivo

	Fate of microfilariae implanted in chambers (pore size)	
Host status and response	0.4	4
In host with latent infection		
Antibodies on cuticle	Yes	Yes
Cells attached to worms	No	Yes
Microfilariae killed	No	Yes
In naive host		
Antibodies on cuticle	No	No
Cells attached to worms	No	No
Microfilariae killed	No	No
In recipient of immune serum		
Antibodies on cuticle	Yes	Yes
Cells attached to worms	No	Yes
Microfilariae killed	No	Yes

(Table 15-4). This may reflect differences in the ability of distinct filarial species to resist potentially toxic molecules released by some types of cells attached to the worms' surface. Alternatively, various host species may use distinct effector mechanisms to control microfilaremia caused by the same parasite. Current evidence is consistent with either interpretation.

Whether in vitro ADCC reactions adequately represent the mechanisms that kill microfilariae in vivo remains a matter of considerable debate. Critics of this concept

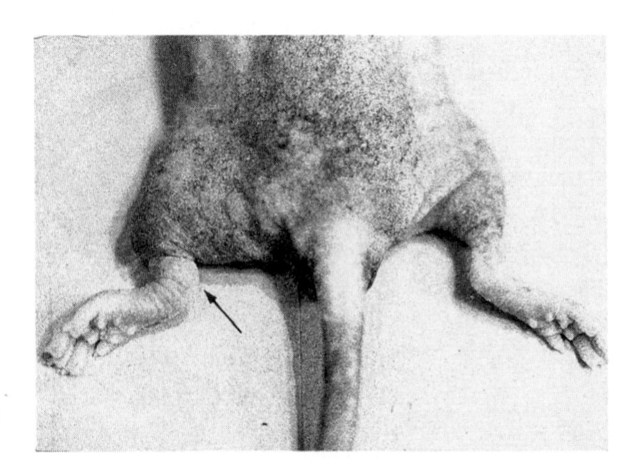

FIGURE 15·2 Subcutaneous lymphedema in nude mice infected with *Brugia pahangi*. Adult worms trigger gross dilatation of lymphatic vessels, which results in functional impairment of lymphatic drainage and accumulation of fluid in interstitial spaces. These changes are reversed when the animals' immune competence is restored by adoptive transfer of T lymphocytes.

TABLE 15·4
In vitro killing of microfilariae by ADCC reactions[1]

Parasite	Host	Antibody class	Effector cell[2]
Dipetalonema viteae	hamster	19S (IgM)	M/multiple
D. viteae	rat	IgE	M/Eo
Litomosoides carinii	rat	IgE	Eo/M
Wucheria bancrofti	man	IgG	multiple
Brugia malayi	man	IgG	M/multiple[3]
Onchocerca volvulus	man	IgG	Eo/neutro
B. pahangi	cat	IgG	Eo/neutro
Dirofilaria immitis	dog	IgM	Neutro

[1] Some reported antibody–effector cell combinations in ADCC reactions to microfilariae; list is not exhaustive.
[2] M = macrophage, Eo = eosinophilic granulocyte, neutro = neutrophilic granulocyte.
[3] Includes platelets.

point out that, at least in the *L. carinii*–albino rat and the *D. viteae*–hamster models, the predominant types of cells that adhere to and "kill" microfilariae in vitro are different from those that appear to operate in vivo. This difference may simply reflect the limitations of in vitro assays. Some cells may not be able to kill microfilariae in vitro because they must be activated by factors secreted by other cell types that are underrepresented in the populations used for in vitro assays. Conversely, in vitro effector cells may not kill parasites in vivo because they are not attracted to their worm target or because other factors, for example, immune complexes, interfere with cell–parasite interactions. Most likely, microfilariae are killed in vivo by the combined effects of several cell types that act sequentially or in concert, although it is not always possible to determine which cells kill the worms and which ones scavenge the dead carcass.

No direct evidence exists to support the popular notion that ADCC-mediated clearance of microfilariae is a process that controls the level of microfilaremia in natural hosts. A proportion of microfilariae isolated from patients with brugian filariasis have immunoglobulin (possibly antibody) on their surface. That these worms are destined to be removed from the circulation is a plausible, but unproven, assumption. On the other hand, the pace at which 19S (IgM) antibodies to the cuticle of *D. viteae* microfilariae become detectable correlates with the duration of microfilaremia in various strains of hamsters. Thus, it appears probable, albeit unproved, that the relatively stable levels of microfilaremia characteristic of natural infections are maintained to a certain extent by the same types of host immune reactions to microfilariae as those that lead to latent filariasis in experimental infections.

Which microfilarial antigens act as targets of such "protective" immune responses remains unknown. Clear-cut associations between the absence of microfilaremia and detectable serum levels of antibodies to surface antigens of microfilariae have been described in many natural and experimental filarial infections. Formal proof is still lacking, however, that surface-directed antibodies mediate microfilarial clearance. In fact, immune sera that reduce microfilaremia in the *B. malayi* – mouse model react with many constituents of microfilariae but appear devoid of antibodies to surface antigens. Other possible target antigens for clearance-mediating antibodies include proteases secreted by the worms (particularly those that might inactivate ADCC-mediating antibodies) and worm-derived modulators of immune effector cells or inhibitors of inflammatory reactions. Experiments to determine which parasite antigens are specifically recognized by sera from amicrofilaremic donors are in progress in many laboratories. So far, differential recognition of microfilarial antigens with relative molecular weights (M_r) of 25 kilodaltons (kD), 70 kD, and 75 kD has been reported. In addition, two monoclonal antibodies that reduce microfilaremia upon adoptive transfer have been described. These react with microfilarial antigens of 70 and 75 kD, and 110 kD, respectively, and should provide invaluable reagents to isolate microfilarial antigens that act as targets of immune responses resulting in the clearance of this parasite stage from the circulation.

Immune Evasion

The mere fact that filarial parasites develop and survive in immunocompetent hosts indicates that they are able to evade or resist host immune responses, either by hampering immune effector mechanisms or by preventing the development of immune responses with protective potential. Several mechanisms by which filarids may evade host immune responses have been identified (Table 15-5).

Immune Evasion by Disguise, or the Invisible Parasite

It would seem natural that microfilariae might blend into their environment by acquiring a surface coat of natural constituents of blood that would shield parasite antigens from host reactions and thereby enhance the worms' survival. Several types of host molecules have been found on the surface of microfilariae.

Albumin can be detected on blood- or tissue-derived microfilariae of some filarial species, but not on microfilariae shed in vitro. The biological role of this molecule in terms of parasite survival is unclear. Host albumin often prevents the radiolabeling of parasite surface antigens, but it does not appear to inhibit the killing of microfilariae by ADCC reactions to these antigens.

The protective role of host blood group substances remains equally in doubt. Isohemagglutinins promote the mixed agglutination of *W. bancrofti* microfilariae with erythrocytes, and blood group A and B determinants have been detected on these parasites (but not on microfilariae of some other species). One report suggests that blood group antigens may not contribute to the survival of filarial worms: immunization of jirds with cat antigens does not alter the recovery or longevity of *B. pahangi* transferred

TABLE 15·5
Possible mechanisms of immune evasion by filarial worms

Disquise of parasite antigens by host molecules (albumin, blood group substances, immunoglobulin)
Polyclonal activation of host immune system
Immune suppression
Inactivation of immune effector mechanisms (complement, antibody, other)
Other mechanisms (surface turnover, innate resistance)

from cats into these jirds. However, these experiments did not rule out the possibility of rapid exchange of cat-derived blood-group determinants for jird determinants once the microfilariae were in the latter's bloodstream. Such exchange can occur in vitro: when microfilariae of *W. bancrofti* are incubated in heterologous sera, blood group determinants present on the worms are rapidly replaced with those present in the serum.

Regardless of their role in immune evasion, blood group substances or other host-derived molecules may insure the continual recirculation of microfilariae by masking surface determinants of the parasite that lack sialic acid, which might otherwise trigger clearance of the worms by the reticuloendothelial system.

Immunoglobulin also can be present on microfilariae recovered from blood or tissues of infected donors but is absent from microfilariae produced in vitro by gravid female worms. In contrast to blood group determinants, immunoglobulins are not adsorbed by microfilariae incubated in vitro in normal human serum. This finding suggests that "adsorbed" immunoglobulin constitutes antimicrofilarial antibody. These antibodies could favor parasite survival if they do not themselves kill the worms but instead prevent a protective type of immune effector mechanism from reacting with its target antigen. Such "blocking" antibodies have been identified in other parasitic infections and in tumor models.

Thus, although several types of host molecules at times can be present on filarial worms, it is far from evident that they contribute to the survival of parasites in vivo. In fact, the true origin of these putative host molecules remains in doubt, because many studies performed so far fail to prove that filariae themselves lack the ability to produce the "adsorbed" host material.

Polyclonal Activation of the Host's Immune System

Many parasitic infections are associated with some degree of polyclonal activation of the host's immune system. This pattern is also true for filariasis: elevated titers of antifilarial antibodies per se account for only a proportion of the increased levels of serum immunoglobulins observed in patients with lymphatic filariasis. Filaria-specific T cell clones augment the in vitro production of specific antibody and of "nonspecific" immunoglobulin by sensitized human B cells. *B. malayi* microfilariae also contain a mitogen that stimulates the B cell helper activity of T lymphocytes from uninfected donors. Thus, both specific parasite antigens and nonspecific mitogenic stimuli may contribute to the hypergammaglobulinemia of lymphatic filariasis.

Polyclonal activation could favor parasite survival in several ways. It could engage the host's immune system in the production of antibodies or the expansion of T cell clones reacting with worm antigens that are irrelevant to the host-parasite relationship to the detriment of immune responses with protective potential. It could generate antibodies that react with "relevant" antigens, yet are unable to mediate effects that benefit the host, for example, blocking antibodies. It could also produce large amounts of immune complexes that could interfere with both the afferent and efferent arms of immune responses to worm antigens. Immune complexes containing parasite antigen can be readily detected in sera from infected patients or animals and inhibit, for example, the in vitro killing of microfilariae by ADCC reactions.

Immune Suppression

Cellular and humoral immune responses to parasite antigens are more vigorous in amicrofilaremic hosts than in those with patent filarial infections. This results in part from the activation of immunosuppressive regulatory circuits of the host's immune system by parasite molecules. Active suppressor T lymphocytes and phagocytic cells presumed to be suppressor macrophages are present in the blood of microfilaremic patients and in the spleen and lymph nodes of infected animals. Sera from microfilaremic donors contain substances that inhibit proliferation of T cells induced by parasite antigens. Cells from microfilaremic donors also produce less interleukin-2 and gamma interferon than lymphocytes from amicrofilaremic individuals upon stimulation with parasite antigens. These interrelated but evidently independent mechanisms suppress immune responses to filarial antigens without affecting the response to antigens from other sources.

In addition to this type of antigen-specific suppression, parasite extracts also inhibit mitogen-induced proliferation of lymphocytes from infected as well as uninfected donors, a form of nonspecific immune suppression. In animal models, antigen-specific and nonspecific modes of immune suppression appear to be mediated by distinct cell types that are differentially distributed among organs and tissues of the immune system.

Precisely what activates suppressor cells in microfilaremic hosts remains unclear. Termination of microfilaremia with drugs restores the ability of treated individuals to react to filarial antigens. It is therefore not surprising that antigenic extracts of microfilariae can induce suppressor T lymphocytes. Such extracts contain proteins of high molecular weight that suppress mitogen-induced lymphocyte proliferation; whether the same parasite materials also inhibit T cell responses to filarial antigens remains to be determined. Immunosuppressive worm products of similar high molecular weight are present in peritoneal exudates of jirds and in sera and lymph of nude mice infected with *B. malayi*, suggesting that such molecules could be active in vivo.

Microfilariae are not the only worm stage that contains immunosuppressive molecules: extracts of adult *B. malayi* also suppress the in vitro reactivity of sensitized T cells to microfilarial antigens. This finding suggests that immune responses to later worm stages may be modulated by cross-reactive determinants present in early developmental stages of filarial helminths. Although immune suppression is most pronounced in microfilaremic hosts, the phenomenon may occur much earlier after the onset of infection than is generally recognized and predate the onset of patency. For example, in the *B. malayi* jird model, immune suppression can first be demonstrated 2 to 3 weeks after infection, when third stage larvae first moult in the mammalian host. The onset of immune suppression in human filariasis is difficult to date, but it appears to start soon after previously unexposed individuals immigrate into an endemic area. Comparison of immune responses to filarial antigens in children and adult natives of endemic areas further suggests that the phenomenon increases with prolonged exposure. In addition, pre- or perinatal exposure to circulating worm antigens may induce neonatal tolerance to filarial antigens in offspring of infected mothers. This possibility might explain the marked differences in the clinical course of lymphatic filariasis in native and immigrant residents of endemic areas, a concept supported by studies in animal models.

Filaria-induced suppression of host immunity could benefit the parasite in several ways. One can speculate that the first few larvae of natural trickle infections suppress their host's immune system in order to allow the development of larvae that subsequently invade the same host. This notion is supported by experimental observations. Immune suppression induced by an established infection could also decrease the host's resistance to reinfection and thereby increase worm burdens. This result has been observed in experimental filariasis. For example, adult *L. carinii* transferred to naive cotton rats are rejected, but they survive in previously infected rats. One must recall, however, that infected individuals rarely display overwhelming parasitemias, so this aspect of immune suppression may be balanced by active sensitization to parasite antigens in the repeatedly infected host. On the other hand, although immune suppression would favor parasite survival, it may also be beneficial to the host by reducing the intensity of immune-mediated components of lymphatic pathology. This has been well documented in the *B. pahangi*–jird model, and will be described later in the chapter.

The role of the nonspecific type of immune suppression in the pathogenesis of lymphatic filariasis is difficult to assess. A priori, this suppression should increase the host's susceptibility to environmental pathogens. In contrast to malaria, a disease in which nonspecific immune suppression is prominent, lymphatic filariasis appears not to be associated with increased morbidity or mortality from common infections. In fact, the severity of malaria is said to be reduced by concomitant filariasis. On the other hand, it has been suggested that streptococcal infections contribute to the pathogenesis of elephantiasis by exacerbating preexisting lymphatic lesions. Whether filaria-induced immune suppression weakens resistance to this pathogen and predisposes to the poly-parasitism commonly observed in endemic areas is unknown. Poor responses to some types of vaccines have been described, but these were noted in both infected and apparently healthy residents of the endemic area.

Inactivation of Immune Effector Mechanisms

Microfilariae of various species activate complement through the alternative pathway, a finding that is consistent with the relatively low amount of sialic acid on their surface. Several types of host cells possess receptors for complement, but these cells might not be able to interact with microfilariae covered with C3 fragments that lack the appropriate polypeptide sequences for recognition. Whether this property is a mechanism of immune evasion could be confirmed by detailed biochemical analysis of the complement fragments generated during the activation of C3 by filarial helminths.

Antibodies that promote the adherence of cells to the sheath of *B. malayi* microfilariae are inactivated when added to the worms in vitro. This process leads to a progressive decline in antibody levels in the supernatant of these cultures and reduces the adherence of cells to microfilariae pretreated with antibody. The Fc portion of such antibody is functionally and/or physically destroyed, perhaps by one of the numerous proteases present in and secreted by microfilariae, while the antigen-binding part of the molecule remains attached to the parasite (Table 15-6). This process could be an extremely effective mechanism by which microfilariae resist ADCC reactions to antigens on their surface. It might explain, for example, why microfilariae covered with "immunoglobulin" continue to circulate in natural hosts. Reexamination of this finding with antisera to defined portions of antibody molecules might very well confirm that the phenomenon of antibody inactivation occurs in vivo as well as in vitro.

Recent studies indicate that filariae contain enzymes that convert arachidonic acid into pharmacologically active products. These results raise the possibility that these

TABLE 15·6
Functional inactivation of antibodies by *B. malayi* microfilariae

Duration of preincubation with antibody (hours)	Antibody portion detectable[1]	
	$F(ab')_2$	Fc
2	Yes	Yes
24	Yes	No

[1] $F(ab')_2$ was detected with a fluorescein-labeled antibody; Fc, with ^{125}I-labeled protein A.

parasites might be able to inactivate host-derived mediators of inflammation. A similar mechanism may explain the relative resistance of *B. malayi* microfilariae to oxygen-derived products generated by host cells that are toxic to other parasites and kill mammalian target cells.

Finally, parasite-mediated induction of isotype-specific helper T cells, such as those that contribute to the high levels of nonspecific IgE present in sera from individuals with helminth infections, might contribute to the survival of filarial worms. Saturation of cell surface receptors with polyclonal IgE directed against nonparasite antigens could effectively prevent the triggering of mast cells by filarial antigens and thereby avert the possible damage to the parasite that would otherwise be caused by the cross-linking of filaria-specific IgE on mast cells by parasite antigens.

Other Mechanisms of Immune Evasion

Several other mechanisms of immune evasion have been suggested, but they remain to be evaluated experimentally. Filariae probably do not show the sort of antigenic variation described for some protozoa, but third- and fourth-stage larvae of *D. immitis*, for example, differ greatly in the structure, charge, and antigenic makeup of their respective cuticles. Thus, the moulting process renders obsolete immune reactions to third stage larval antigens and exposes the host to previously hidden antigenic determinants. Whether rapid turnover of the surface of other parasite stages contributes to their survival is not known. Developmental changes in the properties of microfilarial sheaths have been detected, but there is no conclusive evidence for turnover of the epicuticular surface of filarial helminths between the larval moults and in the adult stage.

In addition to developmental changes in the surface properties of filarial worms, the complex structure of filarial surfaces per se undoubtedly provides a certain degree of innate resistance of these worms to certain types of immune effector mechanisms. Cytolytic T cells are believed to kill by "puncturing" the lipid bilayer membrane of their targets. Controversy exists as to whether filariae are bounded by a "true" membrane, but it is difficult to imagine that T cells would be able to invaginate the filarial cuticle. Similarly, the microfilarial sheath, which is rich in chitinlike material, may confer upon microfilariae an additional degree of innate resistance to immune effector mechanisms.

Finally, an individual's ability to recognize stage-specific antigens must influence the outcome of exposure to filarial worms. So far, none of the genetic factors that influence an individual's ability to control filarial infections has been definitively linked to immune response genes or to genes encoded in the major histocompatibility complex. However, the duration of microfilaremia in animals with subtle defects of the immune system generally exceeds that in their immunocompetent counterparts and varies among genetically different strains of the same animal species. This difference suggests that filarial parasites are able to exploit to their advantage some as yet poorly defined characteristics of immune responses to parasite antigens that differ among members of genetically heterogeneous populations.

Immunopathology of Lymphatic Filariasis

W. bancrofti, *B. malayi*, and *B. timori* cause a broad spectrum of acute and chronic disease manifestations in human beings. Signs of acute filariasis include recurrent episodes of fever accompanied by acute lymphadenitis, retrograde lymphangitis, and transient lymphedema; orchitis, epididymitis, and funiculitis; and acute asthmalike attacks characteristic of the tropical eosinophilia syndrome. Chronic manifestations of lymphatic filariasis consist of persistent lymphedema, elephantiasis, hydroceles, and chyluria. Immune complex–mediated glomerulonephritis, endomyocardial fibrosis, and multijoint arthritis are uncommon sequelae of filarial infections.

Pathogenesis of Lymphatic Lesions

The lymphatic pathology resulting from experimental filarial infections in animal models has been described in detail and appears to be similar to that noted in the surprisingly few available human biopsy and autopsy materials. These descriptive studies have popularized several concepts of the pathogenesis of lymphatic filariasis.

First, it is generally believed that adult worms are the primary cause of lymphatic pathology. Gross lymphatic lesions generally are associated with adult parasites; microfilariae appear to induce little, if any, lymphatic pathology. Viable adult worms in dilated lymphatics are usually not surrounded by inflammatory infiltrates, but tissues containing dead and degenerating worms characteristically show a granulomatous inflammation with partial or complete occlusion of the lymphatic lumen. The lymphatic wall may be thickened; interstitial and intraluminal fibrosis are frequently seen in what are considered to be longstanding lesions.

However, other parasite stages may contribute to the lymphatic pathology of filarial infections. Minor alterations in lymphatics occur as early as 3 days after injection of infective larvae of *B. pahangi* into cats, and continued heavy exposure to infective larvae may exacerbate symptomatic lesions. Studies in other animal models further suggest that moulting fluids, products of preadult worm stages, or substances released by female worms may initiate the observed pathological changes, but this possibility has not yet been tested experimentally. Concurrent bacterial or fungal infections can exacerbate worm-induced lesions in experimental animals, a finding that may be relevant to the situation in endemic areas where individuals at risk often are poorly shod. Precisely what triggers the recurrent episodes of acute adenitis and lymphangitis in human filariasis remains to be determined. It also is unknown if the pathogenesis of these acute events is the same as that of lesions associated with chronic elephantiasis.

In marked contrast to the pathogenesis in lymphatic filariasis, most pathological changes in the skin and eyes of individuals infected with *O. volvulus*, the cause of river blindness in humans, appear to be triggered by the microfilarial stage of this parasite. These lesions have been attributed to immediate type hypersensitivity responses, to immune complex reactions, or to direct "toxic" effects elicited by *O. volvulus* microfilariae. Adult *O. volvulus* induce the formation of subcutaneous onchocercomata. These circumscribed fibrous nodules contain most female worms and consist of variable granulomatous reactions containing macrophages, T lymphocytes, plasma cells, fibroblasts and blood vessels, and a fibrous capsule.

A second concept is that the location of adult filariae determines which anatomical sites are affected. This criterion would explain, for example, why urogenital lesions are common in bancroftian but rare in brugian filariasis. Different species of filarial larvae evidently manifest characteristic tropisms for certain body sites that influence the ultimate localization of the adult worms. It is also thought that the lymphatic lesions caused by any of the three human filariae have the same pathogenesis, irrespective of the clinical manifestations they elicit.

Observations in traditional models of lymphatic filariasis have also led to the widely accepted idea that lymphatic damage and the resulting clinical manifestations of filariasis are caused by host immune responses to antigens released from degenerating adult worms. For example, treatment with immunosuppressive agents reduces the degree of lymphatic pathology. Furthermore, cell-mediated and humoral responses to antigenic extracts of adult worms are generally much higher in symptomatic individuals than in those with asymptomatic microfilaremia. Such descriptive studies provide no direct proof, however, that host reactions to parasite antigens cause the acute or chronic disease manifestations of lymphatic filariasis.

The traditional view that the lymphatic pathology of filariasis is caused almost entirely by host immune responses to dead or dying adult worms is challenged by recent findings in the *B. malayi*–nude mouse model, which indicate that dilation of lymphatics and chronic lymphedema can develop in the absence of detectable host immune responses to filarial antigens. The pathologic changes in nude mice parasitized with adult *B. malayi* are similar to those observed in other host species, except that cellular infiltrates into interstitial tissues and parasitized lymphatics are notably absent. Lymphatics of nude mice harboring adult worms become massively dilated and tortuous, a process that resembles early lymphatic changes in human filariasis and that is reversed by surgical removal of the worms. A minority of these mice develop chronic lymphedema of limbs drained by the affected lymphatics. Lesions of comparable severity are not induced by infections with *B. pahangi* or *B. patei*, an observation reminiscent of the marked geographic differences in the prevalence or severity of chronic lymphatic filariasis in various endemic areas. These findings suggest that adult *B. malayi* trigger lymphatic pathology by the direct action of as-yet-unidentified worm products on host tissues in the absence of filaria-specific immune responses.

With the exception of inflammatory reactions to the parasite, most clinical and pathological manifestations of chronic filariasis — even the secondary hyperkeratotic and fibrotic changes in the skin — also occur in nonfilarial lymphatic disease and lymphedema from any cause. Studies in nonfilarial models of lymphedema suggest that chronic dermal backflow of high-protein lymph is sufficient to cause these lesions. The protein content in dermal lymph obtained from *B. malayi*–infected nude mice is high. If the protein concentration of lymph or hydrocoele fluid were similarly elevated in human filariasis, this concept would provide an attractive alternative to the traditional views on the pathogenesis of chronic lymphatic filariasis. The idea that lymphostasis may be the pathogenic mechanism that leads to lymphedema and elephantiasis in filarial infections is also supported by clinical observations: the creation of lymphovenous shunts to drain lymph accumulated in affected limbs results in a rapid, dramatic regression of advanced filarial elephantiasis, a condition once assumed to be irreversible.

Lymphedema and dermal backflow in nonfilarial conditions can occur without mechanical blockage of lymph channels. The same is true in the *B. malayi*–nude mouse model of filariasis. Most likely, dilatation of lymphatic vessels with concomitant loss of valvular function and flaccid paralysis greatly reduce the ability of lymphatics to drain interstitial fluids. This process might particularly affect tissues in close proximity to adult worms. These tissues appear to be subjected to the activity of enzymes released by the parasites, as indicated, for example, by the increased rate of collagen turnover in tissues of cats parasitized with *B. malayi*.

These studies suggest that adult worms can initiate some components of lymphatic pathology by acting directly on host tissues, but it is clear that host immune responses to parasite antigens aggravate the pathological consequences of filarial infections. Immune reconstitution of nude mice by administration of cells from mice that have rejected a larval inoculum results in an episode of acute lymphangitis and in the death of a proportion of adult *B. malayi* present in the lymphatics of the recipient. These worms become surrounded by inflammatory cells, and lymphatics harboring parasites become partially or completely obstructed by granulomas in which the worms now are embedded, by aggregates of as yet unidentified cells, or by ill-defined lymph thrombi. Such reconstituted mice also display humoral and cell-mediated immune responses to worm antigens.

Additional evidence for a role of host immune responses in the pathogenesis of lymphatic filariasis comes from studies in the *B. pahangi*–jird model. Lymphatic lesions in jirds infected several times by the subcutaneous route are more severe than in singly infected animals. Presensitization of jirds to soluble parasite extracts increases the severity of lymphatic lesions, which are most marked in animals with latent infections and high antibody titers. In this study, presensitization of jirds did not affect adult worm development; hence, the death of larvae or adult *B. pahangi* or the differences in parasite burdens cannot be invoked to account for the increased severity of lesions in the treated animals. On the other hand, a single subcutaneous challenge inoculum causes less severe lymphatic pathology in jirds already carrying an intraperitoneal infection than in previously uninfected jirds. These experiments strongly support the idea that even though immune responses to filarial antigens may not actually initiate the pathologic process, they almost certainly contribute to the lymphatic pathology. They also suggest that the way in which parasite antigens are presented to the host's immune system may have a dramatic impact on the pathological consequences of filarial infections. These concepts are of critical importance to the future development of vaccines or other forms of immunoprophylaxis.

Pathogenesis of the Tropical Pulmonary Eosinophilia (TPE) Syndrome

The pathogenesis of the filaria-induced syndrome of tropical pulmonary eosinophilia (TPE) markedly differs from that of the lymphatic pathology. Microfilariae play virtually no role in inducing lymphatic inflammation, and the continual clearance of microfilariae in the lungs and other organs is generally not associated with definable clinical symptoms or marked tissue reactions. Perhaps this pattern reflects the ability of this parasite stage to suppress its host's immune system, as described earlier. However, microfilariae appear to be the main cause of the TPE syndrome, which consists of recurrent episodes of asthmalike attacks that often occur at night, elevated serum levels of total and filaria-specific IgE, a marked increase in the number of eosinophilic granulocytes in blood and tissues, and nonspecific but characteristic radiographic changes in the lungs. The syndrome occurs in individuals who appear to be hyperreactive to many filarial antigens, but especially to those derived from microfilariae. In particular, levels of IgG and IgE antibodies to microfilarial antigens are much higher in patients with TPE than in those with other filarial disease syndromes.

TPE most likely results from the orderly succession of two types of immune responses to microfilarial antigens. According to this concept, the syndrome starts when microfilariae are trapped in the lungs and are killed in situ. This first step is believed to be mediated by ADCC reactions that require IgG antibodies to surface antigens on the worms, as described earlier. Dying microfilariae then would release internal antigens and allergens that elicit no obvious adverse reactions in the majority of patients with filariasis. In allergic individuals, however, these parasite molecules would combine with filaria-specific IgE antibody. Antigen-induced cross-linking of IgE on pulmonary mast cells would then trigger the release of these cells' wide array of soluble mediators of immediate hypersensitivity reactions. This process results in the clinical symptoms of recurrent bouts of asthmatic attacks. Thus, TPE is believed to be a form of latent filariasis in an allergically sensitized hyperimmune individual.

Why do only some patients with lymphatic filariasis develop TPE? Serum levels of filaria-specific IgE antibodies are much lower in individuals with lymphatic forms of filariasis than in patients with TPE. This finding suggests that IgE responses to parasite allergens are subjected to immunoregulatory control, as has been described for other types of antifilarial immune responses. In addition, sera from patients with lymphatic filariasis contain blocking IgG (mostly IgG4) antibodies that react with the same parasite molecules as those recognized by IgE antibodies. Reaction of IgG blocking antibodies with microfilarial allergens in the fluid phase most likely prevents the interaction of worm products with cell-bound IgE.

Although patients with and without TPE can be segregated on the basis of in vitro reactions to microfilarial antigens, studies performed so far fail to explain the apparent inability of patients with TPE to generate adequate levels of blocking antibodies, and why only a minority of patients with filarial infections develop this syndrome. Perhaps the syndrome occurs only in patients who are superinfected with or exposed to filarial parasites of animals. Filariae of various species share many antigenic determinants; release of common allergens when animal parasites are killed in a nonpermissive host could conceivably trigger the symptoms of TPE in individuals allergically sensitized by prior infection with human parasites. This idea remains unproved, but at least one human volunteer experimentally infected with *B. pahangi* developed symptoms consistent with the TPE syndrome.

Coda

The specters of drug-resistant parasites and insecticide-resistant vectors motivate the immunologist to develop vaccines to prevent infection with or transmission of the many parasites that threaten human well-being. In the case of complex multicellular organisms such as lymphatic filariae, this formidable task is greatly hampered by our outright ignorance in several critical areas. First, we do not yet understand what accounts for the diversity in infection and pathology among humans in an endemic area. No single animal model adequately mimics the complexity of human filariasis, which renders experimental investigation of this phenomenon difficult. Is the "exposed, but noninfected" human innately resistant, in the way mice are nonpermissive hosts for *W. bancrofti*, for instance? If so, what would be the genetic basis of such resistance? Or have these individuals mounted a protective immune response, as is suggested by the differential susceptibility of normal and athymic mice to *B. malayi* infections? If so, is the ability to mount such response genetically linked, or does it reflect a particular history of exposure to the parasite?

In the case of infected individuals, it is still not clear what distinguishes those who develop chronic obstructive lesions. Is the pathology a consequence of an inappropriate immune response, as mimicked by reconstituted nude mice harboring *B. malayi* infections? Or is it triggered by a direct ("toxic") action of adult worms on lymphatic tissues and sustained by backflow of high-protein lymph into the interstitium, as models of nonfilarial lymphedema suggest? Would vaccination exacerbate this pathology, as does presensitization of *B. pahangi*–infected jirds, particularly if the vaccine were only partially effective?

Why do so many microfilaremic individuals remain free of chronic lymphatic pathology? Does this state depend on prenatally induced tolerance, or is there a genetically defined susceptibility to the suppressive effects of filarial antigens? Can transmission of filariasis be decreased by controlling microfilaremia, or would such control strategy result in lymphatic pathology or the TPE syndrome?

TPE has also defied analysis by animal models. It is still not clear why only certain individuals mount the underlying allergic response and fail to make the appropriate complement of blocking antibodies. Is the allergic response filarial antigen-specific, or is it a function of generalized hypersensitivity? Would this outcome be determined by the genetic background of the individual or by the person's exposure history, particularly, exposure to animal filarids or pre- or perinatal sensitization to filarial allergens?

The basic predicament that hinders the timely development of antifilarial vaccines is our failure to understand what kills the parasite in certain individuals only. Many phenomena described in this chapter vary greatly, depending on experimental conditions and the host–parasite combination being studied. In no instance have these observations been correlated with quantitative exposure to defined parasite antigens. Thus, control of infection could be genetically linked or could result from particular exposure patterns. The mechanism could be immune-mediated or could reflect a nonpermissive environment for parasite development. These basic questions can be addressed only by judicious experimentation in animal models in parallel with studies on human filariasis.

Additional Reading

DENHAM, D. A., and P. B. MCGREEVY. 1977. *Brugian* filariasis: Epidemiological and experimental studies. *Adv. Parasitol.* 15:243.

EVERED, D., and S. CLARK (eds.) 1987. *Filariasis. Ciba Foundation Symposium No. 127*, John Wiley and Sons Ltd.

HAQUE, A., and A. CAPRON 1986. Filariasis: Antigens and host-parasite interactions. In *Parasite Antigens*, T. W. Pearson, ed. Marcel Dekker, pp. 317–402.

OGILVIE, B. M., and C. D. MACKENZIE. 1981. Immunology and immunopathology of infections caused by filarial nematodes. In *Parasitic Diseases*, vol. 1: *The Immunology*, J. M. Mansfield, ed. Marcel Dekker, pp. 227–289.

OTTESEN, E. A. 1984. Immunological aspects of lymphatic filariasis and onchocerciasis in man. *Trans. R. Soc. Trop. Med. Hyg.* 78:9.

PHILIPP, M., M. J. WORMS, R. M. MAIZELS, and B. M. OGILVIE. 1984. Rodent models of filariasis. *Cont. Top. Immunobiol.* 12: 275.

PIESSENS, W. F., and C. D. MACKENZIE. 1982. Immunology of lymphatic filariasis and onchocerciasis. In *Immunology of Parasitic Infections*, S. Cohen and K. S. WARREN, eds. Blackwell Scientific Publications, pp. 622–653.

SELKIRK, M. E., D. A. DENHAM, F. PARTONO, I. SUTANTO, and R. M. MAIZELS. 1986. Molecular characterization of antigens of lymphatic filarial parasites. *Parasitology* (Supplement) 91:15.

PART III

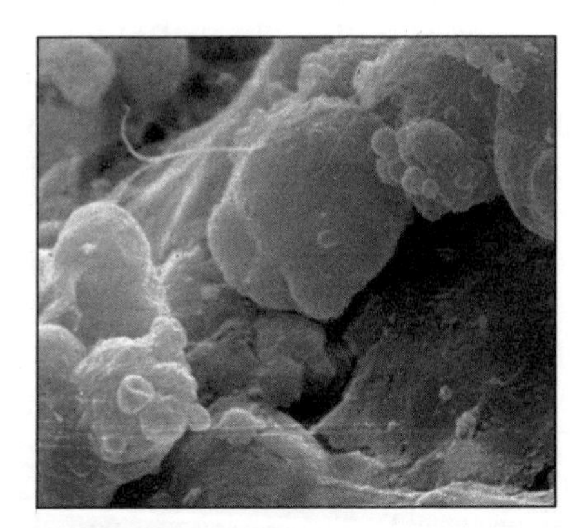

MOLECULAR BIOLOGY OF PARASITES

David J. Wyler

The youngest basic-science discipline applied to the study of parasites and parasitic diseases is molecular biology. The ability to clone parasite genes and express them in *Escherichia coli* or yeast has provided the first glimmer of hope for the development of vaccines against some of the parasites. Before the availability of recombinant DNA technology, there was no conceivable way in which sufficient antigenic material could be produced from any of the medically important parasites for use in vaccines. In only a few years the very rapid application of the new methods led to cloning of plasmodial genes, large-scale antigen production, and phase I and phase II clinical trials of an antisporozoite vaccine (Chapter 16). Although recombinant vaccines against other plasmodial stages and other parasites have not yet progressed this far, their development is underway. From experience with the sporozoite vaccine, it is clear that more than cloning and expressing genes of interest may be required to obtain a successful vaccine. Yet the availability of large amounts of antigenic material and the primary sequence data that can be derived by application of molecular techniques have revolutionized the field. Perhaps the only alternatives to this approach in vaccine development, given the paucity of antigenic material available directly from the parasite, are antiidiotype vaccines (such vaccines employ antibodies [antiidiotypic antibodies] that mimic the structural conformation of an epitope.)

An additional practical application of molecular technology has been in the area of rapid diagnosis. With specific cDNA probes, it is now possible not only to determine the presence of leishmania in skin biopsies, but also to speciate them (Chapter 18). The effectiveness of this diagnostic method is facilitated by the recognition that the extrachromosomal DNA of the kinetoplast contains so-called minicircles, which are highly divergent within the genus, allowing for species discrimination by kinetoplast DNA (kDNA) hybridization techniques. This achievement is an important practical development because conventional diagnostic methods, which require parasite identification by microscopy or isolation of parasites in culture of biopsy material, are cumbersome and often unreliable. Plasmodia also can be detected in blood in this manner (Chapter 16). Although the current level of sensitivity of the molecular method of diagnosis is no greater than that achievable by microscopic examination of blood smears, modifications (that might include amplification by the polymerase chain reaction method) may increase sensitivity. The application of molecular techniques to malaria diagnosis presumably will have greater application in epidemiological surveillance efforts than in clinical practice in areas where malaria is endemic.

Application of molecular biology approaches to the pharmacology of certain parasitic diseases has spurred great excitement and optimism for the idea of developing new antimicrobials by rational design rather than by the traditional and inefficient approach of drug screening. The strategy of this approach is to identify critical enzymes of the parasite that are distinctive from that of the host. Isolating the gene encoding the enzyme, expressing it in *E. coli*, and gleaning the fine structure of the protein by x-ray crystalography—especially its substrate-binding domain—provide the data base (Chapter 19). With this information, drugs might then be designed that can bind avidly to this domain and competitively block access by the substrate. Insights into the genetic basis of drug resistance also have been revealing. For example, resistance to folate antimetabolites in *Leishmania* species has been found to result from extraordinary amplification of the gene that encodes the enzymes that are the targets of such antimicrobial agents (Chapter 18). Molecular pharmacology, as this new discipline is called, could hold great promise for winning the race of medicine against parasites such as *Plasmodia*, which can develop multidrug resistance at a rate that exceeds our ability, by conventional means, to find effective alternative agents. One intriguing strategy is to prepare derivatized oligonucleotides that aneal with specific parasite mRNA, prevent translation, and thereby incapacitate parasite metabolism (Chapter 16).

Although practical application of the new techniques has understandably received the greatest attention, emphasis on basic aspects of the molecular biology of certain parasites has uncovered some unexpected findings of broad biological and evolutionary significance. For example, comparisons of primary ribosomal RNA sequences suggest that the plasmodia causing human malaria are more closely related, in evolutionary terms, to rodent and avian parasites than to parasites causing malaria in other primates (Chapter 16). Similar ribosomal RNA analysis in Kinetoplastida (such as African trypanosomes and *Leishmania*) suggests that these eukaryotes may be among the most distantly divergent, in evolutionary terms, perhaps having branched even before plants (Chapter 17)! Another striking finding has been the unique manner in which these organisms transcribe genetic information: by a noncontiguous construction of mRNA. In contrast to the conventional cis-splicing of most eukaryotes, these protozoa can carry out trans-splicing, a process whereby mRNA transcribed from genome on remote sites on a chromosome—or even on different chromosomes—forms a hybrid mRNA that then is translated (Chapters 17 and 18).

Many interesting biological questions about specific parasites have yet to be subjected to molecular scrutiny. One of the most obvious is gene regulation in the transformation of the parasite from one stage to another. Efforts in this direction have been initiated. For example, expression of the gene for α- and β-tubulin has been found to be expressed as leishmania transform from the intracellular (aflagellate) amastigote stage to the extracellular (flagellate) promastigote stage (Chapter 18). Tubulin is the major protein in promastigotes, perhaps not surprisingly so in view of the presence of an extensive subpellicular microtubular system that includes the flagellum. What triggers gene tubulin gene expression now awaits definition. Along similar lines, an important challenge to the biology of malaria is to determine the molecular basis of gametocytogenesis (the emergence from plasmodial merozoites of gametocytes, the sexual stages; see Chapter 1). Since gametocytes are responsible for initiating transmission in the mosquito, knowledge that could lead to strategies for preventing their genesis in the human host could have distinct practical value. A molecular understanding of the manner whereby specific environmental factors trigger interstage transformation in schistosomes (see chapter 6) is but one example of a worthy goal in molecular helminthology. Indeed, one particularly exciting aspect of parasite molecular biology is that contemplation of parasite cell biology can engender numerous interesting questions that are now, or soon will be, amenable to molecular biological investigation. Furthermore, the parasitic relationship may provide a fertile ground for gaining insights into fundamental evolutionary aspects of gene structure and regulation.

As the molecular biology of parasites matures, it is likely that many important discoveries will be made as a result of asking the right biological questions—with the assistance of some serendipity. Inasmuch as none of the disciplines in modern parasite biology stands alone, new knowledge in the cellular biology of parasites and the immunology of parasitic diseases will no doubt influence perspectives and research directions in molecular biology studies. By the same token, these other areas can be expected to be influenced by progress in basic molecular parasitology.

16

Molecular Biology of *Plasmodium*

■

Thomas F. McCutchan

Protozoan parasites of the genus *Plasmodium* are the causative agent of malaria, a disease that today affects more than 200 million people in the tropics. Of the four human malarias, those due to *Plasmodium falciparum* and *Plasmodium vivax* are the most common. This chapter will focus on the role of molecular biology in the efforts to understand and eradicate malaria and discuss the molecular biology relating to diagnosis, taxonomy, genetics, gene analysis, and vaccine production. The life cycle of *Plasmodium* is reviewed in Chapter 1.

Evolution

Members of the genus *Plasmodium* have historically been grouped according to the hosts that they infect, then subdivided according to morphological and biological characteristics. Thus, for example, the species are classified as primate, rodent, avian, and lizard malarias. This classification scheme implies that the parasites have evolved with their hosts and that there is a greater relatedness among parasites in related hosts than among those greatly separated in evolution. A major exception to this scheme is the human malarial organism *P. falciparum*; this parasite species appears to be more closely related to parasites that cause avian and rodent malarias than to those that cause primate

I thank Dr. Diane Taylor, Dr. Thomas Wellems and Dr. Altaf Lal for supplying material for the figures in this chapter.

malaria. This exception was suggested on the basis of the similarities in the structure and composition of the genomes of *Plasmodia* causing avian and rodent malaria with those of *P. falciparum*. More recently these relationships have been confirmed by comparing the primary sequence of ribosomal RNAs from the different species of malarial parasite. Ribosomal RNA sequence comparisons constitute a generally accepted measure of evolutionary relationship and thereby add weight to these conclusions.

Diagnosis of Malaria

In the standard, time-consuming diagnosis of malaria, a skilled technician does a microscopic examination of blood smears. In many areas diagnosis is omitted altogether; people with malaria-like symptoms are simply treated with antimalarial drugs empirically. The detection of malarial infections in patients by the development and use of a DNA hybridization assay could be a valuable tool in the control of the disease. The success of these assays depends on finding DNA probes that hybridize to the DNA of a particular malarial parasite in a sample of blood and not to the DNA of the host or of other microbes that might be in the same sample. In comparison to molecular diagnosis, microscopy by a skilled technician can detect a concentration of about 20 parasites/1 microliter (μl) of blood. In addition, he or she can determine by microscopy the species of the parasite and degree of anemia. Refining of a molecular diagnosis for malaria to obtain this information will be difficult but clearly possible. The degree of sensitivity that has already been reported for molecular probes approaches 50 parasites per microliter, and increased sensitivity in assays of this type can be expected in the future.

Species-specific probes certainly will be available soon; it is clear that potential diagnostic probes include ones that detect the highly repetitive sequences of *P. falciparum* not found in the other species. It is even possible that similar, although less sensitive, molecular assays may be developed for detecting such other characteristics of a parasite as drug resistance. Perhaps the strength of molecular diagnosis is not in performing better than a skilled microscopist but in allowing a large number of assays to be done with great consistency. A perfected molecular form of diagnosis should allow relatively untrained personnel to do a large number of assays every day with great accuracy.

Plasmodium Genes

This chapter focuses on the three *Plasmodium* genes that have received the greatest attention and discusses reasons for interest in a particular gene as well as the technology involved in its study. An overlying theme in this discussion is how gene structure relates to rapid changes in immunodominant portions of a protein.

The Circumsporozoite Protein

Interest in the circumsporozoite protein (CSP) of malarial parasites is related mainly to the need for a malarial vaccine (see Chapter 9). The important sporozoite antigen, being studied for its ability to elicit a protective response, is a single protein, the circumsporozoite (CS) protein. It covers the entire surface of the sporozoite. The reaction of a fluorescein-tagged anti-CS protein antibody with the sporozoite surface is shown in Figure 16-1. The dominant epitope recognized by antibodies to the CS protein and the one involved in the protective response is included in a single immunodominant region of the protein. The fact that a single antigen could elicit a protective response suggested that a molecular biological approach could provide a source of material for a vaccine. The important gene would be cloned and the product produced by microbes in the laboratory.

Cloning of the circumsporozoite gene Cloning of the CS protein gene presented a particular problem to molecular biologists because of the lack of availability of the sporozoite stage of the parasite. The first CS protein gene was cloned from the simian malaria *P. knowlesi*. To do this, a large number of sporozoites were collected from the

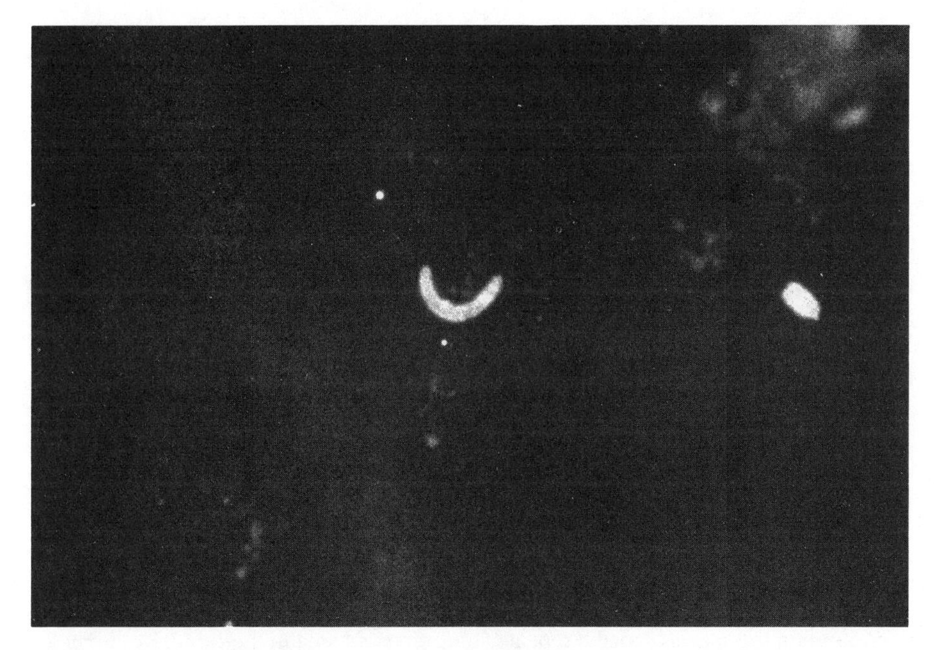

FIGURE 16·1 Immunofluorescent microscopy of a sporozoite with the use of anticircumsporozoite antibody. The distribution of the fluorescinated antibody reveals that the circumsporozoite antigen is distributed over the surface of the parasite. Monoclonal antibodies to the circumsporozoite protein identified this surface component as the immunodominant epitope and facilitated cloning of the gene that encodes the protein.

salivary gland of infected mosquitoes. *P. knowlesi* was chosen because in the laboratory it produces approximately 10 times more sporozoites in an infected mosquito than any of the human malarial species. From these sporozoites, mRNA was isolated and used to make a cDNA library in *E. coli* that contained copies of sporozoite-specific messages. The cDNA containing a fragment of the CS protein gene was identified with a monoclonal antibody to the CS protein. This gene segment was then used to identify the entire CS protein gene. Unfortunately, the *P. knowlesi* gene was of little use in identifying CS genes of other species of malaria, because there were no apparent sequence similarities among genes carried by different species of parasites. It therefore appeared that the most direct approach to isolating genes from different species was to start from the beginning and collect sporozoites from each of the parasites of interest.

Asexual parasites were grown in culture and their DNA was used to prepare an expression library containing intact genes, including those from all stages of the parasite's life cycle. This work was made possible by an earlier discovery indicating that the mung bean nuclease cuts *Plasmodium* DNA in areas flanking genes, thereby reducing the genome to fragments containing intact genes. Bacterial clones containing the CS protein genes were detected by anti-CS protein antibody and standard screening techniques. The clones detected with this approach contained the complete CS protein gene of *P. falciparum*.

Sequence evolution of the circumsporozoite protein The cloning and DNA sequencing of the *P. falciparum* and *P. vivax* CS protein gene allowed comparison of these sequences with that of the *P. knowlesi* gene. Although there were only two small regions of DNA sequence homology among the three, there did appear to be a common structural relationship. Evolutionary conservation of structure at some level was expected among analogous proteins that are presumably responsible for the same functions. For example, concentrations of hydrophobic amino acids at the amino and carboxy terminus of the protein probably serve as signal and anchor regions for the CS proteins. Regions of the three CS proteins that contain a concentration of charged amino acids can also be aligned and probably result from structural restraints on the molecule. The two regions of sequence homology between the *P. falciparum* gene and the other two CS genes are small and flank the repeated or immunodominant portion of the molecule (Figure 16-2). It seems possible that the strong conservation of sequence in these two areas reflected their involvement in some essential parasite function such as reception for liver invasion. The sequence homology between the *P. vivax* and *P. knowlesi* genes is much more extensive and extends over the molecule, except for the central repeating area. In general, at least small regions of conserved sequence in each of the CSP genes flank a species-specific repeating epitope. In Figure 16-2 a schematic of the CS gene shows the position of the highly conserved regions (regions I and II) surrounding the species-specific repeats. This repeated area in the center of the molecule is the immunodominant area of the protein and the subject of the most intensive investigation regarding vaccines. Some typical DNA sequencing data from the analysis of a CS protein gene are shown in Figure 16-3. The data shown in the figure have been read from the arrow on the left in an ascending order to the arrow on the right. The

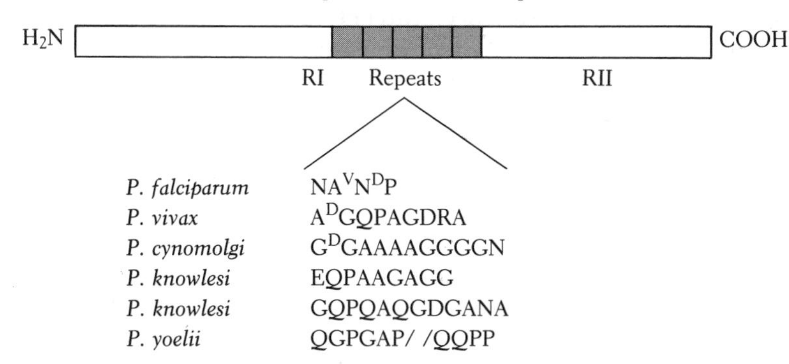

FIGURE 16·2 The immunodominant repetitive epitopes in the circumsporozoite protein of several different species of *Plasmodia*. In contrast to the species specificity of the repeating units, the regions (RI and RII) that flank them are conserved in the genus.

translation of raw data to the gene sequence and then to the protein sequence of the CS repeating epitope is also shown in Figure 16-3. Notice that although the primary DNA sequence varies somewhat from repeat to repeat, only one nucleotide change in the five segments results in an amino acid change. The other changes are in silent third positions, which do not result in the amino acid change.

The extent and rate of variability in the repeat region are important questions in vaccine research. Changes in the immunodominant target would nullify the effectiveness of a vaccine. Perhaps much can be learned from the work done on *P. falciparum* S-antigen, discussed later, which has a repeating immunodominant region similar to that found in the CS protein. Clearly the repeat area of S-antigen varies so rapidly that there are antigenically distinct populations of *P. falciparum*. The CS proteins do not appear to vary as rapidly as the S-antigens, but there are indications of a similar type of drift. The analysis of CS-protein genes from malarial parasites that infect monkeys has

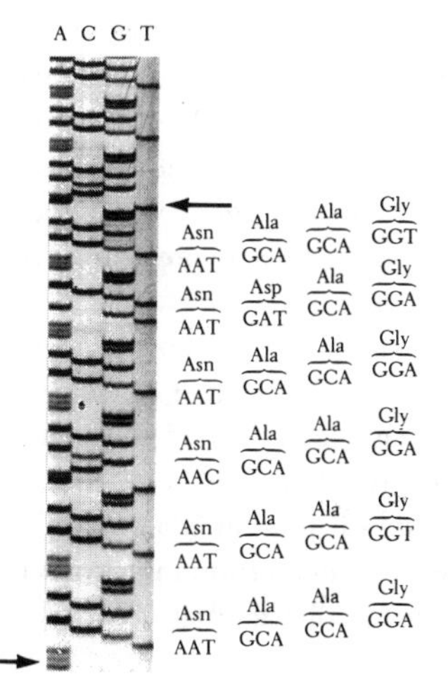

FIGURE 16·3 Representative DNA sequence analysis of a circumsporozoite protein gene. The sequence has been read in ascending order from the arrow on the left to the arrow on the right. Translation of the codons into the amino acid sequence is shown. This type of analysis permits derivation of the primary peptide structure from knowledge of the gene structure.

shown intraspecies variation of the immunodominant region of the protein. Although to date no variation has been seen in isolates of *P. vivax* or *P. falciparum*, this work raises at least two questions. Are there reasons to believe that the *P. falciparum* CS protein gene will be more stable than the CS protein of corresponding monkey malarias? What effect would small subpopulations containing variant CS protein epitopes have on the effectiveness of a vaccine?

The S-Antigen

The S-antigen is a soluble, heat stable protein that is released from the malarial infected erythrocyte as it ruptures. S-antigens circulate in the blood of infected individuals, and the serum of some immune individuals recognize these molecules. Although the expression of a single type of S-antigen is a stable genetic characteristic of a parasite, there are probably a large number of immunologically distinct types. Apparently these immunologically distinct types of S-antigen have a worldwide distribution, since antibody to a mixture of types can be found in a single immune individual and a given type of S-antigen can be found in endemic areas throughout the world. If, as evidence suggests, the frequency of a single type of S-antigen in a particular area varies greatly with time, the dominant form in a single isolated area may vary from year to year.

The S-antigen also may be a target for protective immunity. Studies of patients in The Gambia, West Africa, show a high frequency of different S-antigen serotypes occurring in sequential natural infection, which could indicate that immunity to parasites expressing one S-antigen protect the individual from reinfection by parasites carrying that particular type. Antibodies to S-antigen may then play a role in the gradual development of immunity to malaria. This type of immunity would not extend to protection from the parasites expressing the S-antigen types that the host immune system has not seen before. It may also be that the frequency of individual S-antigens oscillates in a particular area in response to the development of immunity to specific S-antigens by members of the host population.

The general approach that resulted in the isolation of the S-antigen gene was to isolate any gene that produced a product recognized by antibodies of people who are immune to malaria. Individual genes were then characterized and identified by assessing the properties of the protein product of the gene. The S-antigen gene was identified in this manner because the protein produced in the parasite had previously been extensively characterized.

The series of events leading to the cloning of the S-antigen gene proceeded in the following manner. A cDNA library was made with mRNA from a mixture of the asexual stages of *P. falciparum*. The library vector, bacteriophage lambda-gt11, allowed the expression of *P. falciparum* cDNAs as peptides after infection in *E. coli*. Bacterial colonies expressing the *P. falciparum* proteins encoded on the lambda vector were detected with serum from humans that were immune to *P. falciparum* infection. This procedure allowed these researchers to isolate a number of bacterial colonies that contained clones of individual *P. falciparum* genes whose proteins were being recognized by the host immune system. Immunologically specific serum recognizing only a single *P. falciparum* product could then be produced by injecting the bacterially produced product into mice. Having monospecific serum allowed researchers to localize the position of the unknown *P. falciparum* protein on the malarial parasite by an immunofluorescence assay and to characterize its time of expression. It was clear from the immunofluorescence study that one clone from the library produced a product that was synthesized in schizonts, that seemed to be released upon rupture of the infected erythrocyte, and that was isolate-specific. All these features indicated that the clone carried an S-antigen gene. They then used their monospecific serum to identify *P. falciparum* proteins on Western blots. They found that it recognized single proteins of variable size in several cloned lines of parasites. These parasite lines produced S-antigens that were antigenically similar but physically distinguishable by size. By correlating specificity and size they were able to match the serum's specificity to the S-antigen protein.

This series of experiments selected for clones carrying and expressing the most immunogenic portion of the S-antigen molecule. When the cDNA clone of S-antigen was analyzed it was shown to contain a DNA sequence that encoded a sequence of 11 amino acids repeated 23 times. The only nucleotide sequence variation that occurred in the repeating element was located in the third position of codons and did not alter the encoded repeating peptide unit. Although it was clear from this study that there were other areas of the S-antigen molecule that were not composed of this repeat, it was suggested that the repeats were the immunodominant portion of the protein. This work also showed that the nucleotide sequence was isolate-specific. It could be found only in strains of *P. falciparum* expressing one particular antigen type, which indicated that the difference between S-antigen genes was not simply the result of alteration of an existing sequence by point mutation insertions or deletions.

The comparison of the gene sequences of antigenically distinct S-antigens of *P. falciparum* has been detailed. These studies showed that the repeating element of S-antigen was bounded on either side by a nonrepeating sequence. Although the repeating elements of the two genes were different, the nonrepeating sequences were very similar. This pattern is similar to that found in comparisons of CSP genes from different species of *Plasmodium* (Figure 16-2). Hence, S-antigen genes are variable between strains and are therefore varying faster than the species-specific CSP genes. The mode of change, however, appears to be similar. Variation in the S-antigen form does not appear to be occurring by rearrangement of sequences, in that the repeated sequence of one gene cannot be found in the genome of another line of parasite expressing a different S-antigen. Although the precise mechanism that generates diverse forms of S-antigen is not understood, it seems to involve simple sequence alterations as one repeating unit of a gene is spread to other units. For example, any changes to the common theme or base pattern of a repeating element are generally reproduced exactly in several other units in the string of repeats. Since it is unlikely to alter each of several units identically by chance, a mechanism of spreading of single sequence alterations has been proposed. Sequences surrounding the coding portions of the genes were homologous; which indicates that the genes are allelic and that polymorphism exists, because the repeated areas can change more rapidly than their surroundings.

The study of the S-antigen genes has revealed several important facts. (1) The structure of the immunodominant epitope has been elucidated. This information has allowed researchers to compare the S-antigen gene structure with other DNA structures that are constructed similarly. (2) Variations in types of S-antigen are not linked to rearrangement and expression of preexisting sequences, as is the case with mammalian immunoglobulin genes or the trypanosome variant surface antigen. A single parasite therefore does not have an infinite repertoire of changes with regard to the S-antigen. (3) The definition of differences between antigenically distinct types indicates that different S-antigens consist of alleles of a single gene rather than a family of different genes. (4) The mechanism that creates diversity has been partially revealed. We know that the repeating immunodominant epitope is changing faster than the rest of the molecule. The mechanism apparently disseminates local changes in one repeat unit to neighboring units.

The mechanism and rate of change of the S-antigen may give us a valuable lesson by providing an extreme example of the way in which other repetitive antigens can vary. It is unlikely that the accumulation of point mutations in the repeating element of S-antigen occurs any more rapidly than it does in any other repeating element. The question is, Why is the result of this spreading of the initial base change through the S-antigen repeats apparently more frequent? In a population of parasites a great variety of amino acid sequences may be acceptable in S-antigen, whereas sequence variation in the circumsporozoite gene, for example, may be selected against. It also may be that the rate of change is inversely proportional to the amount of host immune pressure. In any case, the different repetitive antigens of *Plasmodium* probably change by similar mechanisms.

Histidine-Rich Proteins

Proteins of unknown function in several species of malarial parasites are characterized by an astonishingly high content of histidine. *Plasmodium lophurae* (an avian parasite) synthesizes one such protein (HRP), which has a histidine content of 73 percent. It may constitute as much as 50 percent of the total protein in asexual forms of the parasite. The protein is found in cytoplasmic granules. One of the things about the HRP that has particularly interested researchers is the report that this antigen protects ducklings against infection with *P. lophurae*. Another parasite, *P. falciparum*, has at least three histidine-rich proteins. One of these proteins (PfHRP-I) is associated with the formation or function of the knoblike protrusions on *P. falciparum* – infected erythrocytes. An example of localization of PfHRP-I on these knoblike protrusions by immunoelectron microscopy is shown in Figure 16-4. In panel A, an antibody specific for the PfHRP-I protein has been labeled with an opaque tag, allowed to react with an intact parasite, and the antibody reactivity localized at the knob is visualized by electron microscopy. Panel B shows a similar experiment using normal mouse sera as antibody; the resulting microscopy shows no localization of antibody over knobs as expected. Panel C repeats the experimental protocol used in panel A, except that the parasite used does not have knobs and hence there is no reactivity. The association between knobs and PfHRP-I is of interest because the knobs mediate cytoadherence of infected erythrocytes to endothelium. Another histidine-rich protein from *P. falciparum* is designated PfHRP-II. It is synthesized throughout the asexual cycle and is released as a soluble protein into culture

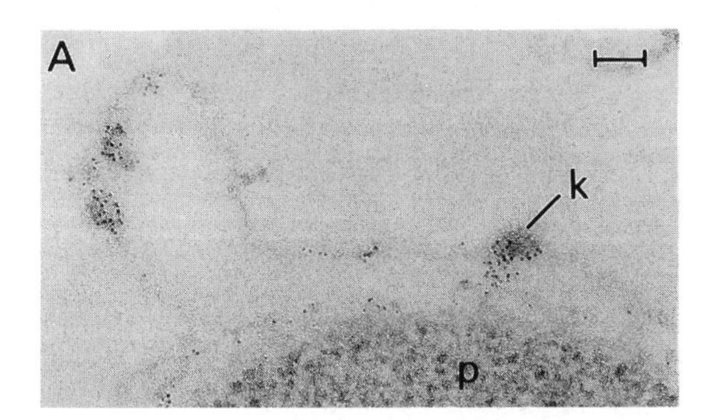

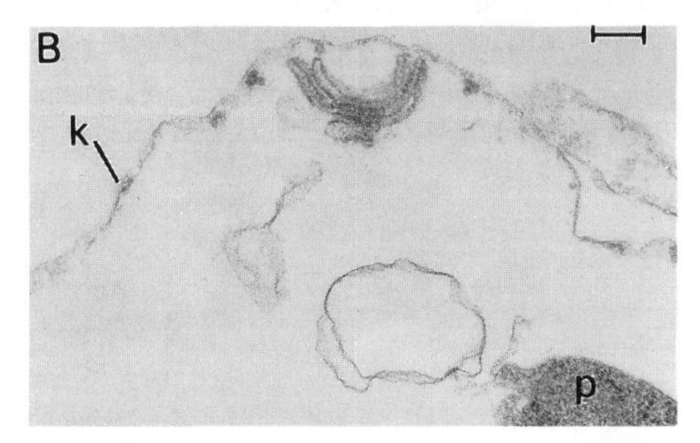

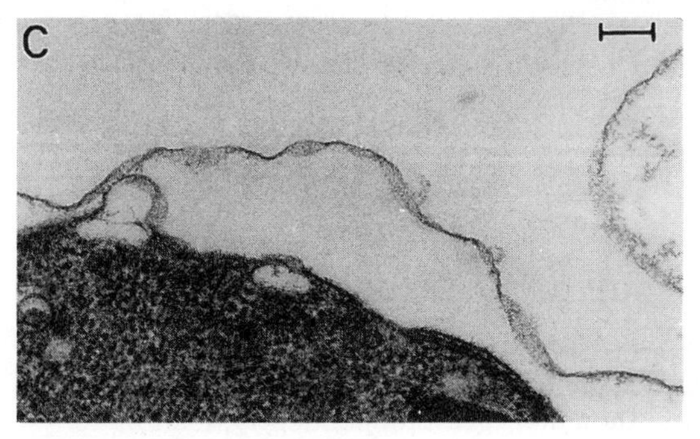

FIGURE 16·4 Immunoelectron microscopy of *P. falciparum* "knobs." A. Antibody to histidine-rich protein has been tagged with dense particles and reacts with the knob (k) of an erythrocyte infected with *P. falciparum* (P). B. Normal mouse serum does not react with knobs, indicating specificity of the interaction in A. C. Antibody to the histidine-rich protein fails to react with erthrocytes infected with a *Plasmodium* that does not induce knob formation.

supernatants from intact cells. There is at least one other histidine-rich protein from *P. falciparum* that remains uncharacterized.

The three histidine-rich protein genes of *P. falciparum* have been isolated. An interesting comparison can be made between the gene sequences of PfHRP-II and PfHRP-III (Figure 16-5). The sequence of the intron and sequence beyond the carboxy terminus of the protein are the same in both genes. Both genes also contain a similar hexapeptide repeat. The hexapeptide repeats vary between genes in only one of six positions. Presumably, one set of repeats arose from the other. Divergence of one region of hexapeptide repeats from the other seems to occur in a uniform way: each repeat unit of the series is altered in the same way. Another set of repeats that are unrelated to the hexapeptide repeats occurs in PfHRP-III. The implication is that both genes were derived from a common precursor by a gene duplication and then diverged independently and assumed different functions. This process, at the very least, establishes the event of gene duplication in *Plasmodium*. It also outlines a mechanism whereby two proteins can share a common immunodominant epitope.

Merozoite Surface Antigens

The merozoite stage of *Plasmodium* invades host erythrocytes to start a new round of asexual replication. The process may involve several independent steps, including (1) initial attachment of the merozoite to the erythrocyte surface; (2) reorientation of the attachment, so that the apical end of the merozoite contacts the surface of the erythrocyte; (3) release of a substance into the erythrocyte from an apical organelle of the merozoite (at this point the shape of the erythrocyte is altered and a vacuole forms

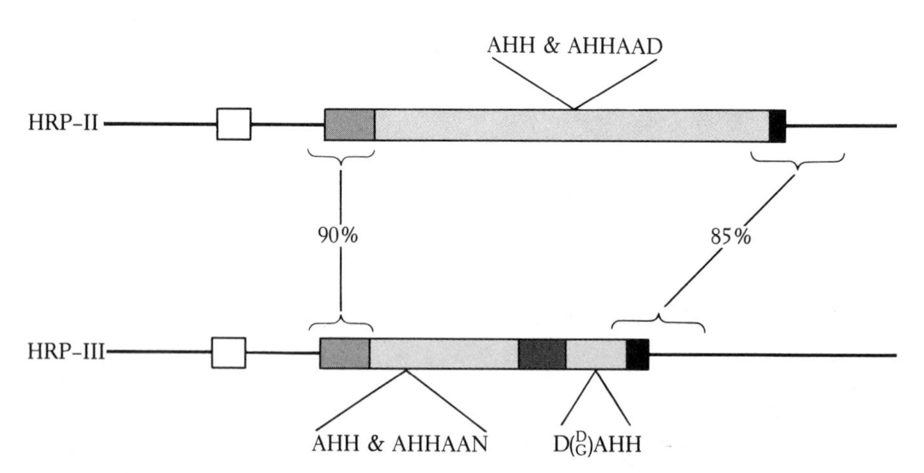

FIGURE 16·5 A comparison of two histidine-rich protein (HRP) genes from *P. falciparum*. It is believed that the two genes were derived from a common precursor by gene duplication and then diverged independently, leading to the differences in the sequence of one region of the hexapeptide repeats shown.

opposite the apical end of the merozoite); and (4) entry of the merozoite into the vacuole. Antibody against epitopes of important antigens involved in any of these steps could be a target for protective immunity. The genes of three *P. falciparum* antigens that are potentially important for vaccine development and are associated with the merozoite stage of development have been cloned. All three can serve as immunogens and give a protective response that blocks merozoite invasion when administered with Freund's complete adjuvant. This does not mean that they would necessarily be useful candidates for a vaccine. It is not known in any of these cases if the protective response is purely antibody mediated, which would be the ideal situation in terms of a vaccine. Furthermore, not all antibody responses would be useful in a vaccine. Ideally, a neutralizing antibody that can block invasion at low titers is needed.

Glycophorin A appears to be a merozoite receptor molecule on the surface of the erythrocyte. A glycophorin A binding protein (GBP) of *P. falciparum* has been identified which is probably involved in the initial binding of the merozoite to the red-cell surface. This protein is on the surface of the merozoite, binds glycophorin, and appears to recognize the surface of the erythrocyte. Antibodies to this protein block invasion of erythrocytes; however, the involvement of this particular protein in reception is unproved. The gene for the GBP has been cloned and the DNA sequenced. The protein contains a 50-amino-acid repeating segment. Preliminary evidence suggests that the structure of the repeat does not vary among isolates of *P. falciparum*.

A merozoite-associated protein has been identified that may be localized to one of the apical organelles involved in erythrocyte invasion. Antibody against this protein stained the surface of ring-infected erythrocytes, but the stain disappeared as the parasite matured. Again, antibodies against the protein block invasion.

A protein of molecular weight (M_r) 195,000 that is synthesized in schizonts is the precursor to three major surface proteins of the merozoite. The function of the proteins is unknown. Preparations of the precursor molecule can induce protective immunity against blood-stage parasites. One of the three proteins (M_r 83,000) is shed when the merozoite invades the red cell. This molecule contains a repeating DNA sequence that may represent an isolate-specific epitope. Two other proteins remain in associated with the ring after invasion. The particular antigenic sites that induce protection have not been defined. Whether or not isolate-specific sites are involved in the protective response also has not been clarified.

Asexual Erythrocyte Parasites

The proteins that are produced by the malarial parasite but reside on the surface of the infected erythrocyte may be an important target of protective immunity. For example, parasite-encoded proteins may be localized in knoblike protrusions on the surface of infected cells that bind to host endothelial cells (a process called cytoadherence). Cytoadherence prevents erythrocytes containing mature parasites from circulating in the peripheral blood and thereby protects them from filtration by the spleen. Analysis of genes whose products are directly involved in this binding might provide a good vaccine target. Antibodies to one of these proteins might block sequestration of parasites and allow them to be cleared from circulation by the spleen.

Sexual-Stage Parasites

Other possible vaccine targets are proteins associated with the sexual cycle of the parasite. It has been shown that the transmission of malaria through the mosquito can be blocked by antisera directed against malarial gametes and zygotes. The specific malarial proteins involved in eliciting such a response have been identified. A vaccine involving these protein products would not prevent the disease directly but would prevent its spread. The hope is that the reduction of transmission of the disease would lead to its eradication.

Genetics of Malaria

The production of an appropriate antigen is one step in producing an effective vaccine (see Chapter 9). Presently four different sources of antigen are being investigated: chemical synthesis, microbial production, vaccinia vaccine, and avirulent *Salmonella* vaccine. Many of the advantages and disadvantages of each approach relate directly to the biology of the vaccine target. The specific example used here is the production of antigen for a circumsporozoite-derived vaccine, and the relationship of approach to CS protein biology. The specific considerations involved in the production of an antigen will of course vary with different vaccine targets.

Chemical synthesis of the repeating epitope of the CS protein appears to be the most direct approach to producing a vaccine antigen. One major objection to this approach, based on work in other systems, is that short, chemically synthesized peptides often do not display an antigenic site in the manner that a complete folded protein would. Initial experiments with chemically synthesized peptides from the repeating epitope of the CS protein indicated that as few as two repeats containing seven amino acids weakly mimicked the antibody-binding site of the complete CS protein. Three repeats or twelve amino acids appear to constitute a complete epitope. This suggested that overall configuration of the CS protein was not necessary to display the appropriate antigenic site and that short, chemically synthesized molecules might be useful in a vaccine. This will not necessarily be the case with other antigens.

Microbial production of antigen based on a recombinant DNA approach has been used by two groups. The *P. knowlesi* CS protein has been produced in yeast cells. A protein that is approximately the size of the precursor CS protein was detected. These researchers report production of 1 microgram (μg) of stable CS protein per 20 milligrams (mg) of total protein. The bacterial fermentation approach is also being investigated. The complete CS protein could not be synthesized in bacteria as it was in the yeast system. Clearly products were rapidly broken down in vivo. But when a smaller segment of the immunodominant region was expressed in the bacterial vector, the products were stable and the yields were nearly 5 percent of total protein, or several thousand times better than the yeast system. These researchers went on to show that antisera raised to their product block hepatoma cell invasion by sporozoites in vitro.

Work directed toward a vaccinia-derived sporozoite vaccine is also proceeding. The gene coding for the CS protein of *P. knowlesi* has been inserted into the *Vaccinia* genome. Cells infected with these recombinant viruses produced a protein that reacted with monoclonal antibodies against the CS protein. Also, rabbits infected with virus produced antibodies against sporozoites. To date, studies of successful vaccinations have not been reported. Some success in producing recombinant vaccinia vaccines for other diseases makes this system look like a promising approach for a malarial vaccine.

Another potential "live vaccine" involves the use of nonvirulent, nonreverting strains of *Salmonella* to deliver antigens. These strains of bacteria could be genetically engineered with any antigen or group of antigens. They grow in the laboratory but cannot grow in the host, because they depend on media supplements that do not exist in host tissue. They apparently exist long enough to elicit an immunogenic response in the host but do not grow and divide. Vaccines potentially could be delivered orally, and there has already been a great deal of testing of the system in the veterinary field. A live bacterial vaccine, however, is another level of increased complexity over the vaccinia system. The fear of loss of control over a "live vaccine" therefore increases.

Although antigen for a sporozoite vaccine can be produced in a number of ways, host immune responses to varied products will not be identical. Thus, a few considerations with regard to antigen production for a sporozoite vaccine relate to the epidemiology of malaria. The first consideration is the dynamics of infection. The sporozoites that are injected into the blood reach the liver within minutes, where they may be protected from the host immune response. If only a single parasite escapes the immune response, the infected individual will contract malaria. Subsequent stages of the parasite's life cycle would not be affected by antibodies directed against sporozoites. Antibody titers must then be held at a very high level to be effective. The immune system must be more than primed to respond: it must be continually producing specific antibodies. The hope for effective long-term protection is that vaccination will produce a lasting response and that there will be natural boosting of this response by the bite of infected mosquitoes. Clearly, the possibility exists that the vaccine response to constructs consisting of less than the entire CS protein may not be boosted by sporozoites contacted in the field. Even the structure of various CS protein may be sufficiently different so that sporozoites from one strain of *P. falciparum* will not boost the response to a CS protein from another strain of *P. falciparum*. In terms of antigen production, these facts favor the live vaccines that use complete gene sequences. The host responses to live vaccines also tend to be more enduring.

Another factor that relates to a sporozoite vaccine is that a purely antibody-mediated response can be protective. This does not exclude the possibility that cell-mediated immunity may be much more effective in protection. Many of the standard arguments against the clean, inexpensive peptide vaccines therefore may not apply in this situation. For example, if only an antibody response is necessary, then the peptide vaccines become attractive even though they would not elicit certain types of T-cell responses.

A final consideration relates to the fact that the group at highest risk is children. Even a short lived vaccine that had to be continually boosted would be of great value in saving lives in this age range. There appears to be no simple definition of the ideal vaccine. Both the live and peptide vaccines have advantages, and a sense of urgency may dictate the parallel development of both.

Drug Resistance

Little work in molecular biology has been done on the origin of drug resistance in malaria. There are two likely candidates for work in this area, pyrimethamine resistance and chloroquine resistance. Pyrimethamine is a chemotherapeutic agent that is used alone in the prevention of malaria and in combination with sulfonamides in the treatment of malaria. This drug inhibits the dihydrofolate reductase (DHFR) of the blood schizonts of avian, rodent, and primate malarias at concentrations that have little effect on the corresponding host enzyme. Some drug-resistant parasites can tolerate levels of the drug at 1000 times this concentration. Although DHFR is a classic example of an amplifiable gene, restriction analysis of DNA from pyrimethamine-susceptible or -resistant lines of parasites showed no obvious gene amplification. Such analysis indicates that resistance is likely to be the result not of an increase in gene copy number but of altered properties of the protein conferring both a different affinity for the drug and an increased stability.

Chloroquine is another commonly used drug in the prevention and treatment of malaria. Most resistant forms that have been studied carry only a two- or threefold level of resistance to the drug, although some greatly increased amounts of resistance have been reported lately. The mechanism of resistance is unknown. Drug resistance seems to be acquired gradually in response to drug pressure. It is assumed that increases in resistance are the result of multiple mutations. The origins of chloroquine resistance may be much more difficult to find than those of pyrimethamine resistance but are of great practical importance.

Contributions of molecular biology to the discovery of new drugs may well be significant in the future. Analysis of proteins such as tubulin may help us focus on central differences between comparable host and parasite enzymes. Potentially, defining these differences will help researchers identify drugs that will selectively kill parasites. Another possibility for the creation of new drugs rests with the use of derivatized oligonucleotides as drugs that will selectively anneal with parasite messenger RNAs and prevent their translation. Neither of these approaches has yielded a new drug to date, but both may have future potential.

Genetics of Malaria

The techniques commonly used to study the genetics of free-living organisms are difficult to apply to malarial parasites, which have a complex cycle involving two hosts; therefore, the amount of classical genetics done in malarial parasites is limited. The number of classical genetic markers is small, and no markers have been reported to be linked. Selectable genetic markers, which would make the recovery of recombinant clones easier, are mainly limited to chloroquine and pyrimethamine resistance. Each genetic cross is an arduous task. Parents with readily definable markers must be cloned and then mated in the mosquito. Progeny of each mating must then be cloned, propagated and analyzed. Even in the face of these difficulties, several important facts have been revealed by classical genetics in malaria. Several markers have been shown to segregate in a normal Mendelian fashion. These studies have also indicated that the blood-stage parasites are haploid and that the diploid stage of the parasite resides in the mosquito after fusion of the gametes to form a zygote.

The use of pulsed-field gradient gel electrophoresis will help resolve some of the genetic questions in the field of malaria. This technique allows researchers to electrophoretically resolve the chromosomes of *P. falciparum* and analyze them individually. Two advances have come from this work to date. A large number of genetically linked markers have been established, and it has been shown that the chromosome sizes of different geographical isolates can vary. The latter may allow the firm demarcation of various taxonomic categories, such as an isolate, strain, and subspecies. Analysis of the parents and progeny of mated parasites will also yield significant information about the genetics of *P. falciparum*.

Coda

Plasmodium is not an especially good model system for studying molecular biology. The complete life cycle of the parasite is difficult to maintain, classical genetic techniques are difficult to apply to the system, and no procedure for transformation has been developed. The research effort has been directed toward understanding and controlling malaria rather than understanding molecular mechanisms in general. The link between understanding molecular mechanisms and understanding the biology of the parasite should, however, not be underestimated. Of considerable importance in this regard is understanding more fully the way in which this parasite evades the host immune system, the amount of natural genetic diversity of the parasites and how it arose, and the evolutionary relationship among the *Plasmodium* species. Most of the genes that have been studied encode antigens that are on the surface of the parasite or are actually released from the parasite. The unique features of these *Plasmodium* genes seem to be the ubiquity of the short repeating epitopes and the frequent sharing of cross-reacting epitopes among different antigens. Clarifying the reasons for these unusual features of *Plasmodium* genes may enhance our understanding of the way that these parasites evade host immunity and may contribute to the production of an effective vaccine.

Additional Reading

BEALE, G. H., R. CARTER, and D. WALLIKER. 1977. Genetics. In *Rodent Malaria*, R. Killick-Kendrick and W. Peters, eds. Academic Press, pp. 213–245.

COCHRANE, A. H., R. S. NUSSENZWEIG, and E. H. NARDIN. 1980. Immunization against sporozoites. In *Malaria*, vol 3, J. P. Kreier, ed. Academic Press, pp. 163–202.

GODSON, G. N. 1985. Molecular approaches to malaria vaccines. *Sci. Am.* 252(5):52–59.

KEMP, D. J., R. L. COPPEL, and R. F. ANDERS. 1987. Repetitive proteins and genes of malaria. *Ann. Rev. Microbiol.* 4:181–208.

MILLER, L. H., R. J. HOWARD, R. CARTER, M. F. GOOD, V. NUSSENZWEIG, and R. S. NUSSENZWEIG. 1986. Research toward malaria vaccines. *Science* 234:1349–1356.

17

Molecular Biology of Trypanosomes

John C. Boothroyd

Through the application of recombinant DNA technology, we have recently gained some insight into the inner workings of the trypanosomes. Few, if any, other biological systems have revealed such a remarkable collection of extraordinary phenomena in such a short period of time. This is a testimony not only to the power of the techniques being used but also to the unusual position these organisms occupy on the evolutionary tree.

This chapter will present an overview of the areas where some of the more interesting findings have been made. As there are many recent reviews on antigenic variation in the African trypanosomes, this topic will receive proportionately less attention. Instead, areas that are just as unusual and interesting but that have not received the same coverage will be emphasized. The chapter is divided into six parts, each representing a functional area of study: genome and genetics, transcription, translation, antigenic variation, mitochondria and respiration, and glycolysis. Most of the work described has been obtained from studies of *Trypanosoma brucei*, but some of the results extend to other species in the order Kinetoplastida. Biochemistry of nucleotide biosynthesis is beyond the scope of this chapter; interested readers are referred to the excellent review noted in the Additional Reading section.

I thank all my colleagues (both in and out of my lab) for sharing their data before publication. I also thank Ms. L. Sargent for secretarial help. Work referred to from this laboratory was supported, in part, by grants from the National Institutes of Health (AI21025) and the MacArthur Foundation. J. C. B. is a Burroughs Wellcome Molecular Parasitology Scholar.

Genome and Genetics

Like most other eukaryotes, all trypanosomes have DNA within two organelles, the nucleus and mitochondrion. The mitochonrial DNA is highly unusual and will be discussed in detail in the section entitled Mitochondria and Respiration.

The genome of *T. brucei* is essentially diploid, with a total complexity of about 4.3×10^7 base pairs (bp). Unfortunately, the chromosomes of trypanosomes do not visibly condense during mitosis (possibly owing to an apparent lack of histone H1); their analysis therefore has awaited the arrival of technologies that do not depend on cytology. Such procedures, termed pulse-field gradient gel electrophoresis, or orthogonal field alternating gel electrophoresis, have recently been developed (Figure 17-1a). These techniques are similar to ordinary agarose gel electrophoresis but are capable of resolving much larger DNA pieces. The method exploits the fact that the time taken for large DNA to reorient in an electrical field that has been rotated through 90 degrees depends on the length of the molecule (although the analogy is biophysically inaccurate, this is akin to the molecules having to repeatedly turn a tight corner). Thus, by switching the orientation of the electrical field from north–south to east–west in repeated pulses (Figure 17-1b), DNA of up to several million base pairs can be resolved.

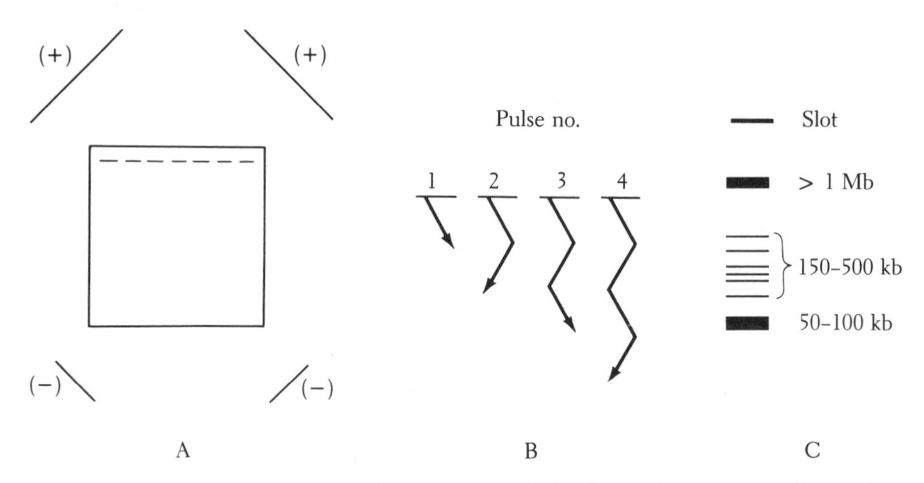

FIGURE 17·1 Chromosome configuration. Part A shows the principle behind orthogonal field alternating gel electrophoresis (OFAGE). A horizontal agarose gel is placed between four electrodes, as shown. Each pair of electrodes (+ and −) is connected to a voltage source in alternating sequence. Under pulse times of a few seconds to many minutes, the DNA slowly migrates into the gel. Ultimately, conventional bands of discretely sized DNA are produced without enormous distortion of a given lane because of the vector sum of the voltage fields, as shown in part B. Part C shows a schematic of trypanosome chromosomes analyzed in this way.

With the use of these techniques, the chromosomes of trypanosomes have been found to fall into four general size classes (Figure 17-1c). The largest is well over 10^6 bp (perhaps even over 10^7 bp) and comprises about 60 percent of the genome. The second set of about 3–6 chromosomes is $1-2 \times 10^6$ bp. Most, if not all, of these two size classes are apparently present at two copies per nucleus. The third class of chromosomes is about $2-7 \times 10^5$ bp. In *T. brucei*, these number about six and are apparently present at only one copy per nucleus. The mini-chromosomes form the fourth and smallest class, with an average size of about $0.5-1.5 \times 10^5$ bp. This class, which is unique to the African trypanosomes, is comprised of the most members, totaling about 100 chromosomes per nucleus (or about 15 percent of the total nuclear DNA). The concept of ploidy is somewhat misleading when applied to members of the mini-chromosome class: they are highly redundant and repetitive, carrying related surface antigen genes, as discussed later.

The chromatin of trypanosomes is apparently conventionally organized in nucleosomes comprised of a histone octamer and about 200 bp of DNA. As yet, however, no H1 histone or equivalent cross-linker has been identified. The chromosome ends (referred to as telomeres) also appear to be conventional, possessing over the last few hundred base pairs a repeated hexanucleotide with occasional nicks. The sequence of this hexanucleotide is similar but not identical to that seen in other protozoa. Preceding this is a subtelomeric sequence that is apparently conserved within the species.

Only a few genes have been mapped to a given chromosome or even class of chromosomes. However, even at this early stage, a trend is clearly emerging, so that "housekeeping genes," or genes involved in routine cellular processes common to most eukaryotic cells, are located on the largest class of chromosomes. These genes are therefore diploid in the trypanosomes. The surface-antigen genes involved in the phenomenon of antigenic variation, on the other hand, are found on all classes of chromosomes, even the mini-chromosomes. Indeed, these smallest chromosomes appears to function exclusively as vehicles for the many surface antigen genes necessary for antigenic variation. This fact also serves to explain their lack in Kinetoplastida species such as *T. cruzi* which do not undergo extensive antigenic variation.

Little is known about the sexual cycle of *T. brucei*, but it is now clear that such a cycle does occur during transmission by the insect vector, the tsetse fly. On the basis of findings to date, it appears that the entire sexual cycle, from meiosis and gametogenesis to zygote formation and subsequent asexual expansion of the population, occurs within the fly. The details of the ploidy of the different stages in this process is still a matter of debate.

During meiosis, it may be that the mini-chromosomes are not evenly segregated (owing to their haploid state in an otherwise diploid nucleus). However, this would not impair the viability of a daughter cell so long as it received sufficient genes to enable it to undergo several cycles of antigenic variation in the mammalian host and thereby survive long enough to be picked up by another fly. Subsequent genetic exchange during the sexual cycle in the fly would enable it to collect a new set of surface antigen genes when the pool is reassorted. There is no definitive evidence for a sexual cycle in the other Kinetoplastida so far studied.

Transcription

One of the most unusual features of the trypanosomes is their mode of transcribing nuclear genes. This process involves the production of a chimeric mRNA by transcription of two segments of DNA that are not physically contiguous.

The first indications of this unusual process came from sequence analysis of the 5′-end of the mRNAs coding for surface antigens. This showed that roughly the 5′-most 35 nucleotides of the mRNAs encoding different antigens were identical and not contiguously encoded with the protein-coding portion of these genes. Synthetic oligonucleotide probes were constructed and used to determine where the mini-exon coding for this spliced leader was located. The results indicated that the mini-exon is part of a 1.35-kilobase (kb) unit (the mini-exon repeat), which is tandemly repeated about 200 times in the genome and is unlinked to the protein-coding exon for these surface antigen genes. In some cases, the mini-exon repeats are on different chromosomes from the surface antigen genes.

It is now known that the mini-exon sequence is present at the 5′-end of most, probably all, mRNAs in all stages of all Kinetoplastida. Although the mini-exon repeat may vary in size and overall composition, the mini-exon sequence is highly conserved. Its exact size in *T. brucei* has recently been shown to be 39 nucleotides plus the usual eukaryotic 7-methyl guanosine in 5′-5′ tri-phosphate linkage. The original estimates of 35 nucleotides were incorrect as a result of the highly modified state of the first four nucleotides, which probably affected the assays used. Given their abundance and presumed functioning in the generation of so many mRNAs, it was not surprising to find that each mini-exon repeat is independently transcribed to give a short, discrete RNA of about 140 nucleotides (in *T. brucei*, shorter in other Kinetoplastida species), the first 39 of which correspond to the mini-exon sequence.

Based on these observations, two basic schemes for the synthesis of chimeric mRNAs were originally proposed. The first involved the use of the med (mini-exon-derived) RNA as a primer in a transcriptional reinitiation event upstream of the protein-coding exon followed by cis-splicing of the resulting molecule. The second model supposed that some sort of bimolecular mechanism operated to combine the mini-exon portion of medRNA with the exon portion of an independently generated transcript from protein-coding genes. Some sort of splicing must be involved because only 39 nucleotides of the medRNA are ultimately found on the mature mRNA. This splicing apparently uses much of the same basic machinery used by higher eukaryotes. This fact is apparent from looking at the sequences located at the exon–intron boundary in the medRNA and intron–exon boundary in the coding gene. These conform to the absolute consensus sequences found in nuclear-encoded genes of all other eukaryotes. Moreover, it has recently been found that the Kinetoplastida have several of the mRNAs involved in cis-splicing in higher eukaryotes.

Recent evidence strongly suggests that the second model, trans-splicing, is correct (Figure 17-2). This model supposes that the medRNA and an acceptor RNA are spliced by a pathway similar to the more conventional cis-splicing, except that the two RNAs

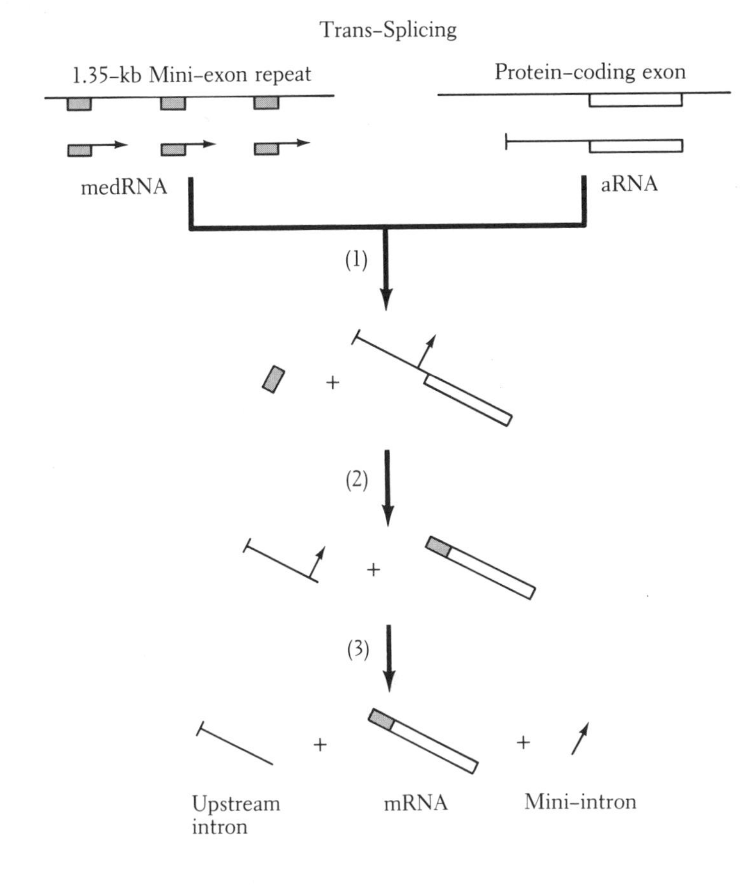

FIGURE 17·2 Model of trans-splicing. The mini-exon repeats direct the synthesis of mini-exon–derived RNA (medRNA); the protein-coding exon directs the transcription of the acceptor RNA (aRNA). The first step in trans-splicing involves formation of a branched species (comprising the intron portion of medRNA (minRNA) and aRNA and free miniexon. In the second step, the two exons are joined so that the two introns are released as a discrete, branched species. The third step comprises the debranching of the branched species into the two component introns. This reaction yields three final products: the upstream intron of aRNA, minRNA, and the mature mRNA. (Reprinted, with permission, from SUTTON, R. E. and J. C. BOOTHROYD. 1986. Evidence for trans-splicing in trypanosomes. *Cell* 47:527–535 © Cell Press.)

are never found as a single, covalently linked precursor. Evidence for this model came from the ability of a debranching activity of HeLa cells to release the tail portion of medRNA (the mini-intron sequence) from higher-molecular-weight RNA. It is now clear that the analogy to cis-splicing is accurate because the linkage between the miniintron and the acceptor intron is a 2′-5′-phosphodiester bond and the trypanosomes have a debranching enzyme with the same activity as the HeLa enzyme.

The unusual process of trans-splicing in Kinetoplastida is critically important to our understanding of the way trypanosome genes are regulated. Trans-splicing may have equally important significance, at least to the molecular biologist, in telling us something of the evolution of RNA splicing, an exciting and contentious subject. This suggestion stems from the fact that trypanosomes are also unusual in having no known conventional (cis) introns in the many genes now studied; that is, no protein-coding exons are interrupted by introns. This total lack is without precedent in eukaryotes and begs the question of whether trypanosomes have somehow removed their introns. An alternative suggestion is that they never had them. If true, this idea argues that trans-splicing predates cis-splicing and that the cis-machinery adapted that of the trans-process for its own purposes.

What is the function of trans-splicing and what is the role of the mini-exon sequence itself? One way in which trans-splicing is apparently used to advantage is in the generation of multiple, capped mRNAs from a single, polycistronic precursor transcript, as illustrated for the α- and β-tubulin genes of *T. brucei* in Figure 17-3. This process would explain the high frequency of trypanosome genes found tandemly arranged on the chromosome, sometimes with the same gene(s) repeated several times. The model proposes that transcription begins at a promoter upstream of the cluster and that through a combination of trans-splicing and polyadenylation, mature, capped mRNAs are generated, each coding for a single polypeptide. By this pathway, each mRNA would acquire a cap, which is believed to be a prerequisite for stability and translatability of eukaryotic mRNAs. (Note that without trans-splicing, acquisition of a cap could not occur by any known pathway, because caps can be added only to transcription initiation sites, probably because of their unique 5′-triphosphate end.)

In addition to the tandem reiteration of many genes, the evidence for this pathway is fourfold.

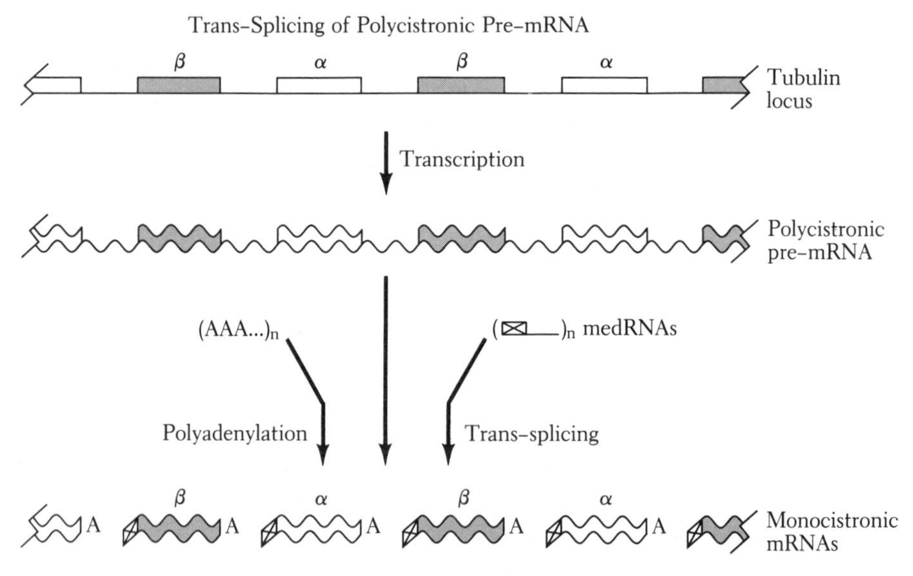

FIGURE 17·3 Model of polycistronic precursor transcripts. The model proposes that a polycistronic RNA is transcribed from a locus composed of multiple coding regions (in this case, the tubulin locus of *T. brucei* encoding both α- and β-exons). Multiple monocistronic mRNAs are generated from this transcript by trans-splicing to yield a capped (medRNA derived) mini-exon sequence at the 5′-end, and by cleavage and polyadenylation to yield the polyadenylated 3′-end. This processing could occur post- or co-transcriptionally, and the order of trans-splicing versus polyadenylation is not known.

1. Heat shock causes the accumulation of polycistronic transcripts.
2. Pulse-chase experiments suggest a precursor–product relationship between polycistronic transcripts and mRNAs.
3. Intergenic regions are transcribed in nuclear run-on assays.
4. Kinetics of UV inactivation indicate some genes are 50 kb downstream of their promoter, with multiple coding regions intervening.

The question of the function of the mini-exon sequence itself is intangible at present because we know of no exception to the rule that all trypanosome mRNAs include this sequence. That such a sequence will prove essential in the translation of these mRNAs seems very likely but tells us little about their function. It would simply confirm that the translational machinery and mini-exon sequence have co-evolved and are now functionally interdependent, just as the bacterial ribosome requires its sequence-specific binding site. However, in the absence of in vitro translation or DNA transformation, it has not been possible to test this idea.

Translation

As in all other eukaryotes, translation in Kinetoplastida is the responsibility of the ribosome, a complex organelle composed of many proteins and RNAs (rRNA). Most of the rRNA in trypanosomes is encoded by a tandemly repeated segment of DNA (about 22 kb per repeat in *T. brucei*; see Figure 17-4). In most eukaryotes the rDNA is transcribed as a single long precursor that is subsequently processed by several cleavages into the 18S, 5.8S, and 28S rRNAs. The Kinetoplastida adhere to this basic pattern, except that the small subunit rRNA (18S-like) is in fact the largest of any known, at about 21S, and the large subunit rRNA (28S-like) is further processed into two medium and several small RNAs. The machinery involved in this processing is not yet identified, but all the cleavage sites have been precisely located. Interestingly, there is no apparent

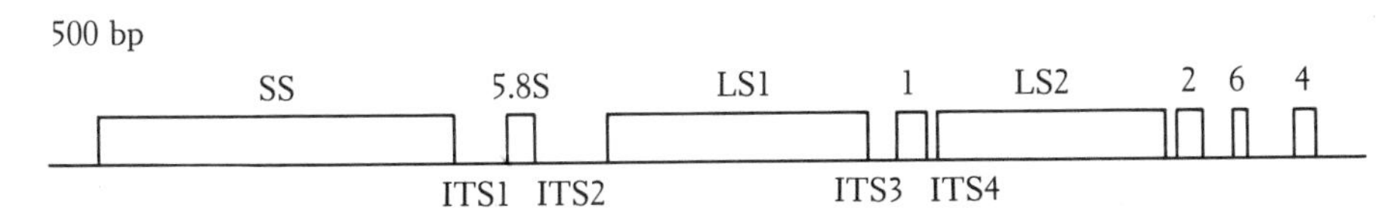

FIGURE 17·4 Organization of ribosomal DNA. A portion of a single rDNA repeat of *T. brucei* is shown. The regions represented in stable RNAs are boxed and identified as small subunit RNA (SS), large subunit RNA (LS1 and LS2), 5.8S RNA, and the small ribosomal RNAs (1, 2, 4, and 6). The intervening transcribed spaers (ITS), which are processed out and rapidly degraded, are also shown. The final component of the rRNA, 5S RNA, is encoded by its own DNA repeat.

commonality in either primary or predicted secondary structure between all sites. The signal for cleavage therefore remains cryptic. Cleavage of the 28S rRNA into two halves is known to occur in a few other eukaryotic systems (e.g., *Drosophila* and *Plasmodium berghei*) but in no other system does this result in the generation of several small, stable RNA fragments that remain associated with the mature ribosome.

Eukaryotic ribosomes generally contain another small RNA with a sedimentation coefficient of about 5S (120 nucleotides). As in other eukaryotes, the trypanosome 5S RNA is encoded by its own tandemly repeated DNA rather than as part of the rDNA repeat described above.

Ribosomal RNA is among the most conserved RNA and as such is frequently used to deduce evolutionary relatedness between organisms. The complete sequence of 5S rRNA for two kinetoplastid species has been obtained. This shows that in addition to being the largest small subunit rRNA yet characterized, the order Kinetoplastida is one of the most distantly diverged (relative to the main eukaryotic line), apparently branching off before even the split into plants, fungi, and animals (Figure 17-5). As we progress further in our understanding of the basic molecular biology of the Kinetoplastida, information on their position on the phylogenetic tree will be increasingly important in enabling us to place new findings in an evolutionary context.

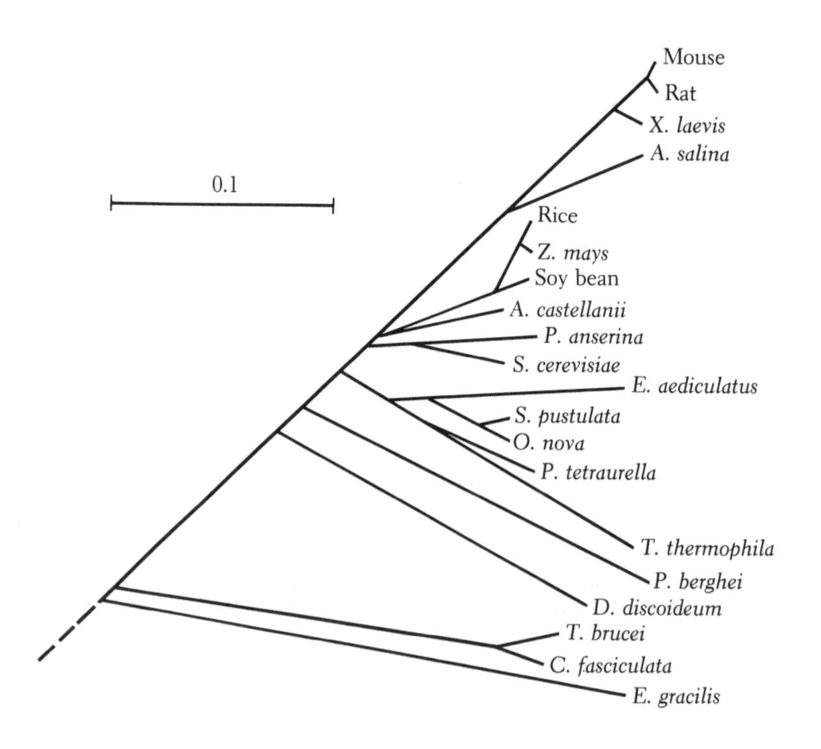

FIGURE 17·5 Placement of trypanosomes on the evolutionary tree. Using the small subunit rRNA sequence as a measure of sequence variation, investigators have devised this phylogenetic tree, showing the placement of the *T. brucei* relative to other organisms, including plasmodia. (Reprinted with permission from GUNDERSON J. H., T. F. MCCUTCHAN, and M. L. SOGIN. 1986. Sequence of the small subunit ribosomal RNA gene expressed in the bloodstream stages of *Plasmodium berghei*; evolutionary implications. *J. Protozool.* 33;525–529.)

Antigenic Variation

This portion of the chapter deals with the extraordinary way in which the African trypanosomes successfully elude the immune system of their mammalian hosts. The biochemistry and cell biology of this antigenic variation is discussed in Chapter 3. The molecular biology will be discussed here.

Antigenic variation describes the ability of the African trypanosomes to alter the antigenic nature of their surface coat. On an individual trypanosome, this coat is generally composed of a single type of abundant glycoprotein molecule known as the variant surface glycoprotein, or VSG. Each parasite contains many hundreds of genes coding for different VSGs. They effect antigenic variation by activating these genes one at a time, so that during the course of an infection initiated by a single parasite, the host is faced with a succession of antigenically distinct populations. No known stimulus triggers this switch; it appears to be spontaneous, occurring approximately once in every 10^6 divisions. Antigenic variation is sustained through the appearance of new antigenic types before the preceding wave has been completely eliminated. It is slow enough, however, to prevent the full repertoire from being revealed to the host too quickly. This prolongment is further ensured by expressing the VSG genes in a nonrandom way, so that for a given strain, there is about a 40 percent chance of one antigenic type being followed by a second, defined type, with the remainder of the switched types largely limited to a small subset of the repertoire.

The molecular basis of this phenomenon is now partially known. First, the expression of a gene is somehow controlled at the level of transcription, with only one VSG gene typically transcribed at any one time. In order for a VSG gene to be active, it must reside in a limited number of expression sites, which are invariably found very near the ends of the trypanosome chromosomes (the telomeres, discussed earlier). A VSG gene enters an expression site by gene conversion (Figure 17-6), a process whereby a gene in one locus (here the expression site) is replaced by a copy of a gene normally found in a different locus. The repertoire of VSG genes in trypanosomes is generally maintained as a collection of basic copy, or BC, genes, which are located at chromosome-internal positions. During gene conversion a duplicate of the basic-copy sequence replaces a gene in an expression site, the new copy being designated an expression linked copy, or ELC. The displaced ELC is permanently lost in this process, but this loss does not generally affect the total repertoire of the parasite, because usually there is a BC corresponding to this ELC at a stable, internal site within the genome (the "white" gene in Figure 17-6). Once in a telomere, the process of activation is unknown. However, several sorts of activation have been described. First, activation can occur directly as a result of the conversion if the target telomere is already active (right side of Figure 17-6). Second, there can be telomere exchange such that the recently created gene is placed on an active chromosome and consequently is turned on. Third, the telomeric copy can in turn convert another, active telomeric locus, resulting in activation. Fourth, there can be no detectable rearrangements; in this case, the new telomeric copy is simply activated in situ, with corresponding switch-off of the previous expression

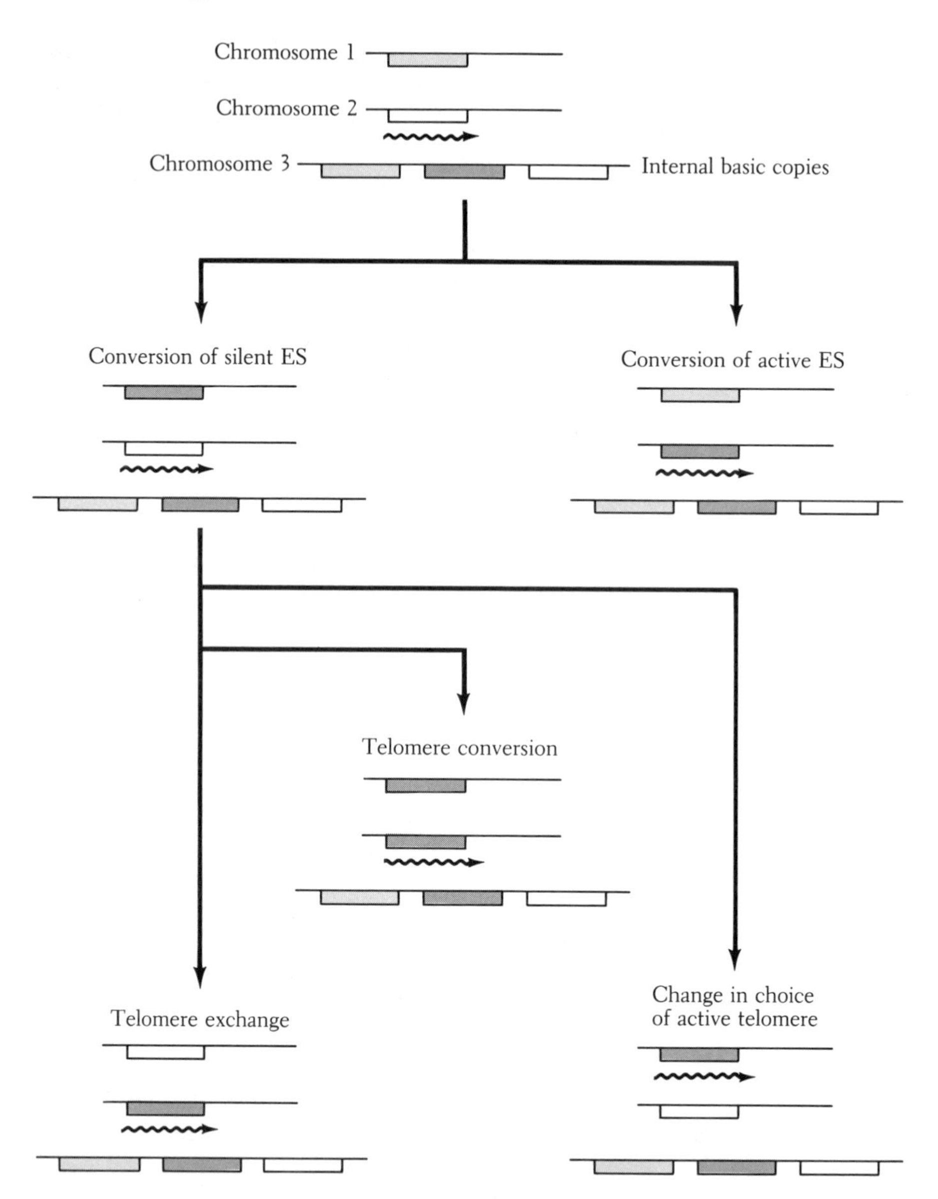

FIGURE 17·6 Models for molecular genetics of antigenic variation. Activation of VSG genes requires that they first be duplicated and transposed into a telomeric expression site. In this drawing, several VSG genes (represented by the boxes) are shown on three chromosomes. The upper drawing represents a parasite expressing the "white" gene in expression site 1 (ES1), as indicated by the arrow. Activation can proceed by a one-step process if the basic copy of the "black" gene is copied into the active ES1, as shown by the right-hand

site. The way in which the trypanosome designates one or another of these expression sites for active transcription is not currently known. The only clue is that the nucleotides of a silent expression site are frequently modified. However, whether this modification is a cause or an effect of expression has not been determined.

As discussed earlier, a given antigenic type is generally succeeded by a discrete subset of the repertoire. This subset seems likely to be partially defined as comprising genes already residing in expression sites. They are, it seems, more likely to be expressed than internal genes which have not yet entered an expression site. The tag identifying the single gene most likely to follow expression of any given gene is not yet known, but it could reside in adjacent sequences rather than in the VSG-coding portion itself.

Mitochondria and Respiration

Trypanosomes have a single mitochondrion that extends over the length of the cell. Like other eukaryotes, many of the mitochondrial proteins are encoded by circular DNA within this organelle. Trypanosomes are exceptional, however, in that their mitochondrial DNA is comprised of two sorts of circular DNA, the maxicircle and minicircle. Together, these two types of DNA comprise a catenated network that, for unknown reasons, is always found near the base of the flagellum. This DNA is easily visualized by basic stains, and before its composition was known, cytologists gave it the name *kinetoplast* for "body associated with motion." As it is found only in the trypanosomes and such closely related protozoa as *Leishmania* and *Crithidia*, it gives its name to the order comprising these genera, the Kinetoplastida. It also gives its name to the mitochondrial DNA of trypanosomes, which is usually referred to as kinetoplast DNA or kDNA.

Despite the unusual structure of its DNA, the fact remains that the mitochondrion must perform its essential function of generating ATP through oxidative respiration. In the trypanosomes, this process is highly regulated. When in the bloodstream of the mammalian host, there is an ample supply of readily converted energy in the form of glucose. The glycolysis necessary to convert the energy in glucose to ATP is handled by an unusual organelle, the glycosome, which will be discussed in greater detail later. Within the actively dividing, long-slender forms of the bloodstream, the mitochondrion is largely quiescent, with no detectable cytochromes or mitochondrial ATP generation. Morphologically, the mitochondrion in this stage is a long, narrow tube with few cristae;

---

pathway. Alternatively, the black gene could convert a silent ES (2), as shown in the left-hand pathway. This would be phenotypically silent with continued expression of the white gene. Following this (by any amount of time), the black gene could be activated by telomere exchange (such that the ES2 becomes activated), telomere conversion (such that a third copy of the black gene is generated, now in the active ES1) or a change in the choice of active telomere with no rearrangements.

biochemically, most of the enzymes and cytochromes necessary for respiration are absent. In the insect vector, the situation is very different, with proline and other amino acids being the major carbon source. In this stage of its life cycle, respiration is fully switched on and the mitochondrion is expanded with many cristae. Transcription of the mitochondrial genome is generally increased in this stage and there is a concomitant increase in the amount of the enzymes and cytochromes necessary for oxidative phosphorylation.

The two sorts of kDNA are very different in both structure and function. The mini-circles are covalently closed circles of 0.7–2.5 kb, depending on the biological species. They have no known coding function and no substantial open-reading frame, but they may be transcribed to give one or more short transcripts. There are about 5000–10,000 minicircles in each kinetoplast network, but these are generally not homogeneous even within a single network. Instead, they are known to be highly variable in sequence, with only short, conserved regions likely to be involved in their replication. Notable exceptions to this variability are *Trypanosoma equiperdum* and *Crithidia fasciculata*, which have nearly homogeneous populations of minicircles.

The other component of kDNA is the maxi-circle, which ranges in size from 20 kb to 38 kb. In other systems this component would be termed the mitochondrial genome. It contains genes for the ribosomal RNAs of the mitochondria (although no such ribosomes have yet been detected), several mitochondrial enzymes or subunits thereof, and some evolutionarily conserved open-reading frames of unknown coding function. As with other mitochondrial genomes, the maxi-circle uses UGA to code for tryptophan (as opposed to using it as a stop codon in the "universal" code).

Comparisons with the mitocondrial genomes of other organisms reveal several properties unique to the Kinetoplastida. First, the putative ribosomal RNAs encoded by the maxi-circle are substantially smaller than those found in any other organism, prokaryote or eukaryote (about 600 and 1150 nucleotides for the 9S and 12S RNAs, respectively). Second, no tRNA genes have yet been identified on any kinetoplastid maxicircle.

The third and most unusual property of the maxi-circle is that several of its genes are "incomplete" yet functional. This paradox describes the fact that transcripts from several of the genes are nonfunctional unless subjected to a remarkable process of uridine addition. This process of RNA editing, which has been observed in three disparate Kinetoplastida species, *T. brucei*, *Leishmania tarentolae*, and *C. fasciculata*, involves the addition of several (over 300 in some cases) uridine residues at precise positions within a transcript both in and out of the protein-coding region. Without such additions, the coding region is not a continuous open-reading frame and may even lack the start codon, AUG (i.e., the corresponding gene encodes only AG at this position, so a U insertion is necessary for functionality). The mechanism of U addition, or even whether it is co- or posttranscriptional, is not yet known. Interestingly, the phenomenon is apparently developmentally regulated in *T. brucei*, so that in the bloodstream forms where respiration is quiescent, the transcripts lack the added Us and are thus nonfunctional. In the insect forms, on the other hand, the transcripts possess the Us and are fully functional, encoding essential mitochondrial enzymes.

Transcription of the maxi-circle is not well understood. No promoters have been definitively identified, and the primary transcription units have not been mapped. The ribosomal RNAs appear to be derived from longer primary transcripts (as is the case in mammalian mitochondria) based on pulse/chase experiments. The situation for the unedited precursors of the protein-encoding mRNAs is not yet known.

The function of the maxi-circle is clear: it encodes several, but not all, of the mitochondrial enzymes, the remainder of these enzymes presumably being encoded in the nucleus. The function of the mini-circle is more enigmatic. Given its variability and lack of open-reading frames, it may be serving as purely structural DNA, perhaps as a scaffolding on which to array the maxi-circle. The structure of such a scaffold could be a stack of parallel, elongated ovals (perhaps supercoiled), like baguettes in a baker's shop. This pattern is consistent with electron micrographs, which show a striated layer with a width equal to the predicted length of one or two mini-circles. At least one of the "sharp turns" could be due to a periodicity in the sequence, wherein a run of adenosines is known to be present at about every 10 nucleotides. Such a nonrandom interval (approximately equivalent to the number of residues in one helix turn) can cause the DNA to bend a few degrees at each adenosine run; the cumulative effect is a substantial turn in a single plane.

Replication of mini-circles first requires that they be freed from the complex by a topoisomerase and then replicated by a Cairns-type (theta) mechanism. The resulting two progeny molecules are reintegrated into the complex, where the nicks that remained after replication are sealed. Maxi-circles, on the other hand, replicate while still part of the network by a rolling-circle mechanism. The linear extension of the rolling circle is cleaved at a discrete site into monomeric units which are not part of the network. The manner by which the linear units are circularized and reattached to the network is not yet known.

The short-stumpy bloodstream forms, which are thought to be the forms infectious to the fly, also have increased levels of maxi-circle transcripts, consistent with the notion that they are preadapting to life in the fly.

Glycosomes

Trypanosomes rely on glycolysis for energy production when in the glucose-rich bloodstream of their mammalian hosts. This process enables them to aerobically metabolize glucose, producing two molecules of the three-carbon pyruvate. This process requires the consumption of four ATP molecules while generating six, resulting in a net gain of only two ATPs per glucose molecule metabolized. Because this process is the major source of ATP to these very motile parasites, the trypanosomes have evolved an unusually efficient system of glycolysis, which metabolizes glucose about 50 times faster than the system in mammals. This efficiency is attained through the enzymes involved being not only highly abundant but also concentrated into discrete, membrane-limited organelles called the glycosomes. These organelles, which are analogous to the microbodies found in other organisms, are about 0.3 microns (μm) across and number 200–300 per cell. They are bounded by a single membrane and constitute about 4 percent of the total volume of the trypanosome and 9 percent of the total cell protein. The nine enzymes involved in glycolysis are the major components of the glycosome. In addition, however, the enzyme machinery involved in other metabolic processes, such as synthesis of ether lipids, beta-oxidation of fatty acids, and biosynthesis of pyrimidines, is also present, although in much lower amounts.

One clear advantage to the compartmentalization of glycolysis is that as the intermediates in the glycolytic pathway are generated, they cannot diffuse too far before being acted on by the next enzyme in the pathway. The intriguing possibility of channeling of the intermediates from one enzyme directly to the next (without diffusion) has proved not to be the case.

The genes coding for the different glycolytic enzymes in trypanosomes are nuclear-encoded (there is no detectable nucleic acid in the glycosome). In general, the genes for the different enzymes are not physically linked. Interestingly, however, like many other nuclear genes in the trypanosomes, some glycolytic enzymes are encoded by two or more tandemly reiterated genes. In at least one case (phosphoglycerate kinase), the individual genes comprising a group are not absolutely identical, with their respective products being localized to the glycosome and cytoplasm, respectively. They are also present in different amounts in the various life stages.

The glycosome enzymes have exceptionally high pIs (9–10) compared with their homologues in the trypanosome cytoplasm and other organisms. For each enzyme, the extra, positively charged residues are found in two topological clusters spaced 40–50 angstroms (Å) apart. It has been suggested that these clusters are the tags that cause these enzymes to be transported into the glycosomes. This intriguing possibility is yet to be tested experimentally.

Coda

Clearly the application of molecular biology to the kinetoplastid organisms has revealed many unusual and unprecedented phenomena. But where do we go from here, apart from filling in the many missing steps in these unusual processes? Determining where trypanosomes fit into the larger picture of evolutionary biology is going to be just as exciting an area for future research. We know they are exceptional in so many processes, but how isolated are they as a group? The trypanosomes promise much in what they can tell us about the evolution of many complex processes whose origins are unclear from studies of more recently evolved systems. All this is in addition to the potential of applying a detailed knowledge of such fundamentally different pathways to effecting new means of controlling this critically important class of parasites. Exciting times await us.

Additional Reading

BORST, P. 1986. How proteins get into microbodies (peroxisomes, glyoxysomes, glycosomes). *Biochem. Biophys. Acta* 286:179–203.

CLAYTON, C. 1988. The molecular biology of the Kinetoplastida. In *Genetic Engineering*, Vol. 7, P. W. J. Rigby, ed. Academic Press, pp. 1–56.

HAMMOND, D. J., and W. E. GUTTERIDGE. 1984. Purine and pyrimidine metabolism in the trypanosomatidae. *Mol. Biochem. Parasitol.* 13:243–261.

RYAN, K. A., T. A. SHAPIRO, C. A. RAUCH, and P. T. ENGLUND. 1988. Replication of kinetoplast DNA in trypanosomes. *Ann. Rev. Microbiol.* 42:339–358.

SIMPSON, L., and J. SHAW. 1989. RNA editing and the mitochondrial cryptogenes of kinetoplastid protozoa. *Cell* 57:355–366.

18

Molecular Biology of *Leishmania*

Dyann F. Wirth

The *Leishmania* organisms are single-celled members of the order Kinetoplastida. The leishmania has two distinct stages in its life cycle, the promastigote stage, in which the parasite is extracellular, flagellated, and motile, and the amastigote stage, in which the parasite is intracellular, nonmotile, and nonflagellated. The promastigote form is found in the insect vector, whereas the amastigote resides in macrophages of the mammalian host. The parasite multiplies in both stages of its life cycle.

Human leishmaniasis is caused by at least 14 different species and subspecies of the genus *Leishmania*. The clinical manifestations of infection and details of host–parasite biology are reviewed elsewhere in this book (see Chapters 5 and 11). Epidemiological studies over the last 25 years have demonstrated that in general leishmaniasis is a zoonotic disease; the parasite is transmitted to humans from a reservoir mammalian host by a phlebotomine sandfly vector during a blood meal. Only a subset of the numerous species of the genus *Leishmania* found to infect mammals also infects and causes disease in humans, presumably because of a combination of factors, including the intrinsic susceptibility of humans and the feeding habits of the sandfly vector. Enormous effort has been devoted to the isolation and characterization leishmanias that infect humans and to the identification of the principal mammalian reservoirs and species of phlebotomine vector. The result of many such studies has been the correlation of particular clinical manifestations with certain species or subspecies of the parasite.

Leishmania species identification is based on a variety of ecological, biological, biochemical, and immunological criteria. Previously, each cultured isolate of the parasite was analyzed by one or more of these criteria and grouped into species and subspecies. Controversies remain as to whether organisms isolated in distant geographic locations but sharing certain common properties belong to the same or distinct subspecies of the genus *Leishmania*. Whether these organisms represent strains of the same subspecies or distinct subspecies cannot be resolved because there is no single generally accepted method for species identification in the genus *Leishmania*. There is no defined sexual stage, so that traditional criteria for species identification cannot be applied. This problem complicates comparison of the disease epidemiology in distinct geographic locations and represents a potential limitation on the transfer of control measures from one geographic location to another.

Recently there has been an interest in molecular mechanisms that might be unique to the parasite and therefore useful either in understanding the parasite itself or certain facets or the host–parasite interaction. The obvious long-term practical goal of such work is the identification of potential new targets for chemotherapeutic attack of the parasite. This area should receive focus because although vaccine work is in progress, both the diversity of *Leishmania* sp. and the potential problems of antigenic change will significantly hamper those developments.

Several questions are of fundamental interest to molecular biologists with regard to the parasite and its life cycle. These include organization, kinetoplast DNA structure and function, developmental control of gene expression, and novel biochemical and molecular processes. Recent work has defined the chromosomal organization of the parasite, but there is no information on the exchange of genetic information in these organisms and the establishment of genetic systems for analysis. This area is just beginning to develop. No sexual stage has been defined for these parasites, and controversy remains concerning even the most basic questions, such as whether the organism is haploid or diploid in either promastigote or amastigote stages. The kinetoplast DNA of the parasite has stimulated particular interest; this fascinating set of DNA molecules is unique to the order Kinetoplastida, and the most extensive molecular characterization of any DNA molecule in the leishmania parasite has been of the kDNA maxi-circle and mini-circle. Control of gene expression in the various stages of the parasite's life cycle has been a focus of research since the parasite exists in vastly different environments within the insect vector and the mammalian host. One major interest for molecular biologists is to determine which adaptive mechanisms the parasite has developed to survive in two distinct environments: the sandfly gut and the host macrophage. Novel biochemical and molecular pathways that are specific for the parasite have attracted attention because their study could lead to new strategies for chemotherapeutic attack.

Genetic Organization and Function

The genome size of *Leishmania donovani* has been determined by Cot analysis and is estimated to be 5×10^7 base pairs (bp) per haploid genome. Recent work using pulse-field gel electrophoresis has demonstrated that the *Leishmania* species have a minimum of 16 chromosomes or chromosome-sized DNA molecules. Cot analysis of the ribosomal RNA demonstrates that there are approximately 150 copies of ribosomal RNA per haploid genome. The organization of the genomic DNA is in the very early stages of analysis.

Separation of large chromosome-sized DNA molecules by the technique of alternating-field gel electrophoresis has allowed the identification of 16 to 23 chromosome-sized DNA molecules in the *Leishmania* species (See Figure 18-1). In general, the chromosomes range in size from 150 kilobases (kb) to greater than 1000 kb. Before this work, it had not been possible to visualize the parasite's chromosomes because, unlike mammalian chromosomes, they are not visible as condensed chromosomes at mitosis. Several laboratories have analyzed chromosomes from different leishmania species and have discovered that there is significant size polymorphism of certain chromosomes between species. This result is not unexpected; different yeast species also show chromosome size polymorphism. In addition, Southern blot analysis demonstrates species-specific variation in the size of specific chromosomes (Figure 18-1).

Several structural genes have now been localized onto specific chromosomes-sized DNA molecules. In this way, a molecular karyotype of several *Leishmania* organisms has been derived. At least five chromosomes have been defined by the location of alpha and beta tubulin genes, rRNA genes, and the gene for the 5′-spliced

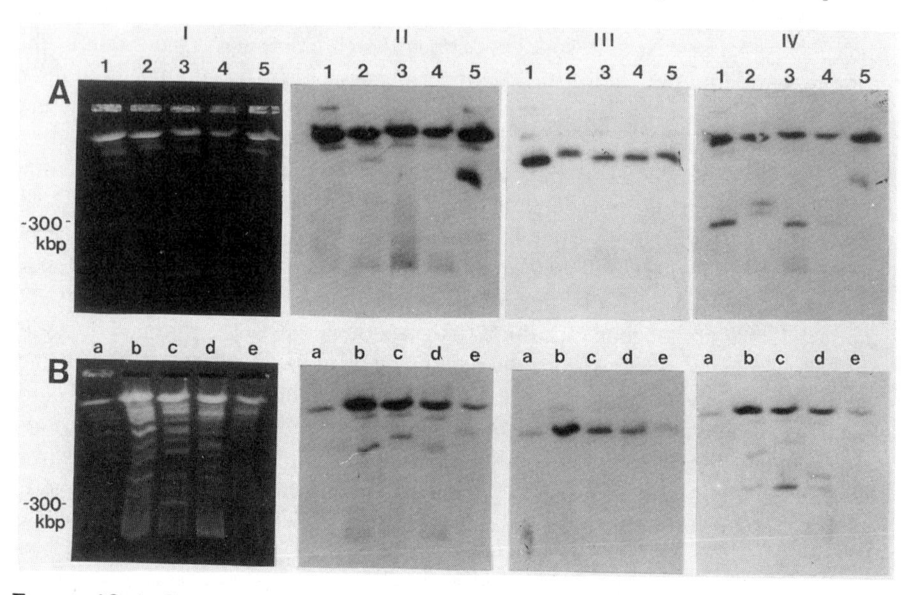

FIGURE 18·1 Survey of molecular karyotypes in a variety of species. Chromosome-sized DNA molecules derived from several *Leishmania* species' promastigotes were separated by orthogonal field alternating gel electrophoresis (OFAGE), pulse time 90 seconds, blotted to nitrocellulose, and probed with the following: plate II, β-tubulin; plate III, α-tubulin (without first washing old hybridization from β-tubulin and 5′-spliced-leader sequence genes); and plate IV, 5′-spliced-leader sequence (without first washing old hybridization for β-tubulin). (A) Lane 1, Josefa; 2, WR563-W50-2-C11a; 3, L11; 4, PH8; 5, M4147. (B) Lane a, 17359; b, H21; c, WR309; d, WR352-C11; e, Khartoum. (From COMEAU, A. M., S. I. MILLER, and D. F. WIRTH. 1986. Chromosome location of four genes in *Leishmania*. *Mol. Biochem. Parasitol.* 21:161–169 by permission of Elsevier Science Publishers BV.)

leader sequence. The mapping of alpha tubulin genes and ribosomal genes demonstrated that each mapped to a distinct large chromosome in a variety of isolates. Both the beta tubulin gene and the 5′-spliced-leader sequence show evidence of size variation and variation in the number of chromosomes with homologous sequences. This pattern indicates a certain plasticity in the arrangement of genes on chromosomes and may have implications with regard to the evolution of the various strains and species.

Structure and Function of Kinetoplast DNA

In addition to the nuclear DNA, leishmanias have an unusual mitochondrial DNA, called the kinetoplast DNA, that comprises 10 percent to 15 percent of the total DNA of the cell. It is composed of two types of circular DNA molecules. The first, referred to as the maxi-circle (20,000 bp long), is the mitochondrial DNA of the parasite; the second type, the mini-circle (800–1200 bp), has no known coding function. The maxi-circle DNA is present in approximately ten copies per kinetoplast; the mini-circle DNA is present in 10,000–50,000 copies.

Replication of both the kDNA maxi-circle and mini-circle has been extensively studied. Elegant work has demonstrated that the maxi-circle replicates by the rolling-circle mechanism, whereas the mini-circle replicates as individual circles, using an RNA primer at the site of the putative origin of mini-circle replication.

Several mitochondrial proteins have been identified by cross-hybridization with genes from yeast and mammalian cells. In addition, DNA sequence analysis of maxi-circle DNA has identified several open reading frames on the maxi-circle DNA. Transcripts of the maxi-circle genes have been identified in both the promastigote and amastigote stages. Developmental regulation of maxi-circle gene expression is less pronounced than in a related kinetoplastid organism, *Trypanosome brucei*, where several maxi-circle transcripts are developmentally regulated in their expression.

Kinetoplast DNA Mini-Circle

One of the most puzzling DNA molecules in nature is the kinetoplast DNA mini-circle. It has been studied extensively in both *Leishmania* and *Trypanosome* species. (both African and South American). It represents approximately 10 percent of the total cell DNA and is repeated 10,000 to 100,000 times per kinetoplast. Its function is unknown; no protein product or transcript has been demonstrated, although many mini-circle sequences contain a small open-reading frame of 30–50 amino acids. A unique 13-nucleotide sequence is contained in all mini-circles sequenced to date and is proposed to be the origin of DNA replication. All mini-circles have a unique structural feature: a DNA sequence that contains a natural bend. This feature was first described because of anomalous migration of the mini-circle DNA in agarose and acrylamide gels. The DNA sequence itself is not conserved, but the structural feature of the bend is conserved. The function of this bend in the DNA is unknown, but one proposal is that it serves as a protein binding site important for the function or structure of the kDNA.

DNA mini-circles appear to undergo relatively rapid DNA sequence divergence, presumably because of the noncoding nature of the DNA. This observation has led to a useful application of molecular biology to the differentiation of *Leishmania* species and subspecies and subsequently to the direct diagnosis of leishmaniasis in patients. In the New World *Leishmania* organisms, the kDNA mini-circle isolated from *L. mexicana* does not share any sequence homology with those isolated from *L. braziliensis*. This difference in DNA sequence has provided the basis for a DNA probe that can distinguish these two *Leishmania* species directly when material obtained from a lesion is applied to nitrocellulose (Figure 18-2). This methodology has now been used to diagnose over 300 patients with cutaneous leishmaniasis offering both a more rapid and specific diagnosis than is currently available with any other methodology.

Work to develop subspecies-specific diagnostic probes and molecular analysis of kDNA have shown species-related restriction site heterogeneity within the kDNA and several groups have cloned species- or isolate-specific kDNA fragments from visceral and Old World cutaneous strains of *Leishmania*. This work demonstrates that within the kDNA network, fragments of DNA exist with different taxonomic specificities. These cloned subspecies-specific DNA probes are important for the development of future diagnostic probes but are also interesting with regard to the kDNA sequence divergence.

Perhaps one of the fascinating molecular questions in kDNA analysis is the mechanism of DNA sequence divergence within the kDNA mini-circle. Within a single mini-circle, there are regions of conserved DNA sequence and regions of highly divergent DNA sequence. The 10,000 mini-circles within a single *Leishmania* organism are heterogeneous and fall into a number of different sequence classes. Within a given sequence class, the mini-circle sequences are identical. Sequence analysis of different classes of mini-circles in *Leishmania tarentole* and *T. brucei* has revealed that mini-circles of different classes share 150 bp of homology but are widely divergent throughout the rest of their length. In addition, in *L. mexicana amazonensis*, sequences with different taxonomic specificities are present in a defined order along the length of a single cloned mini-circle. There appears to be a gradient of sequence divergence on the length of the mini-circle. There is a 230-bp region which is conserved across species boundaries, an adjacent 170-bp region that is found only in the *L. mexicana* complex and a 400 bp region which is highly divergent and found only in a single or small number of *L. mexicana amazonenesis* isolates. The maintenance of homogeneous sequence classes in the face of rapid sequence variation presumably requires a mechanism whereby mutations in a single mini-circle are transmitted to other mini-circles. In

---

FIGURE 18·2 A schematic representation of the method used for the DNA probe diagnosis of cutaneous leishmaniasis. A biopsy of the cutaneous lesion is taken and is divided in half; the tissue is then spotted directly onto a DNA-binding membrane. This membrane is then treated to denature the DNA and is hybridized with the specific probe for either *Leishmania mexicana* or *Leishmania braziliensis*. Specific hybridization is detected by x-ray film. (This diagram was prepared by J. Joseph and W. O. Rogers.)

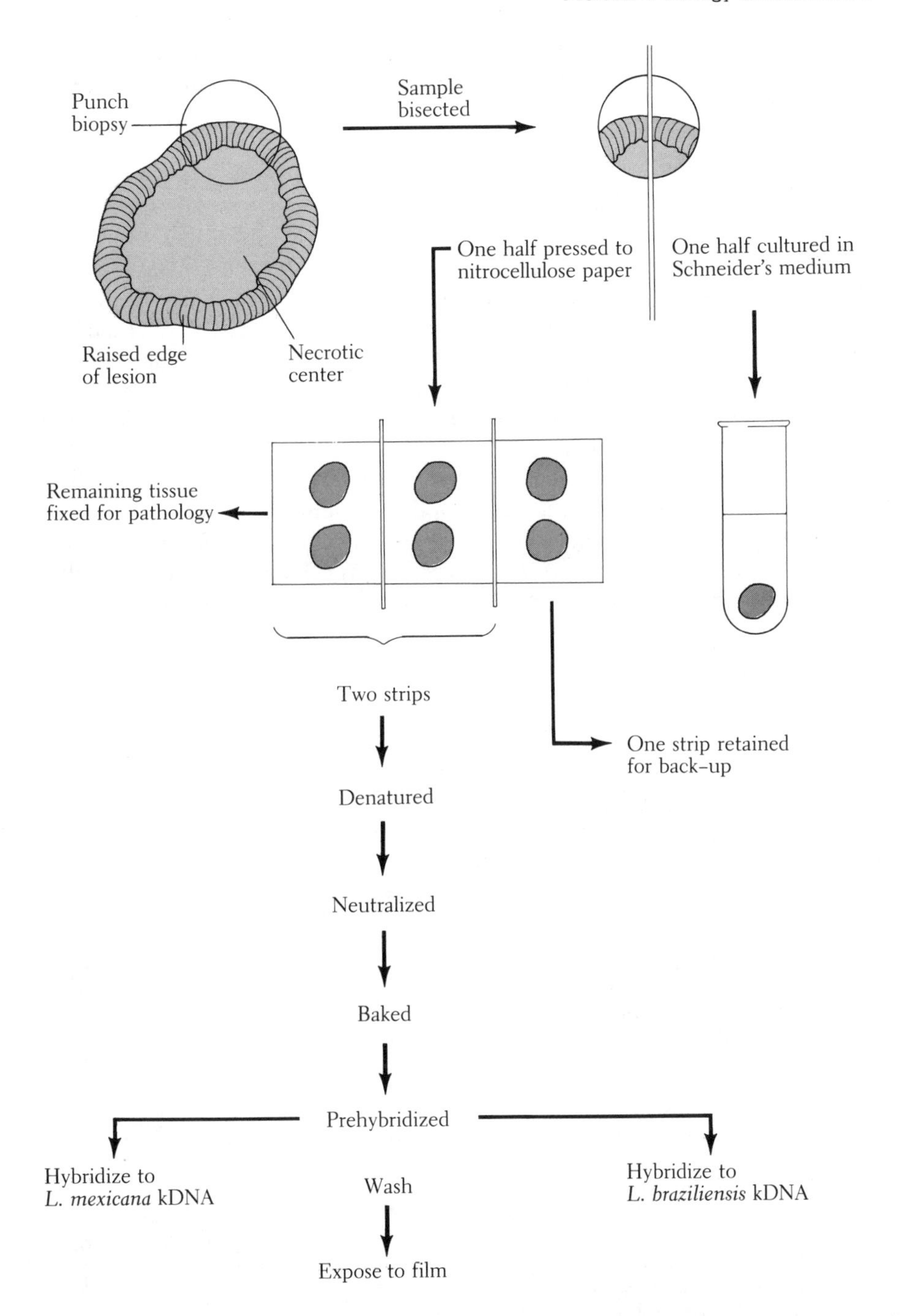
Punch
biopsy
Sample
bisected
Raised edge
of lesion
Necrotic
center
One half pressed to
nitrocellulose paper
One half cultured in
Schneider's medium
Remaining tissue
fixed for pathology
Two strips
One strip retained
for back-up
Denatured
Neutralized
Baked
Prehybridized
Hybridize to
L. mexicana kDNA
Hybridize to
L. braziliensis kDNA
Wash
Expose to film

addition, some mechanism must exist for the generation of rapid sequence divergence in one portion of the mini-circle molecule with maintenance of relatively conserved regions within the same molecule (Figure 18-3).

Developmentally Regulated Gene Expression

To study the developmental regulation of gene expression, it was important to choose the major proteins that are regulated during the transition from amastigote to promastigote. The major biosynthetic change is the increase in the synthesis of alpha and beta tubulin, the major structural proteins of the flagellum and subpellicular microtubules. These proteins can represent up to 25 percent of the total protein synthesized; they therefore represent the major transcriptional products of the parasite during the transition from amastigote to promastigote.

Leishmania species have a novel cytoskeletal structure: organized arrays of microtubules, referred to as subpellicular microtubules, form the basis for the parasite structure. These microtubules are closely opposed to the inner surface of the plasma membrane and completely outline the parasite cytoplasm. Both the amastigote and the promastigote stages contain subpellicular microtubules; the promastigote has in addi-

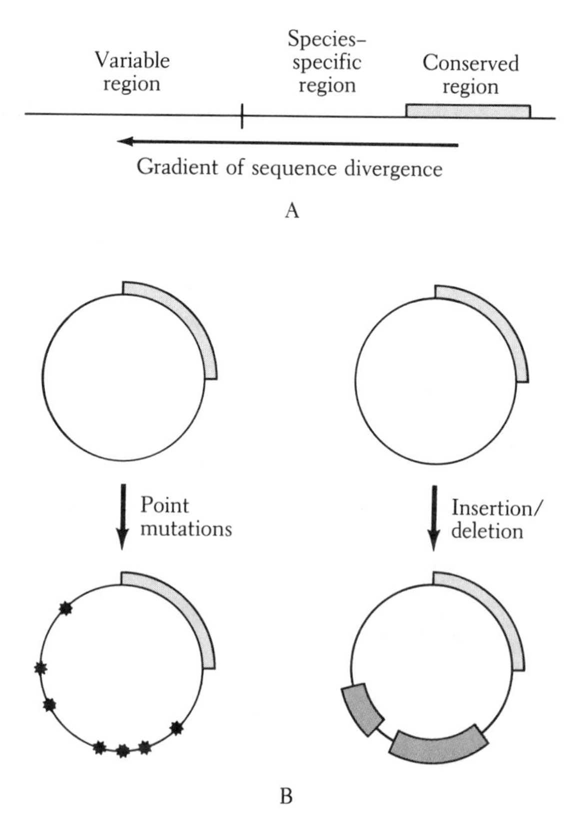

FIGURE 18·3 Models of kDNA sequence divergence. The striped box represents the conserved region of the mini-circle. A. This linear representation of a mini-circle contains regions with different sequence divergence. The conserved region contains sequences shared by all mini-circle molecules. The region directly adjacent to the conserved region is shared by all mini-circles derived from strains of the same *Leishmania* sp. The remainder of the mini-circle sequence (about 50 percent) is highly variable and is often unique to a single isolate. B. Two models help to explain the apparent differences in sequence diversity within a single mini-circle. The first model (I) is that point mutations are accumulated at a rapid rate throughout the molecule but that these mutations are corrected, perhaps by gene conversion, in the conserved region. The greater the distance from the conserved region, the greater is the accumulation of point mutations. The second model (II) proposes the insertion of new sequences by either recombination or rearrangement. DNA sequence comparison of several mini-circles from *L. mexicana amazonensis* is in agreement with the second model, namely, large-scale changes in the DNA sequences rather than accumulation of point mutations. (From ROGERS, W. O., and D. F. WIRTH. 1987. Kinetoplast DNA mini-circles: regions of extensive sequence divergence. *Proc. Natl. Acad. Sci. USA* 84:565–569.)

tion a flagellum with the standard 9 + 2 microtubule arrangement found in flagella from many different organisms. During transition from amastigote to promastigote, the flagellum is assembled, and this process presumably accounts for the large increase in synthesis of the alpha and beta tubulin proteins.

The genes encoding both the alpha and beta tubulin proteins have been cloned and analyzed in *L. enriettii*, a rodent leishmania. Both genes are arranged in separate tandem arrays (Figure 18-4). The alpha tubulin gene is composed of 10–15 copies of a 2-kb identical repeat. All the copies of the alpha tubulin gene are contained on a single chromosome in all *Leishmania* species that have been analyzed. In contrast, the beta tubulin gene, which is approximately 4 kb long, is present both as a tandem array of 10–15 genes on a single chromosome and in separate copies on at least two other

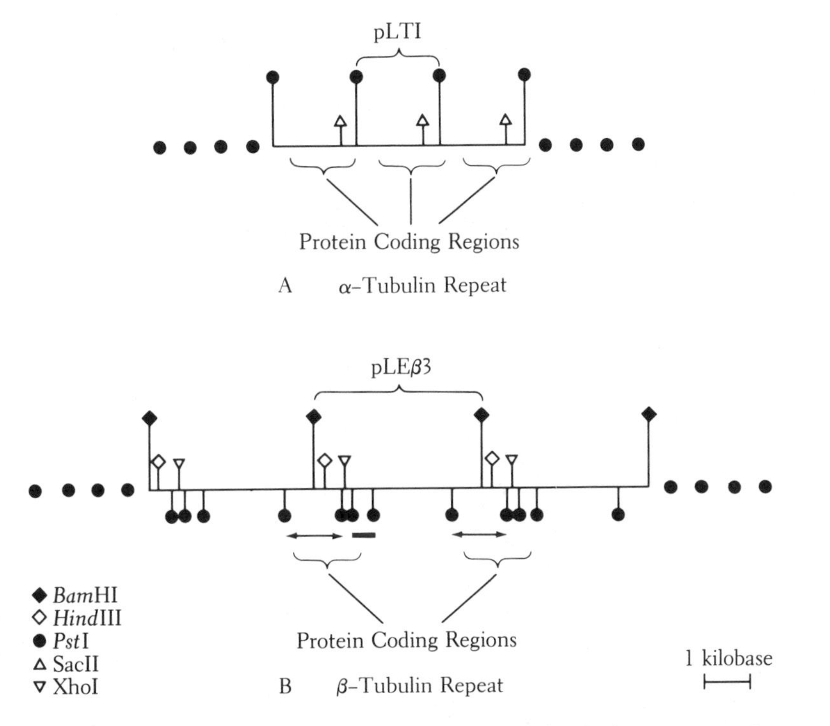

Figure 18·4 Models for the arrangement of α- and β-tubulin genes in the genome of *L. enriettii*. A. The structure of tandemly repeated α-tubulin genes is indicated, along with selected restriction sites. The 2-kb *Pst*I fragment that forms the insert for the α-tubulin genomic clone pLT1 is marked in brackets, and the α-tubulin protein-coding region is delineated by the bar. The direction of transcription is from left to right. B. The structure of the tandemly repeated β-tubulin genes is displayed, along with selected restriction sites. The 4-kb *Bam*HI fragment that constitutes the insert of the β-tubulin genomic clone pLEB3 is indicated by the brackets, and the protein coding region is marked by the narrow bars. The direction of transcription is from right to left. (From LANDFEAR, S. M., and D. F. WIRTH. 1985. Structure of mRNA encoded by tubulin genes in *Leishmania enriettii. Mol. Biochem. Parasitol.* 15:61–82 by permission of Elsevier Science Publishers BV.)

chromosomes (Figure 18-4b). The exact function of the beta tubulin genes not contained in the repeat is unknown. They may be nonfunctional pseudogenes or they may be genes required for special functions, such as mitosis. Why the alpha tubulin gene complex does not have these genes outside the repeat is a curiosity.

The expression of alpha and beta tubulin in *L. enriettii* is controlled at the level of mRNA accumulation. Both of the tubulin mRNAs are present in 8–10-fold greater quantity in the promastigote than the amastigote. In another species, *L. mexicana*, translational control also may play a role in the regulation of tubulin protein expression. Approximate sizes of the alpha and beta tubulin mRNAs were determined by fractionating total RNA on denaturing gels and sequentially hybridizing the same Northern blot with the cloned probes for alpha and beta tubulin. Both mRNAs appear to be slightly greater than 2 kb long, with the beta tubulin slightly larger that the alpha tubulin mRNA. It is interesting to note that although the beta tubulin gene is almost twice as large as the alpha tubulin gene, the final mRNA molecules are very similar in size. The sizes of the tubulin RNAs isolated from promastigotes are identical to those isolated from amastigotes.

Detailed transcriptional analysis of the tubulin mRNAs has revealed that each mRNA is modified at its 5′-end to include the identical 35-nucleotide sequence. This 35-nucleotide sequence is not encoded upstream from the gene, but is spliced onto the structural gene RNA by a novel mechanism, described later, which results in the formation of a hybrid mRNA composed of transcripts from two distinct genes on separate chromosomes. DNA sequence and S1 mapping analysis have led to the identification of putative gene control or RNA processing sequences. The 35-nucleotide sequence is added onto the 5′-end of the structural gene mRNA at the position of a consensus splice site in both the alpha and beta tubulin upstream regions. There is a single splice site in the alpha tubulin and two putative splice sites in the beta tubulin. The most striking feature of the upstream region of both the alpha and beta tubulin genes is a tandem repeat of the dinucleotide AC just upstream from the RNA splicing site. The AC runs occur at different distances from the splice site in the two genes, but their position relative to the actual transcription initiation sites is unknown. Purine-pyrimidine runs of the type observed here may have the potential to form zDNA, but the role of DNA structure in gene expression in these parasites remains to be determined. In the alpha tubulin sequence, there is consensus polyadenylation site at the position of the 3′-end of the mRNA.

Other Protein Genes

Analysis of structural protein genes in *Leishmania* organisms has been limited mainly by the workers in the field. mRNA and DNA sequences homologous to the heat-shock proteins (HSP 70) have been identified both in African trypanosomes and *Leishmania* species. As described next, the genes encoding dihydrofolate reductase/thymidylate synthetase bifunctional enzyme have been isolated from drug-resistant parasites.

Novel Biochemical and Molecular Mechanisms

RNA Processing

The *Leishmania* species have an unusual mechanism of mRNA processing. Each mRNA is a hybrid RNA molecule containing RNA sequences derived from two distinct parts of the genome, in fact from two different chromosomes. Each mRNA contains both a unique sequence, which encodes a structural protein, and a common sequence, which is an identical 35-nucleotide cap at its 5′-end. In the *L. enriettii*, this cap, or spliced leader sequence (SLS), is encoded by a 438-bp gene arranged in a tandem array of approximately 200 copies. These spliced-leader genes are transcribed to give a 85-nucleotide precursor that has the 35-nucleotide cap sequence at its 5′-end. The genes for the SLS are located on a separate and distinct chromosome from that of the alpha and beta tubulins. An analogous gene was originally identified in *Trypanosoma brucei*, and work in that system has demonstrated that this SLS is added onto all mRNAs in the cell (see Chapter 17). The exact function of this SLS is not known. It may be a novel mechanism for the capping of mRNAs, or it may be important as a signal in the translation of proteins. Determination of function must await a transfection system that will allow site-specific mutagenesis of the gene.

One of the major areas of current research is the elucidation of the mechanism of this novel hybrid mRNA formation. DNA sequence and primer extension analysis have demonstrated the presence of consensus splice sites both in the SLS transcript and in the upstream regions of the alpha and beta tubulin genes. There are two basic models for the addition of the SLS to mRNA precursors, the cis-splicing model and the trans-splicing model (see Figure 18-5). In the cis-splicing model, the SLS is covalently attached to the primary transcript of the tubulin gene, either as a primer during transcription or posttranscriptionally by ligation, thereby creating a hybrid intron. Subsequently unimolecular splicing occurs similar to that found in eukaryotic systems. This model predicts that there will be two products of splicing: a mature hybrid mRNA, with the SLS attached to the tubulin structural RNA, and the hybrid intron, consisting of the downstream region of the SLS transcript and the upstream region of the tubulin primary transcript.

Evidence is accumulating that the splicing is more likely to be through a trans-splicing mechanism. In this model, the SLS gene and the tubulin structural gene are independently transcribed and splicing occurs between two separate transcripts. No covalent hybrid intron structure is formed. This model predicts that there will be three products of splicing: the hybrid mRNA, the downstream region of the SLS transcript, and the upstream region of the tubulin gene. The downstream region of the SLS gene has been identified in *L. enriettii*, and recent work in *T. brucei* and *L. enriettii* has demonstrated that incubation with extracts known to debranch intron lariat structures results in the accumulation of the downstream region of the SLS transcript. What remains to be demonstrated is the presence of free upstream transcript from the tubulin

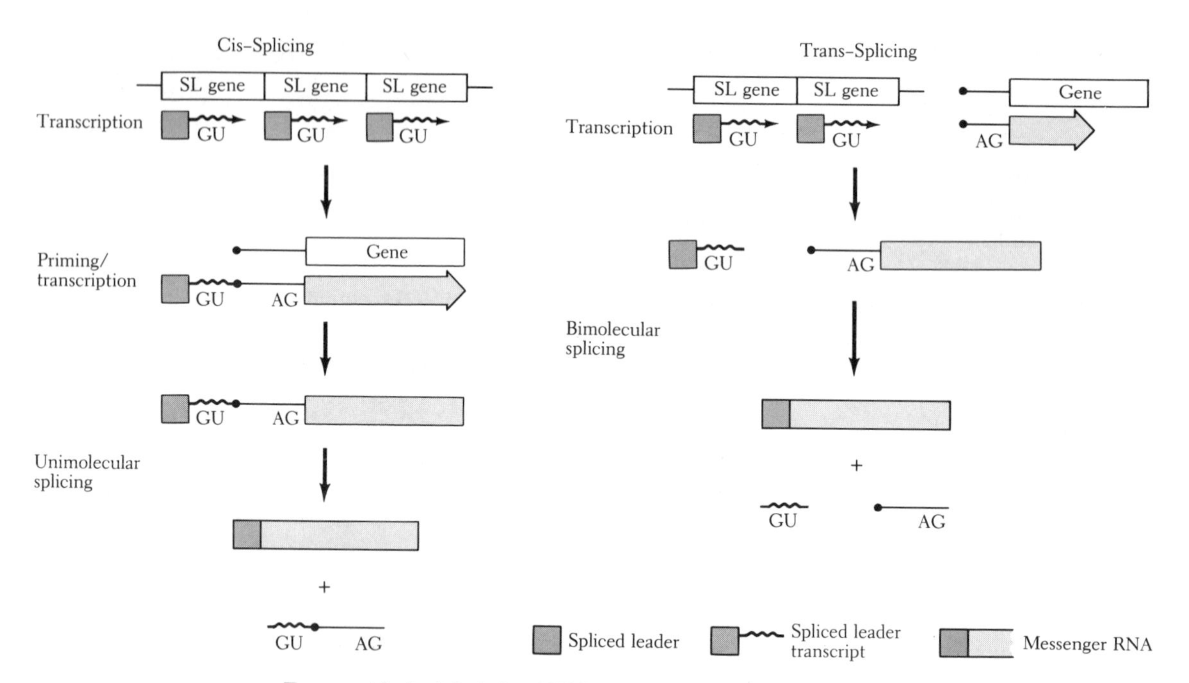

FIGURE 18·5 Models of RNA processing after discontinuous transcription. In cis-splicing, the spliced-leader transcript could prime transcription of a gene, resulting in a precursor with an intron. Unimolecular splicing produces a single RNA by-product containing genomic sequence upstream to the AG consensus splice site and the remainder of the spliced-leader transcript. Alternatively, in cis-splicing, a ligation reaction after two unlinked transcription events could result in a unimolecular splicing event. In trans-splicing, independent transcription events followed by bimolecular splicing produce the two RNA by-products shown and mature mRNA. One RNA by-product contains a spliced-leader genomic sequence, and a separate by-product contains a transcribed genomic sequence from the gene processed to mRNA. Primed transcription after removal of the 3′-end of the spliced-leader transcript would also result in a RNA processing product comprising the free 3′-end of the spliced-leader transcript. GU represents the nucleotides of the consensus eukaryotic splice junction found in the spliced-leader transcript immediately 3′ to the spliced leader. AG represents the nucleotides of the consensus eukaryotic splice junction found immediately 5′ to the splice junction in all trypanosomatid mRNA genes examined to date (Adapted from MILLER, S. I., S. M. LANDFER, and D. F. WIRTH. 1986. Cloning and characterization of a *Leishmania* gene encoding a RNA spliced leader sequence. *Nucl. Acids Res.* 14:7341 by permission of Oxford University Press.)

or any structural gene. Future experiments in this area include the identification of splicing complexes and the isolation of enzymes responsible for this novel RNA processing mechanism.

Drug Resistance

Major areas of interest both for genetic analysis and for the more practical consideration of drug therapy are the mechanisms of drug resistance in the *Leishmania* organisms. Work in several laboratories has demonstrated that gene amplification plays a major role in the development of drug resistance. At least one mechanism of resistance is the overproduction of the target gene product in the resistant parasite (Figure 18-6), a pattern that is similar to the mechanism found in many drug-resistant mammalian cells and that may be an important in vivo step in the development of drug resistance.

Leishmania major parasites have been selected for their resistance to the folic acid analogue methotrexate; in separate experiments, they have been selected for resistance to an inhibitor of thymidylate synthetase, 10-propargyl-5,8-dideazafolate (CB3717). These resistant parasites were selected after stepwise increases in drug concentration. *L. major* has a bifunctional enzyme that contains both DHFR and TS activity, and resistant parasites show an increased activity of the bifunctional enzyme. In mutants selected with methotrexate, this increased enzyme activity is accompanied by the amplification of two regions of the DNA: one of 30 kb, the R region, and the other of 80 kb, the H region. In mutants selected with CB3717, the TS inhibitor, only the R-region DNA is amplified. These amplified DNA regions exist as circular extrachromosomal "factors" demonstrated by migration in chromosome gels and by visualization in the electron microscope. Initially, these circles were unstable in the absence of selective pressure; a decrease in the amount of amplified DNA was accompanied by a

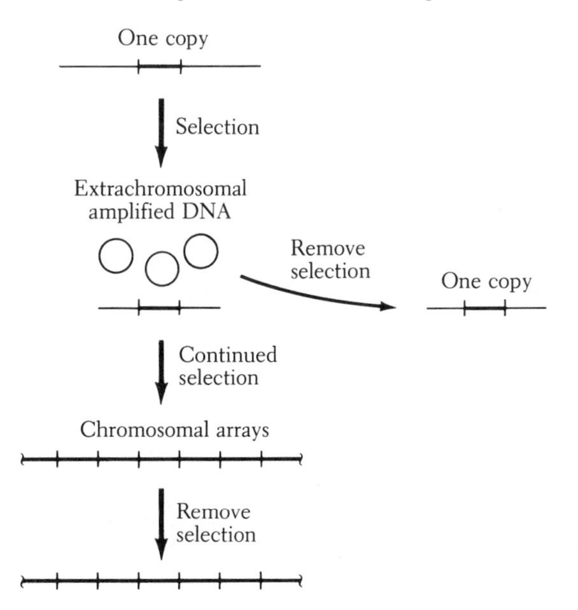

FIGURE 18·6 Summary of the molecular structure of amplified DNA in unstable and stable *Leishmania* lines resistant to methotrexate. (Adapted from BEVERLEY, S. M., et al. 1984. Unstable DNA amplifications in methotrexate-resistant *Leishmania* organisms consist of extrachromosomal circles which relocalize during stabilization. *Cell* 38:431 © Cell Press.)

decrease in the bifunctional enzyme activity. After prolonged selective periods, stable mutants have been isolated. Analysis of both stable and unstable resistant mutants by orthogonal field alternating gel electrophoresis (OFAGE) demonstrates extrachromosomal circles homologous to the R and H regions. Thus, these circles must contain sequences for replication and stable inheritance.

The structural gene for the bifunctional enzyme is located on the 30-kb amplified circle, and its protein sequence is derived from the cDNA sequence and genomic sequences. It is interesting that the DHFR portion of the protein, located at the amino terminus has little primary sequence homology with mammalian or bacterial enzymes, whereas the TS or carboxyl portion of the enzyme has regions of significant homology with thymidylate synthetase isolated from other organisms. The DHFR portion of the protein does share secondary structure homology and known substrate binding sites with other DHFR proteins.

Mutant parasites resistant to tunicamycin, an inhibitor of N-acetylglucosaminyltransferase in the dolichol pathway of protein glycosylation, have been isolated in a similar stepwise manner. These resistant parasites have an increased level of the target enzyme; there is also evidence from a resistance-specific banding pattern of *Hin*dIII-digested, ethidium bromide–stained DNA that resistant parasites have amplified DNA. The gene product of this amplified DNA has yet to be determined, however. It is interesting to note that these resistant mutants have increased virulence.

Gene amplification of enzyme structural genes appears to be at least one important mechanism in the development of drug resistance in *Leishmania* species. An important question that remains to be addressed is the molecular mechanism of this circular amplification and its role in the development of drug resistance in the field.

Coda

The major development necessary to continue molecular genetic analysis of leishmanias is the development of a transfection system to allow the introduction and analysis of modified genes in these parasites. Initial attempts at this work in a related Kinetoplastida species, *Crithidia fasciculata*, involved the introduction of a drug-resistant gene (G418) carried on a plasmid. Resistant parasites were isolated, but rearrangement of the plasmid DNA in these parasites made the interpretation of these results ambiguous.

Several areas of research need to be pursued in order to develop a successful transfection system. First, some mechanism of introducing DNA into the cells is necessary. Work in other eukaryotic systems suggests that electroporation, microinjection, or direct DNA addition should be tried. Second, a selectable marker, either drug resistance or an auxotrophic marker, is necessary. Finally, in order to achieve stable transformation, either an independent origin of replication or an integrating plasmid will be required. Highly repeated genes such as the SLS or tubulin genes may be good candidates for integration and the kDNA molecule or the amplified stable circles may provide the function of the origin of replication.

Once this transfection system is developed, questions of the structure and function relationships of the various putative controlling elements in gene expression can be directly assayed. This system also may help to define important regions of receptors or other virulence factors and could be the breakthrough that will take molecular work with this parasite from description to functional explanation.

19

Molecular Biology of Schistosomes and Filariae

George R. Newport and Nina Agabian

Filariae and schistosomes rank among the most damaging helminth parasites infective to humans. Approximately 300 million people harbor schistosomes, 800,000 of whom die each year; filariae affect 300,000 individuals, and although they account for little mortality, they lead to blinding and disfigurement of significant numbers of people. Moreover, infections by these parasites tend to follow a chronic course that immeasurably saps the vitality and compromises the productivity of millions of children and adults. In the present chapter we review some recent developments in the study of the molecular biology of these two classes of parasites, emphasizing research designed to develop strategies for their eradication.

The study of the molecular biology of parasitic helminths is a new discipline that arose from the opportunity to apply technologies offered by molecular genetics to questions raised by workers investigating the immunology, biochemistry, physiology, and biology of the organisms. Although it is clear that the use of recombinant DNA techniques in the study of schistosomes and filariae will primarily serve as an amplification system for providing quantities of proteins whose availability is otherwise limited, preliminary indications suggest that application of the new technologies also may aid in unraveling basic biological questions.

The authors acknowledge support from the Edna McConnell Clark and John and Catherine MacArthur Foundations for making it possible to write this chapter. N. A. is a Burroughs Wellcome Scholar in Molecular Parasitology. We wish to thank J. Deneris for her helpful suggestions, and S. Craig, A. Dessain, A. Davis, M. Doenhoff, J. McKerrow, G. Mitchell, T. Nielsen, F. Rotman, and A. Sher for sharing information with us before publication.

Vaccine Development

The goal of conferring sterilizing, or even partial, immunity to humans against schistosomes and filariae is obviously a desirable objective, and the search for safe, inexpensive vaccines that can impart this protection is a high priority. Such vaccines would complement current efforts to control these diseases with chemotherapy, vector control, and education. Although scientific progress to date is encouraging, scientific and economic considerations make it currently unclear whether such vaccines can be developed or if a large-scale vaccination program could be implemented. A major motivation to develop vaccines is the realization that in the absence of protective immunity, individuals treated with antihelminthic drugs tend to become reinfected.

The life span of schistosomes and filariae within a vertebrate host is extraordinarily long; schistosomes have been documented to remain viable and fecund for up to 35 years. They stimulate a host immune response of variable intensity, and how they survive host attack has intrigued many scientists (see Chapters 7 and 14).

While investigators for decades have attempted to identify antigens having immunoprophylactic potential, recent molecular and immunological methodologies have provided several powerful approaches to the problem. Different laboratories have tackled the issue in different ways, with some degree of success. Convinced that target antigens should be routinely exposed to the immune system, some workers have proceeded to isolate and genetically express components of the parasites' surface or excretory/secretory antigens, and are testing their prophylactic efficacy in murine models. Others have produced and characterized monoclonal antibodies that were chosen for their ability either to kill worms in vitro or to confer resistance to infection after passive transfer into immunologically naive recipients. Antibodies identified in this manner have then been used as reagents for isolating the protective antigen itself, usually from recombinant DNA sources. Others have concentrated on the identification of antigens recognized by antibody from hosts naturally resistant to infection but not recognized by antibodies of susceptible hosts. The most thorough approach has involved fractionation of parasite extracts, assessment of the protective value of each fraction, and purification of active components. None of the candidate antigens characterized to date has proved capable of engendering sterile immunity, although this failure may reflect suboptimal stimulation of appropriate T-cell populations or of B cells of the right isotype rather than identification of an unsuitable antigen.

Schistosomes

The characterization of protective antigens with monoclonal or monospecific polyclonal antibodies that are cytotoxic to schistosomes has yielded a perplexing catalog of molecules. For example, preliminary data indicate that enzymes of the parasite's glycolytic pathway may represent important vaccine candidates. Adult schistosomes and filariae are anaerobes that use glucose as their sole source of energy. Their consumption of the sugar is prodigious: schistosomes consume the equivalent of 20 percent of their dry weight per hour. The importance of glycolysis to worm survival has been known for many years, and trivalent antimonials (once used as antischistosomal agents) appear to kill the organisms by interfering with the action of phosphofructokinase.

Independent research groups have characterized three apparently protective antigens and identified them as aldolase, glyceraldehyde-3-phosphate dehydrogenase, and triose phosphate isomerase. Curiously, the enzymes catalyze sequential reactions that form the core of the glycolytic pathway. It has recently been demonstrated that aldolase also is a potential vaccine target of *Plasmodium falciparum*. Should these proteins indeed impart protection when used as vaccines, their interplay with the immune system would raise several questions. For example, since these enzymes are not rate-limiting to the pathway (whereas membrane transport and the three kinases are), it is surprising and unclear why immune responses to them have antihelminthic consequences. Furthermore, the fact that the enzymes are normally found in the cytoplasm of eukaryotic cells raises the perplexing question of how the immune system can launch an attack on them in intact organisms. Finally, the amino acid sequences of the three proteins are very similar to host enzymes (Figure 19-1), which raises the concern that vaccines based on these proteins might elicit autoimmune responses.

Another empirically determined class of schistosome vaccine candidates consists of glutathione metabolizing enzymes. One of these proteins was selected on the basis of the observation that mice resistant to *Schistosoma japonicum* infection mount a response to an antigenic enzyme, whereas mice that are susceptible to infection do not. The antigen has been cloned and expressed in bacteria, and has been found to be a 26-kilodalton (kD) isoenzyme of glutathione S-transferase. Unlike the case for the glycolytic enzymes, the protein demonstrates only a modest degree of resemblance to the host homologue. Mice infected with *S. japonicum* produce antibodies that cross-react with the enzyme from *Schistosoma mansoni*, and sequencing of the *S. mansoni* homologue has revealed that they are nearly identical in structure (Figure 19-2) and differ significantly from a 20-kD isoenzyme isolated from *S. mansoni*. The similarity between the enzymes of the two parasite species is much greater than that noted for their heat-shock protein, HSP-70 (which ranks among the phylogenetically most conserved proteins), and is reflected in the sequence homology of 5′ untranslated portions of their RNAs, which begin with an unusual string of polyT. A second isoenzyme of the glutathione S-transferase of *S. mansoni* also has prophylactic potential. The protein has a size of 28 kD, shows little sequence similarity to the 26-kD isoenzyme, and induces production of antibodies that do not cross-react with the 26-kD homologue or with homologues of the host. The enzymes are normally found in the cytoplasm or microsomal fraction of metazoan cells; however, there are indications that they may also be found on the surface of the parasite. The functional role of the schistosome isoenzymes is unknown, but their presence in the tegument suggests that they may play a role in protecting the organisms from oxygen-mediated damage. Protective responses stimulated by the antigen appear to involve the production of specific IgG2a antibodies that direct eosinophil-mediated cytotoxicity (see Chapters 7 and 14).

Aldolase

```
Sm  ald  RHAADESTATMGKRLQQIGVENNEENRRLYRQLLFSADHKLAENISGVILFEETLHQ
RM  ald  IL******GSIA****S**T**T*****F*****LTA*DRVNPC*G*****H***Y*
HL  ald  IL*****VG***N***R*K***T*****QF**I***V*SSINQS*G*****H***Y*

Sm  ald  KSDDGKTLPTLLAERNIIPGIKVDKGVVPLAGTDNETTTQGLDDLASRCAEYWRLGC
RM  ald  *A***RPF*QVIKSKGGVV**************NG********G*SE***Q*KKD*A
HL  ald  KDSQGKLFRNILKEKGIVV***L*Q*G******NK***I****G*SE***Q*KKD*V

Sm  ald  RFAKWRCVLKISSHTPSYLAMLEMLMYL
RM  ald  D**********GE****A**IM*NANV*
HL  ald  D*G***A**R*ADQC**S**IQ*NANA*
```

Glyceraldehyde-3-Phosphate Dehydrogenase

```
         1                                                          60
SCH      GSRAKVGINGFRGIGRLVLRAAFLKNTVDVVSVNDPFIDLEYMVYMIKRDSTHGTFPGEV
HUM      MGKV***VN**GR*****T****NSGK**I*AI*******N*****FQY*****KFH*T*

                                                                    120
SCH      STENGKLKVNGKLISVHCERDPANIPWDKDGAEYVVESTGVFTTIDKAQAHIKNNRAKKV
HUM      KA*****VI**NP*TIFQ***SK*K*GDA***************ME**G**LQGG-**RV

                                                                    180
SCH      IISAPSADAPMFVVGVNENSYEKSMSVVSNASCTTNCLAPLAKVIHDKFEIVEGLMTTVH
HUM      *************M***HEK*DN*LKII*******************N*G**********
```

Figure 19·1 Inferred amino acid sequence of the amino termini of two putative schistosome vaccine target antigens. In the top sequence, the primary sequence of a clone identified as encoding the schistosome aldolase (sch ald) is compared to that of clones encoding rat muscle aldolase (RM ald) and human liver aldolase (HL ald). In the bottom panel, the sequence of the schistosome glyceraldehyde-3-phosphate dehydrogenase is compared to an isoenzyme from human beings. Asterisks represent regions of identity.

Another unexpected schistosome vaccine candidate is paramyosin. The peptide was identified by fractionation of parasite extracts and testing the degree of protection conferred by each fraction in a mouse model. A cDNA encoding part of the schistosome paramyosin has been used to isolate a filarial homologue, and this homologue has been shown to confer partial resistence to filariae in immunized animals. Paramyosin is found

```
Sm26            M A P K L G Y W K I K G L V Q P T R L L L E Y L
Sj26          T K L * I * * * * * * * * * * * * * * * * * * * *
RYB             * P M I * * * * N V R * * T H * I * * * * * * T
RYC             * P G G * P V L H Y F D G R G R M E P I R W L *
Sm28            * A G E H I K V I Y F D G R G R A E S I R M T *

Sm26   (#68) D V K L T Q S M A I L R Y I A D K H N M L G G C P K E R A E I S M L E G A
Sj26   (#68) * * * * * * * * * * I * * * * * * * * * * * * * * * * * * * * * * * * *
RYb    (#67) S R * I * * * N * * M * * L * R * * H L C * E T E E * * I R A D I V * N Q
RYc    (#62) G M * * V * T R * * * N * * * T * Y A L Y * K D M * * * * L * D * Y S E G
Sm28   (#65) V K W M L E * L * * A * * M * K * H * * M * E T D E * Y Y S V E K * I * Q
```

FIGURE 19·2 Inferred amino acid sequences of the amino termini of the 26-kilodalton (kD) glutathione S-transferase of *S. japonicum* (Sj26), the rat Yb isoenzyme (RYb), the rat Yc isoenzyme (RYc), and a 28-kD isoenzyme from *S. mansoni* (Sm28) compared to the 26-kD protein of *S. mansoni* (Sm26). Asterisks represent regions of identity.

in invertebrate catch muscles, where, in association with myosin, it forms complexes that permit strong, protracted muscle contraction, as seen in the adductor muscles of clams. Immunocytochemical studies indicate that the protein is also found in nonfibrillar form within elongate membrane-bound structures within the worm's tegument (see Chapter 6), although it is apparently not exposed on the surface.

Like schistosome myosin, sequenced portions of the parasite's paramyosin indicate that it consists of a series of seven amino acid repeats (a,b,c,d,e,f,g), in which residues a and d are generally hydrophobic, and which are part of a 28-amino-acid unit. Acidic and basic periods within the 28 amino acid units are almost out of phase with each other, so that each segment can be subdivided into zones of alternating charge. In general, it appears that the molecule, like myosin, evolved from a 28-amino-acid peptide whose gene underwent a series of duplications and that the register between the molecule and myosin is mediated by ionic interactions. The amino acid sequence of paramyosin predicts that the molecule forms a classic α helix, where an antiparallel second strand can make periodic contacts at regularly spaced intervals where amino acids having a nonpolar side chain are positioned. Amino acids facing the aqueous environment of paramyosin dimers tend to have hydrophilic side chains. Once the complete amino acid sequences of schistosome myosin and paramyosin are determined, it will become possible to construct and test models on how the two proteins interact. Unlike the case of glutathione S-transferases, paramyosin appears to stimulate T lymphocytes to produce interferon gamma, which activates macrophages to kill larval parasites. Notably, it seems that antibodies play no role in this cytotoxic reaction. The protein also may elicit an inflammatory reaction in sensitized hosts, which could wall off invading larvae.

Another schistosome antigen — myosin — has been cloned and is recognized by antibodies generated by animals that are poorly resistant to infection by sera from susceptible animals. Antibodies against the antigen rise dramatically in patients after chemotherapeutic cure. The protein is unique in being encoded by a single-copy gene (all organisms studied to date have their myosins encoded by a multigene family).

Preliminary indications are that, unlike paramyosin, schistosome myosin is not a promising vaccine candidate.

Taken as a whole, these findings add a new dimension to our understanding of schistosome antigens susceptible to cytotoxic arms of the immune response. Particularly puzzling is the mechanism by which host antigen processing cells manage to recognize cytoplasmic components of living schistosomes and subsequently translate this information into an immune response capable of homing in on molecules enclosed by a plasma membrane. Also surprising is the fact that nearly all potentially protective antigens sequenced to date share significant blocks of homology with host homologues, yet they apparently do not elicit an autoimmune response. Whether this is attributable to special modes of presentation of the antigens by the parasites or to an inherent lack of immunogenicity of evolutionarily conserved domains is unknown, but the question has obvious relevance to vaccine development. The observation that schistosome antigens contain regions whose structure is highly conserved suggests that these may have important functional purposes.

Filariae

As is the case with schistosomes, experimental evidence suggests that it is possible to develop protective immunity to filariae. Several workers have reported that immunization with irradiation-attenuated L_3 larvae can lead to a significant depression of adult worm burden and microfilaremia. Monoclonal antibodies against 70-kD and 75-kD stage-specific surface antigens of *Brugia malayi* have been developed that upon passive transfer lower the microfilaremia of infected animals. Immunization of gerbils with a soluble extract of *B. malayi* induces a 50 percent reduction in adult worm burden and a 80 percent to 100 percent decrease in parasitemia on challenge. The development of resistance correlates with the appearance of IgG antibodies directed at proteins of 25 kD, 42 kD, 60 kD, and 112 kD. Immunization of animals with purified 25 kD and 60 kD proteins leads to their ability to lower the number of circulating microfilariae.

A cDNA encoding the 60-kD antigen has recently been cloned and sequenced. The predicted size of the cDNA's translation product is 63 kD, and the antigen does not have any consensus glycosylation sequences or any hydrophobic domains suggestive of membrane-anchoring sites. Interestingly, the mRNA for the protein begins with a 5′ untranslated leader sequence that is identical to the trans-spliced leader of some actin genes of *Caenorhabditis elegans* (a free-living nematode). The 22-nucleotide spliced leader is present in several locations in the parasite's genome, including (as in *C. elegans*) within the 5s rRNA gene repeat. The spliced leaders are located at the 5′-end of a 109-base pair (bp) transcriptional unit of nonpolyadenylated RNA. The role of the spliced leaders, and why some genes acquire this sequence (as opposed to the trans-spliced leaders in trypanosomes, which is a universal feature of the mRNAs of the organisms) is unclear and awaits the development of appropriate assay systems which can evaluate their significance.

Chemotherapy

The drugs of choice currently used in the treatment of schistosomiasis and filariasis are praziquantel and ivermectin, respectively. Praziquantel is effective in a single oral dose and appears to paralyze the organisms by perturbing their tegument and by affecting calcium transport; the immune system probably plays a role in killing and clearing organisms debilitated by the drug. Praziquantel is also effective against other trematode and cestode parasites but not against nematodes whose cuticle blocks the action of the drug. Verapamil, a slow calcium blocker, also disrupts the schistosome tegument and may prove a less expensive alternative to praziquantel. The semisynthetic antibiotic mixture, ivermectin, similarly affects calcium fluxes at neuromuscular junctions, presumably by increasing the secretion of gamma amino butyrate (GABA) and enhancing the binding of the neurotransmitter to its receptor. Although not a perfect drug, ivermectin is a welcome alternative to diethylcarbamazine, in that it kills microfilariae more slowly, thereby decreasing the possibility of inducing potentially life-threatening Mazzoti-like reactions. The latter are thought to be allergic responses triggered by a sudden appearance of high levels of antigen released by dying microfilariae.

The perceived need to develop new antihelminthic drugs stems from three problems. First, researchers suspect that the parasites will, in time, develop resistance to currently available drugs. Second, current therapeutic regimens are too expensive to apply on a large scale, and reinfection is a constant threat. Finally, in the case of filariasis, available drugs have serious side effects and are not completely effective. It has been recently noted that veterinary use of ivermectin may introduce new ecological problems because the compound is not biodegradable and is toxic to a wide variety of invertebrates.

Combining methods in molecular biology with emerging ones in computer modeling can in theory strengthen and facilitate innovative approaches to the rational development of therapeutic drugs. Antihelminthic drugs currently in use are all products of empirical research, an inefficient method considering that on the average only 1 in 10,000 tested compounds shows any activity. The concept behind rational drug design is to take advantage of biochemical idiosyncrasies that distinguish parasite from host metabolism. This might be accomplished in at least two ways. First, methods in molecular genetics can be used to clone and express active parasite enzymes in more manipulable hosts (yeast and bacteria). If enzymatically active, the enzyme can then be studied by conventional biochemical techniques. If the gene can genetically complement a yeast strain deficient in the enzyme, it is conceivable that the protein's properties can also be studied in vivo. In the other approach, the primary sequence of the protein is determined, and if the three-dimensional structure of a homologue from another organism is known, predictions may be made about the conformation of the parasite peptide. After determination of the probable active site is made, the molecules can be modified by in vitro mutagenesis in order to test the validity of the predicted model. Several laboratories are currently developing algorithms to predict the structure of chemicals that might interfere with the function of active sites of known morphology and chemistry. One group, for example, has developed a model that correctly predicted that methotrexate should inhibit the activity of the dihydrofolate reductase-thymidylate synthetase of leishmania.

One of the better examples of rational identification of drug targets has centered on how purine nucleoside pools are formed in *S. mansoni*. This organism is incapable of synthesizing these compounds de novo and must rely on salvage pathways involving acquisition of precursors from the host. Purines and their respective nucleosides gain entry into the parasites by carrier-mediated mechanisms involving at least five transporters. The surface of the worm contains nucleotidases that may be coupled to nucleoside transport.

Culture media preincubated with adult schistosomes (which regularly slough off their outer surface) are capable of converting adenosine to adenine, inosine, and hypoxanthine, suggesting the existence of a salvage pathway consistent with the scheme outlined in Figure 19-3 and noteworthy in three ways: (1) the organisms lack a nucleotide kinase, so that the initial substrate for nucleic acid synthesis is always a free base; (2) adenosine deaminase activity is developmentally regulated, being absent in larval worms, which therefore cannot interconvert adenosine and guanosine nucleosides; and (3) a single enzyme, hypoxanthine guanidyl phosphoribosyl transferase (HGPRTase), catalyzes the ribosylation of guanidine and hypoxanthine (Figure 19-3). The proposed pathway indicates that the availability of guanosine nucleotides is strictly dependent on the activity of the parasite's HGPRTase. Studies on the schistosome enzyme HGPRTase have shown that it is structurally different from mammalian homologues, which makes it a plausible drug target.

A cDNA encoding a schistosome HGPRTase has recently been cloned and sequenced. The message for the protein is approximately 1.5 kb in size and hybridizes to a cDNA encoding a mammalian counterpart under conditions of moderate stringency. The predicted size of the protein is 26 kD, indicating that the active 105,000-dalton (D) protein is a tetramer. Alignment encoding the predicted amino acid sequence of the cDNA for the schistosome enzyme with that of humans and *Plasmodium falciparum* indicates that the enzyme is more closely related to that of humans (47.9 percent versus 36.9 percent). The areas of greatest identity between the human and schistosome enzyme correspond to the putative purine and phosphoribosyl binding sites (Figure 19-4). Interestingly, despite considerable differences in amino acid sequence the pre-

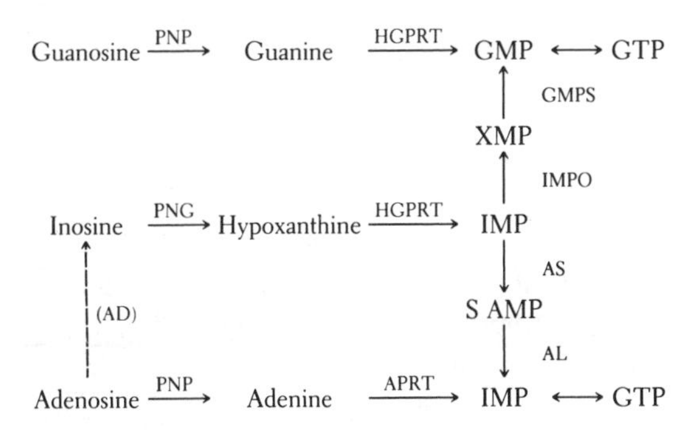

FIGURE 19·3 Purine salvage pathways of schistosomes. Abbreviations: PNP, purine nucleoside phosphorylase; HGPRT, hypoxanthine-guanine phosphoribosyl transferase; GMPS, guanosine monophosphate transferase; IMPD, inosine monophosphate dehydrogenase; AS, adenylosuccinyl AMP synthetase; AL, adenylosuccinyl AMP lyase; AD, adenine deaminase. Solid lines represent general pathways; dashed lines indicate stage-specific ones.

Putative Purine Binding Site

Schistosome	LMCVLKGGFKFLADL . . . . VKSYV
Mouse	AL******Y**F*** . . . . L***C
Human	AL************* . . . . L***C

Putative Phosphoribosylphosphate Binding Site

Schistosome	KDKNVLVVEDIIDTGKTITKL
Mouse	TG****I**********MQT*
Human	TG****I**********MQT*

Figure 19·4 Amino acid sequences of the putative purine and phosphoribosylphosphate-binding sites of the schistosome HGPRTase are compared with those of the enzyme from mice and humans. Asterisks represent points of identity; periods denote the presence of amino acids that do not appear to participate in binding.

dicted α-β-α structure of the two enzymes is very similar and demonstrates considerable differences from that of malarial parasites. The 13 amino acid sequence stretch corresponding to the putative binding sites of the human and schistosome HGPRTase differs by one conservative and one nonconservative substitution, suggesting that development of inhibitors acting at this site will prove difficult.

Several critical proteins of filariae that differ significantly in structure from those of mammals have been identified. These include their respiratory electron pathway complex III, tubulins, chitin, proteinases, amino acyl tRNA synthetases, and glutathione-folic acid metabolizing enzymes. There are indications that the filarial aldolase exerts a rate-limiting action on glycolysis since its activity is lower than that of phosphofructokinase, an enzyme that can catalyze 10 times more substrate than that present at steady-state levels. The filarial aldolase thus appears to be a nonequilibrium regulatory enzyme, and may prove a good chemotherapeutic target; even slight inhibition of such enzymes can greatly reduce the metabolic flow through the pathway. Whether this is also true in malarial and schistosome parasites, where preliminary indications are that the enzyme may prove a vaccine target, is unclear. Derivatives of amoscanate, a noxious compound previously used in the treatment of hookworm and schistosome infections, have been shown to have microfilaricidal action. Part of the activity of these compounds appears to be attributable to inhibition of aminoacyl tRNA synthetases specific for arginine, lysine, and serine. The antimalarial compound mefloquine similarly inhibits filarial aminoacyl tRNA transferases, but not those of mammals, indicating that these enzymes represent a potential drug target.

Developmental Biology and Stage-Specific Gene Expression

Filaria and schistosome worms display a range of intricate developmental plans involving permutations of (1) life in aquatic, vertebrate, and invertebrate habitats; (2) shifts in temperature, ambient gas phases, and available nutrients; and (3) changes in the identity of hostile environmental pressures. The mechanisms whereby the organisms find it possible to transform from one stage to another are poorly understood, and how these are regulated at the level of gene expression is completely unknown. Several laboratories have recently begun to address the issue, primarily by analyzing stage- and tissue-specific transcripts.

The parasites under discussion are genetically programmed to sequentially and specifically survive within widely different habitats. During these transitions, they display both non-stage-specific (glycolysis, DNA replication, transcription and translation, and gluconeogenesis) and stage-specific (respiratory pathways used by larval forms but not adults) metabolic strategies. While the inference of differential gene expression is easy to understand in the case of stage-specific gene expression, many workers have suggested that common metabolic pathways may be mediated by isoenzymes subject to stage-specific appearance. Studies in comparative enzymology have revealed that protein-catalyzed reactions are subject to ionic, pH, and temperature defined modulation. Temperature is particularly important, as it influences rates of reactions, rates of protein denaturation, affinity toward substrate, and pH dependence. These considerations may critically affect the function of near-equilibrium, regulatory, or membrane-bound enzymes (changes in temperature affect membrane fluidity).

How schistosomes and filariae have solved the problem of different stages adapting to different environments (for example, see Chapter 6) is unknown. The solution might involve the expression of separate genomic segments, the synthesis of enzymes that have broad physical optima, or synthesis of developmentally regulated factors that, depending on need, modulate enzyme function. Schistosome glyceraldehyde 3-phosphate dehydrogenase is encoded by a single-copy gene, indicating that the enzyme expressed during development in the snail host derives from the same genetic locus as that expressed during the parasitism of mammalian hosts. Some preliminary observations that deal with the developmental biology of the organisms are discussed next.

Developmentally Regulated Proteins of Unknown Function

Stage-specific gene expression by filariae and schistosomes requires a broad range of intrinsic genetic versatility, allowing the organisms to alternate from a poikilothermic to a homeothermic environment. Although investigations are still limited by technology (e.g., development of systems for genetic analysis, DNA transformation, and in vitro culture systems), examination of the subject has in principle been made more accessible by recent advances in molecular biology.

In one study, an expression cDNA library was screened with serum from schistosome-infected patients. Of 21 cDNA clones expressing parasite antigens, three proved to be developmentally regulated. Northern analyses of adult worm, egg, and cercarial poly A^+ RNA revealed that two of the three stage-specific cDNAs hybridized to mRNAs which change in length during development. While this observation in itself is not particularly unusual, the observation that the two RNAs appear to be transcribed from the same (single copy) gene is significant. How, and to what extent, schistosomes manage to produce multiple transcripts from a single gene is unknown; however, preliminary results indicate that the size variations noted above are due to the use of alternate start sites and alternative RNA processing.

Proteinases

One of the more surprising features of the schistosome proteins that stimulate a mammalian immune response is that most of these are synthesized by invertebrate as well as vertebrate hosts. A notable exception to this trend is the organism's proteinases, which appear to play specialized roles at different phases of development.

Helminth proteinases are interesting for several reasons: in addition to playing a role in parasite nutrition, they may be essential for invasion of mammalian hosts, they may play a role in stimulating inflammatory reactions, they may be important for differentiation, and they may participate in the production of egg yolk. It has been hypothesized that proteases secreted by the parasite or anchored onto its membranes may serve to evade the immune response by hydrolyzing bound immunoglobulin so that the Fc portion, necessary for recognition by effector cells, is removed. As many of these proteases are immunogenic, they may prove to be valuable diagnostic reagents and vaccine candidates.

Two hydrolases, referred to as hemoglobinases, responsible for the digestion of hemoglobin by schistosomes have recently been cloned and expressed in *Escherichia coli*. One of the enzymes is a thiol proteinase that displays enzymatic similarities to vertebrate cathepsin B_2 and is synthesized by cells lining the parasite gut. Products of its activity are hematin, which is regurgitated, and a large number of oligopeptides, which result from digestion of hemoglobin. The fate of the oligopeptides is unknown but it has been hypothesized that they are shuttled into the next anatomical compartment—the vitellaria, in the case of female worms, which consume the bulk of the hemoglobin—where they serve as egg yolk for the soon-to-develop miracidium (see Chapter 6). The enzyme is not essential to the parasites: they can survive for several months in protein-free, chemically defined media. On the other hand, worms maintained in culture quickly lose the ability to deposit eggs, and female schistosomula grown in media containing red blood cells reach sexual maturation, whereas those grown in the absence of the cells do not. Moreover, unisexual female worms, or female worms grown in nonpermissive hosts, contain little pigment in their guts and do not achieve sexual maturity.

The inferred amino acid sequence of one of the cloned cDNAs encoding the enzyme is not homologous to any known protease. This is somewhat surprising as cystinyl proteases from slime molds to humans share recognizable blocks of homology, especially around regions forming the active site. The other shows clear homology to vertebrate cathepsins. Northern analysis indicates that female worms make more of the enzymes than do male worms and that only trace levels are produced during the intramolluscan stage of the life cycle. The enzymes are predictably immunogenic in natural infections, induce a hypersensitivity response, and are one of the more promising substrates for development of an immunodiagnostic assay system.

Another stage- and cell-specific schistosome proteinase has been cloned from a sporocyst cDNA library. The cDNA encodes the cercarial elastase which facilitates the schistosome breach of the human skin barrier, and is arguably the best-characterized macromolecule of a parasitic helminth. It is a calcium-dependent serine protease that is exquisitely designed to fragment the major components of skin. In addition to elastase, the enzyme is capable of digesting keratin, types IV and VIII collagen, proteoglycan, fibronectin, and laminin.

Molecular cloning of the cercarial elastase has provided new information about how the enzyme is regulated, how it functions, and where it is synthesized. Nucleotide-sequence analyses of the cDNA predicts that its mRNA untranslated leader, which is decidedly AT-rich, is unusually long (>220 bp). The translation product of the longest open-reading frame indicates that the protein possesses an amino terminal extension of 27 amino acids not present in the mature enzyme (Figure 19-5). Because the extra amino acids are predominantly hydrophobic, they probably represent a peptide leader that maintains the enzyme in an inactive (zymogen) form. The length of the leader is also consistent with its potential functioning as a signal peptide that directs export of the elastase into secretion granules. Notably, activation of the enzyme is predicted to occur at a site recognized by the enzyme itself, suggesting that once a few zymogen molecules are activated by autocatalysis, a massive burst of autocatalytic activation might occur. The preproenzyme consists of 244 amino acids, which, aside from regions flanking the active site histidine, aspartate, and serine, and regions corresponding to probable protein-folding areas, shares little homology with that of other serine proteases. The mature protein starts with the sequence IsoArgSerGly, which differs radically from the consensus ValValGlyGly, characteristic of serine proteases from invertebrate and vertebrate sources. Together these data suggest that the immune response stimulated by the enzyme should not cross-react with host tissues.

In situ nucleic acid hybridization studies indicate that the cercarial protease mRNA is transcribed by the postacetabular gland cells (also known as the escape glands or penetration glands) during early stages of cercarial development, and studies involving monoclonal antibodies indicate that the inactive form of the enzyme is stored within these cells in secretory vesicles. Southern analyses indicate that the protein is encoded by one or a small number of genes that, because of the presence of intervening sequences, extend over several kilobases.

Understanding the biosynthesis and activation of the schistosome penetration elastase suggests that the potentiation of schistosome penetration of the host skin and its subsequent access to the circulatory system follows this course of events. At the 32-cell stage of cercarial development (when the acetabular cells, which are the largest cells in the organism, first become microscopically obvious) or shortly thereafter, the cercarial elastase gene is transcriptionally activated in acetabuler cells and an mRNA is expressed whose translational product is inactive and contains signals that direct it toward a subset of actabular secretion granules. Elastase mRNA synthesis ceases sometime before full maturation of the cercaria. Upon contact with skin lipids and stimulation by heat, the cercaria secretes the storage granules (see Chapter 6), and somehow a small number of preproelastase molecules become activated, perhaps by specific cleavage by another

```
                                                 −1
SE          MSNRWFVVVVTLFTYCLTFERVSTWL
REI         MLRFLVFASLV*YGHSTQDFPETNAR
REII   MIRTLLLSALVGALSCGYPTYEVQHDVS

       1                  20                  40                  60
SE     IRSGEPVQHPAEFPF-----IAFLTTERTMCTGSLVSTRAVLTAGHCVCSPLPVIRVSFL
REI    VVG*AEARR-NSW*SQISLQYLSGGSWYHT*G*T*IRRNW*M**A***S-----------
REI    VVG*QEASP-NSW*WQVSLQYLSSGKWHHT*G****ANNW****A**IS-----------

                          80                  100                 120
SE     TLRNGAQQGIHHQPSGVKVAPGYMPSCMSARQRRPIAQTLS--GFDIAIVMLAQMVNIQS
REI    SQMTF*VVVGD*NLSQNDGTEQ*VSVQKIMVHPTWNSNNVA-A*Y***LLR***S*T*NN
REII   NS*TYRVLLGR*SLSTSESGSLAVQVSKLVVHEKWN**K**NN*N***L*K**SP*A*T*

                          140                 160                 180
SE     GIRVISLPQPSDIPPPGTGVFIVGTGRDDDND---RDPSRLNGGILKKGRATIMECRHAT
REI    YVQLAV***EGT*LANNNPCY*T*W**TRT*GQLSQTLQQAYLPSVDYSICSSSSYWGS*
REII   K*QTAC**PAGT*LPNNYPCYVT*W**LQT*GAT*DVLQQGRLLVVDYATCSSASWWGSS

                          200                 220                 240
SE     NGNPICVKAGQNFGQLPAPGDSGGPLLPSLQGPVLGVVSHGVTLPNLPDIIVEYASVARM
REI    VKTTTM*C**GDGVRSGCQ*******HCL---VNGQYSV****SFVSSMGCNVSKKPTVF
REII   VKTTNM*C**GDGVTSSCN*******NCQ--ASNGQWQV**I*FFGSTLGCNYPRKPSVF

SE     VDFVRSNI
REI    TRVSAYISWMNNVIAYT
REII   TRVSNYIDWINSVIAKN
```

FIGURE 19·5 The predicted schistosome elastase (SE) amino acid sequence is compared with that of rat pancreatic elastases I and II. Abbreviations: REI, rat pancreatic elastase I; REII, rat pancreatic elastase II. Dashes indicate gaps and asterisks regions of identity. Components of the catalytic triad (histidine 46, aspartate 106, and serine 202) are shown in bold lettering, and the numbering system is arbitrary. The leader sequences are shown on the top lines.

unidentified parasite protease. The remaining share of elastase zymogen would then be activated by autocatalytic processing. The observation that the active enzyme can be isolated from cercariae artificially exposed to linoleic acid suggests that host factors are not important in its activation.

Preliminary data indicate that interference with the activity of cercarial elastase may be of practical value. Monoclonal antibodies that both inhibit the enzyme activity and are cytotoxic to schistosomula have been developed, suggesting that the enzyme may represent a realistic vaccine target. Studies involving the use of chemicals that inhibit the enzyme indicate that the enzyme is essential for successful parasite invasion; parasites whose elastases are irreversibly inactivated do not reach the vasculature and instead die in the skin.

Interestingly, the interstitial fluids of vertebrate skin contain proteins that are broad-spectrum inactivators of serine proteases. These antiproteases, which exist in multiple forms, are synthesized in the liver, from which they are subsequently secreted into the plasma. The reactive center of the inhibitors lies near their carboxy terminus and acts as bait for foreign proteases that covalently bind to it and are thereby inactivated. The finding that these proteins appear to have evolved in an accelerated Darwinian manner has led to the suggestion that they have been selected for their ability to recognize tissue-invasive parasites such as *Pseudomonas* organisms, schistosomes, and filariae. Whether helminth proteases secreted during tissue migration also have evolved in an accelerated manner is unknown, but it is noteworthy that the major protease of *Schistosomatium douthitti*, another digenetic trematode, is completely different from that of *mansoni*.

Schistosomes possess two other proteases that appear to be stage-specific and that are thought to have a specialized function. One is a surface-bound or secreted enzyme that degrades immunoglobulins so that the Fab fragment remains bound to the worm's tegument while the Fc fragment, which is the portion recognized by effector cells, is released into the circulation. The bound immunoglobulin therefore presents a hostlike barrier to the immune system. Some of the peptides generated by degradation of immunoglobulins appear to be immunosuppresive and may therefore further facilitate survival in the vertebrate host. The other enzyme is a protease synthesized within schistosome eggs. Although the function of this protease is unknown, some workers believe that it may play a role in assuring that eggs laid in the circulatory system manage to work their way into the lumen of the intestine.

Less is known about the proteases of filariae. Microfilariae of *Onchocerca volvulus* contain a collagenase that may facilitate their movement through interstitial spaces. The enzyme is immunogenic, shares epitopes with bacterial collagenases, and may act as a virulence factor by degrading host elastin and collagen fibers, thereby leading to loss of elasticity of the skin and the "hanging groin" syndrome. The enzyme also may expose new epitopes of host collagen molecules to the immune system, and therefore lead to autoimmune manifestations. *O. cervicalis* contains a metaloprotease of 60 kD and a serine protease of 30 kD which display collagenase activity. Extracts of *Dirofilaria immitis* microfilariae contain a 22-kD and a 76-kD protease. Proteolytic activity has also been noted in third-stage larvae of *Brugia pahangi* and *B. malayi*. The role of the latter in skin invasion has not been determined.

In general, the observation that tissue-invading helminths possess and secrete collagenolytic or elastinolytic enzymes is consistent with their migratory behavior. These classes of enzymes are notable for having a broad spectrum of substrate specificities; they therefore can degrade a large number of different proteins. The relative importance of these enzymes in the dissolution of tissue barriers, however, has proved difficult to assess. On an equally conjectural plane, it is possible that these enzymes play an additional role in preventing activation of host blood clotting mechanisms by interfering with proteolytic events in the activation of coagulation cascade.

Eggshell Proteins

The schistosome eggshell is a highly resilient, somewhat pliable biopolymer that appears to be composed of proteins cross-linked with each other by quinone bridges. The complex is highly resistant to proteases and contains pores that presumably allow the influx of nutrients required by developing miracidia and that permit the secretion of macromolecules, which lead to granuloma formation (see Chapter 14). Acid hydrolyzates of purified eggshells indicate that they are rich in glycine, aspartate, histidine, and lysine but contain unexpectedly low levels of tyrosine.

The predicted amino acid sequence of a cDNA encoding a probable component of the schistosome eggshell shows areas of considerable similarity to keratin and silkworm chorion proteins. The protein is unusual in that it contains regular repeats of GlyGlyGlyTyr, GlyGlyGlyCys, GlyGlyAsp, GlyGlyAsn, and GlyGlyLys, which indicates that it tends to form β-pleated sheets.[1] The protein is tyrosine-rich, a feature consistent with the probability that the eggshells are cross-linked by quinone tanning, a process involving the hydroxylation of tyrosine, the oxidation of hydroxyl residues on the aromatic ring, and covalent binding of the latter to free amino groups of other amino acids. Production of the eggshell mRNA appears to be mediated by a small gene family that is under transcriptional control and that, unlike silkworm chorion genes, is not amplified during expression.

The predicted amino acid sequence of another cDNA encoding a probable eggshell protein of 48 kD also demonstrates significant identity with chorion proteins. It consists of a series of pentapeptide repeats having a consensus sequence of GlyTyrAspLysTyr. The repeats would be predicted to lead toward a structure consisting of alternating beta turns and beta sheets, and the carboxy terminal portion of the peptide contains a histidine-rich region likely to form an alpha helix. The protein is remarkably rich in tyrosine, a feature that indicates that it is not an abundant component of the eggshell, as hydrolysis of the complex yields little of this amino acid. Treatment of the protein with phenol oxidase results in the formation of covalently linked polymers.

[1] Gly, glycine; Tyr, tyrosine; Cys, cysteine; Asp, aspartate; Asn, asparagine; Lys, lysine.

Epidemiology

Several laboratories have applied techniques of molecular biology toward the development of probes that might be useful in identifying infected mosquito, blackfly, or mollusc vectors or in assessing the occurrence, species, and distribution of larvae within in large areas. The Onchocerchiasis Control Program, for example, has, over the last 13 years, relied on vector-control measures to manage onchocerchiasis. It has assessed its success by assaying the prevalence of infective larvae in the blackfly population over a huge area of Africa. In recognition of the fact that light infections are difficult to detect and that onchocerchiasis has a relatively long prepatent period, the program has not relied on determination of the incidence of infection in the human population.

Repetitive Sequences

The major portion of the genome of eukaryotic cells consists of nontranscribed repetitive sequences whose function is poorly understood. Some workers theorize that these sequences play a structural role, somehow stabilizing the genome and perhaps even regulating gene expression, while others feel that these sequences provide a plasticity that may accelerate evolutionary processes. Repetitive sequences include duplicated pseudogenes, transposons, and satellite DNA. They can be clustered or dispersed throughout unique sequence DNA, and since they lack constraints imposed on functional genes, they are evolutionarily unstable.

In most respects, the composition of helminth genomes has proven similar to that of other invertebrates. Their genome size is on the order of 10^8 bp, and their GC content ranges from 43 percent to 47 percent, compared with roughly 38 percent to 47 percent in mammals and 30 percent to 45 percent in invertebrates (gross exceptions exist). Modified bases have not been found in parasitic helminths, suggesting that transcriptional regulation through methylation may not be of general importance. Although the genome size of parasitic helminths tends to be roughly twice that of free-living relatives (which posess some of the smallest genomes among the metazoa), the fraction consisting of repetitive DNA tends to be lower. Thermal denaturation studies indicate that the increase in genome size of these parasites is due to the introduction of low-copy-number sequences.

Several workers have taken advantage of the inherent variability of repetitive sequences to develop nucleic acid probes that allow the identification of filariae in infected flies or mosquitoes, in blood samples, or in skin biopsies. Since these sequences exist in large copy number, these probes have the potential of being sensitive. Although repetitive sequences have a high rate of drift, they may show considerable homology within a genus, thus limiting their usefulness. In one study, detailed later, this problem was circumvented by constructing oligonucleotide probes that spanned areas of divergence between different species, thus allowing species-specific identification of the parasites.

In the case of *Onchocerca* species, an important question that might be clarified by the use of repetitive DNA probes is whether forest and savannah strains of the parasite have different biological characteristics. The ability to discriminate between the two forms of the parasite also may assist in determining whether outbreaks of infection are due to invasion of forest- or savannah-dwelling flies. Preliminary indications are that the geographically separated strains may lead to different clinical manifestations and that they might be separable on the basis of differences in isoenzyme patterns. Early studies indicate that probes hybridizing to repetitive DNA may help to speciate members of the genus *Onchocerca* and to discriminate between forest and savannah types. The probes are sensitive enough to identify a single infective larva.

In the case of *Brugia* spp., it is felt that development of species-specific probes may further our understanding of the parasite host range. *B. malayi* is thought to infect humans, nonhuman primates, and several other vertebrates, whereas *B. timori* is thought to infect only humans. *B. pahangi* is known to infect domestic animals and is a suspected human pathogen. A genomic clone representing a repetitive sequence of *B. malayi* has been isolated. Twenty-five percent of the parasite's genome consists of repetitive DNA, and the isolated repeat, which consists of a tandemly arrayed 322-bp sequence repeated 30,000 times per haploid genome, accounts for approximately half of this DNA. Under stringent conditions, the probe can easily distinguish *Brugia* from other genera of filariae, but it differs from the major repeat of *B. pahangi* by only 11 percent. As differences between the two repeats are clustered in pockets, workers have designed oligonucleotides that can easily discriminate between the two species.

Ribosomal Genes and their RNAs

Eukaryotic ribosomal RNA genes are typically part of multigene families whose members are tandemly linked (mRNA) in the genome. In general the arrangement follows a plan wherein the gene encoding the 18S RNA subunit lies 5′ to that encoding the 5.8S mRNA subunit, the latter being placed 5′ to that encoding the 28s mRNA subunit. The three subunits are transcribed by RNA polymerase I as a polycistronic unit whose spacer regions are subsequently removed by RNA splicing. Processed ribosomal RNAs are highly conserved in nucleotide sequence, a feature that has assisted in clarification of the taxonomic relationships between helminth parasites. The intergenic regions, on the other hand, vary in length and have evolved more freely. Since these spacer regions represent a homogeneous family of repeated sequences, they may present species-specific differences that can be exploited toward the development of highly discriminating probes.

Polymorphisms in RNA genes also have been exploited as a means of determining the species, sex, and strain of schistosomes within their molluscan host. There exist 17 known species of schistosomes, which are categorized into three groups on the basis of egg morphology. In addition, there appear to exist several strains of the parasite, a phenomenon that may account for differences in infectivity, virulence, isoenzyme patterns, and drug resistance among schistosomes. Although these biological characteristics are unlikely to be related to polymorphisms in the structure of rRNA genes, their fixation within discrete subpopulations may nonetheless allow strain identification and thereby provide a marker that can be correlated later with differences in pathogenicity, longevity, and overall biological fitness.

Coding sequences of helminth rRNA genes also have been used to determine phylogenetic distances between the organisms. Sequence analyses of the 5′-end of the 28s rRNA of various nematodes and schistosomes indicate that the differences in size and sequence of these molecules occur within a few hotspots that have retained the same position relative to the core of all mRNAs examined to date. Preliminary data indicate that nematodes radiated a long time ago: the rRNA sequence of *Nematospiroides dubius* is as similar to that of *O. gibsoni* and *B. pahangi* as it is to that of *S. mansoni*.

Immunodiagnosis

Definitive diagnosis of schistosomiasis rests on the demonstration of parasite ova in stool or urine specimens. Similarly, detection of filariasis depends on the visual detection of microfilariae in blood or skin biopsy. Available methods to detect larval parasites are time consuming and insensitive, especially when applied in large epidemiological surveys—problems that have stimulated the development of immunologically based assay systems. Antigens that have rendered assays reproducible, sensitive, and specific have been identified: four of these have been cloned; two are proteolytic enzymes that are discussed elsewhere in this chapter. Preliminary studies on the antibody response to schistosome peptides indicate that the response is erratic, demonstrates little correlation with clinical status, and persists for a long period after chemotherapeutic cure, thus limiting the information garnered by assay of specific antibodies.

Nevertheless, the search for immunodiagnostic schistosome and filarial antigens has provided significant information relevant to understanding helminth biology. Strangely enough, it appears that immunodominant antigens of schistosomes include a family of macromolecules called heat-shock proteins (HSPs). HSPs are phylogenetically conserved proteins expressed universally by organisms in response to environmental stress. The function of HSPs is poorly understood, although they appear to protect cells from damage and at the same time rid them of denatured macromolecules. Constitutively expressed HSP cognates are thought to play a role in mediating the assembly or disassembly of higher macromolecular structures; in surveying the cytoplasm for denatured, improperly modified, or foreign proteins; and in mediating the insertion of proteins into, or their translocation across, cell membranes. HSP cognates are subject to developmental regulation and are thought by some to be critical factors of differentiation.

One of the major antigens of schistosome eggs is a 40-kD protein referred to as p40. The protein is also found in adult worms and elicits a strong immunological response in over 90 percent of infected humans. The antigen is actually a family of four nearly identical proteins that are differentially expressed during the schistosome life cycle. One form of the peptide is produced by female worms, two in eggs, and four in adult worms.

The sequence of p40, based on the nucleotide sequence of a cDNA, reveals that it is homologous to *Drosophila* and dogfish α-crystallins and a family of eukaryotic small heat-shock proteins (Figure 19-6). The primary sequence of the schistosome protein indicates that it is a descendant of a protein that was ancestral to the functionally different α-crystallins and small HSPs. The amino-terminal portion of the protein consists of a long stretch homologous to the α-crystallins, a central portion homologous to the carboxy terminus of small HSPs, an amino-proximal region homologous to α-crystallins, and a carboxy-terminal tail similar to that of some small HSPs (Figure 19-6). This sequence suggests that the protein is the result of an early intragenic duplication. The function of the protein is unknown, although its similarity to the α-crystallins and small HSPs suggests that it probably forms large aggregates that somehow stabilize cellular structures.

Background studies on the human immune response to *S. mansoni* indicate that although the response is erratic, virtually all infected individuals produce sera that react to the two schistosome proteins, one having a relative molecular weight (Mr) of 70 kD and the other an Mr of 38 kD. Antibodies to these schistosome proteins are not detectable in uninfected individuals or in persons infected with other parasitic helminths (species of *Brugia*, *Onchocerca*, *Chlonorchis*, *Heterobilharzia*, *Ascaris*, *Tri-*

```
Schist p40    GEDGKVHFKVRFDAQGFAPQDINVTSSEN-RVTVHAKKE-TTTD--G
Schist p40'   GSG**RLHVEVAVDPVYK*E*LF*NVDS*-**VVSGRHHKQKS*QH*
α-Cryst A     VRSDRDK*VIFL*VKH*S*E*LT*KVQ*D-F*EIHG*HN-ERQ*DH-
hHSP 27       IRHTADRWR*SL*VNH***DELT*KTKDG-VREITG*H*-ERQ*EH-
dHSP 22       ATVN*DGY*LTL*VKD*S--ELK*KVLDESV*L*E**S*-QQEAEQ*
DHSP 27       PAV**DG*Q*CM*VSQFK*NELT*KVVD*-T*V*EG*H*-ERE*GH-

Schist p40    R-----KCSREFCRMVQLPKSIDDSQLKCRMTDDGVLMLEAPVKV---
Schist p40    *----SSSFA**SQS*AI*ETV*PLSVSAQV-VGNT*V****LEK---
α-Cryst A     -----GYI****H*R*R**SNV*Q*A*S*SLSA**M*TFSG*KIPSGV
hHSP 27       -----GYI****T*K*T**PGV*PT*VSSSLSPE*T*TV***MPKLAT
dHSP 22       -----GYS**H*LGR*V**DGYEADKVSSSLS****TISV*-NPPGV
dHSP 27       -----GMIQ*H*V*K*T***GF*PNEVVSTVSS****T*K**-PPPSK

SCHIST P40    -DQNQSLTLNESGQVAVRPKSDNQIKAVPASQALVAKGVHGLSYVDD
SCHIST P40'   ------------QHAITH
α-Cryst       DAGHSERAIPV*REEKP--SSAPSS
hHSP 27       Q-S*-EI*IPVTFESRAQLGGRSCKIR
dHSP 22       QETLKEREVTIEQTGEPAK**AEEP*DKT***
dHSP 27       EQAKSERIVQIQQTG-PAHLSVKAPAPEAGDGKAENGSGEKMETSK
```

FIGURE 19·6 Regions of homology between schistosome egg antigen p40, bovine α-crystallin A2, *Drosophila* small heat-shock proteins (dHSP) 22 and 27, and human heat-shock protein 27 (hHSP 27). Sequences are aligned. The egg antigen is represented as follows: p40, amino acids 136–262; p40′, amino acids 263–354.

churis, *Ancylostoma*, or *Echinococcus*), indicating that they may prove to be of diagnostic value. The 70-kD schistosome antigen has been identified as an HSP-70 homologue; epitope mapping has localized the immunogenic portion(s) of the molecule at the carboxy terminus of the protein, the least conserved domain of HSP-70s in general (Figure 19-7). Similar properties are displayed by the *S. japonicum* HSP-70 cognate. Oddly, however, the 70-kD schistosome immunogens do not cross-react with each other. As immunization of rabbits with chicken HSP-70 leads to a response that cross-reacts with the homologue of fruit flies and humans, these findings indicate that schistosomes present the protein to the immune system in a manner that avoids an autoimmune response. Along similar lines, it has recently been noted that *P. falciparum* produces two different HSP-70s that are immunodominant and non-cross-reactive. These findings are of interest because they suggest that the parasite species-specific nature of the protective immune response generated by immunization with live-attenuated vaccines is not attributable to fundamentally different processes or targets, but rather to differences in the structure of homologous parasite antigens. It is possible, however, that HSP-70s, molecules whose three-dimensional configuration is unknown, selectively present their hypervariable carboxy termini to the immune system, a property lost when animals are immunized with the denatured protein.

The function of the schistosome HSP-70 is unknown. Curiously, HSP-70 also has been found to be a dominant immunogen of filariae, trypanosomes, plasmodia, and mycobacteria. In other eukaryotes, HSP-70 is considered to be a molecular chaperone that escorts and influences the conformation of proteins destined to become part of, or which are going to be removed from, higher molecular assemblies. HSP-70 also plays a role in the translocation of proteins across cellular membranes. This process is thought to involve hydrophobic interactions between HSP-70 and the protein to be translocated, and it appears to be associated with conformational changes mediated by the binding and hydrolysis of ATP by HSP. HSP-70 has been implicated as playing a central role in the recognition of denatured proteins and in collaboration with ubiquitin, tagging these with signals recognized by lysozomal proteases. Interestingly, synthesis of HSP-70 increases during senescence of fruit flies, perhaps as a result of increases in the production of improperly formed or glycosylated proteins. In light of the finding that the schistosome life expectancy is unusually long for an invertebrate, it would be interesting to see whether the same holds true for them.

The schistosome HSP-70 is found in the tegument of adult worms and in the hatching fluid, surface cells, and nervous system of miracidia. Synthesis of the protein is constitutive in adult worms and sporocysts, although apparently it does not appear to be made by cercariae. The latter nonetheless maintain a previously synthesized pool of the protein. Transforming schistosomula synthesize massive amounts of HSP during a period in which production of nearly every other protein is repressed. The cue for HSP gene transcription in transforming larvae is not heat, as cercariae subjected to heat shock do not synthesize the protein, whereas cercariae placed in mammalian culture media do. Synthesis of other proteins slowly resumes 16-hour posttransformation, eventually plateauing to steady-state levels. Whatever the function of HSP-70, its mRNA is one of the more abundant species in adult worms, accounting for 0.1 percent to 1.0 percent of

```
     445
Sm   E R A L T K D N N L L G K F E L S G I P P A P R G T P Q I E V T F D I D A N G I L N V S A V D K G T G K Q N K I T I
Sj   * * T M * * * * * * * * * * * * * * C * * * * * * * V * * * * * * * * * * * * * * * * * * * * * * A * * * E * * * * *
Hs   * * * M * * * * * * * * * R * * * * * * * * * * * * V * * * * * * * * * * * * * * * * * * * T * T K D S * * * A * * * * *

Sm   T N D K G R L S K E E I E R M V A D A D K Y K A E D E K Q R D R V A S K N S L E S Y V Y T M K Q Q V E G E - L K E K
Sj   * * * * * * * * * * * * D * * I N E * * * * * S * * * * * K N * I C * * * * * * * * * * S * * * S * * * D E M * D *
Hs   * * * * * * * * * * * * Q E * E * * * * * * * * V * * E * * * * * * A * * * * A F N * * A F N * * * S A * * E G * * G *

Sm   I P E S D H - Q V I I S K C E D T I S W L D V H Q S A E K H E T E S K R E E L E K V C A P I I T K V Y Q - A - - - -
SJ   * S * * * R K N R * L * * * * E * * R * M * M N * L * * * E * F * E * K S * * * * * * M * * * * A M N R - * - - - -
Hs   * S * A * K K K * - L D * * Q E V * * * * * A N T L * * * D * F * H * * K * * * Q * * N * * * S G L * * G * - - - -

Sm   - - - - - G G M P G G M H E A S - - G A C G S G - - - - - - - - - - - - - - - - - - - - - - - - - - K G P T I E E V D
Sj   - G G V P S * * * * * * P - - - - - * * * * G * - - - - - - - - - - - - - - - - - - - - - - - - - - * * * * * * * * *
Hs   - - - - - * * - * * * - F G * Q - G P K * * * * - - - - - - - - - - - - - - - - - - - - - - - - - - S * * * * * * * *
```

Figure 19·7 Comparison of the predicted carboxy-terminal amino acid sequences of HSP 70 of *Schistosoma japonicum* (Sj), and *Homo sapiens* (Hs) to that of *Schistosoma mansoni* (Sm). Asterisks denote points of identity with the *S. mansoni* sequence, and dashes represent gaps.

the total message. The observation that the protein is synthesized primarily by tegumental cells suggests that it may play a role in maintaining the integrity of this anatomical compartment.

The observation that despite their similarities, the HSP-70s of *S. mansoni* and *S. japonicum* stimulate production of non-cross-reacting antibodies during the course of natural infection can be taken as a good omen for vaccine development. It seems that in many experimental systems, immunity to the parasites is often specific to the species of sensitizing organism. The response to HSP-70 suggests that this may be due more to interspecific differences in the structure of parasite antigens than to different effector mechanisms or target antigens, and that once a protective antigen is found for one parasite, it may be possible to clone equally active homologues from species that are currently difficult or impossible to maintain in the laboratory. A case in point is the identification of paramyosin as a vaccine target in schistosomes, and subsequent demonstration of the prophylactic potential against filariasis of the nematode homologue.

Coda

Although studies on the molecular biology of schistosomes and filariae are at a stage of infancy, they have significantly contributed to the understanding of both the biology of the organisms and their interaction with their hosts. Many of the parasite antigens have now been assigned a biological function, a significant advance over their previous indexing in the literature on the basis of molecular weight. At the same time, the field has posed a number of new questions to researchers interested in biological and immunological aspects of parasite relationships with their hosts. The answers to these questions are likely to refine efforts to develop vaccines and immunodiagnostic systems. In conjunction with DNA probe technologies, such as those using the polymerase chain reaction, these studies offer the potential to develop highly sensitive probes that could detect and determine the species (and perhaps even strain) of parasites obtained in small biological samples. At a more basic level, the observation that parasitic nematodes possess both cis- and trans-splicing activities makes them a convenient system for studying the mechanics of eukaryotic RNA processing.

Additional Reading

COMLEY, J. C. W., and A. H. W. MENDIS. 1986. Advances in the biochemistry of filariae. *Parasitol. Today* 2:34–37.

PHILIPP, M., T. B. DAVIS, N. STOREY, and C. K. S. CARLOW. 1988. Immunity in filariasis: perspectives for vaccine development. *Ann. Rev. Microbiol.* 42:685–716.

ROLLINSON, D., T. K. WALKER and A. J. G. SIMPSON. 1986. New approaches to schistosome identification. *Parasitol. Today* 2:24–25.

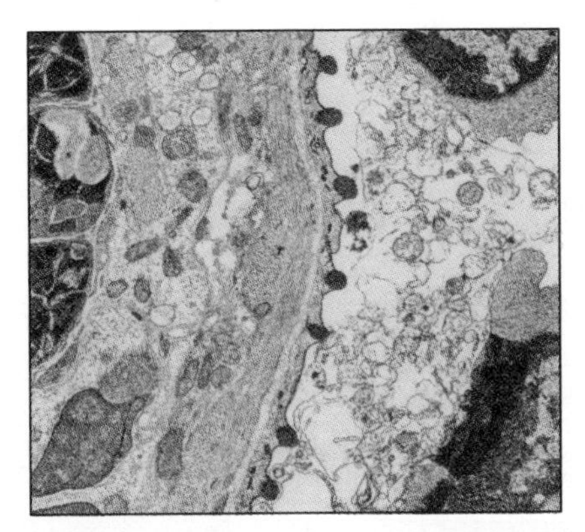

PRIMARY SOURCE PAPERS

Primary Source Papers

Part I Cell Biology

Chapter 1. Cell Biology of *Plasmodium*

ALEY, S. B., M. D. BATES, J. P. TAM, and M. R. HOLLINGDALE. 1986. Synthetic peptides from the cricumsporozoite proteins of *Plasmodium falciparum* and *Plasmodium knowlesi* recognize the human hepatoma cell line HepG2-A16 *in vitro*. *J. Exp. Med.* 164:1915–1922.

BARNWELL, J. W., C. F. OCKENHOUSE, and D. M. KNOWLES. 1985. Monoclonal antibody OKM5 inhibits the *in vitro* binding of *Plasmodium falciparum* infected erythrocytes to monocytes, endothelial and C32 melanoma cells. *J. Immunol.* 135:3494–3497.

CARTER, R., and M. M. NIJHOUT. 1977. Control of gamete formation (exflagellation) in malarial parasites. *Science* 195:407–409.

CARTER, R., and D. C. KAUSHAUL. 1984. Characterization of antigens on mosquito midgut stages of *Plasmodium gallinaceum* III. Changes in zygote surface proteins during transformation to mature ookinete. *Mol. Biochem. Parasitol.* 13:235–241.

COPPEL, R. L., J. G. CULVENOR, A. E. BIANCO, P. E. CREWTHER, H.-D. STAHL, G. V. BROWN, R. F. ANDERS, and D. J. KEMP. 1986. Variable antigen associated with the surface of erythrocytes infected with mature stages of *Plasmodium falciparum*. *Mol. Biochem. Parasitol.* 20:265–277.

DLUZEWSKI, A. R., K. RANGACHARI, R. J. M. WILSON, and W. B. GRATZER. 1983. Properties of red cell ghost preparations susceptible to invasion by malaria parasites. *Parasitology* 87:429–438.

HOLLINGDALE, M. R., E. H. NARDIN, S. THARAVANIJ, A. L. SCHWARTZ and R. S. NUSSENZWEIG. 1984. Inhibition of entry of *Plasmodium falciparum* and *P. vivax* sporozoites into cultured cells: an *in vitro* assay of protective antibodies. *J. Immunol.* 132:909–913.

KUTNER, S., W. V. BREUER, H. GINSBURG, S. D. ALEY, and Z. I. CABANTCHICK. 1985. Characterization of permeation pathways in the plasma membrane of human erythrocytes infected with early stages of *Plasmodium falciparum:* association with parasite development. *J. Cell. Physiol.* 125:521–527.

LYON, J. A., R. H. GELLER, J. D. HAYNES, J. D. CHULAY, and J. L. WEBER. 1986. Epitope map and processing scheme for the 195,000 dalton surface glycoprotein of *Plasmodium falciparum* merozoites deduced from cloned overlapping segments of the gene. *Proc. Natl. Acad. Sci. USA.* 83:2989–2993.

MAZIER, D., R. L. BEAUDOIN, S. MELLOUK, P. DRUILHE, B. TEXIER, J. TROSPER, F. MILTGEN, I. LANDAU, C. PAUL, O. BRANDICOURT, C. GUGUEN-GUILLOUZO, and P. LANGLOIS. 1985. Complete development of hepatic stages of *Plasmodium falciparum in vitro*. *Science* 227:440–442.

MREMA, J. E. K., S. G. LANGRETH, R. C. JOST, R. H. RIECKMANN, and H. G. HEIDRICH. 1982. *Plasmodium falciparum:* isolation and purification of spontaneously released merozoites by nylon membrane seive. *Exp. Parasitol.* 54:285–295.

PERKINS, M. E. 1981. The inhibitory effects of erythrocyte membrane proteins on the *in vitro* invasion of the human malarial parasite into its host cell. *J. Cell. Biol.* 90:563–567.

POLOGE, L., and J. V. RAVETCH. 1986. Chromosomal rearrangement in a *P. falciparum* histidine rich protein gene is associated with knobless phenotype. *Nature (Lond.)* 322:474–477.

PONNUDURAI, T., A. H. W. LENSEN, A. D. E. M. LEEUWENBERG, and J. H. E. TH. MEUWISSEN. 1982. Cultivation of fertile *Plasmodium falciparum* gametocytes in semiautomated systems. I. Static cultures. *Trans. R. Soc. Trop. Med. Hyg.* 76:812–818.

ROBERTS, D. D., J. A. SHERWOOD, S. L. SPITALNIK, L. J. PANTON, R. J. HOWARD, V. M. DIXIT, W. A. FRAZIER, L. H. MILLER, and V. GINSBURG. 1985. Thrombospondin binds falciparum malaria parasitized erythrocytes and may mediate cytoadherence. *Nature (Lond.)* 318:64–66.

SAUL, A., G. LAMONT, W. H. SAWYER, and C. KIDSON. 1984. Decreased deformability in melanesian ovalocytes from Papua, New Guinea. *J. Cell Biol.* 98:1348–1354.

TRAGER, W., and J. B. JENSEN. 1976. Human malaria parasites in continuous culture. *Science* 193:674–675.

Chapter 2. Cell Biology of *Toxoplasma gondii*

AIKAWA, M., Y. KOMATA, T. ASAI, and O. MIDORIKAWA. 1977. Transmission and scanning electron microscopy of host cell entry by *Toxoplasma gondii*. *Am. J. Pathol.* 87:285–296.

CANNING, E. U., and M. ANWAR. 1968. Studies on meiotic division in coccidial and malarial parasites. *J. Protozool.* 15:290–298.

CORNELISSEN, A. W. C. A., J. P. OVERDULVE, and M. VAN DER PLOEG. 1984. Determination of nuclear DNA of five eucoccidian parasites, *Isospora (Toxoplasma) gondii*, *Sarcocystis cruzi*, *Eimeria tenella*, *E. acervulina* and *Plasmodium berghei*, with special reference to gamontogenesis and meiosis in *I. (T.) gondii*. *Parasitology* 88:531–553.

DUBREMETZ, J. F., C. RODRIGUEZ, and E. FERREIRA. 1985. *Toxoplasma gondii:* redistribution of monoclonal antibodies on tachyzoites during host cell invasion. *Exp. Parasitol.* 59:24–32.

FRENKEL, J. K., J. P. DUBEY, and N. L. MILLER. 1970. *Toxoplasma gondii* in cats: fecal stages identified as coccidian oocysts. *Science* 167:893–896.

HUTCHINSON, W. F., J. F. DUNACHIE, J. C. SIIM, and K. WORK. 1970. Coccidian-like nature of *Toxoplasma gondii*. *Br. Med. J.* 1:142–144.

NICHOLS, B. A., and M. L. CHIAPPINO. 1987. Cytoskeleton of *Toxoplasma gondii*. *J. Protozool.* 34:218–227.

NICHOLS, B. A., M. L. CHIAPPINO, and G. R. O'CONNOR. 1983. Secretion from the rhoptries of *Toxoplasma gondii* during host-cell invasion. *J. Ultrastruct. Res.* 83:85–98.

NORRBY, R. 1971. Immunological study on the host cell penetration factor of *Toxoplasma gondii*. *Infect. Immun.* 3:278–286.

OVERDULVE, J. P. 1970. The identity of Toxoplasma Nicolle and Manceaux, 1909, with Isospora Schneider, 1881. *Proceedings Koninklijke Nederlandse Akademie van Wetenschappen, Series C* 73:129–141.

PFEFFERKORN, E. R.. 1978. *Toxoplasma gondii:* the enzymic defect of a mutant resistant to 5-fluorodeoxyuridine. *Exp. Parasitol.* 44:26–35.

PFEFFERKORN, E. R., and L. H. KASPER. 1983. *Toxoplasma gondii:* genetic crosses reveal phenotypic suppression of hydroxyurea resistance by fluorodeoxyuridine resistance. *Exp. Parasitol.* 55:207–218.

PFEFFERKORN, E. R., and L. C. PFEFFERKORN. 1976. *Toxoplasma gondii:* isolation and preliminary characterization of temperature-sensitive mutants. *Exp. Parasitol.* 39:365–376.

PFEFFERKORN, E. R., and L. C. PFEFFERKORN. 1977. *Toxoplasma gondii:* specific labeling of nucleic acids of intracellular parasites in Lesch-Nyhan Cells. *Exp. Parasitol.* 41:95–104.

PFEFFERKORN, E. R., and L. C. PFEFFERKORN. 1978. The biochemical basis for resistance to adenine arabinoside in a mutant of *Toxoplasma gondii*. *J. Parasitol.* 64:486–492.

PFEFFERKORN, E. R., and L. C. PFEFFERKORN. 1979. Quantitative studies of the mutagenesis of *Toxoplasma gondii*. *J. Parasitol.* 65:364–370.

PFEFFERKORN, E. R., and L. C. PFEFFERKORN. 1981. *Toxoplasma gondii:* growth in the absence of host cell protein synthesis. *Exp. Parasitol.* 52:129–136.

PFEFFERKORN, E. R., and L. C. PFEFFERKORN. 1980. *Toxoplasma gondii:* genetic recombination between drug resistant mutants. *Exp. Parasitol.* 50:305–316.

PFEFFERKORN, E. R., L. C. PFEFFERKORN, and E. D. COLBY. 1977. Development of gametes and oocysts in cats fed cysts derived from cloned trophozoites of *Toxoplasma gondii*. *J. Parasitol.* 63:158–159.

SCHWARTZMAN, J. D. 1986. Inhibition of a penetration-enhancing factor of *Toxoplasma gondii* by monoclonal antibodies specific for rhoptries. *Infect. Immun.* 51:760–764.

SCHWARTZMAN, J. D., and E. R. PFEFFERKORN. 1981. Pyrimidine synthesis by intracellular *Toxoplasma gondii*. *J. Parasitol.* 67:150–158.

SCHWARTZMAN, J. D., and E. R. PFEFFERKORN. 1982. *Toxoplasma gondii:* purine synthesis and salvage in mutant host cells and parasites. *Exp. Parasitol.* 53:77–86.

SHEFFIELD, H. G., and M. L. MELTON. 1970. *Toxoplasma gondii:* the oocyst, sporozoite, and infection of cultured cells. *Science* 167:892–893.

SIBLEY, L. D., J. L. KRAHENBUHL, G. M. W. ADAMS, and E. WEIDNER. 1986. *Toxoplasma* modifies macrophage phagosomes by secretion of a vesicular network rich in surface proteins. *J. Cell Biol.* 103:867–874.

TANABE, K., and K. MURAKAMI. 1984. Reduction in the mitochondrial membrane potential of *Toxoplasma gondii* after invasion of host cells. *J. Cell Sci.* 70:73–81.

Chapter 3. Cell Biology of African Trypanosomes

AMAN, R. A., and C. C. WANG. 1986. Absence of substrate channeling in the glycosome of *Trypanosoma brucei*. *Mol. Biochem. Parasitol.* 19:1–10.

BULOW, R., and P. OVERATH. 1986. Purification and characterization of the membrane-form variant surface glycoprotein hydrolase of *Trypanosoma brucei*. *J. Biol. Chem.* 261:11918–11923.

CARDOSO DE ALMEIDA, M. L., and M. J. TURNER. 1983. The membrane form of variant surface glycoproteins of *Trypanosoma brucei*. *Nature* 302:349–352.

CROSS, G. A. M. 1975. Identification, purification and properties of clone-specific glycoprotein antigens constituting the surface coat of *Trypanosoma brucei*. *Parasitology* 71:393–417.

CROSS, G. A. M. 1984. Structure of the variant glycoproteins and surface coat of *Trypanosoma brucei*. *Phil. Trans. R. Soc. Lond. B* 307:3–12.

FERGUSON, M. A. J., M. G. LOW, and G. A. M. CROSS. 1985. Glycosyl-1,2-dimyristylphosphatidylinositol is covalently linked to *Trypanosoma brucei* variant surface glycoproteins. *J. Biol. Chem.* 260:14547–14555.

FREYMANN, D., P. METCALF, J. J. TURNER, and D. WILEY. 1984. The 6Å resolution X-ray structure of a variable surface glycoprotein from *Trypanosoma brucei*. *Nature* 311:167–169.

GALLO, J. M., and B. H. ANDERTON. 1983. A subpopulation of trypanosome microtubules recognized by a monoclonal antibody to tubulin. *Embo J.* 2:479–483.

JENNI, L., S. MARTI, J. SCHWEIZER, B. BETSCHART, R. W. LEPAGE, J. M. WELLS, A. TAIT, P. PAINDAVOINE, E. PAYS, and M. STEINERT. 1986. Hybrid formation between African trypanosomes during cyclical transmission. *Nature* 322:173–175.

KIMMEL, B. E., S. SAMSON, J. WU, R. HINCHBERG, and L. R. YARBROUGH. 1985. Tubulin genes of the African trypanosome *Trypanosoma brucei rhodesiense*. Nucleotide sequence of a 3.7 kb fragment containing genes for alpha and beta tubulin. *Gene* 35:237–248.

LOW, M. G., M. A. J. FERGUSON, A. H. FUTERMAN, and I. SILMAN. 1986. Covalently attached phosphatidylinositol as a hydrophobic anchor for membrane proteins. *Trends Biochem. Sci.* 11:212–215.

MESHNICK, S. R. 1986. Molecular pharmacology of protozoan parasites. *Parasitol. Today* 2:221–223.

OPPERDOES, F. R., P. BAUDHUIN, I. COPPENS, C. DEROE, S. W. EDWARDS, P. J. WEIJERS, and O. MISSET. 1984. Purification, morphometric analysis and characterization of the glycosomes (microbodies) of the protozoan hemoflagellate *Trypanosoma brucei*. *J. Cell Biol.* 98:1178–1184.

OSINGA, K. A., B. W. SWINKEL, W. C. GIBSON, P. BORST, G. H. VEENEMAN, J. H. VAN BOOM, P. A. M. MICHELS, and F. R. OPPERDOES. 1985. Topogenesis of microbody enzymes: a sequence comparison of the genes for the glycosomal (microbody) and cytosolic phosphoglycerate kinases of *Trypanosoma brucei*. *EMBO J.* 4:3811–3817.

REW, R. S., and R. H. FETTERER. 1985. Mode of action of antinematodal drugs. In *Chemotherapy of Parasitic Diseases*, W. C. Campbell and R. S. Rew, eds. Plenum Press, pp. 321–337.

RICE-FICHT, A. C., K. K. CHEN, and J. E. DONELSON. 1981. Sequence homologies near the C-termini of the variable surface glycoproteins of *Trypanosoma brucei*. *Nature* 294:53–57.

RIFKIN, M. R. 1978. Identification of the trypanocidal factor in normal human serum: high density lipoprotein. *Proc. Natl. Acad. Sci.* USA 75:3450–3454.

RIFKIN, M. R. 1984. *Trypanosoma brucei:* biochemical and morphological studies of cytotoxicity caused by normal human serum. *Exp. Parasitol.* 58:81–93.

RUSSELL, D. G., D. MILLER, and K. GULL. 1984. Tubulin heterogeneity in the trypanosoma *Crithidia fasciculata*. *Mol. Cell. Biol.* 4:779–790.

SEEBECK, T., and P. GEHR. 1983. Trypanocidal action of neuroleptic phenothiazines in *Trypanosoma brucei*. *Mol. Biochem. Parasitol.* 9:197–208.

SHAPIRO, S. Z. 1986. *Trypanosoma brucei*: release of variant surface glycoprotein during the parasite life cycle. *Exp. Parasitol.* 61:432–437.

STIEGER, J., and T. SEEBECK. 1986. Monoclonal antibodies against a 60 kDa phenothiazine-binding protein from *Trypanosoma brucei* can discriminate between different trypanosome species. *Mol. Biochem. Parasitol.* 21:37–45.

STIEGER, J., T. WYLER, and T. SEEBECK. 1984. Partial purification and characterization of microtubular protein from *Trypanosoma brucei*. *J. Biol. Chem.* 259:4596–4602.

TAIT, A. 1980. Evidence for diploidy and mating in trypanosomes. *Nature* 287:536–538.

VICKERMAN, K., and T. M. PRESTON. 1976. Comparative cell biology of the kinetoplastid flagellates. In *Biology of the Kinetoplastida*, vol. 1, W. H. R. Lumsden and D. A. Evans, eds. Academic Press, pp. 130–135.

Chapter 4. Cell Biology of *Trypanosoma cruzi*

ARAYO-JORAGE, T. C., E. P. SAMPAIO, and W. DE SOUZA. 1986. *Trypanosoma cruzi:* inhibition of host cell uptake of infective bloodstream forms by alpha-2-macroglobulin. Z. *Parasiten Kd.* 72:323–329.

BOSCHETTI, M. A., M. M. PIRAS, D. HENRIQUEZ, and R. PIRAS. 1987. The interaction of a *Trypanosoma cruzi* surface protein with Vero cells and its relationship with parasite adhesion. *Mol. Biochem. Parasitol.* 24:175–184.

CAVALLESCO, R., and M. E. A. PEREIRA. 1988. Antibody to *Trypanosoma cruzi* neuraminidase infection *in vitro* identifies a subpopulation of trypomastigotes. *J. Immunol.* 140:617-625.

CONTRERAS, V. T., J. M. SALLES, N. THOMAS, C. M. MOREL, and S. GOLDENBERG. 1985. *In vitro* differentiation of *Trypanosoma cruzi* under chemically defined conditions. *Mol. Biochem. Parasitol.* 16:315–327.

JOINER, K., S. HIENY, L. V. KIRCHHOFF, and A. SHER. 1985. gp 72, the 72 kilodalton glycoprotein is the membrane acceptor site for C3 on *Trypanosoma cruzi* epimastigotes. *J. Exp. Med.* 161:1196–1212.

LANAR, D. E., and J. E. MANNING. 1984. Major surface proteins and antigens on the different *in vivo* and *in vitro* forms of *Trypanosoma cruzi*. *Mol. Biochem. Parasitol.* 11:119–131.

NOGUEIRA, N., S. CHAPLAN, and A. COHN. 1980. *Trypanosoma cruzi:* factors modifying ingestion and fate of blood form trypomastigotes. *J. Exp. Med.* 152:447–451.

OUASSI, M. A., J. CORNETTE, and A. CAPRON. 1986. Identification and isolation of *Trypanosoma cruzi* trypomastigote cell surface protein with properties expected of a fibronectin receptor. *Mol. Biochem. Parasitol.* 19:201–211.

OUASSI, M. A., J. CORNETTE, D. AFCHAIN, A. CAPRON, H. GRAS-MASSE, and A. TARTAR. 1986. *Trypanosoma cruzi* infection inhibited by peptides modeled from a fibronectin cell attachment domain. *Science* 234:603–607.

PARODI, A. J., and J. J. CAZZULO. 1982. Protein glycosylation in *Trypanosma cruzi*. Partial characterization of protein-bound oligosaccharides labeled *in vivo*. *J. Biol. Chem.* 257:7641–7645.

PEREIRA, M. E. A., and R. HOFF. 1986. Heterogeneous distribution of neuraminidase activity in strains and clones of *Trypanosoma cruzi* and its possible association with parasite myotropism. *Mol. Biochem. Parasitol.* 20:183–189.

PRIOLI, R. P., J. M. ORDOVAS, I. ROSENBERG, E. J. SCHAEFER, and M. E. A. PEREIRA. 1987. Cruzin, a specific inhibitor of *Trypanosoma cruzi* neuraminidase, is similar to high density lipoprotein. *Science* 238:1417-1419.

PRIOLI, R. P., I. ROSENBERG, and M. E. A. PEREIRA. 1987. Specific inhibition of *Trypanosoma cruzi* neuraminidase by the human plasma glycoprotein cruzin. *Proc. Natl. Acad. Sci. USA*. 84:3097–3101.

PIRAS, M. M., D. HENRIQUEZ, and R. PIRAS. 1987. The effect of fetuin and other sialoglyproteins on the *in vitro* penetration of *Trypanosoma cruzi* trypomastigotes into fibroblastic cells. *Mol. Biochem. Parasitol.* 22:135–143.

PREVIATO, J. O., A. F. B. ANDRADE, M. C. V. PESSOLANI, and L. MENDONCA-PREVIATO. 1985. Incorporation of sialic acid into *Trypanosoma cruzi* macromolecules. A proposal for a new metabolic route. *Mol. Biochem. Parasitol.* 16:85–90.

SCHARFSTEIN, J., M. SCHECHTER, M. SENNA, J. M. PERALTA, L. MENDONCA-PREVIATO, and M. A. MILES. 1986. *Trypanosoma cruzi:* characterization and isolation of a 57/51,000 m.w. surface glycoprotein (GP-57/51) expressed by epimastigotes and bloodstream trypomastigotes. *J. Immunol.* 137:1336–1341.

SHER, A., and D. SNARY. 1982. Specific inhibition of the morphogenesis of *Trypanosoma cruzi* by a monoclonal antibody. *Nature* (Lond.) 300:639–640.

VILLATA, F., and F. KIERSZENBAUM. 1987. Role of membrane N-linked oligosaccharides in host cell interaction with invasive forms of *Trypanosoma cruzi*. *Mol. Biochem. Parasitol.* 22:109–114.

Chapter 5. Cell Biology of *Leishmania*

ALEXANDER, J., and D. G. RUSSELL. 1985. Parasite antigens, their role in protection, diagnosis, and escape: the leishmanias. *Curr. Top. Microbiol. Immunol.* 120:43–67.

ALVING, C. R., E. A. STECK, W. L. CHAPMAN JR., V. B. WAITS, L. D. HENDRICKS, G. M. SWARTZ JR., and W. L. HANSON. 1980. Liposomes in leishmaniasis: Therapeutic effects of antimonial drugs, 8-aminoquinolines and tetracycline. *Life Sci.* 26:2231–2238.

BLACKWELL, J. M., R. A. B. EZEKOWITZ, M. B. ROBERTS, J. Y. CHANNON, R. B. SIM, and S. GORDON. 1985. Macrophage complement and lectin-like receptors bind *Leishmania* in the absence of serum. *J. Exp. Med.* 162:324–331.

BORDIER, C. 1987. The promastigote surface protease of *Leishmania*. *Parasitol. Today* 3:151–153.

CHANG, C. S., and K.-P. CHANG. 1986. Monoclonal antibody affinity purification of a *Leishmania* membrane glycoprotein and its inhibition of *Leishmania*-macrophage binding. *Proc. Natl. Acad. Sci.* USA 83:100–104.

CHAUDHURI, G., and K.-P. CHANG. 1988. Acid protease activity of a major membrane glycoprotein (gp63) from *Leishmania mexicana* promastigotes. *Mol. Biochem. Parasitol.* 27:43-52.

EILAM, Y., J. EL-ON, and D. T. SPIRA. 1985. *Leishmania major:* excreted factor, calcium ion and the survival of amastigotes. *Exp. Parasitol.* 59:161–168.

ETGES, R., J. BOUVIER, and C. BORDIER. 1986. The major surface protein of *Leishmania* promastigotes is a protease. *J. Biol. Chem.* 261:9098–9111.

ETGES, R., J. BOUVIER, and C. BORDIER. 1986. The major surface protein of *Leishmania* promastigotes is anchored in the membrane by a myristic acid-labeled phospholipid. *EMBO J.* 5:597–601.

FAIRLAMB, A. H., P. BLACKBURN, P. ULRICH, B. T. CHAIT, and A. CERAMI. 1985. Trypanothione: a novel bis(glutathionyl) spermidine cofactor for glutathione reductase in trypanosomatids. *Science* 22:1485–1487.

FONG, D., M. WALLACH, J. KEITHLY, P. W. MELERA, and K.-P. CHANG. 1984. Differential expression of mRNAs for tubulin during differentiation of a parasitic protozoan, *Leishmania mexicana*, *Proc. Natl. Acad. Sci.* USA 81:5782–5786.

HANDMAN, E., and J. W. GODING. 1985. The *Leishmania* receptor for macrophages is a lipid-containing glycoconjugate. *EMBO J.* 4:329–336.

KAHL, L. P., and D. MCMAHON-PRATT. 1987. Structural and antigenic characterization of a species- and promastigote-specific *Leishmania mexicana amazonensis* membrane protein. *J. Immunol.* 138:1587–1595.

KINK, J. A., and K.-P. CHANG. 1987. Tunicamycin-resistant *Leishmania mexicana amazonensis:* expression of virulence associated with an increased activity of N-acetylglucosaminyltransferase and amplification of its presumptive gene. *Proc. Natl. Acad. Sci.* USA 84:1253–1257.

LANDFEAR, S. M., and D. F. WIRTH. 1984. Control of tubulin gene expression in the parasitic protozoan *Leishmania enriettii*. *Nature* 309:716–717.

MESHNICK, S. R., and J. W. EATON. 1981. Leishmanial superoxide dismutase. A possible target for chemotherapy. *Biochem. Biophys. Res. Commun.* 102:970–976.

MOSSER, D. M., and P. J. EDELSON. 1987. The third complement (C3) is responsible for the intracellular survival of *Leishmania major*. *Nature* 327:329–331.

MUKKADA, A. J., J. C. MEADE, T. A. GLASER, and P. F. BONVENTRE. 1985. Enhanced metabolism of *Leishmania donovani* amastigotes at acidic pH: an adaptation for intracellular growth. *Science* 229:1099–1101.

PAN, A. 1984. *Leishmania mexicana:* serial cultivation of intracellular stages in a cell-free medium. *Exp. Parasitol.* 58:72–80.

PUPKIS, M. F., L. TETLEY, and G. H. COOMBS. 1986. *Leishmania mexicana:* amastigote hydrolases in unusual lysosomes. *Exp. Parasitol.* 62:29–39.

REMALEY, A. T., D. B. KUHNS, R. E. BASFORD, R. H. GLEW, and S. KAPLAN. 1985. Leishmanial phosphatase blocks neutrophil O_2^- production. *J. Biol. Chem.* 260:880–886.

RUSSELL, D. G., and H. WILHELM. 1986. The involvement of the major surface glycoprotein (gp63) of *Leishmania* promastigotes in attachment to macrophages. *J. Immunol.* 136:2613–2620.

SACKS, D. L., and P. V. PERKINS. 1984. Identification of an infective stage of *Leishmania* promastigotes. *Science* 223:1417–1419.

TURCO, S. J., C. L. JOHNSON, D. L. KING, P. A. ORLANDI, and B. L. WRIGHT. 1987. The structure, localization and function of the lipophosphoglycan of *Leishmania donovani*. In *Host–Parasite Cellular and Molecular Interactions in Protozoal Infections*, K.-P. Chang and D. Snary, eds. NATO ASI Series H: *Cell Biology*, Vol. II. Springer-Verlag, pp. 189–197.

WILLIAMS, K. M., J. B. SACCI, and R. L. ANTHONY. 1986. Identification and recovery of *Leishmania* antigen displayed on the surface membrane of mouse peritoneal macrophages infected *in vitro*. *J. Immunol.* 136:1853–1858.

WILSON, M. E., and R. D. PEARSON. 1986. Evidence that *Leishmania donovani* utilizes a mannose receptor on human mononuclear phagocytes to establish intracellular parasitism. *J. Immunol.* 136:4681–4687.

WOZENCRAFT, A. O., G. SAYERS, and J. M. BLACKWELL. 1986. Macrophages type 3 complement receptors mediate serum-independent binding of *Leishmania donovani*. *J. Exp. Med.* 164:1332–1337.

WYLER, D. J., J. P. SYPEK, and J. A. MCDONALD. 1985. *In vitro* parasite-monocyte interactions in human leishmaniasis: possible role of fibronectin in parasite attachment. *Infect. Immun.* 49:305–311.

ZILBERSTEIN, D., and D. M. DWYER. 1986. Proton motive force-driven transport of D-glucose in the protozoan parasite, *Leishmania donovani*. *Proc. Natl. Acad. Sci. USA* 82:1716–1720.

Chapter 6. Cell Biology of Schistosomes, I. Ultrastructure and Transformations

ALLAN, D., G. PAYARES, and W. H. EVANS. 1987. Phospholipid and fatty acid composition of *Schistosoma mansoni* and of its tegumental membranes. *Mol. Biochem. Parasitol.* 23:123–128.

BASCH, P. F., and J. J. DICONZA. 1974. The miracidium-sporocyst transition in *Schistosoma mansoni*: surface changes *in vitro* with ultrastructural correlation. *J. Parasitol.* 60:935–941.

CORDEIRO, M. N., G. GAZZINELLI, C. A. P. TAVARES, and E. A. FIGUEIREDO. 1984. *Schistosoma mansoni*: relationship between turnover rates of membrane proteins and susceptibility to immune damage of schistosomula. *Mol. Biochem. Parasitol.* 11:23–36.

KUSEL, J. R., and P. E. MACKENZIE. 1975. The measurement of the relative turnover rates of proteins of the surface membranes and other fractions of *Schistosoma mansoni* in culture. *Parasitology* 71:261–273.

LEITCH, B., A. J. PROBERT, and N. W. RUNHAM. 1984. The ultrastructure of the tegument of adult *Schistosoma haematobium*. *Parasitology* 89:71–78.

MCDIARMID, S. S., L. L. DEAN, and R. B. PODESTA. 1983. Sequential removal of outer bilayer and apical plasma membrane from the surface epithelial syncytium. *Mol. Biochem. Parasitol.* 7:141–157.

MCKERROW, J. H., P. JONES, H. SAGE, and S. PINO-HEISS. 1985. Proteinases from invasive larvae of the trematode parasite *Schistosoma mansoni* degrade connective-tissue and basement-membrane macromolecules. *Biochem. J.* 231:47–51.

MCKERROW, J. H., S. PINO-HEISS, R. LINDQUIST, and Z. WERB. 1985. Purification and characterization of an elastinolytic proteinase secreted by cercariae of *Schistosoma mansoni*. *J. Biol. Chem.* 260:3703–3707.

MCLAREN, D. J., D. J. HOCKLEY, O. L. GOLDRING, and J. HAMMOND. 1978. A freeze fracture study of the developing tegumental outer membrane of *Schistosoma mansoni*. *Parasitology* 76:327–348.

MURRELL, K. D., D. W. TAYLOR, W. E. VANNIER, and D. A. DEAN. 1978. *Schistosoma mansoni*: analysis of surface membrane carbohydrates using lectins. *Exp. Parasitol.* 46:247–255.

PAN, C.-T. 1980. The fine structure of the miracidium of *Schistosoma mansoni*. *J. Invertebr. Pathol.* 36:307–372.

PAYARES, G., and A. J. G. SIMPSON. 1985. *Schistosoma mansoni* surface glycoproteins analysis of their expression and antigenicity. *Eur. J. Biochem.* 153:195–201.

PEARCE, E. J., D. ZILBERSTEIN, S. L. JAMES, and A. SHER. 1986. Kinetic correlation of the acquisition of resistance to immune attack in schistosomula of *Schistosoma mansoni* with a developmental change in membrane potential. *Mol. Biochem. Parasitol.* 21:259–267.

ROBERTS, S. M., A. N. MCGREGOR, M. VOJVOID, E. WELLS, J. E. CRABTREE, and R. A. WILSON. 1982. Tegument surface membranes of adult *Schistosoma mansoni*: development of a method for their isolation. *Mol. Biochem. Parasitol.* 9:105–127.

RUPPEL, A., and D. J. MCLAREN. 1986. *Schistosoma mansoni*: surface membrane stability *in vitro* and *in vivo*. *Exp. Parasitol.* 62:223–226.

SAMUELSON, J. C., J. J. QUINN, and J. P. CAULFIELD. 1984. Hatching, chemokinesis, and transformation of miracidia of *Schistosoma mansoni*. *J. Parasitol.* 70:321–331.

SAMUELSON, J. C., and J. P. CAULFIELD. 1986. Cercarial glycocalyx of *Schistosoma mansoni* activates human complement. *Infect. Immun.* 51:181–186.

SHORT, R. B. 1983. Sex and the single schistosome. *J. Parasitol.* 69:4–22.

SMYTH, J. D., and D. W. HALTON. 1983. *Physiology of Trematodes*. 2nd Edition. Cambridge University Press, 446 pp.

STIREWALT, M. A., C. E. COUSIN, and C. H. DORSEY. 1983. *Schistosoma mansoni*: stimulus and transformation of cercariae into schistosomules. *Exp. Parasitol.* 56:358–368.

VIAL, H. J., G. TORPIER, M. L. ANCELIN, and A. CAPRON. 1985. Renewal of the membrane complex of *Schistosoma mansoni* is closely associated with lipid metabolism. *Mol. Biochem. Parasitol.* 17:118–203.

YOUNG, B. W., and R. B. PODESTA. 1985. Uptake and incorporation of choline by *Schistosoma mansoni* adults. *Mol. Biochem. Parasitol.* 15:105–114.

Chapter 7. Cell Biology of Schistosomes, II. Tegumental Membranes and Their Interaction with Human Blood Cells

ALLAN, D., G. PAYARES, and W. H. EVANS. 1987. The phospholipid and fatty acid composition of *Schistosoma mansoni* and of its purified tegumental membranes. *Mol. Biochem. Parasitol.* 23:123–128.

CAULFIELD, J. P., and C. M. L. CIANCI. 1985. Human erythrocytes adhering to schistosomula of *Schistosoma mansoni* lyse and fail to transfer membrane components to the parasite. *J. Cell Biol.* 101:158–166.

CAULFIELD, J. P., A. HEIN, G. MOSER, and A. SHER. 1981. Light and electron microscopic appearance of rat peritoneal mast cells adhering to schistosomula of *Schistosoma mansoni* by means of complement or antibody. *J. Parasitol.* 67:776–783.

CAULFIELD, J. P., G. KORMAN, A. E. BUTTERWORTH, M. HOGAN, and J. R. DAVID. 1980. The adherence of human neutrophils and eosinophils to schistosomula: evidence for membrane fusion between cells and parasites. *J. Cell Biol.* 86:46–63.

CAULFIELD, J. P., G. KORMAN, A. E. BUTTERWORTH, M. HOGAN, and J. R. DAVID. 1980. Partial and complete detachment of neutrophils and eosinophils from schistosomula: evidence for the establishment of continuity between fused and normal parasite membrane. *J. Cell Biol.* 86:64–76.

CAULFIELD, J. P., G. KORMAN, and J. C. SAMUELSON. 1982. Human neutrophils endocytose multivalent ligands from the surface of schistosomula of *S. mansoni* before membrane fusion. *J. Cell Biol.* 94:370–378.

CAULFIELD, J. P., H. L. LENZI, P. ELSAS, and A. J. DESSEIN. 1985. Ultrastructure of the attack of eosinophils stimulated by blood mononuclear cell products on schistosomula of *Schistosoma mansoni*. *Am. J. Pathol.* 120:380–390.

CLEGG, J. A. 1972. The schistosome surface in relation to the parasitism. *Symp. Soc. Parasitol.* 10:23–72.

GOLAN, D. E., C. S. BROWN, C. M. L. CIANCI, S. T. FURLONG, and J. P. CAULFIELD. 1986. Schistosomula of *Schistosoma mansoni* use lysophosphatidylcholine to lyse adherent human red blood cells and immobilize red cell membrane components. *J. Cell Biol.* 103:819–828.

MEYER, F., H. MEYER, and E. BUEDING. 1970. Lipid metabolism in the parasitic and free-living flatworms, *Schistosoma mansoni* and *Dugesia dorotocephala*. *Biochim. Biophys. Acta* 210:257–266.

PEARCE, E. J., P. F. BASCH, and A. SHER. 1986. Evidence that the reduced surface antigenicity of developing *Schistosoma mansoni* schistosomula is due to antigen shedding rather than host molecule acquisition. *Parasite Immunol.* 8:79–89.

SAMUELSON, J. C., and J. P. CAULFIELD. 1982. Loss of covalently labeled glycoproteins and glycolipids from the surface of newly transformed schistosomula of *S. mansoni*. *J. Cell Biol.* 94:363–369.

SAMUELSON, J. C., J. P. CAULFIELD, and J. R. DAVID. 1982. Schistosomula of *Schistosoma mansoni* clear concanavalin A from their surface by sloughing. *J. Cell Biol.* 94:355–362.

SHER, A., B. F. HALL, and M. A. VADAS. 1978. Acquisition of murine major histocompatibility complex gene products by schistosomula of *Schistosoma mansoni*. *J. Cell Biol.* 148:46–57.

SIMPSON, A. J. G., D. SINGER, T. F. MCCUTCHAN, D. L. SACKS, and A. SHER. 1983. Evidence that the schistosome MHC antigens are not synthesized by the parasite but are acquired from the host as intact glycoproteins. *J. Immunol.* 131:3315–3330.

YOUNG, B. W., and R. B. PODESTA. 1982. Major phospholipids and phosphatidylcholine synthesis in adult *Schistosoma mansoni*. *Mol. Biochem. Parasitol.* 5:165–172.

YOUNG, B. W., and R. B. PODESTA. 1984. Phospholipids of cercariae and adult *Schistosoma mansoni*. *J. Parasitol.* 70:447–448.

Chapter 8. Cell Biology of *Entamoeba histolytica* and Immunology of Amebiasis

ARROYO, R., and E. OROZCO. 1987. Localization and identification of an *Entamoeba histolytica* adhesin. *Mol. Biochem. Parasitol.* 23:151–158.

BRACHA, R., and D. MIRELMAN. 1984. Virulence of *Entamoeba histolytica* trophozoites: effects of bacteria, microaerobic conditions and metronidazole. *J. Exp. Med.* 160:353–368.

CHADEE, K., W. A. PETRI JR., D. J. INNES, and J. I. RAVDIN. 1987. Rat and human colonic mucins bind to and inhibit the adherence lectin of *Entamoeba histolytica*. *J. Clin. Invest.* 80:1245–1254.

DIAMOND, L. S., D. R. HARLOW, and C. C. CUNNICK. 1978. A new medium for the axenic cultivation of *Entamoeba histolytica* and other *Entamoeba*. *Trans. R. Soc. Trop. Med. Hyg.* 72:431–432.

JOYCE, M. P., and J. I. RAVDIN. 1988. Antigens of *Entamoeba histolytica* recognized by immune sera from liver abscess patients. *Am. J. Trop. Med. Hyg.* 38:74–80.

KEENE, W. E., G. P. MATTHEW, S. ALLEN, and J. H. MCKERROW. 1986. The major neutral proteinase of *Entamoeba histolytica*. *J. Exp. Med.* 163:536–549.

KOBILER, D., and D. MIRELMAN. 1980. Lectin activity in *Entamoeba histolytica* trophozoites. *Infect. Immun.* 29:221–225.

LONG-KRUG, S. A., R. M. HYSMITH, K. J. FISCHER, and J. I. RAVDIN. 1985. The phospholipase A enzymes of *Entamoeba histolytica*: description and subcellular localization. *J. Infect. Dis.* 152:536–541.

LUSHBAUGH, W. B., A. F. HOFBAUER, and F. E. PITTMAN. 1985. Purification of cathepsin B activity of *Entamoeba histolytica* toxin. *Exp. Parasitol.* 59:328–336.

LYNCH, E. C., I. ROSENBERG, and C. GITLER. 1985. An ion channel forming protein produced by *Entamoeba histolytica*. *EMBO J.* 1:801–804.

MCGOWAN, K., A. KANE, N. ASARKOF, J. WICKS, V. GUERINA, J. KELLUM, S. BARON, A. GINTZLER, and M. DONOWITZ. 1983. *Entamoeba histolytica* causes intestinal secretion: role of serotonin. *Science* 221:762–764.

MARTINEZ-PALOMO, A., A. GONZALEZ-ROBLES, B. CHAVEZ, E. OROZCO, S. FERNANDEZ-CASTELO, and A. CERVANTES. 1985. Structural basis of the cytolytic mechanisms of *Entamoeba histolytica*. *J. Protozool.* 32:166–175.

MEZA, I., F. CAZARES, J. L. ROSALES-ENCINA, P. TALAMAS-ROHANA, and M. ROJKIND. 1987. Use of antibodies to characterize a 220-kilodalton surface protein from *Entamoeba histolytica*. *J. Infect. Dis.* 156:798–805.

MIRELMAN, D., R. BRACHA, and P. G. SARGEAUNT. 1984. *Entamoeba histolytica*: virulence enhancement of isoenzyme-stable parasites. *Exp. Parasitol.* 57:172–177.

MIRELMAN, D., R. BRACHA, A. WEXLER, and A. CHAYON. 1986. Changes in isoenzyme patterns of a cloned culture of nonpathogenic *Entamoeba histolytica* during axenization. *Infect. Immun.* 54:827–834.

MUNOZ, M. D. L., J. CALDERON, and M. ROJKIND. 1982. The collagenase of *Entamoeba histolytica*. *J. Exp. Med.* 155:42–51.

OROZCO, E., G. GUARNEROS, A. MARTINEZ-PALOMO, and T. SANCHEZ. 1983. *Entamoeba histolytica*. Phagocytosis as a virulence factor. *J. Exp. Med.* 158:1511–1521.

PETRI, W. A., M. P. JOYCE, J. BROMAN, R. D. SMITH, C. F. MURPHY, and J. I. RAVDIN. 1987. Recognition of the galactose or N-acetyl-D-galactosamine binding lectin of *Entamoeba histolytica* by human immune sera. *Infect. Immun.* 55:2327–2331.

PETRI, W. A., JR., R. D. SMITH, P. H. SCHLESINGER, and J. I. RAVDIN. 1987. Isolation of the galactose-binding lectin which mediates the in vitro adherence of *Entamoeba histolytica*. *J. Clin. Invest.* 80:1238–1244.

RAVDIN, J. I., B. Y. CROFT, and R. L. GUERRANT. 1980. Cytopathogenic mechanisms of *Entamoeba histolytica*. *J. Exp. Med.* 152:377–390.

RAVDIN, J. I., and R. L. GUERRANT. 1981. The role of adherence in cytopathogenic mechanisms of *Entamoeba histolytica*. Study with mammalian tissue culture cells and human erythrocytes. *J. Clin. Invest.* 68:1305–1313.

RAVDIN, J. I., J. E. JOHN, L. I. JOHNSTON, D. J. INNES, and R. L. GUERRANT. 1985. Adherence of *Entamoeba histolytica* trophozoites to rat and human colonic mucosa. *Infect. Immun.* 48:292–297.

RAVDIN, J. I., C. F. MURPHY, R. L. GUERRANT, and S. A. LONG-KRUG. 1985. Effect of calcium and phospholipase A antagonists on the cytopathogenicity of *Entamoeba histolytica*. *J. Infect. Dis.* 152:542–549.

RAVDIN, J. I., C. F. MURPHY, R. A. SALATA, R. L. GUERRANT, and E. L. HEWLETT. 1985. N-acetyl-D-galactosamine-inhibitable adherence lectin of *Entamoeba histolytica*. I. Partial purification and relation to amoebic virulence *in vitro*. *J. Infect. Dis.* 151:804–815.

RAVDIN, J. I., W. A. PETRI JR., C. F. MURPHY, and R. D. SMITH. 1986. Production of mouse monoclonal antibodies which inhibit in vitro adherence of *Entamoeba histolytica* trophozoites. *Infect. Immun.* 53:1–5.

RAVDIN, J. I., P. H. SCHLESINGER, C. F. MURPHY, I. Y. GLUZMAN, and D. J. KROGSTAD. 1986. Acid intracellular vesicles and the cytolysis of mammalian target cells by *Entamoeba histolytica* trophozoites. *J. Protozool.* 33:478–486.

ROSALES-ENCINA, J. L., I. MEZA, A. LOPEZ-DELEON, P. TALAMAS-ROHANA, and M. ROJKIND. 1987. Isolation of a 220-kilodalton protein with lectin properties from a virulent strain of *Entamoeba histolytica*. *J. Infect. Dis.* 156:790–797.

ROSENBERG, I., and C. GITLER. 1985. Subcellular fractionation of amoebapore and plasma membrane components of *Entamoeba histolytica* using self-generating Percoll gradients. *Mol. Biochem. Parasitol.* 14:231.

SALATA, R. A., and J. I. RAVDIN. 1985. The interaction of human neutrophils and *Entamoeba histolytica* increases cytopathogenicity for liver cell monolayers. *J. Infect. Dis.* 154:19–26.

SALATA, R. A., and J. I. RAVDIN. 1986. The interaction of human neutrophils and *Entamoeba histolytica* increases cytopathogenicity for liver cell monolayers. *J. Infect. Dis.* 154:19–26.

SARGEAUNT, P. G., J. E. WILLIAMS, and J. D. GRENE. 1978. The differentiation of invasive and non-invasive *Entamoeba histolytica* by isoenzyme electrophoresis. *Trans. R. Soc. Trop. Med. Hyg.* 72:519–521.

YOUNG, J. D.-E., T. M. YOUNG, L. P. LU, J. C. UNKELESS, and Z. A. COHN. 1982. Characterization of a membrane pore-forming protein from *Entamoeba histolytica*. *J. Exp. Med.* 156:1677–1690.

Part II Immunology

Chapter 9. Immunology of Malaria

BALLOU, W. R., S. L. HOFFMAN, J. A. SHERWOOD, M. R. HOLLINGDALE, F. A. NEVA, W. T. HOCKMEYER, D. M. GORDON, I. SCHNEIDER, R. A. WIRTZ, J. F. YOUNG, G. F. WASSERMAN, P. REEVE, C. L. DIGGS, and J. D. CHULAY. 1987. Safety and efficacy of a recombinant DNA *Plasmodium falciparum* sporozoite vaccine. *Lancet* 1:1277–1281.

BRAKE, D. A., W. P. WEIDANZ, and C. A. LONG. 1986. Antigen-specific interleukin 2-propagated T lymphocytes confer assistance to a murine malarial parasite, *Plasmodium chabaudi-adami*. *J. Immunol.* 137:347–352.

CARTER, R., L. H. MILLER, J. RENER, D. C. KAUSHAL, N. KUMAR, P. M. GRAVES, C. A. GROTENDORST, R. W. GWARDZ, C. FRENCH, and D. WIRTH. 1984. Target antigens in malaria transmission blocking immunity. *Phil. Trans. R. Soc. Lond B* 307:201–213.

CLARK, I. A., N. H. HUNT, G. A. BUTCHER, and W. B. COWDEN. 1987. Inhibition of murine malaria *(Plasmodium chabaudi) in vivo* by recombinant interferon-γ or tumor necrosis factor, and its enhancement by butylated hydroxyanisole. *J. Immunol.* 139:3493–3496.

CLYDE, D. F., V. C. MCCARTHY, R. M. MILLER, and R. B. HORNICK. 1973. Specificity of protection of man immunized against sporozoite-induced falciparum malaria. *Am. J. Med. Sci.* 266:398–403.

COHEN, S., I. A. MCGREGOR, and S. P. CARRINGTON. 1961. Gamma globulin and acquired immunity to human malaria. *Nature* (Lond) 192:733–737.

COLLINS, W. E., R. F. ANDERS, M. PAPAIOANOU, G. H. CAMPBELL, G. V. BROWN, D. J. KEMP, R. L. COPPEL, J. C. SKINNER, P. M. ANDRYSIAK, J. M. FAVOLORO, L. M. CORCORAN, J. R. BRODERSON, G. F. MITCHELL, and C. C. CAMPBELL. 1986. Immunization of Aotus monkeys with recombinant proteins of an erythrocyte surface antigen of *Plasmodium falciparum*. *Nature* (Lond) 323:259–262.

COPPEL, R. L., A. F. COWMAN, R. F. ANDERS, A. E. BIANCO, R. B. SAINT, K. R. LINGELBACH, D. J. KEMP, and G. V. BROWN. 1984. Immune sera recognize on erythrocytes *Plasmodium falciparum* antigen composed of repeated amino acid sequences. *Nature* (Lond) 310:789–792.

DAME, J. B., J. L. WILLIAMS, T. F. MCCUTCHAN, J. L. WEBER, R. A. WIRTZ, W. T. HOCKMEYER, W. L. MALOY, J. D. HAYNES, I. SCHNEIDER, D. ROBERTS, G. S. SANDERS, E. P. REDDY, C. L. DIGGS, and L. H. MILLER. 1984. Structure of the gene encoding the immunodominant surface antigen on the sporozoite of the human malaria parasite *Plasmodium falciparum*. *Science* 225:593–599.

DEL GUIDICE, G., J. A. COOPER, J. MERINO, A. S. VERDINI, A. PESSI, A. R. TOGNA, H. D. ENGERS, G. CORRADIN, and P.-H. LAMBERT. 1986. The antibody response in mice to carrier-free synthetic polymers of *Plasmodium falciparum* circumsporozoite repetitive epitope is I-A^b restricted: possible implications for malaria vaccines. *J. Immunol.* 137:2952–2960.

DESOWITZ, R. S., and R. RAYBOURNE. 1985. Perinatal immune priming in malaria: antigen-induced blastogenesis and adoptive transfer of resistance by splenocytes from rats born of *Plasmodium berghei* infected females. *Parasite Immunol.* 7:451–456.

EGAN, J. E., J. L. WEBER, W. R. BALLOU, M. R. HOLLINGDALE, W. R. MAJARIAN, D. M. GORDON, M. L. MALOY, S. L. HOFFMAN, R. A. WIRTZ, I. SCHNEIDER, G. R. WOOLLETT, J. F. YOUNG, and W. T. HOCKMEYER. 1987. Efficacy of murine malaria sporozoite vaccines: implications for human vaccine development. *Science* 236:453–456.

FERREIRA, A., L. SCHOFIELD, V. ENEA, H. SCHELLEKENS, P. VAN DER MEIDE, W. E. COLLINS, R. S. NUSSENZWEIG, and V. NESSENZWEIG. 1986. Inhibition of development of exoerythrocytic forms of malaria parasites by γ-interferon. *Science* 232:881–884.

GOOD, M. F., J. A. BERZOFSKY, and L. H. MILLER. 1988. The T cell response to the malaria circumsporozite protein: an immunological approach to vaccine development. *Ann. Rev. Immunol.* 6:663–688.

GRAU, G. E., L. F. FJARDO, P. F. PIGUET, B. ALLET, P. H. LAMBERT, and P. VASSALLI. 1987. Tumor necrosis factor (cachectin) as an essential mediator in murine cerebral malaria. *Science* 237:1210–1212.

HERRINGTON, D. A., D. F. CLYDE, G. LOGONSKY, M. CORTESIA, J. R. MURPHY, J. DAVID, S. BAGAR, A. M. FELIX, E. P. HEIMER, D. GILLESSEN, E. NARDIN, R. S. NUSSENZWEIG, V. NUSSENZWEIG, M. R. HOLLINGDALE, and M. M. LEVINE. 1987. Safety and immunogenicity in man of a synthetic peptide malaria vaccine against *Plasmodium falciparum* sporozoites. *Nature* (Lond) 328:257–259.

HO, M., H. K. WEBSTER, S. LOOAREESUWAN, W. SUPANARANOND, R. E. PHILLIPS, P. CHATHAVANICH, and D. A. WARRELL. 1986. Antigen-specific immunosuppression in human malaria due to *Plasmodium falciparum*. *J. Infect. Dis.* 153:763–771.

HOFFMANN, S. L., C. N. OSTER, C. V. PLOWE, G. R. WOOLLETT, J. C. BEIER, J. D. CHULAY, R. A. WIRTZ, M. R. HOLLINGDALE, and M. MUGAMBI. 1987. Naturally acquired antibodies to sporozoites do not prevent malaria: vaccine development implications. *Science* 237:639–642.

HOWARD, R. J. 1984. Antigenic variation of bloodstage malaria parasites. *Phil. Trans. R. Soc. Lond B* 307:141–158.

JENSEN, J. B., S. L. HOFFMAN, M. T. BOLAND, M. A. S. AKOOD, L. W. LAUGHLIN, L. KURIAWAN, and H. A. MARWOTO. 1984. Comparison of immunity to malaria in Sudan and Indonesia: crisis form versus merozoite invasion inhibition. *Proc. Natl. Acad. Sci. USA* 81:922–925.

LOCKYER, M. J., and R. T. SCHWARZ. 1987. Strain variation in the circumsporozoite protein gene of *Plasmodium falciparum*. *Mol. Biochem. Parasitol.* 22:101–108.

MCGREGOR, I. A., H. M. GILLES, J. H. WALTERS, A. H. DAVIES, and F. A. PEARSON. 1956. Effects of heavy and repeated malarial infections in Gambian infants and children: effects of erythrocytic parasitization. *Br. Med. J.* 2:686–692.

NUSSENZWEIG, R. S., and V. NUSSENZWEIG. 1984. Development of sporozoite vaccines. *Phil. Trans. R. Soc. Lond B* 307:117–128.

ORJIH, A. V., and R. S. NUSSENZWEIG. 1979. *Plasmodium berghei*: suppression of antibody response to sporozoite stage by acute blood stage infection. *Clin. Exp. Immunol.* 38:1–8.

PATARROYO, M. E., R. AMADOR, P. CLAVIJO, A. MORENO, F. GUZMAN, P. ROMERO, R. TASCON, A. FRANCO, L. A. MURILLO, G. PONTON, and G. TRUJILLO. 1988. A synthetic vaccine protects humans against challenge with asexual blood stages of *Plasmodium falciparum* malaria. *Nature* (Lond) 332:158–161.

PHILLIPS, R. E., and D. A. WARRELL. 1986. The pathophysiology of severe falciparum malaria. *Parasitol. Today* 2:271–282.

SKAMENE, E., M. STEVENSON, and S. LEMIEUX. 1983. Murine malaria: dissociation of natural killer (NK) cell activity and resistance to *Plasmodium chabaudi*. *Parasite Immunol.* 5:557–565.

TROYE-BLOMBERG, M., G. ANDERSSON, M. STOCZKOWSKA, R. SHABO, P. ROMERO, E. PATARROYO, H. WIGZELL, and P. PERLMANN. 1985. Production of IL-2 and IFN-γ by T cells from malaria patients in response to *Plasmodium falciparum* or erythrocyte antigens *in vitro*. *J. Immunol.* 135:3498–3504.

UDEINYA, I. J., L. H. MILLER, I. A. MCGREGOR, and J. B. JENSEN. 1983. *Plasmodium falciparum* strain-specific antibody blocks binding of infected erythrocytes to amelanotic melanoma cells. *Nature* (Lond) 303:429–431.

WEISS, W. R., M. SEDEGAH, R. L. BEAUDOIN, L. H. MILLER, and M. F. GOOD. 1988. $CD8^+$ T cells (cytotoxic/supressors) are required for protection in mice immunized with malaria sporozoites. *Proc. Natl. Acad. Sci. USA* 85:573–576.

WYLER, D. J. 1979. Cellular aspects of immunoregulation in malaria. *Bull. WHO* 57(suppl):239–243.

WYLER, D. J. 1983. The spleen in malaria. In *Malaria and the Red Cell*, D. Evered and J. Whelan, eds. Ciba Foundation Symposium No. 94. London: Pitman Books Ltd., pp. 98–116.

Chapter 10. Immunology of Toxoplasmosis

DESMONTS, G., and J. COURVEUR. 1979. Congenital Toxoplasmosis. A prospective study of the offspring of 542 women who acquired toxoplasmosis during pregnancy. Pathophysiology of congenital disease. In *Sixth European Congress of Perinatal Medicine*,

Vienna, 1978, K. Baumgarten, O. Thalhammer, and A. Pollack, eds. George Thieme, pp. 51–60.

DUBEY, J. P., N. L. MILLER, and J. K. FRENKEL. 1970. The *Toxoplasma gondii* oocyst from cat feces. *J. Exp. Med.* 132:636–662.

DUBEY, J. P., and J. K. FRENKEL. 1972. Cyst-induced toxoplasmosis in cats. *J. Protozool.* 19:155–177.

FRENKEL, J. K., and L. JACOBS. 1958. Ocular toxoplasmosis. *Arch. Ophthalmol.* 59:260.

FRENKEL, J. K. 1953. Host, strain and treatment variation as factors in the pathogenesis of toxoplasmosis. *Am. J. Trop. Med. Hyg.* 2:390–411.

FRENKEL, J. K., B. M. NELSON, and J. ARIAS-STELLA. 1975. Immunosuppression and toxoplasmic encephalitis. *Hum. Pathol.* 6:97–111.

HANDMAN, E., J. W. GODING, and J. S. REMINGTON. 1980. Detection and characterization of membrane antigens of *Toxoplasma gondii*. *J. Immunol.* 124:2578–2583.

HAUSER, W. E., JR., S. D. SHARMA, and J. S. REMINGTON. 1982. Natural killer cells induced by acute and chronic toxoplasma infection. *Cell. Immunol.* 69:3330–3346.

HUGHES, H. P. A., C. A. SPEER, J. E. KYLE, and J. P. DUBEY. 1987. Activation of murine macrophages and a bovine monocyte cell line by bovine lymphokines to kill the intracellular pathogens *Eimeria bovis* and *Toxoplasma gondii*. *Infect. Immun.* 55:784–791.

HUTCHISON, W. M., R. M. PITTILO, S. J. BALL, and J. C. SLIM. 1980. Scanning electron microscopy of the cat small intestine and mucosal alteration observed during *Toxoplasma gondii* infection. *Ann. Trop. Med. Parasitol.* 74:427–437.

JONES, T. C., K. A. BIENZ, and P. ERO. 1986. *In vitro* cultivation of *Toxoplasma gondii* cysts in astrocytes in the presence of gamma interferon. *Infect. Immun.* 51:147–156.

KASPER, L. H., M. S. BRADLEY, and E. R. PFEFFERKORN. 1984. Identification of stage-specific sporozoite antigens of *Toxoplasma gondii* by monoclonal antibodies. *J. Immunol.* 132:443–449.

MCCABE, R. E., B. J. LUFT, and J. S. REMINGTON. 1984. Effect of murine interferon gamma on murine toxoplasmosis. *J. Infect. Dis.* 140:961–962.

MCLEOD, R., and D. MACK. 1986. Secretory IgA specific for *Toxoplasma gondii*. *J. Immunol.* 136:2640–2643.

MCLEOD, R., H. CONEN, and R. ESTES. 1984. Immune responses to ingested *Toxoplasma*. Description of a mouse model of *Toxoplasma* acquired by ingestion. *J. Infect. Dis.* 149:234–244.

MCLEOD, R., and R. ESTES. 1985. Role of lymphocyte blastogenesis to *T. gondii* antigens in containment of chronic latent *T. gondii* infections in humans. *Clin. Exp. Immunol.* 62:24–30.

PFEFFERKORN, E. R. 1984. Interferon blocks the growth of *Toxoplasma gondii* in human fibroblasts by inducing the host cells to degrade tryptophan. *Proc. Natl. Acad. Sci. USA.* 81:1908–1912.

PINON, J. M., H. THOANNES, and N. GRUSON. 1985. An enzyme-linked immunofiltration assay used to compare infant and maternal antibody profiles in toxoplasmosis. *J. Immunol. Methods* 77:15–23.

REYES, L., and J. K. FRENKEL. 1987. Specific and nonspecific mediation of protective immunity to *Toxoplasma gondii*. *Infect. Immun.* 55:856–863.

SHARMA, S. D., J. K. MULLENAX, F. G. ARAUJO, H. A. ERLICH, and J. S. REMINGTON. 1983. Western blot analysis of antigens of *Toxoplasma gondii* recognized by human IgM and IgG antibodies. *J. Immunol.* 131:1977–1983.

SHARMA, S. D., F. G. ARAUJO, and J. S. REMINGTON. 1985. *Toxoplasma* antigen isolated by affinity chromatography with monoclonal antibody protects mice against lethal infection with *Toxoplasma gondii*. *J. Immunol.* 133:2818–2820.

SHARMA, S. D., J. M. HOFFLIN, and J. S. REMINGTON. 1986. In vivo recombinant interleukin 2 administration enhances survival against a lethal challenge with *Toxoplasma gondii*. *J. Immunol.* 135:4160–4163.

WARE, P. L., and L. H. KASPER. 1987. Strain-specific antigens of *Toxoplasma gondii*. *Infect. Immun.* 55:778–783.

WILSON, C. B., and J. E. HAAS. 1984. Cellular defenses against *Toxoplasma gondii* in newborns. *J. Clin. Invest.* 73:1606–1616.

Chapter 11. Immunology of Leishmaniasis

BADARO, R., T. C. JONES, R. LORENCO, B. J. CERF, D. SAMPAIO, E. M. CARVALHO, H. ROCHA, R. TEIXEIRA, and W. D. JOHNSON, JR. 1986. A prospective study of visceral leishmaniasis in an endemic area in Brazil. *J. Infect. Dis.* 154:639–649.

BLACKWELL, J., D. M. PRATT, and J. SHAW. 1986. Molecular biology of *Leishmania*. *Parasitol. Today* 2:45–53.

BLACKWELL, J. M. 1985. Receptors and recognition mechanisms of *Leishmania* species. *Trans. R. Soc. Trop. Med. Hyg.* 79:606–612.

BLACKWELL, J. M., B. ROBERTS, and J. ALEXANDER. 1985. Response of BALB/c mice to leishmanial infection. *Curr. Top. Microbiol. Immunol.* 122:97–106.

CARVALHO, E. M., R. BADARO, S. G. REED, T. C. JONES, and W. D. JOHNSON, JR. 1985. Absence of gamma interferon and interleukin 2 production during active visceral leishmaniasis. *J. Clin. Invest.* 76:2066–2069.

GREENBLATT, C. L. 1980. The present and future of vaccination for cutaneous leishmaniasis. In *Progress in Clinical and Biological Research*, Vol 47. *New Developments with Human and Veterinary Vaccines*. Alan R. Liss, pp. 259–285.

HANDMAN, E., L. F. SCHNUR, T. W. SPITHILL, and G. F. MITCHELL. 1986. Passive transfer of *Leishmania* lipopolysaccharide confers parasite survival in macrophages. *J. Immunol.* 137:3608–3613.

HARRISON, L. H., T. G. NAIDU, J. S. DREW, J. E. ALENCAR, and R. D. PEARSON. 1986. Reciprocal relationships between undernutrition and the parasitic disease, visceral leishmaniasis. *Rev. Infect. Dis.* 8:447–453.

HEINZEL, F. P., M. D. SADICK, B. J. HOLADAY, R. L. COFFMAN, and R. M. LOCKSLEY. 1989. Reciprocal expression of interferon-γ or interleukin 4 during the resolution of murine leishmaniasis. Evidence for expansion of distinct helper T cell subsets. *J. Exp. Med.* 169:59–72.

HOOVER, D. L., D. S. FINBLOOM, R. M. CRAWFORD, C. A. NACY, M. GILBREATH, and M. S. MELTZER. 1986. A lymphokine distinct from interferon-γ that activates human monocytes to kill *Leishmania donovani in vitro*. *J. Immunol.* 136:1329–1333.

LIEW, F. Y., C. HALE, and J. G. HOWARD. 1985. Prophylactic immunization against experimental leishmaniasis. IV. Subcutaneous immunization prevents the induction of protective immunity against fatal *Leishmania major* infection. *J. Immunol.* 135:2095–2101.

LIEW, F. Y., A. SINGLETON, E. CILLARI, and J. G. HOWARD. 1985. Prophylactic immunization against experimental leishmaniasis. V. Mechanism of the anti-protective blocking effect induced by subcutaneous immunization against *Leishmania major* infection. *J. Immunol.* 135:2102–2107.

MILON, G., R. G. TITUS, J.-C. CEROTTINI, G. MARCHAL, and J. A. LOUIS. 1986. Higher frequency of *Leishmania major*-specific L3T4+ T cells in susceptible BALB/c as compared with resistant CBA mice. *J. Immunol.* 136:1467–1471.

MURRAY, H. W., B. Y. RUBIN, and C. D. ROTHERMEL. 1983. Killing of intracellular *Leishmania donovani* by lymphokine-stimulated human mononuclear phagocytes. Evidence that interferon-γ is the activating lymphokine. *J. Clin. Invest.* 72:1506–1510.

MURRAY, H. W., J. J. STERN, K. WELTE, B. Y. RUBIN, S. M. CARRIERO, and C. F. NATHAN. 1987. Experimental visceral leishmaniasis: production of interluekin 2 and interferon-γ, tissue immune reaction, and response to treatment with interleukin-2 and interferon-γ. *J. Immunol.* 138:2290–2297.

NICKOL, A. D., and P. F. BONVENTRE. 1985. Visceral leishmaniasis in congenic mice of susceptible and resistant phenotypes: T-lymphocytes mediated immunosuppression. *Infect. Immun.* 50:169–174.

PETERSON, E. A., F. A. NEVA, A. BARRAL, R. CORREA-CORONAS, H. BOGAERT-DIAZ, D. MARINEZ, and F. E. WARD. 1984. Monocyte suppression of antigen-specific lymphocyte responses in diffuse cutaneous leishmaniasis patients from the Dominican Republic. *J. Immunol.* 132:2603–2606.

REINER, N. E., and C. J. MALEMUD. 1985. Arachidonic acid metabolism by murine peritoneal macrophages infected with *Leishmania donovani: in vitro* evidence for parasite-induced alterations in cyclooxygenase and lipoxygenase pathways. *J. Immunol.* 134:556–563.

REINER, N. E., W. NG, and W. R. MCMASTER. 1987. Parasite-accessory cell interactions in murine leishmaniasis. II. *Leishmania donovani* suppresses macrophage expression of class I and class II major histocompatibility complex gene products. *J. Immunol.* 138:1926–1932.

REMALEY, A. T., R. H. GLEW, D. B. KUHNS, and R. E. BASFORD. 1985. *Leishmania donovani:* surface membrane acid phosphatase blocks neutrophil oxidative metabolite production. *Exp. Parasitol.* 60:331–341.

SACKS, D. L., S. L. LAL, S. N. SHRIVASTAVA, J. BLACKWELL, and F. A. NEVA. 1987. An analysis of T cell responsiveness in Indian kala-azar. *J. Immunol.* 138:908–913.

SACKS, D. L., P. A. SCOTT, R. ASOFSKY, and F. A. SHER. 1984. Cutaneous leishmaniasis in anti-IgM-treated mice: enhanced resistance due to functional depletion of a B cell-dependent T cell involved in the suppressor pathway. *J. Immunol.* 132:2072–2077.

SACKS, D. L., S. HIENY, and A. SHER. 1985. Identification of cell surface carbohydrate and antigenic changes between noninfective and infective developmental stages of *Leishmania major* promastigotes. *J. Immunol.* 135:564–569.

SCOTT, P. 1985. Impaired macrophage leishmanicidal activity at cutaneous temperature. *Parasite Immunol.* 7:277–288.

SCOTT, P., P. NATOVITZ, R. L. COFFMAN, E. PEARCE, and A. SHER. 1988. Immunoregulation of cutaneous leishmaniasis. T cell lines that transfer protective immunity or exacerbation belong to different T helper subsets and respond to distinct parasite antigens. *J. Exp. Med.* 168:1675–1684.

SYPEK, J. P., and D. J. WYLER. 1985. Cell contact-mediated macrophage activation for antileishmanial defense: mapping of the genetic restriction to the I region of the MHC. *Clin. Exp. Immunol.* 62:449–457.

TITUS, R. G., R. CEREDIG, J.-C. CEROTTINI, and J. A. LOUIS. 1985. Therapeutic effect of anti-L3T4 monoclonal antibody GK1.5 on cutaneous leishmaniasis in genetically susceptible BALB/c mice. *J. Immunol.* 135:2108–2114.

TITUS, R. G., G. C. LIMA, H. D. ENGERS, and J. A. LOUIS. 1984. Exacerbation of murine cutaneous leishmaniasis by adoptive transfer of parasite-specific T cell populations capable of mediating *Leishmania major*-specific delayed-type hypersensitivity. *J. Immunol.* 133:1594–1600.

WILLIAMS, K. M., J. B. SACCI, and R. L. ANTHONY. 1986. Identification and recovery of leishmania antigen displayed on the surface membrane of mouse peritoneal macrophages *in vitro*. *J. Immunol.* 136:1853–1858.

WOZENCRAFT, A. O., G. SAYERS, and J. M. BLACKWELL. 1986. Macrophage type 3 complement receptors mediate serum-independent binding of *Leishmania donovani*. *J. Exp. Med.* 164:1332–1337.

WYLER, D. J., D. I. BELLER, and J. P. SYPEK. 1987. Macrophage activation for antileishmanial defense by an apparently novel mechanism. *J. Immunol.* 138:1246–1249.

Chapter 12. Immunology of African Trypanosomiasis

CROSS, G. A. M. 1975. Identification, purification, and properties of clone-specific glycoprotein antigens constituting the surface coat of *Trypanosoma brucei*. *Parasitology* 71:393–417.

DEGEE, A. L. W., and J. M. MANSFIELD. 1984. Genetics of resistance to the African trypanosomes. IV. Resistance of radiation chimeras to *Trypanosoma rhodesiense* infection. *Cell. Immunol.* 87:85–91.

DEGEE, A. L. W., R. F. LEVINE, and J. M. MANSFIELD. 1988. Genetics of resistance to the African trypanosomes. VI. Heredity of resistance and VSG-specific immune responses. *J. Immunol.* 140:283–288.

DEGEE, A. L. W., G. SONNENFELD, and J. M. MANSFIELD. 1985. Genetics of resistance to the African trypanosomes. V. Qualitative and quantitative differences in interferon levels among resistant and susceptible mouse strains. *J. Immunol.* 134:2723–2726.

DEMPSEY, W. L., and J. M. MANSFIELD. 1983a. Lymphocyte function in experimental African trypanosomiasis. V. Role of antibody and the mononuclear phagocyte system in variant specific immunity. *J. Immunol.* 130:405–411.

DEMPSEY, W. L., and J. M. MANSFIELD. 1983b. Lymphocyte function in experimental African trypanosomiasis. VI. Parasite specific immunosuppression. *J. Immunol.* 130:2896–2898.

DONELSON, J. E., and A. C. RICE-FICHT. 1985. Molecular biology of trypanosome antigenic variation. *Microbiol. Rev.* 49:107–125.

GROSSKINSKY, C. M., and B. A. ASKONAS. 1981. Macrophages as primary target cells and mediators of immune dysfunction in African trypanosomiasis. *Infect. Immun.* 33:149.

INVERSO, J. A., and J. M. MANSFIELD. 1983. Genetics of resistance to the African trypanosomes. II. Differences in virulence associated with VSSA expression among clones of *Trypanosoma rhodesiense*. *J. Immunol.* 130:412–417.

INVERSO, J. A., A. L. W. DEGEE, and J. M. MANSFIELD. 1988. Genetics of resistance to the African trypanosomes. VII. Trypanosome virulence is not linked to VSG expression. *J. Immunol.* 140:289–293.

LEVINE, R. F., and J. M. MANSFIELD. 1984. Genetics of resistance to the African trypanosomes. III. Variant specific antibody responses of H-2 compatible resistant and susceptible mice. *J. Immunol.* 133:1564–1569.

MURRAY, M., J. C. M. TRAIL, C. E. DAVIS, and S. J. BLACK. 1984. Genetic resistance to African trypanosomiasis. *J. Infect. Dis.* 149:311–319.

PAULNOCK, D. M., C. SMITH, and J. M. MANSFIELD. 1988. Antigen presenting cell function in African trypanosomiasis. In *Antigen Presenting Cells: Diversity, Differentiation, and Regulation*. J. Tew and L. Schook, eds. Alan R. Liss. pp. 135–143.

SACKS, D. L., K. M. ESSER, and A. SHER. 1982. Immunization of mice against African trypanosomiasis using anti-idiotypic antibodies. *J. Exp. Med.* 155:1108–1119.

SACKS, D. L., G. BANCROFT, W. H. EVANS, and B. A. ASKONAS. 1982. Incubation of trypanosome-derived mitogenic and immunosuppressive products with peritoneal macrophages allows recovery of biological activities from soluble parasite fractions. *Infect. Immun.* 36:160–168.

SACKS, D. L. 1984. Induction of protective immunity using antiidiotypic antibodies: immunization against experimental African trypanosomiasis. In *Idiotypy in Biology and Medicine*, H. Kohler, J. Urbain, and P.-A. Cazenave, eds. Academic Press, p. 401–416.

SEED, J. R. 1978. Competition among serologically different clones of *Trypanosoma brucei gambiense in vivo*. *J. Protozool.* 25:526–529.

TIZARD, I., K. H. NIELSEN, J. R. SEED, and J. E. HALL. 1978. Biologically active products from African trypanosomes. *Microbiol. Rev.* 42:661–681.

TURNER, M. J. 1985. The biochemistry of the surface antigens of the African trypanosomes. *Br. Med. Bull.* 41:137–143.

VICKERMAN, K., and A. G. LUCKINS. 1969. Localization of variable antigens in the surface coat of *Trypanosoma brucei* using ferritin conjugated antibody. *Nature* (Lond.) 224:1125–1126.

WELLHAUSEN, S. R., and J. M. MANSFIELD. 1979. Lymphocyte function in experimental African trypanosomiasis. II. Splenic suppressor cell activity. *J. Immunol.* 122:818–824.

YAMAMOTO, K., M. ONODERA, K. KATO, M. KAKINUMA, T. KIMURA, and F. F. RICHARDS. 1985. Involvement of suppressor cells induced with membrane fractions of trypanosomes in immunosuppression of trypanosomiasis. *Parasite Immunol.* 7:95–106.

Chapter 13. Immunology of Chagas' Disease

HAREL-BELLAN, A., M. JOSKOWICA, D. FREDELIZI, and H. EISEN. 1983. Modification of T-cell proliferation and interleukin 2 production in mice infected with *Trypanosoma cruzi*. *Proc. Natl. Acad. Sci. USA.* 80:3466–3469.

KIERSZENBAUM, F., C. A. GOTTLIEB, and D. B. BUDZKO. 1981. Antibody-independent, natural resistance of birds to *Trypanosoma cruzi* infection. *J. Parasitol.* 67:656–660.

KIERSZENBAUM, F., and H. M. GHARPURE. 1983. Killing of circulating forms of *Trypanosoma cruzi* by lymphoid cells from acutely and chronically infected mice. *Int. J. Parasitol.* 13:377–381.

KIERSZENBAUM, F., and G. SONNENFELD. 1984. *β*-interferon inhibits cell infection by *Trypanosoma cruzi*. *J. Immunol.* 132:905–908.

KIPNIS, T. L., A. U. KRETTLI, and W. DIAS DA SILVA. 1985. Transformation of trypomastigote forms of *Trypanosoma cruzi* into activators of alternative complement pathway by immune IgG fragments. *Scand. J. Immunol.* 22:217–226.

KRETTLI, A. U., J. R. CANÇADO, and Z. BRENER. 1982. Effect of specific chemotherapy on the levels of lytic antibodies in Chagas' disease. *Trans. R. Soc. Trop. Med. Hyg.* 76:334–340.

KRETTLI, A. U., and Z. BRENER. 1982. Resistance against *Trypanosoma cruzi* associated to anti-living trypomastigote antibodies. *J. Immunol.* 128:2009–2012.

MARTINS, M. S., L. HUDSON, A. U. KRETTLI, J. R. CANÇADO, and Z. BRENER. 1985. Human and mouse sera recognize the same polypeptide associated with immunological resistance to *Trypanosoma cruzi* infection. *Clin. Exp. Immunol.* 61:343–350.

MINOPRIO, P. M., H. EISEN, L. FORNI, M. R. D'IMPERIO, M. JOSKOWICZ, and A. COUTINHO. 1986. Polyclonal lymphocyte responses to murine *Trypanosoma cruzi* infection. I. Quantitation of both T- and B-cell responses. *Scand. J. Immunol.* 24:661–668.

NOGUEIRA, N., S. CHAPLAN, M. REESINK, J. TYDINGS, and Z. A. COHN. 1981. *Trypanosoma cruzi:* induction of microbicidal activity in human mononuclear phagocytes. *J. Immunol.* 128:2142–2146.

PLATA, F., F. G. PONS, and H. EISEN. 1984. Antigenic polymorphism of *Trypanosoma cruzi:* clonal analysis of trypomastigote surface antigens. *Eur. J. Immunol.* 14:392–299.

RIBEIRO DOS SANTOS, R., J. O. MARQUEZ, C. C. VON GAL FURTADO, J. C. RAMOS DE OLIVEIRA, A. R. MARTINS, and F. KÖBERLE. 1979. Antibodies against neurons in chronic Chagas' disease. *Trop. Med. Parasitol.* 30:19–23.

RIBEIRO DOS SANTOS, R., and L. HUDSON. 1980. *Trypanosoma cruzi:* immunological consequences of parasite modification of host cells. *Clin. Exp. Immunol.* 40:36–41.

REED, S. G. 1983. Spleen cell mediated suppression of IgG productions to a non-parasite antigen during chronic *Trypanosoma cruzi* infection in mice. *J. Immunol.* 131:1978–1982.

SCOTT, M. T., and D. SNARY. 1979. Protective immunization of mice using cell surface glycoprotein from *Trypanosoma cruzi*. *Nature* (Lond.) 282:73–74.

SCOTT, M. T., and M. GOSS-SAMPSON. 1984. Restricted IgG isotype profiles in *T. cruzi* infected mice and Chagas' disease patients. *Clin. Exp. Immunol.* 58:372–279.

SNARY, D. 1985. Biochemistry of surface antigens of *Trypanosoma cruzi*. *Br. Med. J.* 41:144–148.

SNARY, D., J. E. FLINT, J. N. WOOD, M. T. SCOTT, M. D. CHAPMAN, J. DODD, T. M. JESSELL, and M. A. MILES. 1983. A monoclonal antibody with specificity for *Trypanosoma cruzi*, central and peripheral neurones and glia. *Clin. Exp. Immunol.* 54:617–624.

WOOD, J. N., L. HUDSON, T. M. JESSELL, and M. YAMAMOTO. 1982. A monoclonal antibody defining antigenic determinants on subpopulations of mammalian neurons and *Trypanosoma cruzi* parasites. *Nature* (Lond.) 296:34–38.

Chapter 14. Immunology of Schistosomiasis

BALLOUL, J. M., J.-M. GRZYCH, R. J. PIERCE, and A. CAPRON. 1987. A purified 28,000 dalton protein from *Schistosoma mansoni* adult worms protects rats and mice against experimental schistosomiasis. *J. Immunol.* 138:3448–3453.

BICKLE, Q. D., B. J. ANDREWS, and M. G. TAYLOR. 1986. *Schistosoma mansoni:* characterization of two protective monoclonal antibodies. *Parasite Immunol.* 8:95–107.

BUTTERWORTH, A. E., M. CAPRON, J. S. CORDINGLEY, P. R. DALTON, D. W. DUNNE, H. C. KARIUKI, G. KIMANI, D. KOECH, M. MUGAMBI, J. H. OUMA, M. A. PRENTICE, B. A. RICHARDSON, T. K.

ARAP SIONGOK, R. F. STURROCK, and D. W. TAYLOR. 1985. Immunity after treatment of human schistosomiasis mansoni. II. Identification of resistant individuals, and analysis of their immune responses. *Trans. R. Soc. Trop. Med. Hyg.* 79:393–408.

BUTTERWORTH, A. E., R. BENSTED-SMITH, A. CAPRON, M. CAPRON, P. R. DALTON, D. W. DUNNE, J. M. GRZYCH, H. C. KARIUKI, J. KHALIFE, D. KOECH, M. MUGAMBI, J. H. OUMA, T. K. ARAP SIONGOK, and R. F. STURROCK. 1987. Immunity in human schistosomiasis mansoni: prevention by blocking antibodies of the expression of immunity in young children. *Parasitology* 94:281–300.

COLLEY, D. G., A. A. GARCIA, J. R. LAMBERTUCCI, J. C. PARRA, N. KATZ, R. S. ROCHA, and G. GAZZINELLI. 1986. Immune response during human schistosomiasis: XII. Differential responsiveness in patients with hepatosplenic disease. *Am. J. Trop. Med. Hyg.* 35:793–802.

CORREA-OLIVEIRA, R., S. L. JAMES, D. MCCALL, and A. SHER. 1986. Identification of a genetic locus, Rsm-1, controlling protective immunity against *Schistosoma mansoni*. *J. Immunol.* 137:2014–2019.

GRZYCH, J. M., M. CAPRON, P. H. LAMBERT, C. DISSOUS, S. TORRES, and A. CAPRON. 1985. An anti-idiotype vaccine against experimental schistosomiasis. *Nature* (Lond.) 316:74–76.

GRYZCH, J. M., C. DISSOUS, M. CAPRON, S. TORRES, P. H. LAMBERT, and A. CAPRON. 1987. *Schistosoma mansoni* shares with keyhole limpet haemocyanin a protective carbohydrate epitope. *J. Exp. Med.* 165:865–878.

HAGAN, P., H. A. WILKINS, U. J. BLUMENTHAL, R. J. HAYES, and B. M. GREENWOOD. 1985. Eosinophilia and resistance to *Schistosoma haematobium* in man. *Parasite Immunol.* 7:625–632.

HARN, D. A., M. MITSUYAMA, and J. R. DAVID. 1984. *Schistosoma mansoni*. Anti-egg monoclonal antibodies protect against cercarial challenge *in vivo*. *J. Exp. Med.* 159:1371–1387.

JAMES, S. L. 1986. Induction of protective immunity against *Schistosoma mansoni* by a nonliving vaccine. III. Correlation of resistance with induction of activated larvacidal macrophages. *J. Immunol.* 136:3872–3877.

JAMES, S. L. 1986. Activated macrophages as effector cells of protective immunity to schistosomiasis. *Immunol. Res.* 4:139–148.

KHALIFE, J., M. CAPRON, A. CAPRON, J. H. GRZYCH, A. E. BUTTERWORTH, D. W. DUNNE, and J. H. OUMA. 1986. Immunity in human schistosomiasis mansoni. Regulation of protective immune mechanisms by IgM blocking antibodies. *J. Exp. Med.* 164:1626–1640.

LANAR, D. E., E. J. PEARCE, S. L. JAMES, and A. SHER. 1986. Identification of paramyosin as schistosome antigen recognized by intradermally vaccinated mice. *Science* 234:593–596.

MCLAREN, D. J. 1980. *Schistosoma mansoni*: the parasite surface in relation to host immunity. In *Tropical Medicine Research Studies*, Vol. 1, K. N. Brown, ed. John Wiley.

MANGOLD, B. L., and D. A. DEAN. 1986. Passive transfer with serum and IgG antibodies of irradiated cercaria induced resistance against *Schistosoma mansoni* in mice. *J. Immunol.* 136:2644–2648.

PEARCE, E. J., P. F. BASCH, and A. SHER. 1986. Evidence that the reduced surface antigenicity of developing *Schistosoma mansoni* schistosomula is due to antigen shedding rather than host molecule acquisition. *Parasite Immunol.* 8:79–94.

PEARCE, E. J., S. L. JAMES, J. DALTON, A. BARRALL, C. RAMOS, M. STRAND, and A. SHER. 1986. Immunochemical characterization and purification of Sm-97, a *Schistosoma mansoni* antigen monospecifically recognized by antibodies from mice protectively immunized with a non-living vaccine. *J. Immunol.* 137:3593–3600.

SIMPSON, A. J. G., and S. R. SMITHERS. 1985. Schistosomes: surface egg and circulating antigens. *Curr. Top. Microbiol. Immunol.* 120:205–239.

SMITH, M., and J. A. CLEGG. 1985. Vaccination against *Schistosoma mansoni* with purified surface antigens. *Science* 227:535–538.

SMITH, D. B., K. M. DAVERN, P. G. BOARD, W. U. TIU, E. G. GARCIA, and G. F. MITCHELL. 1986. Mr 26,000 antigen of *Schistosoma japonicum* recognized by resistant WEHI 129/J mice is a parasite glutathione-S-transferase. *Proc. Natl. Acad. Sci. USA* 83:8703–8707.

WARREN, K. S. 1982. The secret of the immunopathogenesis of schistosomiasis: *in vivo* models. *Immunol. Rev.* 61:189–213.

WILKINS, H. A., P. H. GOLL, T. F. DE C MARSHALL, and P. J. MOORE. 1984. Dynamics of *Schistosoma haematobium* infection in a Gambian community. III. Acquisition and loss of infection. *Trans. R. Soc. Trop. Med. Hyg.* 78:227–232.

WILKINS, H. A., U. J. BLUMENTHAL, R. J. HAYES, and S. TULLOCH. 1987. Resistance to reinfection after treatment of urinary schistosomiasis. *Trans. R. Soc. Trop. Med. Hyg.* 81:29–35.

WYLER, D. J. 1983. Regulation of fibroblast functions by products of schistosomal egg granulomas: potential role in the pathogenesis of hepatic fibrosis. In *Cytopathology of Parasitic Diseases*, D. Evered and G. M. Collins, eds. Ciba Foundation Symposium No. 99. London: Pitman Books Ltd., pp. 190–206.

Chapter 15. Immunology of Lymphatic Filariasis

AGGARWAL, A., W. CUNA, C. HAQUE, C. DISSOUS, and A. CAPRON. 1985. Resistance against *Brugia malayi* microfilariae induced by a monoclonal antibody which promotes killing by macrophages and recognizes surface antigen(s). *Immunology* 54:655–663.

BAGAI, R. C., and D. SUBRAHMANYAM. 1968. Studies on the host–parasite relation in albino rats infected with *Litomosoides carinii*. *Am. J. Trop. Med. Hyg.* 17:833–839.

CANLAS, M. M., A. A. WADEE, L. LAMONTAGNE, and W. F. PIESSENS. 1984. A monoclonal antibody to surface antigens on microfilariae of *Brugia malayi* reduces microfilaremia in infected jirds. *Am. J. Trop. Med. Hyg.* 33:420–424.

FUHRMAN, J. A., S. S. URIOSTE, B. HAMILL, A. SPIELMAN, and W. F. PIESSENS. 1987. Functional and antigenic maturation of *Brugia malayi* microfilariae. *Am. J. Trop. Med. Hyg.* 36:17–21.

HOWELLS, R. E. 1987. Dynamics of the filarial surface. In *Filariasis*, Ciba Foundation Symposium No. 127, John Wiley and Sons Ltd. pp. 94–102.

KAZURA, J. W., H. CICIRELLO, and J. W. MCCALL. 1986. Induction of protection against *Brugia malayi* in jirds by microfilarial antigens. *J. Immunol.* 136:1422–1426.

KAZURA, J. W., H. G. CICIRELLO, and K. P. FORSYTH. 1986. Differential recognition of a protective filarial antigen by antibodies from humans with Bancroftian filariasis. *J. Clin. Invest.* 77:1985–1992.

KLEI, T. R., F. M. ENRIGHT, D. P. BLANCHARD, and S. A. UHL. 1982. Effects of presensitization on the development of lymphatic lesions in *Brugia pahangi*–infected jirds. *Am. J. Trop. Med. Hyg.* 31:280–291.

LAMMIE, P. J., and S. P. KATZ. 1983. Immunoregulation in experimental filariasis. II. Responses to parasite and nonparasite antigens in jirds with *Brugia pahangi*. *J. Immunol.* 130:1386–1389.

MACKENZIE, C. D., and J. F. WILLIAMS. 1987. Host responses in onchocerciasis. In *Onchocerciasis*, G. S. Nelson and C. D. Mackenzie, eds. Academic Press, pp. 237–269.

NUTMAN, T. B., A. S. WITHERS, and E. A. OTTESEN. 1985. In vitro parasite antigen-induced antibody responses in human helminth infections. *J. Immunol.* 135:2794–2799.

NUTMAN, T. B., V. KUMARASWAMI, and E. A. OTTESEN. 1987. Parasite-specific anergy in human filariasis. Insights after analysis of parasite antigen-driven lymphokine production. *J. Clin. Invest.* 79:1516–1523.

OTTESEN, E. A., V. KUMARASWAMI, R. PARANJAPE, et al. 1981. Naturally occurring blocking antibodies modulate immediate hypersensitivity responses in human filariasis. *J. Immunol.* 127:2014–2020.

PIESSENS, W. F., F. PARTONO, S. L. HOFFMAN, et al. 1982. Antigen-specific suppressor T lymphocytes in human lymphatic filariasis. *N. Engl. J. Med.* 307:144–148.

PIESSENS, W. F., A. A. WADEE, and L. KURNIAWAN. 1987. Regulation of immune responses in lymphatic filariasis. In *Filariasis*, Ciba Foundation Symposium No. 127. John Wiley and Sons Ltd, pp. 164–172.

SCHACHER, J. F., and P. F. SAYHOUN. 1967. A chronological study of the histopathology of filarial disease in cats and dogs, caused by *Brugia pahangi*. *Trans. R. Soc. Trop. Med. Hyg.* 61:234–243.

SIM, B. K., B. H. KWA, and J. W. MAK. 1982. Immune responses in human *Brugia malayi* infections: serum-dependent cell-mediated destruction of infective larvae in vitro. *Trans. R. Soc. Trop. Med. Hyg.* 76:362–370.

VICKERY, A. C., and A. L. VINCENT. 1984. Immunity to *Brugia pahangi* in athymic nude and normal mice: eosinophilia, antibody and hypersensitivity responses. *Parasite Immunol.* 6:545–559.

VICKERY, A. C., A. L. VINCENT, and W. A. SODEMAN. 1983. Effect of reconstitution on resistance to *Brugia pahangi* in congenitally athymic nude mice. *Parasitology* 69:478–485.

VICKERY, A. C., J. K. NAYAR, and K. H. ALBERTINE. 1985. Differential pathogenicity of *Brugia malayi*, *B. patei* and *B. pahangi* in immunodeficient nude mice. *Acta Trop.* 42:353–363.

WADEE, A. A., and W. F. PIESSENS. 1986. Partial purification of a T cell mitogen from microfilariae of *Brugia malayi*. *Am. J. Trop. Med. Hyg.* 35:141–147.

WEIL, G. J., K. G. POWERS, E. L. PARBUONI, B. R. LINE, R. D. FURROW, and E. A. OTTESEN. 1982. *Dirofilaria immitis*. 6. Antimicrofilarial immunity in experimental filariasis. *Am. J. Trop. Med. Hyg.* 31:477–485.

WEIL, G. J., R. HUSSAIN, V. KUMARASWAMI, S. P. TRIPATHY, K. S. PHILLIPS, and E. A. OTTESEN. 1980. Prenatal allergic sensitization to helminth antigens in offspring of parasite-infected mothers. *J. Clin. Invest.* 71:1124–1129.

WEISS, N., and M. TANNER. 1979. Studies of *Dipetalonema viteae* (Filarioidea). 3. Antibody-dependent cell-mediated destruction of microfilariae in vivo. *Trop. Med. Parasitol.* 30:73–80.

YATES, J. A., and G. I. HIGASHI. 1985. *Brugia malayi:* Vaccination of jirds with ^{60}cobalt-attenuated infective stage larvae protects against homologous challenge. *Am. J. Trop. Med. Hyg.* 34:1132–1137.

Part III Molecular Biology

Chapter 16. Molecular Biology of *Plasmodium*

FERONE, R., J. J. BURCHALL, and G. H. HITCHINGS. 1969. *Plasmodium berghei* dihydrofolate reductase. Isolation, properties and inhibition by antifolates. *Mol. Pharmacol.* 5:49–59.

KILEJIAN, A. 1977. Histidine-rich protein as a model malaria vaccine. *Science* 202:922–924.

KILEJIAN, A. 1979. Characterization as a protein correlated with the production of knob-like protrusions on membranes of erythrocytes infected with *Plasmodium falciparum*. *Proc. Natl. Acad. Sci. USA* 76:4650–4653.

MCCUTCHAN, T. F., J. B. DAME, L. H. MILLER, and J. BARNWELL. 1984. Evolutionary relatedness of *Plasmodium* species as determined by the structure of DNA. *Science* 225:808–811.

MCCUTCHAN, T. F., J. L. HANSEN, J. S. DAME, and J. A. MULLINS. 1984. Mung bean nuclease cleaves *Plasmodium* genomic DNA at sites before and after genes. *Science* 225:625–628.

PERKINS, M. E. 1984. Surface proteins of *Plasmodium falciparum* merozoites binding to the erythrocyte receptors, glycophorin. *J. Exp. Med.* 160:788–798.

PERLMANN, H., K. BERZINS, M. WAHLGREN, J. CARLSON, A. BJORKMAN, M. E. PATARROYO, and P. PERLMANN. 1984. Antibodies in malaria sera to parasite antigens in the membrane of erythrocytes infected with early asexual stage *Plasmodium falciparum*. *J. Exp. Med.* 159:1686–1704.

RAVETCH, J. V., J. P. KOCHAN, and M. PERKINS. 1985. Isolation of the gene for a glycophorin-binding protein implicated in erythrocyte invasion by a malaria parasite. *Science* 227:1593–1597.

SHARMA, S., and G. N. GODSON. 1985. Expression of the major surface antigen of *Plasmodium knowlesi* sporozoites in yeast. *Science* 228:879–882.

SMITH, B. P., M. REINA-GUERRA, S. K. HOISETH, B. A. STOCKER, F. HABASHA, E. JOHNSON, and F. MERRITT, Aromatic-dependent *Salmonella typhimurium* as modified live vaccines for calves. *Am. J. Vet. Res.* 45:59–66.

SMITH, G. L., G. N. GODSON, V. NUSSENZWEIG, R. S. NUSSENZWEIG, J. BARNWELL, and B. MOSS. 1984. *Plasmodium knowlesi* sporozoite antigen: expression by infectious recombinant vaccinia virus. *Science* 224:397–399.

WELLEMS, T. E., and R. J. HOWARD. 1986. Homologous genes encode two distinct histidine-rich proteins in a cloned isolate of *Plasmodium falciparum*. *Proc. Natl. Acad. Sci. USA* 83:6065–6062.

WILSON, R. J. M., I. A. MCGREGOR, and P. J. HALL. 1975. Persistence and recurrence of S-antigen in *Plasmodium falciparum* infections in man. *Trans. R. Soc. Trop. Med. Hyg.* 69:460.

YOUNG, R. A., and R. W. DAVIS. 1983. Efficient isolation of genes by using antibody probes. *Proc. Natl. Acad. Sci. USA* 80:1194–1198.

Chapter 17. Molecular Biology of African Trypanosomes

BENNE, R., J. VANDENBURG, J. P. J. BRAKENHOFF, P. SLOOF, J. V. VANBOOM, and M. C. TROMP. 1986. Major transcript of the frame-shifted coxII gene from trypanosome mitochondria contains four nucleotides that are not encoded in the DNA. *Cell* 46:819–826.

BERNARDS, A., N. VANHARTEN-LOOSBROEK, and P. BORST. 1984. Modification of telomeric DNA in *Trypanosoma brucei:* a role in antigenic variation? *Nucl. Acids Res.* 12:4153–4170.

BOOTHROYD, J. C., and G. A. M. CROSS. 1982. Transcripts coding for different variant surface glycoproteins of *Trypanosoma brucei* have a short, identical exon at their 5′- end. *Gene* 20:281–289.

BORST, P. 1986. Discontinuous transcription and antigenic variation in trypanosomes. *Ann. Rev. Biochem.* 55:701–732.

BORST, P. 1986. How proteins get into microbodies (peroxisomes, glyoxysomes, glycosomes). *Biochim. Biophys. Acta* 286:179–203.

CAMPBELL, D. A., D. A. THORNTON, and J. C. BOOTHROYD. 1984. Apparent discontinuous transcription of *Trypanosoma brucei* variant surface antigen genes. *Nature* 311:350–355.

DE LANGE, T., and P. BORST. 1982. Genomic environment of the expression-linked extra copies of genes for surface antigens of *Trypanosoma brucei* resembles the end of a chromosome. *Nature* (Lond.) 299:451–453.

DE LANGE, T., A. Y. C. LIU, L. H. T. VAN DER PLOEG, P. BORST, M. C. TROMP, and J. H. VAN BOOM. 1983. Tandem repetition of the 5′ miniexon of variant surface glycoprotein genes: a multiple promoter for VSG gene transcription? *Cell* 34:891–900.

FEAGIN, J. E., D. P. JASMER, and K. STUART. 1987. Developmentally regulated addition of nucleotides within apocytochrome b transcripts in *Trypanosoma brucei*. *Cell* 49:337–345.

GIBSON, W. C., and M. A. MILES. 1986. The karyotype and ploidy of *Trypanosoma cruzi*. *EMBO J.* 5:1299–1305.

GIBSON, W. C., K. A. OSINGA, P. A. M. MICHELS, and P. BORST. 1985. Trypanosomes of subgenus *Trypanozoon* are diploid for housekeeping genes. *Mol. Biochem. Parasitol.* 16:231–242.

GUNDERSON, J. H., T. F. MCCUTCHAN, and M. L. SOGIN. 1986. Sequence of the small subunit ribosomal RNA gene expressed in the bloodstream stages of *Plasmodium berghei:* evolutionary implications. *J. Protozool.* 33:525–529.

HAJDUK, S. L., C. R. CARMERON, J. D. BARRY, and K. VICKERMAN. 1981. Antigenic variation in cyclically transmitted *Trypanosoma brucei*. I. Variable antigen type composition of metacyclic trypanosome populations from the salivary glands of *Glossina morsitans*. *Parasitology* 83:595–607.

HAJDUK, S. L., V. A. KLEIN, and P. T. ENGLUND. 1984. Replication of kinetoplast DNA maxicircles. *Cell* 36:483–492.

HAJDUK, S. L. and K. VICKERMAN. 1981. Antigenic variation in cyclically transmitted *Trypanosoma brucei*. Variable antigen type composition of the first parasitaemia in mice bitten by trypanosome-infected *Glossina morsitans*. *Parasitology* 83:609–621.

HASAN, G., M. J. TURNER, and J. S. CORDINGLEY. 1984. Ribosomal RNA genes of *T. brucei:* mapping the regions specifying the six small ribosomal RNA. *Gene* 27:75–86.

HOEIJMAKERS, J. H. J., A. C. C. FRASCH, A. BERNARDS, P. BORST, and G. A. M. CROSS. 1980. Novel expression-linked copies of the genes for variant surface antigens in trypanosomes. *Nature* (Lond.) 284:78–80.

KOOTER, J., T. DE LANGE, and P. BORST. 1984. Discontinuous synthesis of mRNA in trypanosomes. *EMBO J.* 3:2387–2392.

LONGACRE, S., and H. EISEN. 1986. Expression of whole and hybrid genes in *Trypanosoma equiperdum* antigenic variation. *EMBO J.* 5:1057–1063.

LONGACRE, S., U. HIBNER, A. RAIBAUD, H. EISEN, T. BALTZ, C. GIROUD, and D. BALTZ. 1983. DNA rearrangement and antigenic variation of *Trypanosoma equiperdum:* multiple expression-linked sites in independent isolates of trypanosomes expressing the same antigen. *Mol. Cell. Biol.* 3:399–409.

MURPHY, W. J., K. P. WATKINS, and N. AGABIAN. 1986. Identification of a novel Y branch structure as an intermediate in trypanosome mRNA processing: evidence for trans splicing. *Cell* 47:517–525.

NELSON, R. G., M. PARSONS, P. J. BARR, K. STUART, M. SELKIRK, and N. AGABIAN. 1983. Sequences homologous to the variant antigen mRNA spliced leader are located in tandem repeats and variable orphons in *Trypanosoma brucei. Cell* 34:901–909.

NELSON, R. G., M. PARSONS, M. SELKIRK, G. NEWPORT, P. J. BARR, and N. AGABIAN. 1984. Sequences homologous to the variant antigen mRNA spliced leader are present in Trypanosomatidae which do not undergo antigenic variation. *Nature* (Lond.) 308:665–667.

OPPERDOES, F. R., P. BORST, and W. LEENE. 1977. Localization of glycerol-3-phosphate oxidase in the mitochondrion and particulate NAD^+-linked glycerol-3-phosphate dehydrogenase in the microbodies of the bloodstream form of *Trypanosoma brucei. Eur. J. Biochem.* 76:29–39.

OSINGA, K. A., B. W. SWINKELS, W. C. GIBSON, P. BORST, G. H. VEENEMAN, J. H. VANBOOM, P. A. M. MICHELS, and F. R. OPPERDOES. 1985. Topogenesis of microbody enzymes: a sequence comparison of the genes for the glycosomal (microbody) and cytosolic phosphoglycerate kinases of *Trypanosoma brucei. EMBO J.* 4:3811–3817.

PAINDAVOINE, P., F. ZAMPETTI-BOSSELER, E. PAYS, J. SCHWEIZER, M. GUYAUX, L. JENNI, and M. STEINERT. 1986. Trypanosome hybrids generated in tsetse flies by nuclear fusion. *EMBO J.* 5:3631–3636.

PAYS, E., M. F. DELAUW, M. LAURENT, and M. STEINERT. 1984. Possible DNA modification in GC dinucleotides of *Trypanosoma brucei* telomeric sequences: relationship with antigen gene transcription. *Nucl. Acids Res.* 12:5235–5247.

PAYS, E., M. GUYAUX, D. AERTS, N. VAN MEIRVENNE, and M. STEINERT. 1985. Telomeric reciprocal recombination as a possible mechanism for antigenic variation in trypanosomes. *Nature* (Lond.) 316:562–564.

PAYS, E., S. VAN ASSEL, M. LAURENT, M. DARVILLE, T. VERVOORT, N. VAN MEIRVENNE, and M. STEINERT. 1983. Gene conversion as a mechanism for antigenic variation in trypanosomes. *Cell* 34:371–381.

SCHNARE, M. N., J. C. COLLINGS, and M. W. GRAY. 1986. Structure and Evolution of the small subunit ribosomal RNA gene of *Crithidia fasciculata. Curr. Genet.* 10:405–410.

SIMPSON, A. M., N. NECKELMANN, V. F. DE LA CRUZ, M. L. MUHICH, and L. SIMPSON. 1985. Mapping and 5′ end determination of kinetoplast maxicircle gene transcripts from *Leishmania tarentolae. Nucl. Acids Res.* 13:5977–5993.

SUTTON, R. E., and J. C. BOOTHROYD. 1986. Evidence for trans splicing in trypanosomes. *Cell* 47:527–535.

TSCHUDI, C., F. F. RICHARDS, and E. ULLU. 1986. The U2 RNA analogue of *Trypanosoma brucei gambiense:* implications for a splicing mechanism in trypanosomes. *Nucl. Acids Res.* 22:8893–8903.

VAN DER PLOEG, L. H. T., A. Y. C. LIU, P. A. M. MICHELS, T. DE LANGE, P. BORST, K. MAJUMDER, H. WEBER, and G. H. VEENEMAN. 1982. RNA splicing is required to make the messenger RNA for a variant surface antigen in trypanosomes. *Nucl. Acids Res.* 10:3591–3604.

VAN DER PLOEG, L. H. T. 1986. Discontinuous transcription and splicing in trypanosomes. *Cell* 47:479–480.

VAN DER PLOEG, L. H. T., D. C. SCHWARTZ, C. R. CANTOR, and P. BORST. 1984. Antigenic variation in *Trypanosoma brucei*, analyzed by electrophoretic separation of chromosome-sized DNA molecules. *Cell* 37:77–74.

VICKERMAN, K. 1985. Developmental cycles and biology of pathogenic trypanosomes. *Br. Med. Bull.* 41:105–114.

WILLIAMS, R. O., J. R. YOUNG, and P. A. O. MAJIWA. 1979. Genomic rearrangements correlated with antigenic variation in *Trypanosoma brucei*. *Nature* (Lond.) 282:847–849.

Chapter 18. Molecular Biology of *Leishmania*

BEVERLEY, S. M., J. A. CODERRE, D. V. SANTI, and R. T. SCHIMKE. 1984. Unstable DNA amplifications in methotrexate-resistant *Leishmania* consist of extrachromosomal circles which relocalize during stabilization. *Cell* 38(2):431–439.

CARLE, G. E., and M. V. OLSON. 1987. Orthogonal-field-alternation gel electrophoresis. *Methods Enzymol.* 155:468–482.

COMEAU, A. M., S. I. MILLER, and D. F. WIRTH. 1986. Chromosome location of four genes in *Leishmania*. *Mol. Biochem. Parasitol.* 21(2):161–169.

DETKE, S., G. CHAUDHURI, J. A. KINK, and K. P. CHANG. 1988. DNA amplification in tunicamycin-resistant *Leishmania mexicana*. Multicopies of a single 63-kilobase supercoiled molecule and their expression. *J. Biol. Chem.* 263(7):3418–3424.

FEAGIN, J. E., J. M. SHAW, L. SIMPSON, and K. STUART. 1988. Creation of AUG initiation codons by addition of uridines within cytochrome b transcripts of kinetoplastids. *Proc. Natl. Acad. Sci. USA.* 85(2):539–543.

GARVEY, E. P., and D. V. SANTI. 1986. Stable amplified DNA in drug-resistant *Leishmania* exists as extrachromosomal circles. *Science* 233(4763):535–540.

KIDANE, G. Z., D. HUGHES, and L. SIMPSON. 1984. Sequence heterogeneity and anomalous electrophoretic mobility of kinetoplast minicircle DNA from *Leishmania tarentolae*. *Gene* 27(3):265–277.

LAINSON, R. 1983. The American leishmaniases: some observations on their ecology and epidemiology. *Trans. R. Soc. Trop. Med. Hyg.* 77(5):569–596.

LANDFEAR, S. M., D. MCMAHON-PRATT, and D. F. WIRTH. 1983. Tandem arrangement of tubulin genes in the protozoan parasite *Leishmania enriettii*. *Mol. Cell. Biol* 3(6):1070–1076.

LOPES, U. G., and D. F. WIRTH. 1986. Identification of visceral *Leishmania* species with cloned sequences of kinetoplast DNA. *Mol. Biochem. Parasitol.* 20(1):77–84.

MILLER, S. I., S. M. LANDFEAR, and D. F. WIRTH. 1986. Cloning and characterization of a *Leishmania* gene encoding a RNA spliced leader sequence. *Nucl. Acids Res.* 14(18):7341–7360.

MURPHY, W. J., K. P. WATKINS, and N. AGABIAN. 1986. Identification of a novel Y branch structure as an intermediate in trypanosome mRNA processing: evidence for trans splicing. *Cell* 47(4):517–525.

ROGERS, W. O., and D. F. WIRTH. 1987. Kinetoplast DNA minicircles: regions of extensive sequence divergence. *Proc. Natl. Acad. Sci. USA.* 84(2):565–569.

ROGERS, W. O., and D. F. WIRTH. 1988. Generation of sequence diversity in the kinetoplast DNA minicircles of *Leishmania mexicana amazonensis*. *Mol. Biochem. Parasitol.* 30(1):1–8.

SACKS, D. L., and P. V. PERKINS. 1984. Identification of an infective stage of *Leishmania* promastigotes. *Science* 223(4643):1417–1419.

SAMARAS, N., and T. W. SPITHILL. 1987. Molecular karyotype of five species of *Leishmania* and analysis of gene locations and chromosomal rearrangements. *Mol. Biochem. Parasitol.* 25(3):279–291.

SCHWARTZ, D. C., and C. R. CANTOR. 1984. Separation of yeast chromosome-sized DNAs by pulsed field gradient gel electrophoresis. *Cell* 37(1):67–75.

SIMPSON, L., N. NECKELMANN, V. F. DE LA CRUZ, A. M. SIMPSON, J. E. FEAGIN, D. P. JASMER, and J. E. STUART. 1987. Comparison of the maxicircle (mitochondrial) genomes of *Leishmania tarentolae* and *Trypanosoma brucei* at the level of nucleotide sequence. *J. Biol. Chem.* 262(13):6182–6196.

SUTTON, R. E., and J. C. BOOTHROYD. 1988. Trypanosome trans-splicing utilizes 2′-5′ branches and a corresponding debranching activity. *EMBO J.* 7(5):1431–1437.

VAN DER PLOEG, L. H., S. H. GIANNINI, and C. R. CANTOR. 1985. Heat shock genes: regulatory role for differentiation in parasitic protozoa. *Science* 228(4706):1443–1446.

WIRTH, D. F., and D. M. PRATT. 1982. Rapid identification of *Leishmania* species by specific hybridization of kinetoplast DNA in cutaneous lesions. *Proc. Natl. Acad. Sci. USA.* 79(22):6999–7003.

WIRTH, D. F., W. O. ROGERS, R. BARKER, JR., H. DOURADO, L. SUESEBANG, and B. ALBUQUERQUE. 1986. Leishmaniasis and malaria: new tools for epidemiologic analysis. *Science* 234(4779):975–979.

Chapter 19. Molecular Biology of Schistosomes and Filariae

BALLOUL, J. M., P. SONDERMEYER, D. DREYER, M. CAPRON, J. M. GRZYCH, R. J. PIERCE, D. CARVALLO, J. P. LECOCQ, and A. CAPRON. 1987. Molecular cloning of a protective antigen of schistosomes. *Nature* (Lond.) 326:149–153.

CORDINGLEY, J. S. 1987. Trematode eggshells: novel protein biopolymers. *Parasitol. Today* 3:341–344.

DAVIS, A., R. BLANTON, F. ROTTMAN, R. MAURER, and A. MAHMOUD. 1986. Isolation of cDNA clones for differentially expressed genes of the human parasite *Schistosoma mansoni*. *Proc. Natl. Acad. Sci. USA* 83:5534–5538.

DAVIS, A. H., J. NANDURI, and D. C. WATSON. 1987. Cloning and gene expression of *Schistosoma mansoni* protease. *J. Biol. Chem.* 262:12851–12855.

ERTTMAN, K. D., T. R. UNNASCH, B. M. GREENE, E. J. ALBIEZ, J. BOATENG, A. M. DENKE, J. J. FERRARONI, M. KARAM, H. SCHULZ-KEY, and P. N. WILLIAMS. 1987. A DNA sequence specific for forest form *Onchocerca volvulus*. *Nature* (Lond.) 327:415–417.

HEDSTROM, R., J. CULPEPPER, R. HARRISON, N. AGABIAN, and G. NEWPORT. 1987. A major immunogen in *Schistosoma mansoni* infections is homologous to the heat-shock protein HSP70. *J. Exp. Med.* 165:1430–1435.

HEDSTROM, R., J. CULPEPPER, V. SCHINSKI, N. AGABIAN, and G. NEWPORT. 1988. Schistosome heat-shock proteins are immunologically distinct host-like antigens. *Mol. Biochem. Parasitol.* 29:275–282.

LANAR, D. E., E. J. PEARCE, S. L. JAMES, and A. SHER. 1986. Identification of paramyosin as schistosome antigen recognized by intradermally vaccinated mice. *Science* 234:593–596.

MATSUMOTO, Y., G. PERRY, R. J. C. LEVINE, R. BLANTON, A. A. F. MAHMOUD, and M. AIKAWA. 1988. Paramyosin and actin in schistosomal teguments. *Nature* (Lond.) 333:76–78.

NENE, V, D. W. DUNNE, K. S. JOHNSON, D. W. TAYLOR, and J. S. CORDINGLEY. 1986. Sequence and expression of a major egg antigen from *Schistosoma mansoni*: homologies to heat-shock proteins and alpha crystallins. *Mol. Biochem. Parasitol.* 21:179–188.

NEWPORT, G. R., R. A. HARRISON, J. MCKERROW, P. TARR, J. KALLESTAD, and N. AGABIAN. 1987. Molecular cloning of *Schistosoma mansoni* myosin. *Mol. Biochem. Parasitol.* 26:29–38.

NEWPORT, G. R., J. MCKERROW, R. HEDSTROM, M. PETITT, L. MCGARRIGLE, P. BARR, and N. AGABIAN. 1988. Molecular cloning and analysis of the proteinase that facilitates infection by schistosome parasites. *J. Biol. Chem.* 263:13179–13184.

NIELSEN, T. W., P. A. MARONEY, R. G. GOODWIN, K. G. PERRINE, J. A. DENKER, J. NANDURI, and J. W. KAZURA. 1988. Cloning and characterization of a potentially protective antigen in lymphatic filariasis. *Proc. Natl. Acad. Sci. USA* 85:3604–3607.

PETRALANDA, I., L. YARZABAL, and W. F. PIESSENS. 1986. Studies on a filarial antigen with collagenase activity. *Mol. Biochem. Parasitol.* 19:51–59.

PIESSENS, W. F., L. A. MCREYNOLDS, and S. A. WILLIAMS. 1987. Highly repeated DNA sequences as species-specific probes for *Brugia*. *Parasitol. Today* 3:378–379.

QU, L. H., N. HARDMAN, L. GILL, L. CHAPPELL, M. NICOLOSO, and J.-P. BACHELLERIE. 1986. Phylogeny of helminths determined by rRNA sequence comparison. *Mol. Biochem. Parasitol.* 20:93–99.

SEARCY, D. G., and A. J. MACINNIS. 1970. Measurements by DNA renaturation of the genetic basis of parasitic reduction. *Evolution* 24:796–806.

SELKIRK, M. E., P. J. RUTHERFORD, D. A. DENHAM, F. PARTONO, and R. M. MAIZELS. 1987. Cloned antigen genes of *Brugia* filarial parasites. *Biochem. Soc. Symp.* 53:91–102.

SENFT, A. W., W. B. GIBLER, and J. J. GUTERMAN. 1986. Influence of calcium-perturbing agents on schistosomes: comparison of the effects of praziquantel and verapamil on worm tegument. *J. Exp. Zool.* 239:25–36.

SMITH, D. B., K. M. DAVERN, P. G. BOARD, W. Y. TIU, E. G. GARCIA, and G. F. MITCHELL. 1986. The Mr 26,000 antigen of *Schistosoma japonicum* recognized by resistant WEHI mice is a parasite glutathione S-transferase. *Proc. Natl. Acad. Sci. USA* 83:8703–8707.

UNNASCH, T. R. 1987. DNA probes to identify *Onchocerca volvulus*. *Parasitol. Today* 3:377–378.

Index